Metal Ions in Life Sciences

Volume 28

Metal Ions and the Route to Life

Guest Editors

Wolfgang Nitschke
CNRS, BIP (UMR 7281)
Aix-Marseille-University
Marseille, France
nitschke@imm.cnrs.fr

Simon Duval
CNRS, BIP (UMR 7281)
Aix Marseille University
Marseille, France
sduval@imm.cnrs.fr

Series Editors

Astrid Sigel, Helmut Sigel
Department of Chemistry
Inorganic Chemistry
University of Basel
St. Johannsring 19
CH-4056 Basel
Switzerland
astrid.sigel@unibas.ch
helmut.sigel@unibas.ch

Eva Freisinger, Roland K. O. Sigel
Department of Chemistry
University of Zurich
Winterthurerstrasse 190
CH-8057 Zurich
Switzerland
freisinger@chem.uzh.ch
roland.sigel@chem.uzh.ch

CRC Press
Taylor & Francis Group
Boca Raton London New York

CRC Press is an imprint of the
Taylor & Francis Group, an **informa** business

Cover illustration: Prepared by Wolfgang Nitschke and Simon Duval. The metals highlighted in the periodic table form the structural matrices (illustrated by the chemical garden shown in the background) hosting nascent life, catalyze crucial anabolic reactions (exemplified by the mixed-metal cluster of the enzyme acetyl CoA-synthase) and are indispensable in the conversion of environmental free energy into intracellular chemical free energy (schematically shown in the bioenergetically active cell). The image of the chemical garden is courtesy C. Truong, chapter 8.

First edition published 2026
by CRC Press
2385 NW Executive Center Drive, Suite 320, Boca Raton FL 33431

and by CRC Press
4 Park Square, Milton Park, Abingdon, Oxon, OX14 4RN

CRC Press is an imprint of Taylor & Francis Group, LLC

ISBN: 9781032606156 (hbk)
ISBN: 9781032606255 (pbk)
ISBN: 9781003459910 (ebk)

ISSN: 1559-0836 e-ISSN: 1868-0402

DOI: 10.1201/9781003459910

Typeset in Times
by Deanta Global Publishing Services, Chennai, India

About the Editors

Wolfgang Nitschke obtained his Diploma in Physics and his PhD in Biochemistry from the University of Regensburg (Germany). He then worked as a postdoctoral fellow (on grants from the Deutsche Forschungsgemeinschaft, the CEA and EMBO) with A. W. (Bill) Rutherford, Commissariat à l'Energie Atomique (Saclay, France) and with Pierre Joliot, Institut de Biologie Physico-Chimique (Paris/France), followed by a position as Assistant Professor at the Albert-Ludwigs-University's Institut für Biologie (Freiburg/Germany). In 1995, he obtained a permanent position at the CNRS in the Laboratory of Bioenergetics and Protein Engineering (Marseille/France) where he founded and headed the research group "Evolution of Bioenergetics". This group investigates the bioenergetic processes of phylogenetically diverse microorganisms with the aim of elucidating evolutionary pathways from early to extant life and of discerning common, conserved patterns informing on the bioenergetic processes likely to have operated in the earliest forms of life on our planet. Over the last five years, the group has extended its research activities toward characterizing mineral-based processes potentially representing abiotic precursors of crucial reaction schemes in living systems.

Simon Duval is a researcher at the French National Centre for Scientific Research (CNRS) in Marseille, France, since 2018, where he has been part of a team focused on the evolution of bioenergetics. He earned his PhD from Aix-Marseille University, where he studied the redox properties and evolution of molybdenum enzymes, in particular, arsenite oxidase.

Following his PhD, Simon pursued several postdoctoral projects, mostly investigating electron transfer in metalloproteins. Notably, he spent two years at Utah State University (Logan, USA) with Professor Lance Seefeldt, where he explored the properties of Fe-nitrogenase using spectroscopic and biochemical methods.

Simon's expertise in biochemical, biophysical, and electrochemical techniques applied to metalloproteins enabled him to launch an innovative project at the CNRS. This project investigates minerals as catalytic analogs of metalloenzymes, aiming at substantiating their potential roles in emergence-of-life processes. In 2021, he was involved in organizing the European Bioenergetics Conference (EBEC 2021, Aix en Provence), an international congress dedicated to this field.

Over the past five years, his research has increasingly focused on the origins of life, particularly the role of metals in converting free energy within a chemiosmotic prebiotic system. He has developed a unique multidisciplinary approach, combining insights from bioenergetics and mineralogy to study metal ions in minerals and their potential to drive free energy formation at life's emergence.

Astrid Sigel has studied languages; she was an Editor of the *Metal Ions in Biological Systems (MIBS)* series (until Volume 44) and also of the *Handbook on Toxicity of Inorganic Compounds* (1988), the *Handbook on Metals in Clinical and Analytical Chemistry* (1994; both with H. G. Seiler and H.S.), and on the *Handbook on Metalloproteins* (2001; with Ivano Bertini and H.S.). She is also an Editor of the *MILS* series from Volume 1 on, and she co-authored more than 50 papers on topics in Bioinorganic Chemistry.

Helmut Sigel is Emeritus Professor (2003) of Inorganic Chemistry at the University of Basel, Switzerland. He is a Co-editor of the series *Metal Ions in Biological Systems* (1973–2005; 44 volumes) as well as of the Sigels' new series *Metal Ions in Life Sciences* (since 2006). He also co-edited three handbooks and published over 350 articles on metal ion complexes of nucleotides, amino acids, coenzymes, and other bio-ligands. Together with Ivano Bertini, Harry B. Gray, and Bo G. Malmström, he founded (1983) the International Conferences on Biological Inorganic Chemistry (ICBICs). He lectured worldwide and was named *Protagonist in Chemistry* (2002) by *Inorganica Chimica Acta* (issue 339). Among Endowed Lectureships, appointments as Visiting Professor (e.g., Austria, China, Japan, Kuwait, UK), and further honors, he received the P. Ray Award (Indian Chemical Society, of which he is also a Honorary Fellow), the Alfred Werner Award (Swiss Chemical Society), and a Doctor of Science honoris causa degree (Kalyani University, India). He is also an Honorary Member of SBIC (Society of Biological Inorganic Chemistry).

Eva Freisinger is Associate Professor of Bioinorganic Chemistry and Chemical Biology at the Department of Chemistry, University of Zurich (appointed in 2018). She earned her PhD in 2000 from the University of Dortmund, Germany, under the supervision of Bernhard Lippert, followed by a three-year postdoctoral stay with Caroline Kisker at SUNY Stony Brook, USA. Since 2003, she has conducted independent research at the University of Zurich, where she held an SNSF-funded *Förderungsprofessur* from 2008 to 2014 and received her Habilitation in Bioinorganic Chemistry in 2014. Her research focuses on plant metallothioneins, with additional interests in the sequence-specific modification of nucleic acids. She co-chaired the 12th European Biological Inorganic Chemistry Conference (EuroBIC-12, Zurich, 2014) and the 19th International Conference on Biological Inorganic Chemistry (ICBIC-19, Interlaken, 2019), both together with Roland Sigel. In addition to serving on several international advisory boards, she has been Secretary of the EuroBIC conference series since 2014 and was Co-Director of the Department of Chemistry in Zurich from 2000-2024. She joined the *Metal Ions in Life Sciences* (MILS) editorial team starting with Volume 18.

Roland K. O. Sigel is Full Professor (2016) of Chemistry at the University of Zürich, Switzerland. In the same year, he became Vice Dean of Studies (BSc/ MSc), and in 2017, he was elected Dean of the Faculty of Science. From 2003 to 2008, he was endowed with a Förderungsprofessur of the Swiss National Science Foundation, and he is the recipient of an ERC Starting Grant 2010. He received his doctoral degree summa cum laude (1999) from the University of Dortmund, Germany, working with Bernhard Lippert. Thereafter he spent nearly three years at Columbia University (now Yale University), New York, USA, with Anna Marie Pyle. During the six years abroad, he received several prestigious fellowships from various sources, and he was awarded the EuroBIC Medal in 2008 and the Alfred Werner Prize (SCS) in 2009. Between 2015 and 2019, he was the Secretary of the Society of Biological Inorganic Chemistry (SBIC), and since 2018, he has been the Secretary of the International Conferences on Biological Inorganic Chemistry (ICBICs). His research focuses on the structural and functional role of metal ions in ribozymes, especially group II introns, regulatory RNAs, and on related topics. He is also an Editor of Volumes 43 and 44 of the *MIBS* series and of the *MILS* series from Volume 1 on.

Historical Development and Perspectives of the Series

*Metal Ions in Life Sciences**

It is an old wisdom that metals are indispensable for life. Indeed, several of them, like sodium, potassium, and calcium, are easily discovered in living matter. However, the role of metals and their impact on life remained largely hidden until inorganic chemistry and coordination chemistry experienced a pronounced revival in the 1950s. The experimental and theoretical tools created in this period and their application to biochemical problems led to the development of the field or discipline now known as *Bioinorganic Chemistry*, *Inorganic Biochemistry*, or more recently also often addressed as *Biological Inorganic Chemistry*.

By 1970, *Bioinorganic Chemistry* was established and further promoted by the book series *Metal Ions in Biological Systems* founded in 1973 (edited by H.S., who was soon joined by A.S.) and published by Marcel Dekker, Inc., New York, for more than 30 years. After this company ceased to be a family endeavor and its acquisition by another company, we decided, after having edited 44 volumes of the *MIBS* series (the last two together with R.K.O.S.) to launch a new and broader minded series to cover today's needs in the *Life Sciences*. Therefore, the Sigels' new series is entitled

"Metal Ions in Life Sciences".

After publication of 22 volumes (since 2006), we are happy to join forces from Volume 23 on in this still growing endeavor with Taylor & Francis, London, UK, a most experienced Publisher in the *Sciences*.

The development of *Biological Inorganic Chemistry* during the past 40 years was and still is driven by several factors; among these are (i) attempts to reveal the interplay between metal ions and hormones or vitamins, etc.; (ii) efforts regarding the understanding of accumulation, transport, metabolism, and toxicity of metal ions; (iii) the development and application of metal-based drugs; (iv) biomimetic syntheses with the aim to understand biological processes as well as to create efficient catalysts; (v) the determination of high-resolution structures of proteins, nucleic acids, and other biomolecules; (vi) the utilization of power- ful spectroscopic tools allowing studies of structures and dynamics; and (vii) more recently, the widespread use of macromolecular engineering to create new biologically relevant structures at will. All this and more is reflected in the volumes of the series *Metal Ions in Life Sciences*.

* Reproduced with some alterations by permission of John Wiley & Sons, Ltd., Chichester, UK (copyright 2006) from pages v and vi of Volume 1 of the series *Metal Ions in Life Sciences* (MILS-1).

The importance of metal ions to the vital functions of living organisms, hence, to their health and well-being, is nowadays well accepted. However, in spite of all the progress made, we are still only at the brink of understanding these processes. Therefore, the series *Metal Ions in Life Sciences* links coordination chemistry and biochemistry in their widest sense. Despite the evident expectation that a great deal of future outstanding discoveries will be made in the interdisciplinary areas of science, there are still "language" barriers between the historically separate spheres of chemistry, biology, medicine, and physics. Thus, it is one of the aims of this series to catalyze mutual "understanding."

It is our hope that *Metal Ions in Life Sciences* continues to prove a stimulus for new activities in the fascinating "field" of *Biological Inorganic Chemistry*. If so, it will well serve its purpose and be a rewarding result for the efforts spent by the authors.

Astrid Sigel and Helmut Sigel
Department of Chemistry, Inorganic Chemistry
University of Basel, CH-4056 Basel, Switzerland

Eva Freisinger and Roland K. O. Sigel
Department of Chemistry
University of Zürich, CH-8057 Zürich, Switzerland

October 2005 and March 2023

Preface to Volume 28

Metal Ions and the Route to Life

In 1906, more than a century ago, the French biologist Stéphane Leduc posited life's origin to be rooted in morphogenic processes involving metal-bearing minerals, driven by diffusional processes and osmotic effects. Leduc was led to this hypothesis by comparing morphologies and growth properties observed in the so-called chemical gardens (known to chemists and alchemists since the mid-17th century; for an example of a chemical garden, see the cover figure of this volume) to those of living entities such as plants or fungi. Leduc's ideas were at best disregarded but frequently even ridiculed by his contemporaries and by the origin-of-life field ever since. From the times of Berzelius and the birth of organic chemistry in the early 19th century, life was considered to depend on organic molecules crucially and what could be more antithetical than hydrocarbons and metals or minerals? Following this mindset, origin-of-life research throughout the second half of the 20th century was preoccupied with elucidating possible prebiotic sources for these organics.

Meanwhile, however, biology, and particularly biochemistry, had begun to appreciate the paramount roles of metal ions in life's processes. The former "trace elements" turned into main players in living entities from simple bacteria to humans, and the entire *Metal Ions in Life Sciences* series indeed bears witness to this upheaval in our perception of what is important for life to exist.

This emerging awareness of metal ions' importance to life laid the groundwork for a fundamental calling into question the organics-centered origin-of-life scenarios, a paradigm shift in progress over the last two decades. The aim of this volume of the MILS-series is (a) to summarize the biochemical and thermodynamics arguments which feed this paradigm shift, (b) to highlight minerals (and their reactivities) which are currently discussed as promising candidates for mediating the transition from inanimate matter to life, and (c) to outline potential scenarios for life's emergence from inorganic settings which are thermodynamically viable and biologically plausible through their similarities to characteristics of extant life on Earth.

Chapter 1 discusses the thermodynamic and bioenergetic logic for why metal ions and not organic molecules must be considered an absolute necessity for life to emerge. Building on this logic, it is suggested that, in addition to previously considered minerals, layered metal-bearing, nanocrystalline minerals, the so-called anionic clays, deserve particular attention. Finally, the ramifications of recognizing metal ions as life's main players for the concepts of autopoiesis and habitability are discussed.

The catalytic and non-catalytic roles of metal ions in certain types of metabolism and biogeochemical cycles are addressed in Chapter 2. Specific types of metal ions are seen to be associated with individual geochemical transformations. This chapter also outlines major metal-mediated emergence-of-life scenarios based on iron or zinc minerals.

Chapter 3 presents an exhaustive description of the ensemble of metalloproteins encountered in living entities from prokaryotes to complex organisms. The involvement of this multitude of metalloenzymes in a plethora of metabolic pathways perfectly illustrates the paramount role metal ions play in living cells.

The outstanding importance of specific metalloenzymes in mediating between life and its environment is highlighted in Chapter 4. In this chapter, varying availabilities of certain metal ions through not only the paleogeochemical eons but also their heterogeneous presence in different habitats are shown to impact both life's long-term evolutionary trajectory and the potential to adapt to variable environmental conditions. This chapter also raises the problem of repercussions for the concept of habitabilities of other bodies in the universe.

From Chapter 5 onwards, the focus shifts from biology to inorganic chemistry/geochemistry/mineralogy. Chapter 5 describes the versatile electron transfer properties of inorganic nanoparticles (mainly iron oxides and sulfides) and their catalytic activities suggesting that they may have served as prebiotic oxidoreductases. Conceivable evolutionary transitions from such nanocrystalline metal-bearing minerals to actual metalloproteins are discussed. As shown in this chapter, these nanocrystalline minerals are involved in essential geochemical cycles to the present day.

Since the most pertinent functions of prebiotic metal ions and minerals likely reside in performing a plethora of redox reactions, many reaction intermediates will necessarily feature unpaired electrons. Electron paramagnetic resonance (EPR) spectroscopy therefore potentially represents a powerful tool to characterize electronic states and reaction schemes of such minerals during catalytic turnover. Chapter 6 makes the case for recruiting EPR as a technique to study prebiotic reactions mediated by metal ions and by complexes between metals and organics.

Chapter 7 (and to a certain degree also Chapter 5) showcases anionic clays and their unique redox properties and catalytic abilities. This chapter introduces the (Fe^{2+}/Fe^{3+} oxyhydroxy) mineral green rust (taken up again in Chapters 8 and 9) as well as the manganese-based birnessites and details their electronic structures and reactivities. This chapter furthermore highlights the use of electrochemical techniques in obtaining the redox properties of these types of minerals.

Chapter 8 zooms in onto the mineral green rust and its naturally occurring form, fougerite. This chapter presents recently elucidated geochemical conversions catalyzed by green rust. The reactants for these processes are abundant in and around submarine alkaline hydrothermal vents, the chapter then develops a scenario for the emergence of an abiotic form of free energy conversion in such geochemical settings which reproduces the bioenergetic principles operating in extant life.

In Chapter 9, a corresponding scenario is put forward, however, mainly motivated by paleogeochemical observations and thermodynamic considerations. The author acknowledges lines of criticism brought up against his initial hypothesis proposed more than 20 years ago and lays out an updated version of this hypothesis taking into account the points of critique and the wealth of paleogeochemical and mineralogical results which have accumulated over these two decades.

Summing up, the chapters assembled in volume 28 consolidate the notion that the primordial role of metal ions in life's emergence needs to be taken seriously in order to arrive at hypotheses which yield entities actually resembling extant life and which do make thermodynamic sense! As should transpire from the ensemble of the chapters, Leduc certainly missed several essential ingredients for rationalizing the emergence of life, for example, the role of redox gradients, the need for free energy-converting mechanisms and anabolic, organics-generating reaction schemes. Still, his scenario can safely count as impressively visionary and prescient, proposed more than a century before our times. The entire volume 28 emphasizes that the research communities dealing by and large with metal ions are bound to play crucial roles in unraveling the specifics of life's emergence on our planet, roles they may have considered denied due to the prevalence of the organic molecule-centered mindset in the origin-of-life field.

Contents

Contributors to Volume 28

Laura M. Barge
NASA Jet Propulsion Laboratory
California Institute of Technology
Pasadena, CA, USA

Bernardo Barosa
Department of Biology
University of Naples Federico II
Naples, Italy

Deborah Bastoni
Department of Biology
University of Naples Federico II
Naples, Italy

Jacopo Brusca
Department of Biology
University of Naples Federico II
Naples, Italy
and
Department of Environmental Science, Informatics and Statistics
University of Venice Ca' Foscari
Venice, Italy

Martina Cascone
Department of Biology
University of Naples Federico II
Naples, Italy

Vidhya Chakrapani
Howard P. Isermann Department of Chemical and Biological Engineering
Rensselaer Polytechnic Institute
Troy, NY, USA

Adriana Clouet
CNRS, BIP (UMR 7281)
Aix Marseille University
Marseille, France

Angelina Cordone
Department of Biology
University of Naples Federico II
Naples, Italy

Simon Duval
CNRS, BIP (UMR 7281)
Aix Marseille University
Marseille, France

Orion Farr
CNRS, CINaM (UMR 7325)
CNRS, BIP (UMR 7281)
Aix Marseille University
Marseille, France

Gabriella Gallo
Department of Biology
University of Naples Federico II
Naples, Italy

Nil Gaudu
CNRS, BIP (UMR 7281)
Aix Marseille University
Marseille, France

Donato Giovannelli
Department of Biology
University of Naples Federico II
Naples, Italy
and
Marine Chemistry & Geochemistry Department – Woods Hole Oceanographic Institution
Woods Hole, MA, USA
and
Institute for Marine Biological and Biotechnological Resources
National Research Council of Italy (CNR-IRBIM)
Ancona, Italy

and
Department of Marine and Coastal Science
Rutgers University
New Brunswick, NJ, USA
and
Earth-Life Science Institute, ELSI
Tokyo Institute of Technology
Tokyo, Japan

Xiaolan Huang
Center for Clean Water Technology, School of Marine and Atmospheric Sciences
Stony Brook University
Stony Brook, NY, USA

Bhanu P. Jagilinki
Department of Chemistry and Biochemistry
University of Oklahoma
Norman, OK, USA

John-Paul Jones
NASA Jet Propulsion Laboratory
California Institute of Technology
Pasadena, CA

Antonio Longo
Department of Biology
University of Naples Federico II
Naples, Italy
and
Department of Integrative Marine Ecology
Stazione Zoologica Anton Dohrn
Naples, Italy

Wolfgang Maret
Department of Nutritional Sciences
School of Life Course and Population Sciences
Faculty of Life Sciences and Medicine
King's College London
London, UK

Flavia Migliaccio
Department of Biology
University of Naples Federico II
Naples, Italy

Florence Mus
Department of Chemistry and Biochemistry
University of Oklahoma
Norman, OK, USA

Wolfgang Nitschke
CNRS, BIP (UMR 7281)
Aix-Marseille-University
Marseille, France

Paul H. Oyala
Division of Chemistry and Chemical Engineering
California Institute of Technology
Pasadena, CA, USA

John W. Peters
Department of Chemistry and Biochemistry
University of Oklahoma
Norman, OK, USA

Laura E. Rodriguez
Lunar and Planetary Institute
Universities Space Research Association
Houston, TX, USA

Michael J. Russell
Dipartimento di Chimica
Università degli Studi di Torino
Torino, Italy

Luca Tonietti
Department of Biology
University of Naples Federico II
and
Department of Science and Technology
University of Naples Parthenope
and

INAF-OAC
Osservatorio Astronomico di Capodimonte
Naples, Italy

Chloé Truong
CNRS, BIP (UMR 7281)
Aix Marseille University
Marseille, France

Seneca J. Velling
NASA Jet Propulsion Laboratory
Division of Engineering & Applied Science
Kavli Nanoscience Institute
California Institute of Technology
Pasadena, CA, USA

Jessica M. Weber
NASA Jet Propulsion Laboratory
California Institute of Technology
Pasadena, CA, USA

Handbooks and Book Series Published and (Co-) edited by the SIGELs

"Handbook on Toxicity of Inorganic Compounds" (ISBN: 0–8247-7727-1) Eds H. G. Seiler, H. Sigel, A. Sigel; Dekker, Inc.; New York; 1988; 1069 pp

"Handbook on Metals in Clinical and Analytical Chemistry" (ISBN: 0–8247–9094-4) Eds Hans G. Seiler, Astrid Sigel, Helmut Sigel; Dekker, Inc.; New York, Basel, Hong Kong; 1994; 753 pp

"Handbook on Metalloproteins" (ISBN: 0–8247–0520–3) Eds I. Bertini, A. Sigel, H. Sigel; Marcel Dekker, Inc.; New York, Basel; 2001; 1182 pp

Metal Ions in Biological Systems

Volumes 1–44

https://www.routledge.com/Metal-Ions-in-Biological-Systems/book-series/IHCMEIOBISY

(see also the website given below)

Metal Ions in Life Sciences

Volumes 1–28

Details about all books (series) edited by the SIGELs, including the Guest Editors, can be found at http://www.bioinorganic-chemistry.org/mils

Volumes published by Taylor & Francis

Volume 23: Molecular Bio-Sensors and the Role of Metal Ions
Guest-edited by Thomas J. Meade

Volume 24: Targeted Metallo-Drugs: Design, Development, and Modes of Action
Guest-edited by Etelka Farkas and Celine J. Marmion

Volume 25: Modern Avenues in Metal-Nucleic Acid Chemistry
Guest-edited by Jens Müller and Bernhard Lippert

Volume 26: Metal-Containing Molecules and Nanomaterials: From Diagnosis to Therapy
Guest-edited by Ana de Bettencourt-Dias

Volume 27: Lanthanide and Other Transition Metal Ion Complexes and Nanoparticles in Magnetic Resonance Imaging
Guest-edited by Carlos F.G.C. Geraldes

Volume 28: Metal Ions and the Route to Life
Guest-edited by Wolfgang Nitschke and Simon Duval

For more information about this series, please visit: www.routledge.com/Metal-Ions-in-Life-Sciences-Series/book-series/CRCMILS

1 The Role of Metal Ions in the Emergence of Life

A Call for a Change in Paradigm

Wolfgang Nitschke
BIP (UMR 7281), CNRS, Aix-Marseille-University, 13009 Marseille, France
nitschke@imm.cnrs.fr

Simon Duval
BIP (UMR 7281), CNRS, Aix-Marseille-University, 13009 Marseille, France
sduval@imm.cnrs.fr

CONTENTS

DOI: 10.1201/9781003459910-1

ABSTRACT

Organic molecules are generally regarded as the quintessential prerequisites for life to emerge on our planet and elsewhere in the cosmos. This mindset is likely guided by the perception that extant life on Earth is all about amino acids, nucleotides, sugar compounds, and lipids, i.e. heteroatom-containing hydrocarbons. This view, however, neglects more than a century of research revealing the indispensable roles played by metal ions in all sorts of crucial biological processes. Without metal ions, extant life simply would cease to exist! It therefore isn't far-fetched to extrapolate the ineluctable necessity for the implication of metal ions from modern life all the way back to life's very inception. In this chapter, we try to put metal ions to the center stage in the transition from inanimate matter to living entities, a position they may, to our minds, rightfully claim. In addition to pointing out that the supposed primacy of organic matter is a mirage that disappears when numbers of agents rather than "weight ratios" are taken into account, we argue that metal ions and not organics are best-suited to initiate life's most fundamental process, free energy conversion. We then discuss the repercussions that a metal-centered emergence-of-life paradigm entails for the concepts of autopoiesis and habitability.

Like-minded colleagues from the biological sciences, (geo)chemistry, and mineralogy have accepted to contribute to this volume and bolster its message by re-assessing the implications of metal ions in life's mechanisms (Chapters 2–4), presenting "life-like" catalytic properties of metal ions in certain minerals (Chapters 5–8) and elaborating potential scenarios for the emergence of life relying on the agency of metal ions (Chapters 8 and 9).

KEYWORDS

metal-ion-based emergence of life, catalysis, free energy conversion, second law of thermodynamics, redox reactions, bioenergetics, anionic clays, autopoiesis, habitability

1 THE MISTAKEN NOTION OF LIFE BEING DEFINED BY ORGANIC MOLECULES

The processes characterizing life are commonly encapsulated in the term "biochemistry". Now open any textbook on biochemistry, and you will find in-depth introductions into the physico-chemical properties of amino acids, lipids, and nucleic acids; explanations of how they form proteins, biomembranes, and RNA/DNA superstructures; and finally descriptions of how enzymes (that is, proteins and protein assemblies) catalyze the chemical reactions that make cells look "alive". You may be led to believe that the actual definition of catalysis is "what an enzyme does". That said, you may be led to this impression only unless you are an inorganic chemist, a mineralogist, a chemical engineer, in short, not a biologist. In these disciplines, the archetypal catalysts, i.e. agents that decrease the activation barriers impeding chemical reactions from swiftly attaining thermodynamic equilibrium, are metals! Checking back within the above mentioned biochemistry textbooks, we will well find a few pages dealing with the role of metals in the catalysis of certain enzymes. However, these metals will be treated more like welcome helpers to "improve" certain reaction efficiencies rather than as catalytic agents per se.

As we will summarize in the next section and as is discussed in detail in the contributions by Wolfgang Maret (Chapter 2, [1]), Bhanu Jagilinki et al. (Chapter 3, [2]), and Donato Giovannelli et al. (Chapter 4, [3]), the standard perception gained from the biochemistry textbooks is arguably erroneous: Rather than being nice "additives" enhancing the catalytic capabilities of enzymes, metal ions in many (possibly most) cases run the show, meaning "no metal, no catalysis" and exclusively metal-ion-catalyzed processes are at the heart of what makes life possible, that is, energy metabolism, a topic which we will deal with in more detail in Section 1.2.

1.1 Abundances: The Gram-Per-Dry-Weight Fallacy Exposed

The very inception of organic chemistry, at the turn from the eighteenth to the nineteenth century, was motivated by the fact that the first chemists trying to understand animate matter found that living things were made up of a substance appearing to be substantially different from inanimate matter (for example, represented by rocks). When elements became recognized by early chemists, this substance turned out to be mainly hydrocarbons containing additional "heteroatoms" which principally are nitrogen, oxygen, sulfur, and phosphorus. Indeed, living matter is made up almost entirely (about 99% of dry weight) of these "CHONPS" elements (alternatively referred to as CHNOPS or SPONCH) [1], and since the days of the earliest "organic chemists" in the beginning of the nineteenth century, the essence of what makes life possible consequently was considered to lie in the "organic" molecules. However, the devil in this assessment is in the detail of the "almost entirely" and in straightforwardly equating "weight/mass" with importance. The remaining percent of dry weight at first sight may appear insignificant but a closer look suggests otherwise. As touched upon in Section 1.1, the entity catalyzing a given biochemical reaction isn't a single carbon, oxygen, or nitrogen atom, but an enzyme. Such an enzyme, however, is built from an extraordinarily high number of CHONPS atoms. For example, the bacterium *Escherichia coli* needs more than 6,000 carbons, 1,800 oxygens, and 1,700 nitrogens to build the catalytic subunit of its respiratory metalloenzyme nitrate reductase (Nar). The catalytic activity of this enzyme, by contrast, stands or falls by a single transition metal atom, the molybdenum ion! The excess of more than four orders of magnitude of "architectural" CHONPS atoms over the actually catalyzing metal ion in the Nar-enzyme is arguably an extreme case resulting from the exceptionally high molecular weight of this enzyme's catalytic subunit. For most metalloenzymes this ratio nevertheless still exceeds three orders of magnitude, a fact which shines a very different light on the above mentioned 99%-share of CHONPSes in a living cell: Life needs (by numbers) 1,000 times more CHONPS atoms than transition metals to make a functional metalloenzyme. Even taking into account that CHONPS elements are on average 4–5 times lighter than the bulk of the biologically most common transition metals (e.g. Fe, Cu, Co, Ni), we are forced to conclude that the 99%-to-1% dry-weight-ratio actually corresponds to a relative importance (in numbers) of metal ions exceeding those of CHONPS elements by roughly a factor of 2. Organo-centric biochemists may counter that all the above may well apply to metalloproteins but that these are only a small subgroup of the world of enzymes. This actually isn't true. The relative contribution of metalloproteins to the protein-total is estimated to exceed 50% [4] and, even more strikingly, there seems to be no metabolic pathway that doesn't depend on at least one metalloenzyme [5]. The mind-boggling multitude of metal-ion-catalyzed biochemical reactions in all forms of life on Earth is spelled out in the contribution by Bhanu Jagilinki et al. (Chapter 3).

Metal ions' preeminent role as biological catalysts becomes even more manifest if one bears in mind that CHONPSes not only form enzymes but also lipids, sugars, nucleotides, and so on. We have recently explicitly compared numbers (rather than dry-weight-ratios) of metal ions and the respective major organic units (proteins, RNA, DNA, lipids, sugars, monomers) of a living cell [6]. As illustrated in Figure 1, by numbers, the transition metal ions are the most abundant catalytically active entities in living cells rather than mere "trace elements". The biochemically most relevant transition metals iron, nickel, manganese, copper, cobalt, and zinc outnumber enzymes by a factor of 4 and even iron by itself is more abundant than proteins (Figure 1). The towering excess of transition metals over proteins as catalytic agents is in reality even more pronounced since obviously not every protein is an enzyme. Chapters 2 and 3 further detail the crucial roles of metal ions in extant life while Chapter 4 discusses how metal ions and their availability have influenced the evolutionary trajectory of life on our planet.

Obviously, lipids and nucleotides are also crucial in making life possible, and their roles almost certainly cannot be taken over by (or reduced to) metal ions. The important role of these two classes of organics notwithstanding, we maintain that, when it comes to metabolic processes, metal ions are the lead actors. Why is that so?

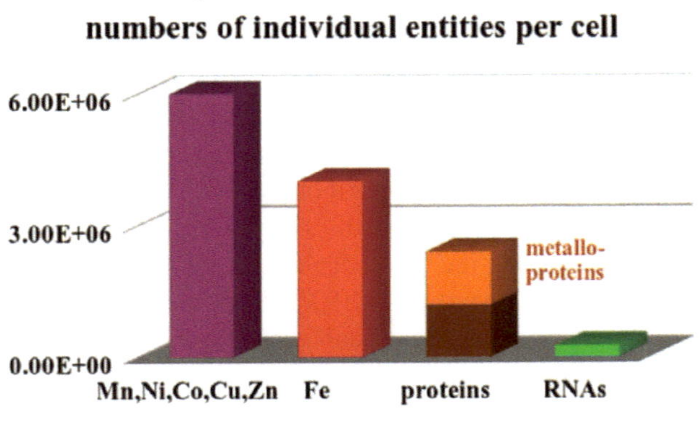

FIGURE 1 Bar chart emphasizing the large excess in numbers of transition metal ions in an average cell over the "organic entities" such as proteins and RNAs (data from the B10NUMB3R5 website, https://bionumbers.hms.harvard.edu) [7] and from [8]). Apart from lipids, the number of all other organic entities would be imperceptible on the scale of this histogram.

1.2 The Reason for the Primacy of Transition Metals: Life Is Redox!

The working-out of thermodynamics during the 19th century (culminating in Boltzmann's theory of statistical mechanics) generated an explanatory tension between the physical/physico-chemical description of the natural world and the life sciences. While systems belonging to the former realm were recognized to tend toward increasing disorder (i.e. to obey the second law of thermodynamics), living entities seemingly defied all thermodynamic rules by generating order from disorder (for a more detailed description of the historical background, see [6]). Vitalism, a now discredited hypothesis stipulating that organic molecules are endowed with an anti-thermodynamic, intrinsically "life-generating", force, was the 19th century's bid at easing this conceptual tension. The knot was finally undone in 1944 by Erwin Schrödinger's recognition that the phenomenon of life can be seamlessly integrated into the framework of thermodynamics if it was understood as a subsystem of a larger one (i.e. living entities plus their environment) which itself was in a state of thermodynamic disequilibrium [9]. Life can under these circumstances take the entire system closer to equilibrium while in the process transforming part of the order of the larger system into its own low entropy [6,9]. Given the biological knowledge at the time Schrödinger wrote his book *What Is Life?* [9], the idea of life embedded in and living off a bath of low entropy necessarily had to remain quite abstract with respect to what this environmental disequilibrium actually was. Elucidation of its nature had to await the 1960s and the deciphering of the process of "oxidative phosphorylation" [10,11]. Efforts to unravel the functioning of initially mitochondria and then chloroplasts marked the launch of a new field of research: "Bioenergetics". Initially confined to the mentioned powerhouse organelles of animal and plant cells, bioenergetics rapidly widened its scope with the study of new and ever more exotic prokaryotes discovered by microbiologists during the second half of the 20th century. The take-home message of several decades of research into the bioenergetic processes of the microbial world can be put in a nutshell: Life universally feeds its entropy decrease from electrochemical disequilibria [12,13] as emphasized in Chapter 4. These redox tensions are in the vast majority of organisms directly provided by the environment in the form of reducing and oxidizing substrates but can, for the case of photosynthetic species, also be generated internally through photochemical reactions [6,12,13].

This recognition provides a straightforward and simple rationalization for why the functional significance of metal ions vastly surpasses that of CHONPS elements. CHONPSes are redox inactive in the range of ambient potentials relevant to bioenergetics. Only very specific molecular assemblies of CHONPS elements (mainly heteroatomic aromats such as flavins, nicotinamides, quinones, and the likes) can cycle between oxidized and reduced states within this range. By contrast, basically every transition metal ion can!

As mentioned, certain organics (e.g. nucleotides) doubtlessly also are indispensable for making life possible. Nevertheless, Schrödinger's thermodynamic insight puts energy metabolism firmly above and prior to all other processes of a living cell and thus rationalizes a primacy of metal ions in life's fundamental mechanisms.

2 FROM EXTANT ORGANISMS BACK TO LIFE'S EMERGENCE

While it may only just be possible to convince the biochemist of a primordial role of metal ions in the functioning of extant life on Earth, the vast majority of researchers in the origin-of-life (OoL) field will remain unmoved, arguing that nascent life necessarily was based on simple organic molecules. Too ingrained in the collective mindset not only of the OoL community but also of the general public is the notion of the "primordial soup of organic molecules" from which life is supposed to emerge spontaneously (if given enough time), on our planet just as on any other planetary body presenting suitable (understood as "comparable" [14]) conditions. In this narrative, mixtures of simple organic molecules step-by-step bootstrapped themselves out of a high entropy state toward increasing complexity until finally arriving at something vaguely reminiscent of primitive cellular structures.

2.1 The Epistemological Problem of "Nascent Life as We Don't Know It"

The notion that earliest life had almost nothing (apart from the CHONPSes, of course) in common with present-day organisms is close to foundational to the vast majority of OoL scenarios. From the "RNA world" [15] via the iron-sulfur world [16] to the "Zinc world" [17] (for a discussion of the latter two, see Chapter 2), emergent life has been depicted as truly alien and almost unrecognizable to the modern microbiologist. Inevitably, invoking alien beginnings entails the necessity for radical changes on the way from fledgling life, estimated to have taken off on planet Earth more than 3.8 (but less than 4.2) billion years ago, to the last universal common ancestor (LUCA), commonly dated to at least 3.5 billion years in the past. Incidentally, we draw the reader's attention to the remarkable fact that over more than 3 billion years of evolution from the LUCA to the present day, the makeup of a living cell hasn't changed fundamentally.

There is of course no empirically based argument refuting a succession of evolutionary upheavals as envisaged in such "alien-life-world" scenarios (as they might be called summarily). However, they pose the epistemological problem of whether such hypotheses, as refined as they may come, can count as "knowledge". Since by definition (by virtue of their constituent evolutionary "quantum leaps") they cannot be empirically falsified and thus fail Karl Popper's criteria [18] but also fare badly when confronted with Occam's razor [19]. However, is there any scenario out there which is more parsimonious in unprovable events and ad hoc assumptions? As we will discuss in the following, such scenarios do indeed exist, and they are all intimately linked to metal ions being instrumental in accomplishing the transition from inanimate matter to life.

2.2 Did Redox Tensions and Transition Metal Ions Spawn Life?

Even Erwin Schrödinger was puzzled by the apparent paradox that in extant life, order is only produced from order while life's origin seemed to imply the generation of order from disorder (chapter 7 in [9]). As mentioned, Schrödinger could not have anticipated that bioenergetics would soon after elucidate the nature of the "order" feeding life, that is, redox disequilibria, nor that this type of free energy was predominantly geochemical, therefore pre-dating the emergence of life and was available to nascent life right from the beginning. And so obviously were metal and transition metal ions, the embodiment of inorganic matter!

Scenarios for life's emergence envisaging an evolutionary trajectory from life's inception to the present day with a constant underlying principle of energy metabolism are therefore conceivable. Nascent life may have been fed by the same kind of free energy as extant organisms, and the main players in the process of converting environmental free energy into life's ordering may have been transition metal ions then just as now. The epistemological appeal of such an evolutionary trajectory has been recognized by Frederick Donnan already more than a century ago [20] (see the introductory section of Chapter 9). A number of scenarios following this lead, and avoiding the problem of abrupt radical changes to varying degrees, have been put forward over the last decades [21–30].

For all the reasons mentioned earlier, we consider it overdue to comprehensively challenge the paradigm of organic molecules being the essence of life and in particular of life's appearance on our planet and to explore the possibilities of a "metal-mediated" emergence of life from diverse angles. Evidently, the "Metal Ions in Life Sciences" series provides the natural and ideal forum for such an endeavor, and we are grateful to the series editors for giving us the opportunity to assemble contributions fleshing out the claims sketched above.

The first three contributions in this volume reassess the role of metal ions in extant life paving the ground for the remainder of the volume which addresses different aspects of how metal ions may have been involved in the appearance of life on Earth.

These putative pioneering metal ions almost certainly were not coordinated by organic molecules but were harbored within certain minerals. Resemblances between the catalytic feats accomplished by such minerals and biochemical reaction schemes have been discussed already several decades ago. Günter Wächtershäuser and colleagues have reported a host of bio-relevant reactions occurring on the surfaces of pyrite crystals [31] (and see Chapter 2 of this volume [1]). Another mineral-based scenario for life's emergence, envisaging Lost-City-type submarine alkaline hydrothermal vents as life's birthplace, focused on the minerals greigite and mackinawite rather than pyrite as primordial catalysts [21]. Just as for pyrite, a number of biologically important reactions have been shown to be catalyzed by these two minerals [27,32–34] and, intriguingly, similarities between the structural arrangements of their metal ions and those of the metalloenzymes performing the corresponding biochemical reactions have been noticed [27–30]. Pyrite, greigite, and mackinawite and their putative roles in bringing about the emergence of life have been extensively discussed over the last decades [31,35,36]. These are (among others) addressed in Chapters 5, 6, and 9 [37–39] of this book. However, we have chosen to also highlight in this volume a specific class of minerals, to our minds so far neglected and vastly underrated by the OoL community, the metal-bearing layered minerals, also called anionic clays. Chapters 5–9 [37–41] address the intriguing (in particular redox-catalytic) properties pertinent to OoL topics of a host of anionic clays with Chapters 8 [41] and 9 [39] further venturing into more elaborate and detailed proposals for how these properties may have been instrumental in getting life going.

3 METAL-BEARING LAYERED MINERALS: A LONG-OVERLOOKED CLASS COMES TO THE FORE

While pyrite and greigite are 3D-solid crystals, the anionic clays consist of 2D layers made up from positively charged metal ions covalently bridged by (most frequently) oxyhydroxides or (more rarely) by sulfides (e.g. mackinawite). The parallel stacking of these sheets is mainly stabilized by anions occupying the interlayer spaces (hence the attribute "anionic" to the term "clay") together with water molecules or simply by water molecules alone.

Figure 2 illustrates the alternating arrangement of metal-containing layers and interlayer spaces in a "generic" representative of this class of minerals. Variations in metal content (i.e. Al, Mg, Ni, Fe etc.), nature of the metal-ion-connecting bridges (i.e. O(H) or S(H)), and composition of the interlayers yield a zoo of anionic clays

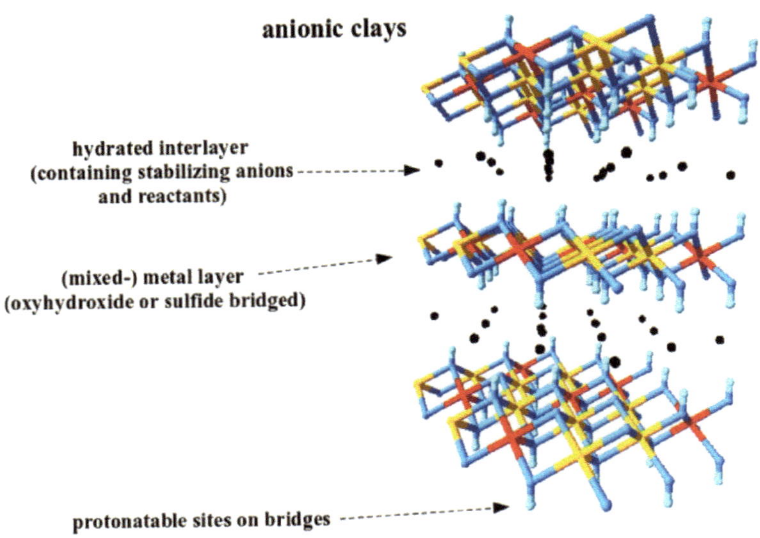

FIGURE 2 Schematic representation of the layered structuring of anionic clays with their characteristic alternating arrangement of positively charged metal-ion-containing sheets and (charge-compensating and hence stabilizing) negatively charged interlayers formed by all sorts of anions. Specific examples of such anionic clays are presented in Chapters 5 and 7–9.

(e.g. hydrotalcite, amakinite, fougerite, brucite, tochilinite, mackinawite, and so on), some of which are presented in detail in Chapters 5 to 9. Just as "true" (i.e. cationic) clays, these layered metal-ion-containing minerals do not feature fixed structures but swell and shrink as a function of water content, type of interlayer anions or redox state of their metal constituents. The majority of these anionic clays are highly unstable under aerobic conditions and therefore difficult to handle and investigate, likely rationalizing why they have failed to attract the due attention of the OoL community so far. In the following section as well as to extensive detail in Chapters 5 and 7–9, the case will be made for the necessity to reappreciate the potential role of anionic clays, and in particular, of the redox-active ones, in life's emergence.

3.1 The Appeal of Layered Minerals to Emergence-of-Life Issues

Layered minerals and in particular the iron-containing layered double oxyhydroxides (LDHs) have been thoroughly studied by the material sciences since more than half a century [42]. Ironically, the pioneers in research on anionic clays have already very early on suggested a potential involvement of these LDHs in getting life started [42], a concept that unfortunately sunk into oblivion for several decades. Yet, there are numerous reasons to take the idea seriously.

- Empirical evidence increasingly indicates that the layered anionic clays were abundantly present in the early planet, as elaborated by M.J. Russell in Chapter 9 [39].
- The fact that reactants can enter the interstitial galleries between the metal-ion-containing sheets implies that the catalytically active surface area of layered minerals is orders of magnitude larger than that of the solid minerals such as pyrite or greigite, a property extensively exploited by the mineral sciences for industrial purposes [43].
- Cationic clays are considered intriguing candidates for assisting prebiotic organic chemistry by favorably pre-ordering reactants via the lattice of charges within sheets [44,45]. The same property obviously is a feature of the anionic clays with the only difference that the opposite charges on reactants will be involved in the 2D ordering. In this vein, Rasmussen et al. have proposed that the anionic clay greenalite may have served as a "template for the assembly and alignment of linear RNA-type molecules" [46].
- Water activity in the interlayers is substantially reduced as compared to bulk aqueous solutions displacing a host of chemical reactions toward the products (in particular those wherein products are prone to hydrolyze). This property is again exploited in industrial use of clays [47].
- The redox-active forms of anionic clays (mostly containing iron) perform a bewildering variety of bio-relevant chemical transformations (see Chapters 5, 8, and 9).
- The individual crystals of many clay minerals grow mainly to nanoscopic sizes and are therefore size-compatible with the predominant catalytic entities of extant life (enzymes), avoiding the necessity of finding excuses for

the transition from macroscopic (e.g. pyrite) 3D solid minerals to the nano-world of a living cell.

3.2 ... And Then There Is the Issue of Free Energy Conversion ...

Schrödinger's affirmation (informed by the second law of thermodynamics as discussed in Sections 1.2 and 2.2) that life must be feeding off an environmental source of free energy is frequently misinterpreted as meaning that life's processes are fundamentally exergonic with their exergonicity being maintained by a constant influx of free energy. However, exergonic processes take a system toward thermodynamic equilibrium which is obviously not what occurs in living entities, the incarnations of systems extremely far from equilibrium. Almost all metabolic processes are indeed highly endergonic rather than exergonic! For example, the biosynthesis of amino acids from simple precursor organics is energetically extremely "uphill", and only infinitesimal quantities of the products (the amino acids) are present once the reaction has reached equilibrium. Of course, there is no direct way how redox tensions, i.e. life's environmental free energy (as discussed in Section 1.2), might render these pathways exergonic. Rather, they are coupled to another chemical reaction scheme, the hydrolysis of polyphosphates (just as is the majority of metabolic reactions). In living cells, the polyphosphate/phosphate ratios are driven to extreme out-of-equilibrium values, conferring an astronomical exergonicity to the hydrolysis reaction of these polyphosphates to yield inorganic phosphate [11]. It is the extreme exergonicity of this reaction scheme which, if coupled to metabolic processes (such as the above mentioned amino acid biosynthesis pathways), drives the latter "uphill".

Rather than being directly driven by environmental redox-free energy, metabolic processes are thus rendered exergonic by their coupling to a different, intracellular, type of free energy, resulting from highly out-of-equilibrium polyphosphate to phosphate ratios (in extant life mainly in the form of ATP/ADP, ATP/AMP, or ADP/AMP couples). Life thus exists due to the fact that there is an intracellular "master-disequilibrium" fueling the entire downstream formation of spatial and temporal organization that characterizes a living cell [48]. This chemical master disequilibrium is constantly maintained at a very high level via a process converting the environmental redox-free energy into the intracellular, chemical, type of free energy. This process of course is bioenergetics (see Section 1.2 and [11,49]). Combining Schrödinger's understanding of the link between life and the second law of thermodynamics with the appreciation of the fundamental difference between the environmental disequilibrium and the intracellular one, it was pointed out that it is ultimately this conversion of ambient redox tensions into polyphosphate/phosphate disequilibria that allows life to exist today [12,13,50]. As we have argued in Section 2.1, it is both intellectually tempting and scientifically parsimonious to assume that this same conversion process thermodynamically drove life into existence and then persisted to the present day (see also [6]).

The appeal of metal ions as prebiotic catalysts for the redox reactions involved in bioenergetic processes has been discussed in Section 2.2 and is argued in detail in Chapter 4 [3]. However, since "catalysts" by definition only facilitate evolution toward thermodynamic equilibrium, these redox reactions can only be part of the story. In addition, there must have been mechanisms which actually did convert disequilibria to get life started! Such reaction steps are considered to involve "Brownian-ratchet"-type mechanisms, that is, fluctuation-induced endergonic transitions rendered irreversible by the driving disequilibrium [48,51,52], very much as those underlying the functioning of extant systems that generate polyphosphate/phosphate disequilibria such as the enzyme ATP synthase.

To allow Brownian-ratchet mechanisms to occur, the mediating agents must possess some degree of conformational flexibility, potentially even coupled to protonation/deprotonation events [13]. Obviously, 3D-solid minerals are unlikely to present such conformational softness. By contrast, the layered minerals highlighted in Chapters 5, 7, 8, and 9 of this volume feature a host of properties making them promising candidates for performing Brownian-ratchet-type reaction schemes, further adding to their attractiveness as essential agents driving the transition from inanimate matter to life.

The contributions by Michael J. Russell (Chapter 9 [39]) and Orion Farr et al. (Chapter 8 [41]) specifically address the topic of how layered minerals may have operated as disequilibrium converters in nascent life.

4 REPERCUSSIONS FOR THE CONCEPT OF AUTOPOIESIS

The question of how to define life has haunted the natural sciences and their precursor, natural philosophy, over the millennia (for a discussion see [6]). In the early 1970s, Maturana and Varela [53] proposed a theory which proved highly influential ever since in attempts to conceptualize the phenomenon of life and, consequently, to rationalize its inception: Autopoiesis. In autopoiesis, a living entity is understood as being defined by two distinct types of closure, a topological one (corresponding to the cytoplasmic membrane) and a functional one called "organizational closure". Organizational (or metabolic) closure is considered to be the very basis of a perceived "autonomy" of living cells and is frequently condensed to the (equally perceived) property that all agents necessary to produce all the other agents involved in cellular processes are produced within the cell itself. These foundational principles of Maturana and Varela's autopoiesis reappear in other theoretical attempts to define life, some of which are essentially contemporary with the publication of the theory of autopoiesis, such as Kauffman's "autocatalytic cycles" [54], Rosen's "(M,R) systems" [55], or Gánti's "chemotons" [56]. For example, Rosen's definition of (M,R) systems as "closed to efficient cause" means that enzymes (the agents of catalysis) are themselves produced by enzymes, echoing autopoiesis' principle of organizational closure. Taking the principle of organizational closure for granted, reflections on the ways how life might have come into being were over the last half century inexorably nudged toward theories gravitating around "self-organizing" molecular systems and autocatalytic networks based on, as to be expected, organic molecules.

The very notion of life's autonomy, dear to autopoiesis and the related concepts, has been criticized based on the fact that cells obviously need to exchange matter with the environment (raw materials and waste), let alone that they can't exist without a constant influx of free energy. The illusion of an autonomous existence of a living entity obviously is unsparingly shattered within milliseconds to minutes (depending on the type of organism) of being deprived of an external source of free energy.

However, the appreciation of the role of metal ions in extant life as conveyed by Chapters 2–4 of this volume [1–3] deals an even heavier blow to the core-tenet of autopoiesis. In extant life, roughly half of the catalysts in an average cell are not produced by enzymes but their catalytic agents, the metal ions, are imported ready-made from the environment. In the overwhelming majority of cases, these metal ion catalysts (which may also be multi-metal clusters) perform roughly the same kinds of catalyzes outside (if possibly less efficiently and less specifically) as they do within the cell, clad in their polypeptide coats, as addressed in Chapters 5, 8, and 9 [37,39,41]. Intriguingly, the share of metalloenzymes involved in bioenergetic free energy conversion vastly exceeds the rough 50% found in the overall population of enzymes and even attains 100% in certain organisms. This implies that in nascent life, prior to coded replication, the fundamental process of energy metabolism (see Sections 1.2, 2.2, and 3.2 as well as [6]) may well have been performed exclusively by metal ions sourced from the environment in blatant disregard of the doctrine of operational closure.

Of course, topological closure is the prerequisite for operational closure to be conceivable. If, according to the considerations outlined above, the principle of operational closure is at best inessential, if not obsolete, what about the concept of topological closure? While the concept of operational closure may be dismissed as a misapprehension resulting from the neglect of the role of metal-ion-based catalysis, the existence of a cytoplasmic membrane "enclosing" living cells simply is an empirical fact observed in all extant life on our planet. However, this cytoplasmic membrane, i.e. the confinement thought to provide topological closure, actually isn't a barrier to most of the agents central to the theory of autopoiesis but is a busy hub transporting atoms and molecules, even multiprotein enzyme complexes, from the outer world to the inside and back.

By contrast, all cytoplasmic membranes have one common property: They are dielectric barriers allowing the (bioenergetics-mediated) build-up of cation-motive gradients which are the universal transducers of environmental free energy into chemical free energy driving all metabolic processes within cells [6]. In stark contrast to the vision promoted by autopoiesis and related concepts, the cytoplasmic membrane does NOT insulate the cell from the outside world but instead represents the indispensable transducer of free energy allowing intracellular low entropy to be maintained by disequilibria in the environment.

Autopoiesis' emphasis on an autonomy of living systems set these systems apart from the "outer world". Taking the paramount role of metal-ion-mediated processes in life seriously and recognizing the cytoplasmic membrane as a coupling device rather than a separating enclosure, on the contrary puts life firmly back into the physical world and draws attention to the extreme inter-dependence of living entities and their environment.

5 THE ROUTE TO LIFE ON EARTH ... AND ELSEWHERE? COROLLARIES FOR "HABITABILITY"

The contributions in this volume suggest that the role of metal ions in the emergence of life is not simply underestimated, it is misconceived. There is indeed a strong case to be made for the notion that metal ions weren't merely ingredients in getting life started, they were the indispensable operators catalyzing crucial "pre-biochemical" reactions (Chapters 2 [3] and 4–9 [5,37–41]) and mediating the conversion of environmental free energy into the elaborate ordering of living entities (Chapters 8 [41] and 9 [37]). Metal ions were arguably the "conditio sine qua non" for the possibility of life to appear on planet Earth.

As further emphasized by all contributions in this volume, while there is some degree of overlap, each individual metal ion excels in a specific range of reaction schemes. It therefore seems highly likely that the emergence of life on planet Earth could only occur in environments supplying the full panoply of catalytically relevant metal ions, comprising (most prominently) iron but also nickel, zinc, and cobalt as well as (potentially) molybdenum and tungsten. Metal-centered emergence-of-life scenarios therefore frequently consider hydrothermal vents with their naturally metal-rich effluents as very promising nurseries for nascent life (see Chapters 4–9).

Astrobiology (also referred to as Exobiology) is a discipline in desperate search for its conjectured research object. Despite strictly no empirical evidence so far for life elsewhere in the universe, astrobiology has over the last decades become a flourishing field enjoying significant attention from the general public. Unable to put actual alien life under the microscope or through the mass spectrometer, astrobiology has focused much of its scope on predicting under which environmental conditions, and thus on which planetary bodies (including moons) life might have appeared. Such "habitability" research aims at extrapolating the conditions deemed essential for the emergence of life on Earth onto other planetary bodies and thereby home in on the most promising targets for further investigation. The "NASA astrobiology roadmap" [14] stipulates as mayor criteria for habitability the presence of (a) "extended regions of liquid water", (b) favorable conditions for assembling organic molecules, as well as (c) an energy source suitable for driving metabolic processes. Obviously, this list reflects the mindset exposed in Section 1 that life is all about organic molecules performing biochemical reactions in aqueous solution, all that enhanced by a pinch of (undefined) "energy" (for a discussion of the distinction between energy and free energy, see [6,50]).

If we take the messages conveyed by the contributions in this volume seriously, a stark reappraisal of the criteria for habitability is ineluctable. Firstly, organic molecules have more recently been found in space almost everywhere they were looked for, from other planets and moons in our solar system [57–60] through asteroids and meteorites [61–65] to interstellar clouds [66]. Apart from the stars themselves and more extreme cosmic bodies such as neutron stars, black holes, and the likes, organics seem to be trivial ingredients of the universe.

This, however, is not the case for metals! Unlike on our planet where transition metal ions were present dissolved in the primeval oceans to get life started, metals and in particular their heavier representatives in the fifth and sixth period of the periodic table are far from evenly available in the universe (see also the corresponding discussion in Chapter 4 [3]).

5.1 A Detour into Nucleosynthesis

Big Bang nucleosynthesis in the nascent universe (supposed to have occurred for about 20 minutes following the Big Bang) produced only helium nuclei from protons together with traces of lithium [67]. Matter in the early universe therefore was essentially restricted to hydrogen and helium while the appearance of heavier elements had to await stellar nucleosynthesis. The ignition of nuclear fusion reactions in nascent stars starting at about 400 million years after the Big Bang began breeding heavier nuclei in these stars' interiors. In addition, the extremely high energies reached during explosive events, such as in supernova explosions and neutron star mergers, allow for further mechanisms of nucleosynthesis to operate [67,68]. The present picture for the origin of elements heavier than lithium views the elements in the third and fourth period (the latter containing iron and most of the biologically relevant transition metals) as produced during core collapse and type Ia supernovae (type Ia supernovae arise from binary star systems) and as distributed into their stellar neighborhoods through the explosion-associated shock waves. By contrast, the quasi-totality of fifth and sixth period elements (including molybdenum and tungsten, see Section 6) are either bred during the late stages of stellar evolution of low- to middle-mass stars (during the so-called "asymptotic giant branch" or AGB-phase) or cataclysmically generated in neutron star merger events. AGB stars expel their outer layers (charged with heavy elements through convection from the inner layers) into the circumstellar medium to form so-called planetary nebulae [67]. Explosive termination of the life cycle of stellar bodies such as supernovae and neutron star mergers are comparatively rare and the distribution of the bulk of the transition metals (including their fifth and sixth period representatives) is therefore expected to be less homogeneous than that of the light elements (containing the CHONPSes). Corresponding inhomogeneities are indeed observed in our parent galaxy via spectral analysis of the photospheres of representative stars [69,70].

To avoid confusion, we point out here that in the terminology of astronomers and cosmologists, all elements heavier than hydrogen and helium (i.e., those provided by the Big Bang) are called "metals". Of course, throughout this chapter (and in the entire volume), the meaning of the term "metal" is as defined by inorganic chemistry!

5.2 A Detour into Planetary System Formation

While the diverse elements present within individual protoplanetary clouds (ultimately collapsing to planetary star systems) are likely to initially be relatively homogeneously distributed (but differing between protoplanetary clouds, see previous section), the series of processes leading to planets and their moons in such systems tend to spatially differentiate abundances of elements as a function of their masses. During early accretion, the higher melting point temperatures of the heavier elements from silicon upwards [71] allowed their condensation into dust grains and on into planetesimals relatively close to the parent stars while the lighter, "volatile", elements were driven outwards by the solar wind and condensed further away from the stars, beyond the so-called soot- and frost-lines [72]. Thus-formed planetesimals therefore come with very dissimilar constitutions as manifested in our present day solar system by the compositional diversity of meteorites ranging from iron-rich representatives to the carbonaceous chondrites enriched in organics and with much lower contributions of iron minerals. Planets formed in the inner parts of the planetary systems are thus enriched in heavier elements, while the more distant (gas-) planets mainly contain low-atomic-mass elements. Delivery of further mass to planetary embryos from diverse parts of the planetary systems via asteroid and comet impacts only partially compensate the initial inhomogeneous distribution of elements leaving for example the planetary bodies in our own parent solar system with a considerable imbalance in the ratios of "lighter" (here defined as excluding the transition metals, that is, up to the 3rd period of the periodic table) to heavier elements. Dissimilar trajectories of differentiation of these elements within planetary bodies after their formation further exacerbate the uneven availability of elements from the fourth period upwards in hypothetical life-bearing environments (i.e. featuring liquid water and sources of free energy). For example, the two largest moons of our own solar system, Ganymede and Titan (moons of Jupiter and Saturn, respectively), are frequently cited as potentially harboring life in subsurface oceans of liquid water [73]. Ganymede's interior is considered to be fully differentiated containing an iron-rich core that gives rise to a magnetic field [74,75] while Titan's core appears to be low in iron and mainly made up from silica (SiO_2) [75,76]. For Saturn's moon Enceladus, the Cassini mission was able to analyze chemicals ejected in cyrovolcanic plumes which are considered to be sourced from its subsurface ocean. Tellingly, these samples contained (apart from a suite of light elements and simple organics) only solid grains made up from silica while iron-bearing silicate minerals (such as olivine) were conspicuously absent [77,78]. All this indicates that strongly differing availabilities of iron (and other transition metals) are to be expected within these subsurface oceans.

5.3 Availability of Metal Ions as a Criterion for Habitability

As all contributors to this volume of MILS try to convey, metal ions and in particular transition metal ions appear to have been indispensable agents permitting the emergence of life on our planet. As discussed in Section 1.2, this likely comes down to the fact that the sources of free energy feeding terrestrial life are redox tensions and transition metals are natural redox catalysts. Neglecting the option that life elsewhere in the universe might be driven into existence by other types of free energy (resulting in phenomena which we would likely have difficulties identifying as "living"), we hold that the presence of elements in the fourth period (and possibly above, see Section 6) needs to be taken into account when assessing the habitability of a given candidate outside of planet Earth. As outlined in the two previous sections, such a presence is non-trivial on other bodies in our solar system and even more so in other planetary systems of the universe. Figure 3 represents a juxtaposition of the traditional set of criteria for habitability as for example laid down in NASA's astrobiology roadmap [14] and the revised set suggested by the arguments developed in the contributions of this volume.

NASA's Europa Clipper mission scheduled to arrive at the Jovian moon Europa in 2030 carries the surface dust analyzer (SUDA) instrument designed to provide a detailed dataset on the elemental composition of samples again thought to be sourced from a subsurface ocean of liquid water [79]. These data will provide a better picture concerning the presence or absence of iron and other transition metals in this subsurface ocean and thus allow to assess the likelihood for processes resembling those having engendered life on the primordial Earth to occur in Europa's subsurface ocean.

The recognition that metal ions rather than organic molecules are key to an emergence of "life as we know it", together with the apprehension that the presence of these heavier elements is far from being a given across the universe (see Sections 5.1 and 5.2 as well as Chapter 4 [3]), further shrink the (recently already downgraded [80]) probabilities for habitable planets in our galaxy. Yet, there is one place in the universe for which we can be sure that all criteria formulated above (Figure 3) are formidably fulfilled, that is, planet Earth. Not only did our home planet in its infancy provide "habitats" (as discussed in Chapters 4, 5, 8, and 9) teeming with metal ions

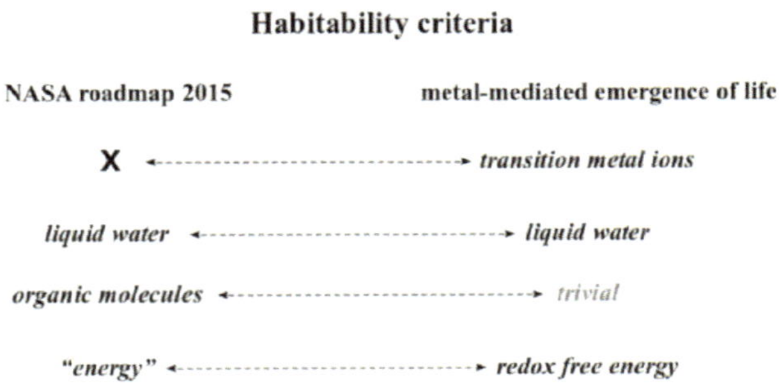

FIGURE 3 Contrasting lists of criteria for habitability based either (left scheme) on OoL scenarios by and large affiliated with the "primordial soup of organics" hypothesis or (right scheme) on the arguments developed in Sections 1 and 2 and corroborated by the entire collection of chapters in this volume. The necessity for the presence of liquid water is maintained in our modified scheme partly due to its role in allowing the formation of anionic clays and partly due to its unique impact on redox reactions as detailed in [13].

in general and transition metals in particular, ready to process the environmental redox tensions occurring in these same habitats (see Section 2.2), but it also saw to it that conditions satisfying the habitability criteria of Figure 3 persisted over almost 4 billion years. Ongoing volcanism and plate tectonics, non-trivial processes enabled by a number of idiosyncrasies proper to our planet [81], secured a wide distribution of all the necessary metal ions in almost all environments throughout the eons, thereby allowing the continuation of the phenomenon "life" on our mother planet to the present day.

6 OUTLOOK: FIFTH AND SIXTH PERIOD ELEMENTS – DISPENSABLE, HELPFUL, OR OUTRIGHT INELUCTABLE

As discussed in Section 5.1, the distribution of elements in the fifth and sixth periods of the periodic table likely is highly inhomogeneous in planetary systems and in the universe in general. Chapters 2 to 4 of this volume [1–3] mention two specific elements in the fifth and sixth periods that play important roles in extant life on Earth, i.e. tungsten and molybdenum. While tungsten appears to be limited to only a few cases of metalloenzymes in bacteria and archaea, molybdenum occurs in a plethora of enzymes, taking on extremely diverse catalytic tasks and operating over the entire range of redox potentials accessible to life [82]. As judged by molecular phylogeny, many of these molybdoenzyme families seem to have been part of the enzyme inventory of the last universal common ancestor (LUCA) [83]. The authors of this chapter have consequently proposed in the past [84] that molybdenum-catalyzed processes likely were indispensable ingredients for getting life started. We have furthermore argued that the paramount role of molybdenum and tungsten comes down to their extraordinary redox flexibilities, surpassing those of all other transition metal ions, and to their exceptional ability to catalyze the redox conversions of notoriously metastable 2-electron compounds such as CO_2 or nitrate [85]. It goes without saying that a strong requirement for molybdenum and/or tungsten to get life (here and elsewhere) off the (rocky) ground would further narrow the range of potentially habitable environments in the universe.

Apart from the three chapters dealing with the biological systems [1–3], there is no mentioning of molybdenum and tungsten in this volume of MiLS. This is not due to a negative bias in setting up the list of contributions but due to a real lack of work addressing Mo/W-containing minerals and their possible roles in bringing forth life. This is all the more astonishing if one considers the prominent attention paid to Mo-containing nanomaterials (e.g. MoO_2 nanosheets) in the search for so-called "advanced materials" [86]. As the finishing words to our introductory contribution to this volume of MILS, we will therefore appeal to colleagues convinced of metal ions' importance in the emergence of life to not neglect the study of molybdenum- or tungsten-containing minerals!

ACKNOWLEDGMENTS

We sincerely thank all the contributors to this volume who have actually honored their initial commitments and who have performed an excellent job in supporting the general message of this volume from very diverse angles. We are furthermore deeply indebted to Hilary Lafoe and Varalika Kathuria of Taylor & Francis for their superhuman patience, unwavering helpfulness, resourcefulness, and kindness in assisting us to get this project over the finishing line.

SD's and WN's research pertaining to the topic of this volume was supported by the French Agence Nationale pour la Recherche (grant no. ANR-22-CE30-0035-01). This work furthermore received support from the French government under the France 2030 investment plan, as part of the initiative d'Excellence d'Aix Marseille Université (A*MIDEX -AMX-21-PEP-039).

Special thanks are due to Mike Russell (University of Turin, Italy) who introduced us to emergence-of-life problems over two decades ago and who stimulates our thinking ever since and to Elbert Branscomb (University of Urbana/Champaign, USA) who kindly but firmly puts us back on track whenever we get lost in the maze of thermodynamics.

ABBREVIATIONS

2D/3D	two-/three-dimensional
Å	Ångström unit
AGB	asymptotic giant branch (of the Hertzsprung-Russell diagram)
AMP/ADP/ATP	adenosine mono/di/tri-phosphate
CHONPS	acronym for the set of elements comprising carbon, hydrogen, oxygen, nitrogen, phosphorus and sulfur (also referred to as CHNOPS or SPONCH)
GR	Green Rust
LDH	layered double hydroxide
LUCA	last universal common ancestor (of bacteria and archaea)
Nar	nitrate reductase (EC number 1.7.5.1)
NASA	National Aeronautics and Space Administration
OoL	origin of life
SUDA	surface dust analyzer
XRD	X-ray diffraction

REFERENCES

1. W. Maret, Chapter 2 of this book.
2. B. P. Jagilinki, F. Mus, J. W. Peters, Chapter 3 of this book.
3. D. Giovannelli, M. Cascone, G. Gallo, F. Migliaccio, B Barosa, D. Bastoni, J. Brusca, A. Longo, L. Tonietti, A. Cordone, Chapter 4 of this book.
4. A. J. Thomson, H. B. Gray, *Curr. Opin. Chem. Biol.* **1998**, *2*,155–158.

5. E. Monosson, *Evolution in a toxic world*, Island Press, Washington DC/USA, 2012.
6. W. Nitschke, O. Farr, N. Gaudu, C. Truong, F. Guyot, M. J. Russell, S. Duval, *Life* **2024**, *14*, 607.
7. R. Milo, R. Phillips, N. Orme, *Cell biology by the numbers*, Garland Science, Taylor & Francis Group, 2016. (https://bionumbers.hms.harvard.edu)
8. T. D. Brock, M. T. Madigan, J. M. Martinko, J. Parker, *Brock biology of microorganisms,* Prentice-Hall, Upper Saddle River/NJ, 2003.
9. W. Schrödinger, *What Is life? The physical aspect of the living cell*, Cambridge University Press, London/UK, 1944.
10. P. Mitchell, *Nature* **1961**, *191*, 144–148.
11. D. G. Nicholls, S. J. Ferguson, *Bioenergetics, 4th ed.*, Academic Press, Amsterdam/Netherlands, 2013.
12. B. Schoepp-Cothenet, R. van Lis, A. Atteia, F. Baymann, L. Capowiez, A.-L. Ducluzeau, S. Duval, F. ten Brink, M. J. Russell, W. Nitschke, *Biochim. Biophys. Acta Bioenerg.* **2013**, *1827*, 79–93.
13. W. Nitschke, B. Schoepp-Cothenet, S. Duval, K. Zuchan, O. Farr, F. Baymann, F. Panico, A. Minguzzi, E. Branscomb, M. J. Russell, *Electrochem. Sci. Adv.* **2022**, *3*, e2100192. https://doi.org/10.1002/elsa.202100192.
14. L. Hays, *NASA astrobiology strategy*, Washington/USA, 2015. http://astrobiology.nasa.gov/uploads/filer_public/01/28/01283266-e401-4dcb-8e05-3918b21edb79/nasa_astrobiology_strategy_2015_151008.pdf.
15. P. Forterre, *Biochimie* **2005**, *87*, 793–803.
16. G. Wächtershäuser, *Assembling pieces of the puzzle*, in *The molecular origins of life,* Ed. A. Brack, Cambridge University Press, Cambridge/UK, 1998, pp. 206–218.
17. A. Mulkidjanian, *Biol. Direct* **2009**, *4*, 1–39.
18. K. Popper, *The logic of scientific discovery*, Routledge, London/UK 2002 (first published as *Logik der Forschung*, Springer, 1934).
19. F. Gross, *Philos. Sci.* **2019**, *86*, 1134–1145.
20. F. G. Donnan, *J. Chem. Educ.* **1928**, *5*, 1558–1570.
21. M. J. Russell, W. Martin, *Trends Biochem. Sci.* **2004**, *29*, 358–363.
22. N. Lane, *BioEssays* **2017**, *39*, 1600217.
23. E. Camprubi, S. F. Jordan, R. Vasiliadou, N. Lane, *IUBMB Life* **2017**, *69*, 373–381.
24. V. Sojo, B. Herschy, A. Whicher, E. Camprubi, N. Lane, *Astrobiology* **2016**, *16*, 181–197.
25. Y. Li, N. Kitadai, R. Nakamura, *Life* **2018**, *8*, 46.
26. A. D. Goldman, J. M. Weber, D. E. LaRowe, L. M. Barge, *Proc. Natl. Acad. Sci. USA*. **2023**, *120*, e2210924120.
27. L. M. White, R. Bhartia, G. D. Stucky, I. Kanik, M. J. Russell, *Earth Planet. Sci. Lett.* **2015**, *430*, 105–114.
28. M. J. Russell, A. J. Hall, *J. Geol. Soc. Lond.* **1997**, *154*, 377–402.
29. W. Nitschke, S. E. McGlynn, E. J. Milner-White, M. J. Russell, *Biochim. Biophys. Acta Bioenerg.* **2013**, *1827*, 871–881.
30. E. L. Shock, E. S. Boyd, *Elements* **2015**, *11*, 395–401.
31. G.. Wächtershäuser, *Pure Appl. Chem.* **1993**, *65*, 1343–1348.
32. Y. Li, *Minerals as prebiotic catalysts for chemical evolution towards the origin of life.* IntechOpen, Rijeka, Croatia, 2022.
33. M. Grosch, M. T. Stiebritz, R. Bolney, M. Winkler, E. Jückstock, H. Busch, S. Peters, A. F. Siegle, J. van Slageren, M. Ribbe, Y. Hu, O. Trapp, C. Robl, W. Weigand, *ChemSystemsChem* **2022**, *4*, e202200025.

34. Takahagi, W., Okada, S., Matsui, Y., Ono, S., Takai, K., Takahashi, Y., & Kitadai, N. (2023). *Proc. Natl. Acad. Sci. U.S.A.* **2023**, *120*, e2303302120.
35. N. Lane, J. F. Allen, W. Martin, *BioEssays* **2010**, *32*, 271–280.
36. B. Herschy, A. Whicher, E. Camprubi, C. Watson, L. Dartnell, J. Ward, J. R. G. Evans, N. Lane, *J. Mol. Evol.* **2014**, *79*, 213–227.
37. X. Huang, Chapter 5 of this book.
38. S. J. Velling, L. E. Rodriguez, P. H. Oyala, J.-P. Jones, L. M. Barge, J. M. Weber, Chapter 6 of this book.
39. M. J. Russell, Chapter 9 of this book.
40. V. Chakrapani, Chapter 7 of this book.
41. O. Farr, N. Gaudu, A. Clouet, M. J. Russell, C. Truong, Chapter 8 of this book.
42. J. D. Bernal, *Proc. Phys. Soc. Lond.* **1947**, *62*, 537–558.
43. H. H. Murray, *Appl. Clay Sci.* **1991**, *5*, 379–395.
44. O. El Samrout, M. Fabbiani, G. Berlier, J. F. Lambert, G. Martra, *Langmuir* **2022**, *38*, 15516–15525.
45. J. T. Kloprogge, H. Hartman, *Life* **2022**, *12*, 259.
46. B. Rasmussen, J. R. Muhling, W. W. Fischer, *Astrobiology* **2021**, *21*, 246–259.
47. G. Nagendrappa, *Appl. Clay Sci.* **2011**, *53*, 106–138.
48. E. Branscomb, M. J. Russell, *Biochim. Biophys. Acta Bioenerg.* **2013**, *1827*, 62–78.
49. B. Guérin, *Bioénergétique*, EDP Sciences, Les Ulis/France, 2004.
50. E. Branscomb, **2023**, preprint. https://arxiv.org/submit/5240040/pdf
51. R. D. Astumian, M. Bier, *Phys. Rev. Lett.* **1994**, *72*, 1766.
52. E. Branscomb, T. Biancalani, N. Goldenfeld, M. J. Russell, *Phys. Rep.* **2017**, *677*, 1–60.
53. M. Maturana, F. Varela, *Autopoiesis and cognition: The realization of the living*, Boston Studies in the Philosophy and History of Sciences, Reichel, Dordrecht, 1971.
54. S. A. Kauffman, *J. Cybern.* **1971**, *1*, 71–96.
55. R. Rosen, *Life itself: A comprehensive inquiry into the nature, origin and fabrication of life*, Columbia University Press, Columbia, 1991.
56. T. Gánti, *Chemoton theory: Theory of living systems*, Springer, New York/USA, 2003 (originally published in Hungarian in 1971).
57. J. I. Lunine, S. M. Hörst, *Rend. Lincei.* **2011**, *22*, 183–189.
58. J. Spacek, P. Rimmer, G. E. Owens, S. R. Cady, D. Sharma, S. A. Benner, *ACS Earth and Space Chem.* **2023**, *8*, 89–98.
59. A. H. Ansari, *Front. Astron. Space Sci.* **2023**, *10*, 1075052.
60. D. P. Cruikshank, C. K. Materese, Y. J. Pendleton, P. J. Boston, W. M. Grundy, B. Schmitt, C. M. Lisse, K. D. Runyon, J. T. Keane, R. A. Beyer, M. E. Summers, F. Scipioni, S. A. Stern, C. M. Dalle Ore, C. B. Olkin, L. A. Young, K. Ennico, H. A. Weaver, V. J. Bray, *Astrobiology* **2019**, *19*, 831–848.
61. K. Altwegg, H. Balsiger, A. Bar-Nun, J. J. Berthelier, A. Bieler, P. Bochsler, C. Briois, U. Calmonte, M. R. Combi, H. Cottin, J. De Keyser, F. Dhooghe, B. Fiethe, S. A. Fuselier, S. Gasc, T. I. Gombosi, K. C. Hansen, M. Haessig, A. Jäckel, E. Kopp, A. Korth, L. Le Roy, U. Mall, B. Marty, O. Mousis, T. Owen, H. Rème, M. Rubin, T. Sémon, C.-Y. Tzou, J. Hunter Waite, P. Wurz, *Sci. Adv.* **2016**, *2*, e1600285.
62. Y. Oba, Y. Takano, Y. Furukawa, T. Koga, D. P. Glavin, J. P. Dworkin, H. Naraoka, *Nat. Commun.* **2022**, *13*, 2008.
63. Y. Oba, T. Koga, Y. Takano, N. O. Ogawa, N. Ohkouchi, K. Sasaki, H. Sato, D. P. Glavin, J. P. Dworkin, H. Naraoka, S. Tachibana, H. Yurimoto, T. Nakamura, T. Noguchi, R. Okazaki, H. Yabuta, K. Sakamoto, T. Yada, M. Nishimura, A. Nakato, A.

Miyazaki, K. Yogata, M. Abe, T. Okada, T. Usui, M. Yoshikawa, T. Saiki, S. Tanaka, F. Terui, S. Nakazawa, S.-i. Watanabe, Y. Tsuda, Hayabusa2-initial-analysis SOM team, *Nat. Commun.* **2023**, *14*, 1292.

64. C. Potiszil, M. Yamanaka, C. Sakaguchi, T. Ota, H. Kitagawa, T. Kunihiro, R. Tanaka, K. Kobayashi, E. Nakamura, *Life* **2023**, *13*, 1448.
65. S. S. Zeichner, J. C. Aponte, S. Bhattacharjee, G. Dong, A. E. Hofmann, J. P. Dworkin, J. E. Elsila, H. V. Graham, J. M. Eiler, *Science* **2023**, *382*, 1411–1416.
66. M. Guélin, J. Cernicharo, *Front. Astron. Space Sci.* **2022**, *9*, 787567.
67. A. Arcones, F.-K.Thielemann, *Astron. Astrophys. Rev.* **2023**, *31*, 1–109.
68. H. Schatz, *Elements* **2010**, *6*, 13–17.
69. S. Pedicelli, G. Bono, B. Lemasle, P. François, M. Groenewegen, J. Lub, J. W. Pel, D. Laney, A. Piersimoni, M. Romaniello, R. Buonanno, F. Caputo, S. Cassisi, F. Castelli, S. Leurini, A. Pietrinferni, F. Primas, J. Pritchard, *Astron. Astrophys.* **209**, *504*, 81–86.
70. T. Bensby, *Proc. Int. Astron. Union* **2013**, *9*, 17–27.
71. H. Palme, K. Lodders, A. Jones, *Solar system abundances of the elements*, in *Solar system abundances of the elements*, Volume 2 of *Treatise on Geochemistry,* Eds. H. D. Holland, K. K. Turekian, Elsevier, 2014, pp. 15–36.
72. Vernazza *Planetesimals: Early Differentiation and Consequences for Planets*, Eds. L. T. Elkins-Tanton, B. P. Weiss, Cambridge University Press, Cambridge, 2016.
73. M. Jebbar, K. Hickman-Lewis, B. Cavalazzi, R. S. Taubner, S. K. M. Rittmann, A. Antunes, *Space Sci. Rev.* **2020**, *216*, 10.
74. F. Sohl, T. Spohn, D. Breuer, K. Nagel, *Icarus* **2002**, *157*, 104–119.
75. H. Hussmann, C. Sotin, J. I. Lunine, *Interiors and evolution of icy satellites*, in *Treatise on geophysics*, Ed. G. Schubert, Elsevier B.V., Berlin, Germany, 2015, pp. 605–635.
76. C. Sotin, K. Kalousová, G. Tobie, *Annu. Rev. Earth Planet. Sci.* **2021**, *49*, 579–607.
77. H.-W. Hsu, F. Postberg, Y. Sekine, T. Shibuya, S. Kempf, M. Horányi, A. Juhász, N. Altobelli, K. Suzuki, Y. Masaki, T. Kuwatani, S. Tachibana, S-i Sirono, G Moragas-Klostermeyer, R. Srama, *Nature* **2015**, *519*, 507–513.
78. F. Postberg, R. N. Clark, C. J. Hansen, A. J. Coates, C. M. Dalle Ore, F. Scipioni, M. M. Hedman, J. H. Waite, *Plume and surface composition of enceladus*, in *Enceladus and the icy moons of saturn*, Eds. P. M. Schenk, R. N. Clark, C. J. Howett, A. J. Verbiscer, J. H. Waite, University of Arizona Press, 2018, pp. 129–162.
79. T. M. Becker, M. Y. Zolotov, M. S. Gudipati, M.S., J. M. Soderblom, M. A. McGrath, B. L. Henderson, M. M. Hedman, M. Choukroun, R. N. Clark, C. Chivers, N. S. Wolfenbarger, C. R. Glein, J. C. Castillo-Rogez, O. Mousis, K. M. Scanlan, S. Diniega, F. P. Seelos, W. Goode, F. Postberg, C. Grima, H.-W. Hsu, L. Roth, S. K. Trumbo, K. E. Miller, the Europa Clipper Composition Working Group, Exploring the Composition of Europa with the Upcoming Europa Clipper Mission. *Space Sci. Rev.* **2024,** *220*, 49. https://doi.org/10.1007/s11214-024-01069-y.
80. M. Scherf, H. Lammer, L. Sproß, *Astrobiology* **2024**, *24*, e916–e1061.
81. M. Bjornerud, *Turning to stone. Discovering the subtle wisdom of rocks*, Headline Publishing Group, London/UK, 2024.
82. S. Grimaldi, B. Schoepp-Cothenet, P. Ceccaldi, B. Guigliarelli, A. Magalon, *Biochim. Biophys. Acta Bioenerg.* **2013**, *1827*, 1048–1085.
83. S. Duval, A.-L. Ducluzeau, W. Nitschke, B. Schoepp-Cothenet, *BMC Evol. Biol.* **2008**, *8*, 1–13.

84. B. Schoepp-Cothenet, R. van Lis, P. Philippot, A. Magalon, M. J. Russell, W. Nitschke, *Sci. Rep.* **2012**, *2*, 263.
85. S. Duval, J. M. Santini, D. Lemaire, F. Chaspoul, M. J. Russell, S. Grimaldi, W. Nitschke, B. Schoepp-Cothenet, *Biochim. Biophys. Acta Bioenerg.* **2016**, *1857*, 1353–1362.
86. I. A. De Castro, R. S. Datta, J. Z. Ou, A. Castellanos-Gomez, S. Sriram, T. Daeneke, K. Kalantar-zadeh, *Adv. Mater.* **2017**, *29*, 1701619.

2 Metal Ion Catalysis in the Abiotic and Biotic Phases of the Development of Living Organisms

Wolfgang Maret
Department of Nutritional Sciences, School of Life Course and Population Sciences, Faculty of Life Sciences and Medicine, King's College London, London, UK
wolfgang.maret@kcl.ac.uk

CONTENTS

ABSTRACT

There is no life without the biological roles of metal ions. Living organisms use at least 25 chemical elements from periods one to six in the periodic system for essential functions. Non-metal elements are the primary focus in the life sciences taught in organic chemistry, biochemistry, and molecular biology. They are collectively referred to by the acronym SPONCH, which stands for Sulfur, Phosphorus, Oxygen, Nitrogen, Carbon, and Hydrogen. The majority of essential chemical elements, however, are metals. Their inorganic chemistry receives correspondingly less attention. The fact that they are minor elements should not distract from their

DOI: 10.1201/9781003459910-2

overall significance. Metal ions play a major role in catalysis, a fundamental aspect of kinetically controlled life. Several of them (V, Mo, Mn, Fe, Co, Ni, Cu, Zn) act as catalysts in reactions that are crucial for intermediary and energy metabolism, the bioleaching, assimilatory processes involving gases to supply the biosphere with the non-metal elements (C, N, O, S, H) and to return them in dissimilatory processes as gases or in biodeposition/biomineralization to the geosphere. In this capacity, metal ions determine biogeochemical cycles and thus the ecology on Earth. Metal ions also control the structure of some biological macromolecules, in particular proteins and nucleic acid, have a function in biological regulation, maintain electrochemical gradients through membranes, and participate in information transfer as intracellular and extracellular signaling ions, including neurotransmission. When the extracellular availability of metal ions is too low, some living organisms have developed strategies to control their acquisition with metallophores. Metallophores and other chelating agents also sequester metal ions in order to make them unavailable to organisms that compete for them. And metal-chelating molecules are employed for intoxicating invading organisms with metal ions, or, in an altruistic way, to assist kin organisms in acquiring the necessary metal ions. All these aspects of an extracellular inorganic biochemistry are important for the ecology of life. With their unique chemical capabilities, metal ions such as iron play a central role in heterogeneous and homogeneous catalysis in prebiotic chemistry and the biochemistry of living organisms, which has been shaped by geological events that provided the chemical and physical environments conducive to the development of life.

KEYWORDS

metal ions, catalysis, bioinorganic chemistry, metallomics

1 THE BIOLOGICAL PERIODIC SYSTEM OF THE ELEMENTS

In the life sciences, biochemistry and molecular biology are taught with a primary focus on organic chemistry but mostly overlook the many aspects of inorganic chemistry as a critical aspect of life, a schism that was pointed out over 30 years ago [1]. The reason for this preoccupation with organic chemistry is that carbon (C) is the core element of life in the building blocks of biomolecules, including nucleobases, amino acids, carbohydrates, and lipids, and in their metabolic pathways. The other "macroelements" are hydrogen (H), nitrogen (N), and oxygen (O). Together with carbon they constitute 92% of the biomass of cells. Two additional chemical elements, phosphorus (P) and sulfur (S), have more specialized roles: sulfur, in redox reactions and interactions with metal ions, and phosphorus, as an energy currency in the anhydride bonds in ATP and ADP (adenosine tri- and diphosphate, respectively) and in biological control. Phosphorus and sulfur add another 4% to the biomass of cells. These six elements are all non-metals and are summarized under the mnemonic acronym CHNOPS or SPONCH for the symbols of the chemical elements. The acronym also lists the elements in the order of decreasing or increasing percentage of biomass for the elements except for H, which is positioned between N and P. Also, hydrogen is on the left side of the periodic system of the elements, where all the metals are positioned. Indeed, under extreme physical conditions, hydrogen behaves as a metal [2]. When the roles of phosphorus and sulfur in biochemistry became known, George Wald, the 1967 Nobel Laureate in Physiology or Medicine, called it "life in the second and third period" [3]. Now, 60 years later, we know that the chemical concepts of life need to include the fourth, fifth, and even the sixth period of the periodic system of the elements. Examples are selenium, bromine, and many metals from the third series in the fourth period, iodine, molybdenum, and cadmium in the fifth period, and tungsten and lanthanides in the sixth period. A considerable bias is introduced in the life sciences with a focus primarily on the SPONCH elements, without considering that at least 14 additional chemical elements are essential for humans. They include ten metals: Na, K, Mg, Ca (1.5% of biomass if bones in humans are included), Mo, Mn, Fe, Co, Cu, Zn as well as the metalloid Se and the non-metals Cl, Br, I, notwithstanding that yet additional elements are essential for other forms of life. For example, silicon (Si) and boron (B) are essential for plants, and cadmium (Cd), tungsten (W), vanadium (V), and nickel (Ni) are essential for other organisms. Most likely, our knowledge of which additional chemical elements are essential for different forms of life is incomplete as we keep discovering essential roles, e.g., bromine for humans in 2014 [4], and the biochemistry of lanthanides/rare earth elements for microorganisms [5]. Thus, much different from a focus on the six SPONCH non-metal macroelements, a greater part of the periodic system of the chemical elements is necessary for life and should be discussed in the wider context of biochemistry [6]. Some of the chemical elements occur at low amounts and accordingly are "microelements" (trace elements, <0.1% of biomass). However, this aspect of their quantity should not be taken as a manifestation of their quality, i.e., their importance. An understanding of biochemistry, the origin of life, its evolution, and the ecology of living organisms is not possible without knowing the roles of the metal ions. Metal ions are critical for life and made life possible in the first place.

Furthermore, analytical techniques with high sensitivity have demonstrated that almost all the chemical elements are found in living organisms [7]. The presence of these additional, nonessential elements has functional implications, too. They can have beneficial or toxic actions. If they are not essential, organisms do not have the molecular mechanisms to control them effectively and as a consequence their accumulation affects the function of the essential elements and other biological functions. All of this knowledge calls for studying biochemistry with bioinorganic as well as bioorganic chemistry. The field of metallomics, an integrated biometal science, addresses this deficiency and complements the fields focusing on the other building blocks of life: proteins (proteomics), DNA/RNA (genomics), lipids (lipidomics), and carbohydrates (glycomics) [8].

2 METAL IONS IN HOMOGENEOUS BIOCATALYSIS

Life depends on nutrients and energy supply and is kinetically controlled [9]. Kinetics, i.e. the factors that determine the rates of chemical reactions, is important for making the essential chemical elements available for living organisms and performing all the reactions in metabolism at the ambient temperatures and pressures at which organisms can live and thrive. For humans, ambient conditions mean atmospheric pressure and a body temperature of about 37°C. However, some organisms thrive under rather different conditions. Extremophiles can live at high (<120°C, thermophiles) or low (>−25°C, psychrophiles) temperatures, at pressures of 1100 bar, at extremes of pH (pH<1 or pH>11), at high salinity, or even at high levels of radiation. The reaction rates of biochemical processes must be sufficiently fast and commensurate with the specific requirements of life. In the control of reaction rates, chemical catalysis is a crucial aspect of the (bio)chemistry of life, its origin, evolution, and maintenance. Metal ions have a major role as catalysts. They have been selected for their superiority in carrying out specific biochemical tasks [1]. Catalysts do not affect the equilibrium of a reaction. Instead, they enhance reaction rates by lowering the activation energy. Biocatalysts effect a significant enhancement of reaction rates. Carbonic anhydrase, a zinc enzyme, may serve as an example of the catalytic prowess of metalloenzymes. It provides a million-fold enhancement of the reaction between carbon dioxide and water to form hydrogen carbonate Eq. (1) with a turnover of 10^6 substrate molecules per second.

$$CO_2 + H_2O \rightarrow HCO_3^- + H^+ \tag{1}$$

Embedding the metal ion as an inorganic cofactor in a protein (metalloenzyme) or RNA (metalloribozyme) matrix adds functionalities such as controlling the metal ion concentration, activating the metal ion for its specific functions, selective recognition of a substrate, exquisite stereochemical control of some reactions, and other functions linked to the interactions of the biocatalyst with other biomolecules. The proximity of reactive groups in the active site of enzymes generates high effective concentrations of reactants intramolecularly. This increase in order compared to intermolecular interactions decreases entropy and is a major factor for rate enhancement.

Catalysis without a biomolecule also occurs in biology, e.g., in the Fenton chemistry of redox-active iron and copper ions in the formation of hydroxyl radicals Eq. (2), albeit it is largely unwanted.

$$H_2O_2 \rightarrow OH\cdot + OH^- \tag{2}$$

Another example is the nickel (Ni^{2+})-assisted hydrolysis of peptides, which is an aspect of the toxic actions of this metal ion that is not essential for humans [10,11]. Also, iron and vitamin B6 catalyze the non-enzymatic production of hydrogen sulfide from cysteine in blood [12].

Whether metal ion catalysts with low molecular ligands have a role in biology is poorly explored. Coenzymes are a case in point. Cu^{2+}/ATP complexes are catalysts in Diels-Alder cyclizations in organic synthesis, implying that reactions of metal ions with low molecular weight ligands may have played a role in the evolution of life [13]. Indeed, Fe^{3+} is the favored metal catalyst in the phosphoryl transfer from acetyl phosphate to ADP to form ATP, providing a possible scenario for the prebiotic formation of this universal energy currency, its role in protometabolism, and later as a coenzyme [14]. Coenzymes based on ribonucleotides have been shown to be synthesized abiotically in a prebiotic chemistry on a mineral surface and are thought to be remnants of early life [15]. For some of the metal ions, cofactor biochemistry is used as a way of handling the metal ion, modulating their chemistry, and generating interactions with proteins. The cofactor controls metal ion availability, holds on to the metal ion, i.e., making the metal ion essentially non-exchanging, and provides specific chemical properties. Examples are the tetrapyrroles of heme (Fe), corrin (Co), chlorophyll (Mg), hydrocorphin (Ni), the pincer nucleotide pyridinium-3,5-bis-thiocarboxylic acid (Ni), and pterin (Mo). In the case of vitamin B12, the primary attention is usually given to the organic cofactor, i.e., the vitamin, when in fact it is the trace metal cobalt that performs the essential chemistry. It is also an example of how some forms of life, including humans, became reliant on other organisms for the synthesis of a cofactor carrying a particular metal ion.

Metal ions are typical catalysts due to their unique chemical properties and thus provide considerable functionality, but they are not the only catalysts, as many enzymes catalyze reactions without using a metal ion as a cofactor. Nevertheless, it is estimated that 40% of all proteins interact with a metal ion cofactor [16]. For enzymes, almost every second enzyme uses a metal ion cofactor. Metal ions have special catalytic properties that cannot be achieved without them. One important aspect is that they can interact with gases and catalyze redox chemistry to convert the gases to products that living organisms can use. This process is critically important for supplying living organisms with the non-metal elements. The efficiency of biocatalysis is evident for nitrogen fixation. In the industrial process, the Haber-Bosch process of synthesizing ammonia, pressures of 200 bar and temperatures of 500°C are needed despite using an iron (or other metal) catalyst. In nature, nitrogenase (requiring Fe, V, and Mo) catalyzes this process under ambient atmospheric conditions.

$$N_2 + 3\,H_2 \rightarrow 2\,NH_3 \tag{3}$$

3 THE SELECTION AND ACQUISITION OF METAL IONS

The biological selection of metal ions depends on a variety of factors. Metal ions must have sufficient strength to bind to a biomolecule. For divalent metal ions of the third series the binding strength is determined by the Irving-Williams series, i.e., $\mathrm{Mn{<}Fe{<}Co{<}Ni{<}Cu{>}Zn}$ [17]. This thermodynamic principle, in addition to kinetic factors, controls the metal ion concentrations in cells and ascertains that each metal ion can function within a certain range of concentrations without interfering with the other metal ions. An extensive intracellular biochemistry exists for controlling metal ions and their availability. It includes many proteins that are transporters (importers and exporters) on the cytoplasmic membrane and membranes of organelles, serve in storage, transport in the cytoplasm, serve as metallochaperones, and act as sensors (transcriptional regulators). For zinc alone, at least about 50 proteins control and coordinate its functions. Another issue is the dependence of the availability of the metal ion on environmental conditions. While some metalloenzymes are specific regarding the metal they use, there is interchangeability for others. While carbonic anhydrase is a zinc enzyme in most cases, a cadmium carbonic anhydrase has also been found in marine diatoms in environments where very little zinc is available [18].

Other examples of using different metal ions for the same purpose include the use of Fe, Mo, and V in nitrogenases, and different metal ions in ribonucleotide reductase and the ribosome [19]. The superoxide dismutases have been examined in detail in this regard. The enzymes use Cu (with or without Zn in the active site), Ni, Mn, or Fe. Some are cambialistic; i.e., they can use either Mn or Fe, but there is also a case of evolutionary changes of residues near the active site to confer specificity for Mn [20]. Thus, metal ion usage can change with the chemical environment, and the preferred metal is context specific.

Even when the availability of metal ions is severely restricted, evolution found solutions for acquisition. For example, the solubility product of $Fe(OH)_3$ is 10^{-39}. It means that not a single iron ion is available from this hydroxide in a liter of fluid. To make this vital element available, microorganisms and plants synthesize chelating agents with commensurate binding constants for Fe(III) and secrete them into the extracellular milieu to scavenge iron ions. In the case of iron, the chelating agents are called siderophores (iron carriers); in the case of copper and zinc, chalkophores and zincophores, respectively; or in the case of metal carriers yet to be coined names, generally metallophores [21]. Living organisms use extracellular biochemistry to engage in chemical warfare with attacks, counterattacks, and defenses against invaders and competitors by either increasing or decreasing the availability of metal ions to maintain their welfare in aquatic or terrestrial ecological niches. To control mutualistic, cambialistic, or parasitic symbiosis with other organisms, a remarkably rich suite of several bioactive molecules is employed. Specific receptors on cells bind the metallophore and internalize the metal complex, followed by harvesting the metal ion through several mechanisms. Thus, while the metal ions must have been available originally for the organism to evolve, later in evolution, organisms developed survival mechanisms that made them independent of the low availability of some metal ions in ecological niches.

4 BIOGEOCHEMICAL CYCLES

Metal ion catalysis has fundamental roles in the reactions that generate the energy for life, in energy conversions, and in interconnected biogeochemical cycles of the SPONCH elements, in handling their gases, some of which are chemically inert, in assimilatory reactions that make the element available or in dissimilatory reactions that use the chemistry of the element in oxidoreduction but do not take up the element into the organism, and in handling molecule fragments such as radicals. Except for phosphorus, metal ions form chemical bonds with all these chemical elements and are involved in their redox chemistry. Metal ions are a key link between the biosphere and the geosphere, and biogeochemical elemental cycles of the metals and non-metals are coupled.

The following list – in reverse order, i.e., CHNOPS – gives exemplary chemical equations with an emphasis on showing the formation of the chemical elements in metal ion-catalyzed reactions.

C $CO_2 + 2\ H_2 \rightarrow C + 2\ H_2O$ (speculative). Metal ions such as Cu, Ni, Fe, Co, and Zn are used in the biological chemistry of the gases of carbon, i.e., CH_4, CO, CO_2. In addition, Mo and W are used in formate dehydrogenases that form CO_2.

H $2\ H^+ + 2\ e^- \rightarrow H_2$ (reversible)

Hydrogenases use Fe and Ni.

N $2\ NO_3^- + 10\ e^- + 12\ H^+ \rightarrow N_2 + 6\ H_2O$ (sum of four reactions catalyzed by nitrate, nitrite, nitric oxide, and nitrous oxide reductases). Anaerobic ammonium oxidation (anammox) also forms elemental nitrogen:

$$NH_4^+ + NO_2^- \rightarrow N_2 + 2\ H_2O$$

Nitrogenases involved in nitrogen fixation use Mo/Fe, V/Fe, or only Fe.

O $2\ H_2O \rightarrow O_2 + 4\ e^- + 4\ H^+$ (photosynthesis). The oxygen evolving complex (OEC) in photosystem II uses Mn and Ca in a Mn_4O_5Ca cluster. Mg is also used, while Fe and Cu are used in the reduction of O_2 to H_2O in the respiratory chain. Mo is used in oxygen (and sulfur) transfer reactions.

P Elemental phosphorus apparently is not formed by living organisms. However, lithoautotrophic, sulfate-reducing bacteria from marine segments employ phosphite (redox state +III) as an energy source and produce phosphate (redox state +V) [22].

$$4\ HPO_3^{2-} + SO_4^{2-} + H^+ \rightarrow 4\ HPO_4^{2-} + HS^-$$

Thus, phosphorus redox biochemistry has a role in biogeochemical cycles. A significant diversity of microbes performing dissimilatory phosphite oxidation under anoxic conditions has been observed [23], including microbes that fix and reduce CO_2. It is thought that the microbes evolved early in the earth's history with a role in carbon fixation and making phosphate available [24].

S $CO_2 + 4\ H_2S + O_2 \rightarrow CH_2O + 4\ S^0 + 3\ H_2O$

As a congener of oxygen in the periodic system of elements there is an even richer biological redox chemistry for sulfur than for oxygen. Biology makes use of the redox states from −II to +VI. Selenium, the congener of sulfur, adds to this redox chemistry with different redox potentials.

Sulfur is important as a ligand donor atom for thiophilic ("sulfur-loving") metal ions such as Fe, Cu, and Zn. For redox-inert metal ions such as Zn, the sulfur ligand confers redox activity on the coordination, a property that is important for the distribution of zinc ions and thus their functions [25].

The biogeochemical sulfur cycle also has a critical role in the evolution of life and in providing fuel for microorganisms through elemental sulfur or sulfate reduction, sulfate reduction, and sulfide oxidation, and in affecting the solubility of metal minerals and the availability of metal ions. It evolved together with the changing redox environment of the earth. Sulfide oxidation is thought to have occurred in the Archean eon of geological history while the thiosulfate pathway developed after the Great Oxygenation Event [26].

A direct connection with the inorganic world – aside from the reminder that we as humans need essential metal ions – is lithotrophy, literally "rock eating", the dependence of life on inorganic substrates including metal ions. Autotrophic organisms such as plants, algae, and many bacteria depend on minerals as nutrients, carbon dioxide as a source of carbon, and water for making organic compounds. They use sunlight as a source of energy if they are photolithotrophs or the reducing power of inorganic substrates if they are chemolithotrophs. Such reduced substrates include elemental hydrogen (H_2), elemental sulfur, hydrogen sulfide (H_2S), Fe^{2+} (photoferrotrophs if energy from sunlight is used), ammonia (NH_3) or nitrite (NO_2^-) (anammox, anaerobic ammonia oxidation), or methane (CH_4) (methanotrophs). Chemolithotrophy is restricted to microorganisms (bacteria and archaea). Metal ions are used either directly as substrates for generating energy or they are used in the enzymes that convert the inorganic substrates. The energy obtained from oxidation of the reduced substrate is then used for biosynthesis. Heterotrophs then use the oxygen formed by autotrophic organisms and reduce it to water in respiration. This oxygen cycle is also coupled to a carbon cycle in the production of carbon dioxide by animals and the fixation of carbon dioxide by plants. Important for sustaining a healthy ecology is that in these cycles some living organisms use greenhouse gases, i.e., carbon dioxide, nitrous oxide (N_2O) and methane. Formation of greenhouse gases also was important for the warming of Earth in its early history to establish conditions to sustain life.

5 ZINC AND IRON

Among the more than 15 metal ions essential for various living organisms, the remainder of the discussion will focus on iron and zinc because of their prime importance as catalysts in the origin and evolution of life. Both metal ions have a relatively high affinity for sulfur, in addition to interacting with nitrogen and oxygen donor atoms. Iron, cobalt, and nickel also form metal-carbon bonds, a feature that is crucial for primordial chemistry and biochemistry, as six of the nine known metal-carbon bonds are found in the last universal common ancestor (LUCA) of all the three domains of life (bacteria, archaea, and eukarya) [27].

Zinc(II) ions are not redox active in biology and therefore differ from the other redox-active metal ions in the third series in the periodic system of the elements. In this capacity Zn is related to the other redox-inert metal ions of Mg and Ca. Zinc is a cofactor of 10% of human proteins. Over 3000 human proteins contain zinc as a catalytic and/or structural cofactor [28]. This count does not include a yet uncounted number of transient interactions of zinc with proteins as a regulatory cofactor. At least 1000 enzymes contain zinc as a catalytic cofactor. A similarly large number of proteins contain structural zinc sites with at least two sulfur donor ligands. Though there are different classes of coordination sites and domains, many of these proteins are generally referred to as zinc fingers. Some zinc finger proteins contain as many as 30 zinc ions. Structural zinc sites enlarge the conformational space of proteins and their interactions considerably. The zinc enzymes include at least six of the seven classes of enzymes (oxidoreductases, transferases, hydrolases, lyases, isomerases, ligases, and translocases), thus showing the ability of zinc to catalyze many different enzymatic reactions. The Zn^{2+} ion is an excellent Lewis acid (electron withdrawing). It can polarize a substrate directly or activate a bound water molecule for nucleophilic attack on the substrate. It can also use a bound water molecule as an acid/base catalyst. The multitude of functions makes zinc a type 2 nutrient, that is, a nutrient of general importance [29].

Iron, in contrast, is redox-active and a type 1 nutrient with more specific yet crucially important functions. It is a cofactor of 2% of human proteins, among which 35% bind the iron ion directly, 48% bind the heme cofactor, and 17% are Fe-S proteins. Thus, the count of iron proteins is significantly less than that of zinc proteins [30]. In addition to the Fe(II)/Fe(III) redox chemistry in many iron enzymes and non-enzymatic iron catalysis, the redox activity of iron also includes higher oxidation states (Fe(IV) and Fe(V)) in some iron enzymes [31].

6 NON-CATALYTIC FUNCTIONS OF METAL IONS

While the article focuses mainly on the paramount importance of metal ions in biocatalysis, metal ions also have other roles. They organize and maintain the structure of macromolecules, in particular proteins and ribonucleic acids, and remarkably, metal ions such as Ca^{2+} and Zn^{2+} are used as messengers in information transfer in intracellular and intercellular signaling [32]. Each metal ion has its own bioinorganic coordination chemistry and biology for controlling its homeostasis. It includes membrane transporters, cellular transporters, storage proteins, chaperone proteins that safeguard the metal ions for insertion into the correct metalloproteins, and sensor proteins that regulate gene expression in a metal-dependent way. For zinc alone, at least 40 human proteins are dedicated to these tasks. Zinc is a structural cofactor for many proteins, determining the 3D architecture, making domains for interactions with other proteins, nucleic acids, and yet other ligands, and binding between proteins to bridge them. Metal ions also have functions in biomineralization and possibly in magnetoreception in the form of magnetite.

7 HETEROGENEOUS CATALYSIS: IRON AND ZINC IN THE ORIGIN OF LIFE AND EVOLUTIONARY SCENARIOS

After the brief exegesis of bioinorganic facts in the scientific literature, pieces of the puzzle in the origin of life and evolutionary scenarios will be discussed, with a primary focus on the ions of iron and zinc and sulfur as their donor ligands. The other metal ions in the third series of the periodic system of the elements also have important, albeit more specialized, roles, and there is some interdependence. For example, copper is needed in the redox chemistry of Fe(III)/Fe(II) in iron metabolism.

Life developed over the four eons of geological time. LUCA, suggested to be a chemolithotroph/anaerobic acetogen and encoding for about 2600 proteins, is placed at 4.2 billion years (Ga) in the Hadean (4–4.5 Ga ago), less than 200 million years after the earth formed [33]. The first fossils are discovered at the borderline to the Archean (2.5-4 Ga ago). The Great Oxygenation Event that increased atmospheric oxygen significantly took place at 2.3 Ga in the Proterozoic (0.54 to 2.5 Ga ago) and saw eukarya and multicellular life develop. And lastly, at the transition to the Phanerozoic (starting 0.54 Ga ago) era, plants and the earliest animals emerged. After the Cambrian Explosion at the end of this period all major animal phyla can be detected in the fossil record. Flowers, birds, and primates are traced back to 0.9 Ga. Thus, living organisms developed early, just about 0.5 Ga after the earth and the oceans formed, and have a 4.1-billion-year history of evolution. The Great Oxygenation Event changed the redox environment on Earth. It is also thought that a transition to a euxinic (anoxic and sulfidic) environment in the deep ocean at 1.8 Ga occurred, followed by the oxygenation of deep waters. The transitions are linked to the availability of metal ions for organisms and consequently how living organisms selected the chemical elements and evolved. The changing metal geochemistry is reflected in metal biochemistry. Thus, metalloproteomes of eukarya demonstrate the increased use of zinc [34]. Because a sulfidic environment does not make enough zinc available, it suggests that eukarya evolved in an oxic environment while prokarya evolved in an anoxic environment [34].

I will discuss neither how metal ion usage in the evolutionary tree of life is being traced with metalloproteomics methods nor scenarios of evolution between FUCA (the first universal common ancestor) and LUCA. Instead, I will elaborate on how inorganic chemistry and organic chemistry are interrelated in the abiogenic phase in the origin of life, emphasizing the role of metal ions in heterogeneous catalysis on mineral surfaces before the role of metal ions in homogeneous catalysis of enzymes and ribozymes became the mainstay of cellular biology. Heterogeneous catalysis on surfaces of solids such as minerals or clays has distinct features in terms of providing stereochemical templates, serving as a reservoir for heat exchange, and having activators present or activating chemicals through adsorption. Like enzymes and ribozymes, surfaces provide a proximity effect, namely being able to accumulate substrates. Since the rate of chemical reactions depends on concentrations, it leads to rate enhancements, which is the essence of catalysis. One of the characteristics of biomolecules is their chirality. Amino acids in proteins are all L-enantiomers, while sugars are generally D-isomers. It has been demonstrated that metal ions in catalysis can provide the energetic difference needed for the selection of a particular stereoisomer [35].

7.1 The Fe/S World

The role of iron (ferrous) disulfide (FeS_2, pyrite – fool's gold) in the origin of life was discussed in articles by Günther Wächtershäuser as early as 1988 [36]. In his iron-sulfur (Fe/S) world theory he postulates that the reducing power of iron sulfide was harnessed in a hydrothermal vent (volcanic) environment to synthesize organic compounds found in metabolism. The canonical [4Fe-4S] and [2Fe-2S] iron/sulfur (FeS) clusters, in which the iron ions are bridged by sulfide (S^{2-}) ligands and bound to the thiolate group of cysteines, are employed for catalytic and structural purposes in many Fe/S proteins in now living organisms [37]. They are ancient motifs since many FeS clusters have been detected in LUCA in addition to LUCA's dependence on transition metal ions, corrins (Co), molybdopterin (Mo), and selenium [38,39]. Fe/S proteins catalyze redox reactions, including fundamental processes such as carbon dioxide fixation. Since FeS clusters can form abiotically [40], they are thought to be among the oldest catalysts and important for primordial chemistry and protometabolism. FeS clusters illustrate the relationship between geochemistry and biochemistry and the significance of heterogeneous catalysis on surfaces, namely the deposition of iron and sulfur in the extracellular bioformation of pyrite and the harnessing of iron and sulfur in the bioleaching of pyrite. Pyrite is the most abundant mineral sulfide in the Earth's crust and the most stable iron sulfide among the others formed, i.e., pyrrhotite ($Fe_{1-x}S$), greigite (Fe_3S_4), and mackinawite (FeS), and soluble Fe-S clusters. The underlying processes in the conversions of these forms are important for biogeochemical cycles of iron, microbial ecology, and sulfur and iron acquisition pathways [41]. Sulfate-reducing bacteria (SRB) produce H_2S and polysulfides (S_n^{2-}) and participate in pyrite biodeposition (bioformation of FeS_2) in submarine hydrothermal vents [42]. This pyritization involves the conversion of less stable FeS forms to FeS_2 and is considered a key energy conversion process in the evolution of life. In contrast to microbial sulfate reduction, mineralization of pyrite is also thought to occur from reduced organic sulfur in sediments that are low in sulfate [43]. Pyrite can also be formed from sulfidation of biogenic magnetite (Fe_3O_4) [44]. Bioleaching of FeS_2, on the other hand, has been observed in methanogens that reduce pyrite to produce aqueous ferrous sulfide clusters (FeS_{aq}) to serve in iron and sulfur uptake [45]. The archaeon *Pyrococcus furiosus* stores iron and sulfur in thioferrate nanoparticles in the IssA protein [46,47]. This storage form, i.e., ferric sulfide, $[FeS_2]^-$, is different from the better-known storage of iron as FeOOH (ferric oxyhydroxide) in ferritin. Both proteins contain a few thousand iron atoms. The IssA monomer can reconstitute [4Fe-4S] clusters *in vitro*. The transition from an extracellular to an intracellular biochemistry is seen in the observation of extracellular FeS clusters. The *Mycobacterium tuberculosis* permease OppABCD requires a [4Fe-4S] cluster [48]. The secretion of the SufC protein of the bacterial FeS cluster biosynthetic system SUF suggests additional significance of biosynthesized FeS clusters and their machinery in *Streptococcus pneumoniae* [49]. A physical interface has developed between abiotic metal ion catalysis and biocatalysis. The so-called mineral-respiring bacteria use extracellular electron transfer (EET) or direct interspecies electron transfer (DIET) with flavins, cytochromes, and nanowires. A proteomic analysis of the extracellular polymeric substances (EPSs) of *Geobacter sulfurreducens* identified FeS cluster proteins that are thought to be involved in EET [50].

In addition to catalysis on surfaces, the iron made available in solution in these reactions has been shown to catalyze the synthesis of important precursors of the molecules of life and to be involved in reaction sequences of protometabolism. For example, iron-rich meteoritic and/or volcanic particles catalyze the synthesis of key precursors of life's building blocks from the fixation of CO_2 [51]. Formation of pyruvate, the end product of glycolysis and the substrate for making acetyl-CoA to enter the citric acid cycle, formate, and acetate and other products can be formed from H_2 and CO_2 in iron/nickel-catalyzed reactions [52]. Also, iron(II) ions can catalyze the formation of nine of the major metabolites of the citric acid cycle, a central metabolic and energetic pathway [53].

7.2 The Zn/S World

It was also hypothesized that life originated in a "zinc world", placing ZnS (sphalerite) in the middle of the abiogenic scenario on primeval earth. ZnS "edifices", deposits of mineral with a hydrothermal origin, are thought to have had a role in primordial earth based on the three postulates of how they work in heterogeneous catalysis and other ways: (i) using solar radiation to drive CO_2 reduction for making biochemical building blocks – the connection to the "carbon world", (ii) serving as templates for making the polymers, and (iii) protecting the formed compounds from photodissociation by absorbing radiation [54,55]. ZnS is produced by dissimilatory sulfate reducers with iron(II) ion catalysis in environments that are low in oxygen and zinc. The bacteria left environmental imprints in sulfide ore deposits as early as 3.4 Ga ago in geological history [56]. Bioleaching from ZnS also occurs by sulfide oxidation that requires Fe(III) catalysis. Two pathways are involved: the thiosulfate pathway, which uses iron, and the polysulfide pathway, which uses iron and copper. Usage of the pathways depends on the solubility of the metal sulfide at low pH [57,58]. The polysulfide mechanism provides protons and is used for acid-soluble metal sulfides whereas the thiosulfate mechanism is used for acid-insoluble metal sulfides.

8 CONCLUSIONS

Great strides have been made in constructing the tree of life over the remarkable period of 4 billion years with beginnings relatively shortly after the earth and sufficient water was formed. Remarkable progress also has been made in our understanding of how the organic molecules of life and their precursors could have been synthesized abiotically under the then prevailing physical and chemical conditions on Earth and how life may have emerged [59,60]. An inevitable conclusion is that metal ions are indispensable in heterogeneous catalysis on surfaces of minerals, clays, or other soil types in scenarios that re-construct the primeval chemistry in the origin of life. Likewise, in simple inorganic complexes, complexes with biological ligands and organic cofactors, and in metalloenzymes and metalloribozymes, metal ions have fundamental roles in homogeneous catalysis in the biotic phase of life, its evolution, and maintenance. Metal ions and some metalloids and non-metal chemical elements are a fifth branch of building blocks of life. Constructing scenarios of how life developed requires knowledge in several disciplines, in particular a combination of the earth sciences and life sciences. Such a combination is rarely part of curricula in higher education. Paradoxically, metal ions are not in the purview of most rudimentary biochemistry teaching [61]. In mostly organic chemistry-focused biochemistry, teaching should include inorganic chemistry of the chemical elements with disciplines such as elemental biology vis-à-vis molecular biology and metallomics [62]. An appropriate quote on this aspect would be "there is no understanding of any cellular system of any organism without insight into biological inorganic as well as organic chemistry" [63]. Bridging these "rifts in this fractured research field" needs to be pursued in interdisciplinary studies and research, and it is therefore important that we transcend "conventional disciplinary boundaries" for progress in assembling the "pieces of a bigger puzzle" and disseminating the findings [64]. Understanding the geologic record is one aspect. Knowledge from the geological and geochemical sciences such as rock and fossil dating and understanding the changing composition of the atmosphere, hydrosphere, and lithosphere is the basis for determining the chemical and physical conditions of prebiotic chemistry, protometabolism, and ensuing biochemistry of living organisms. Scenarios for ecology must fit the prevalent physical parameters (pH, redox potentials, temperature, radiation) that determine the availability of chemical elements, their reactions, and the mechanisms of their acquisition and usage that living organisms have developed. Naturally, a lot of work in this field is retrospective by constructing plausible scenarios based on known chemistry and biochemistry and recognizing primordial features in extant biochemistry. We should not exclude the possibility that at one point in time the pervasive theme of catalytic roles of metal ions included other metal ions that are no longer used. Our thinking usually follows one line of evolution. Including the possibility that there were parallel lines with diversification as well as decimation in the tree of life could reconcile different frameworks such as the emergence of life in volcanic hydrothermal vents either on Earth or in the oceans [65]. Addressing questions about the origin and evolution of life fosters research in inorganic and organic chemistry, in the biological and geological sciences and many other specific disciplines and advances our knowledge with tangible benefits in unexpected discoveries and applications. An intellectually most stimulating question is whether the same or other inorganic chemical frameworks in the evolution of autocatalytic and self-organizing systems and possibly life occur or occurred on other planets in our or other solar systems. Lift off!

ABBREVIATIONS AND DEFINITIONS

LUCA last universal common ancestor
SUF sulfur assimilation system

ACKNOWLEDGMENT

WM acknowledges support from COST Action, FeSimmChemNet. This article is based upon work from COST Action FeSimmChemNet, CA21115, supported by COST (European Cooperation in Science and Technology).

REFERENCES

1. J. J. R. Fraústo da Silva, R. J. P. Williams, *The biological chemistry of the elements.* Clarendon Press, Oxford 1991, 2nd ed., Oxford University Press 2001.
2. S. Desgreniers, *Nature* **2020**, *577*, 626–627.
3. G. Wald, *Life in the second and third periods; or why phosphorus and sulfur for high-energy bonds?* In: M. Kasha, B. Pullman (eds.), *Horizons in Biochemistry*, Academic Press 1962, pp. 127–142.
4. A. Pol, T. R. M. Barends, A. Dietl, *Environ. Microbiol.* **2014**, *16*, 255–264.
5. A. S. McCall, C. F. Cummings, G. Bhave, A. Vancore, A. Page-McCaw, B. G. Hudson, *Cell* **2014**, *157*, 1380–1392.
6. W. Maret, *Metallomics. A primer of integrated biometal sciences.* Imperial College Press, London, 2016.
7. H. Haraguchi, *J. Anal. At. Spectr.* **2004**, *19*, 5–14.
8. W. Maret, *Metallomics* **2022**, *14*, mfac051.
9. R. J. P. Williams, J. J. R. Fraústo da Silva, *J. Theor. Biol.* **2003**, *220*, 323–343.
10. A. A. Karaczyn, W. Bal, S. L. North, R. M. Bare, V. M. Hoang, R. J. Fisher, K. S. Kasprzak, *Chem. Res. Toxicol.* **2003**, *16*, 1555–1559.
11. E. I. Podobas, D. Gutowska-Owsiak, S. Moretti, J. Poznański, M. Kulińczak, M. Grynberg, A. Gruca, A. Bonna, D. Płonka, G. Ogg, W. Bal, *Front. Mol. Biosci.* **2022**, *9*:828674.
12. J. Yang, P. Minkler, D. Grove, R. Wang, B. Willard, R. Dweik, C. Hine, *Commun. Biol.* **2019**, *2*, 194.
13. C. Wang, Q. Qi, W. Li, J. Dang, M. Hao, S. Lv, X. Dong, Y. Gu, P. Wu, W. Zhang, Y. Chen, J. S. Hartig, *Nat. Commun.* **2020**, *11*, 4792.
14. S. Pinna, C. Kunz, A. Halpern, S. A. Harrison, S. F. Jordan, J. Ward, F. Werner, N. Lane, *PLoS Biol.* **2022**, *20* (10), e3001432.
15. A. D. Goldman, B. Kacar, *J. Mol. Evol.* **2021**, *89*, 127–133.
16. C. Andreini, I. Bertini, G. Cavallaro, G. L. Holliday, J. M. Thornton, *J. Biol. Inorg. Chem.* **2008**, *13*, 1205–1218.
17. H. Irving, R. J. P. Williams, *Nature* **1948**, *162*, 746–747.
18. T. W. Lane, F. M. M. Morel, *Proc. Natl. Acad. Sci. USA* **2000**, *97*, 4627–4631.
19. D. G. J. Smethurst, N. Shcherbik, *J. Biol. Chem.* **2021**, *297*, 101374.
20. K. M. Sendra, A. Barwinska-Sendra, E. S. Mackenzie, A. Baslé, T. E. Kehl-Fie, K. J. Waldron, *Nat. Ecol. Evol.* **2023**, *7* (5), 732–744.
21. W. Maret, *Natural Product Comm.* **2024**, *19* (8), 1–13.
22. B. Schink, M. Friedrich, *Nature* **2000**, *406*, 27.
23. S. D. Ewens, A. F. S. Gomberg, T. P. Barnum, M. A. Borton, H. K. Carlson, K. C. Wrighton , J. D. Coates, *Proc. Natl. Acad. Sci. USA* **2021**, *118*, e2020024118.
24. C. R. Walton, S. Ewens, J. D. Coates, R. E. Blake, N. J. Planavsky, C. Reinhard, P. Ju, J. Hao, M. A. Pasek, *Nat. Geosci.* **2023**, *16*, 399–409.
25. W. Maret, B. L. Vallee, *Proc. Natl. Acad. Sci. USA* **1998**, *95*, 3478–3482.
26. K. Mateos, G. Chappell, A. Klos, B. Le, J. Boden, E. Stüeken, R. Anderson, *Sci. Adv.* **2023**, 9, eade4847.

27. W. F. Martin, *Trends Biochem. Sci.* **2019**, *9*, 807–818.
28. C. Andreini, L. Banci, I. Bertini, A. Rosato, *J. Proteome Res.* **2006**, *5*, 196–201.
29. J. C. King, *Am. J. Clin. Nutr.* **2011**, *94*, 679S–684S.
30. C. Andreini, V. Putignano, A. Rosato, L. Banci, *Metallomics* **2018**, *10*, 1223–1231.
31. J. Hohenberger, K. Ray, K. Meyer, *Nat. Commun.* **2012**, *3*, 720.
32. W. Maret, *Regulation of cellular zinc ions and their signaling functions*, in: *Zinc signaling*, 2nd ed., T. Fukada, T. Kambe, eds., Springer Nature Singapore Ltd., 2019, pp. 5–22.
33. E. R. R. Moody, S. Álvarez-Carretero, T. A. Mahendrarajah, J. W. Clark, H. C. Betts, N. Dombrowski, L. L. Szánthó, R. A. Boyle, S. Daines, X. Chen, N. Lane, Z. Yang, G. A. Shields, G. J. Szöllősi, A. Spang, D. Pisani, T. A. Williams, T. M. Lenton, P. C. J. Donoghue, *Nat. Ecol. Evol.* **2024**, *9*, 1654–1666.
34. C. L. Dupont, S. Yang, B. Palenik, P. E. Bourne, *Proc. Natl. Acad. Sci. USA* **2006**, *103*, 17822–17827.
35. J. A. Cowan, R. J. Furnstahl, *ACS Earth Space Chem.* **2022**, *6* (11), 2575–2581.
36. G. Wächtershäuser, *Prog. Biophys. Mol. Biol.* **1992**, *58*, 85–201.
37. C. Vallières, O. Benoit, O. Guittet, M.-E. Huang, M. Lepoivre, M.-P. Golinelli-Cohen, L. Vernis, *Metallomics* **2024**, *16* (5), mfae025.
38. M. Weiss, F. Sousa, N. Mrnjavac, S. Neukirchen, M. Roettger, S. Nelson-Sathi, W. F. Martin, *Nat. Microbiol.* **2016**, *1*, 16116.
39. P. S. Garcia, F. D'Angelo, S. Ollagnier de Choudens, M. Dussouchaud, E. Bouveret, S. Gribaldo, F. Barras, *Nat. Ecol. Evol.* **2022**, *6*, 1564–1572.
40. S. F. Jordan, J. Ioannou, H. Rammu, A. Halpern, L. K. Bogart, M. Ahn, R. Vasiliadou, J. Christodoulou, A. Maréchal, N. Lane, *Nat. Commun.* **2021**, *12*, 5925.
41. A. Kappler, C. Bryce, M. Mansor, U. Lueder, J. M. Byrne, E. D. Swanner, *Nat. Rev. Microbiol.* **2021**, *19*, 360.
42. F. Jia, X. Lei, Y. Yan, Y. Su, H. Zhou, H. Wei, Y. Yuan, C. Zou, X. Shi, C. Yang, *Sci. Rep.* **2023**, *13*, 11650.
43. C. Wei, S. Yin, A. Kappler, S. Tao, D. Zhu, *Fundamental Research (FMRE)*, 2023, in press https://doi.org/10.1016/j.fmre.2023.08.003.
44. E. Runge, M. Mansor, T. H. Chiu, J. Shuster, S. Fischer, A. Kappler, J.-P. Duda, *Comm. Earth Environ.* **2024**, *5*, 252.
45. R. L. Spietz, D. Payne, R. Szilagyi, E. S. Boyd, *Trends Microbiol.* **2022**, *30*, 1072–1083.
46. S. W. Vali, D. K. Haja, R. A. Brand, M. W. W. Adams, P. A. Lindahl, *J. Biol. Chem.* **2021**, *296*, 100710.
47. B. Vaccaro, S. Clarkson, J. Holden, D. W. Lee, C. H. Wu, F. L. Poole II, J. J. H. Cotelesage, M. J. Hackett, S. Mohebbi, J. Sun, H. Li, M. K. Johnson, G. N. George, M. W. W. Adams, *Nat. Commun.* **2017**, *8*, 16110.
48. X. Yang, T. Hu, J. Liang, Z. Xiong, Z. Lin, Y. Zhao, X. Zhou, Y. Gao, S. Sun, X. Yang, L. W. Guddat, H. Yang, Z. Rao, B. Zhang, *Nat. Struct. Mol. Biol.* **2024**, *31*, 1072–1082.
49. Y. Yasui, S. Hirayama, T. Hiyoshi, T. Isono, H. Domon, T. Maekawa, K. Tabeta, Y. Terao, *Microorganisms* **2023**, *11*, 2969.
50. G. Yang, X. Xia, W. Nie, B. Qin, T. Hou, A. Lin, S. Yao, L. Zhuang, *Environ. Res.* **2024**, *245*, 118038.
51. S. Peters, D. A. Semenov, R. Hochleitner, O. Trapp, *Sci. Rep.* **2023**, *13*, 6843.
52. T. Beyazay, K. S. Belthle, C. Farès, M. Preiner, J. Moran, W. F. Martin, H. Tüysüz, *Nat. Commun.* **2023**, *14* (1), 570.
53. K. B. Muchowska, S. J. Varma, J. Moran, *Nature* **2019**, *569*, 104–111.
54. A. Y. Mulkidjanian, *Biol. Direct* **2009**, *4*, 24.
55. A. Y. Mulkidjanian, *Biol. Direct* **2009**, *4*, 27.

56. M. Labrenz, G. K. Druschel, T. Thomsen-Ebert, B. Gilbert, S. A. Welch, K. M. Kemner, G. A. Logan, R. E. Summons, G. De Stasio, P. L. Bond, B. Lai, S. D. Kelly, J. F. Banfield, *Science* **2000**, *290*, 1744–1747.
57. T. Rohwerder, T. Gehrke, K. Kinzler, W. Sand, *Appl. Microbiol. Biotechnol.* **2003**, *63*, 239–248.
58. M. Vera, A. Schippers, S. Hedrich, W. Sand, *Appl. Microbiol. Biotechnol.* **2022**, *106*, 6933–6952.
59. L. Zinke, Abiotic reactions on early earth, *Nat. Rev. Earth Environ.* **2023**, *5*, 602.
60. S. Duval, K. Zuchan, F. Baymann, B. Schoepp-Cothenet, E. Branscomb, M. J. Russel, W. Nitschke, *Met. Ions Life Sci.* **2021**, *21*, 135–157.
61. W. Maret, P. Blower, *Biochem. Mol. Biol. Educ.* **2022**, *50*, 283–289.
62. W. Maret, *Appl. Sci.* **2021**, *11*, 10846.
63. R. J. P Williams, *Inorg. Chim. Acta* **2003**, *356*, 27–40.
64. N. Lane, J. G. Xavier, *Nature* **2024**, *626*, 948–951.
65. S. J. Gould, *Wonderful life*, W. W. Norton & Co, New York and London, 1989.

3 Metals and Life
An Introduction

Bhanu P. Jagilinki, Florence Mus and John W. Peters
Department of Chemistry and Biochemistry, University of Oklahoma, Norman, OK 73019, USA
jw.peters@ou.edu

CONTENTS

DOI: 10.1201/9781003459910-3

ABSTRACT

Metals are ubiquitous across all biological kingdoms, from prokaryotes to eukaryotes. A diverse array of metals from nearly all classified groups play crucial roles in biology, conferring both structurally and functionally. Transition metals, in particular, are pivotal in driving almost all biochemical reactions, significantly influencing the global recycling of elements such as carbon, nitrogen, and sulfur. This chapter emphasizes the integral role of a diverse set of metals in sustaining life, detailing their occurrence, involvement in biochemical reactions, and broader implications for health and disease.

KEYWORDS

metals, metalloenzymes, transition, metalloenzymes, redox reaction, life, cofactors, catalysis

1 INTRODUCTION

Metals are critical to life across all kingdoms. From the simplest prokaryotes to the most complex eukaryotes, metals play a vital role in sustaining life and facilitating numerous biological processes. The diverse array of metals, including those from the alkali, alkaline earth, transition, and non-transition metal groups, contribute to a myriad of functions within living organisms, underscoring the essential nature of their involvement [1]. In the essence of life, several biomolecules, especially proteins and sometimes nucleotides, interact with metals to mediate biological reactions or confer structural stability, which are pivotal for the precise regulation and execution of various metabolic processes. This chapter describes several important metal-dependent biomolecules categorized by the type of primary metal center and their positioning in the periodic table (Figure 1).

Metals from the alkali group, such as sodium and potassium, are indispensable for maintaining cellular homeostasis and facilitating signal transduction. Sodium ions, for instance, generate action potentials in nerve cells, which are fundamental for neuronal communication. On the other hand, potassium ions play a critical role in maintaining the membrane potential of cells, which is essential for various physiological processes, including muscle contraction and heart function. Alkaline earth metals, like calcium and magnesium, also play significant roles in biological systems. Magnesium ions, in addition to their role in stabilizing nucleotides, are vital for the functioning of several enzymes, including those involved in energy production and DNA synthesis. Calcium ions are key players in signal transduction pathways and are involved in muscle contraction, blood clotting, and neurotransmitter release. Transition metals, such as iron, copper, and manganese, are the most versatile and functionally diverse. Iron, as part of the heme group in hemoglobin and myoglobin, is critical for oxygen transport and storage. Iron-sulfur clusters are involved in electron transfer in various metabolic pathways, such as cellular respiration and hydrogen metabolism [2]. Copper, another essential transition metal, is a key component of cytochrome c oxidase, an enzyme in the mitochondrial electron transport chain that drives adenosine triphosphate (ATP) production [3]. Zinc is crucial for the catalytic activity in several enzymes, including DNA polymerases and proteases, and plays a structural role in zinc finger proteins, which regulate gene expression.

The importance of metals in biology is further highlighted by the fact that metal ion imbalances can lead to various diseases [4]. For example, iron deficiency can result in anemia, while excess iron can cause hemochromatosis. Similarly, zinc

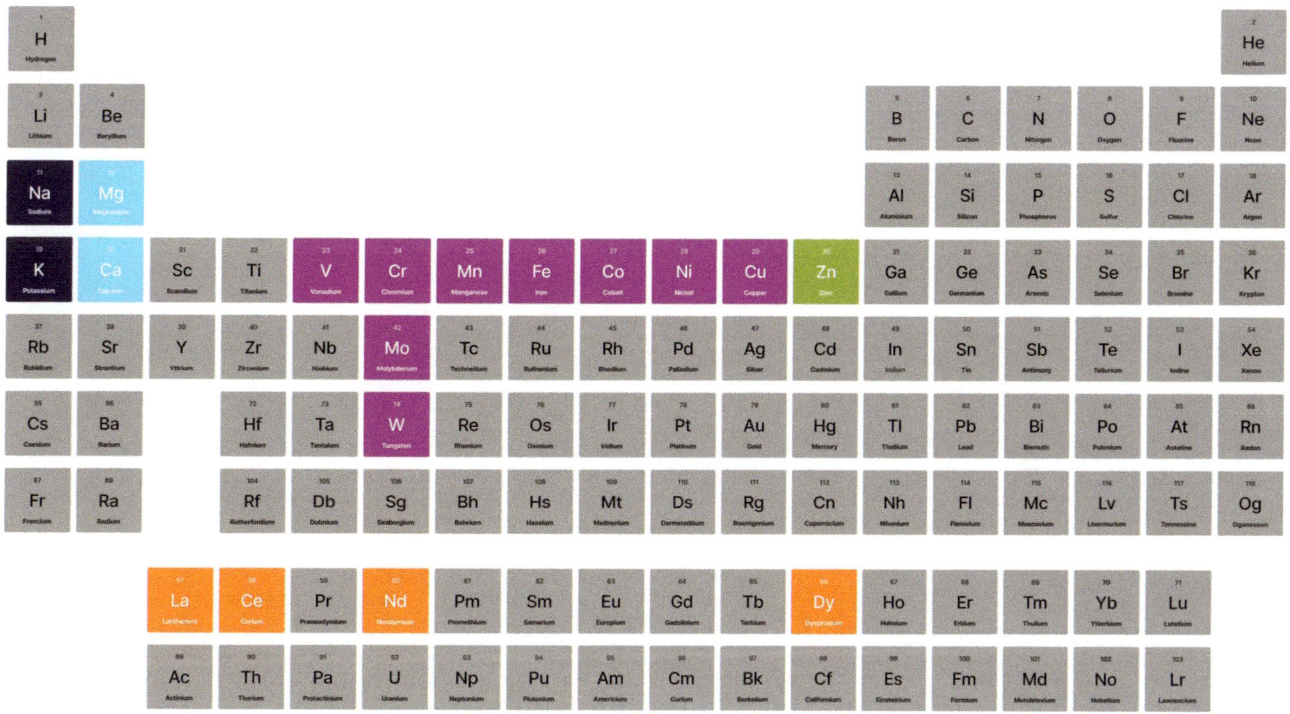

FIGURE 1 This chapter covers the highlighted metals that are biologically relevant. Color scheme: Alkali metals – blue; alkaline earth metals – cyan; transition metals – magenta; non-transition d-block metals – green; lanthanides – orange. (This periodic table has been made using a free interactive template from www.figma.com).

deficiency can impair immune function, and copper imbalances are associated with disorders like Wilson's disease. Understanding the diversity of metals in life provides insight into the fundamental mechanisms of biology and has practical implications for medicine and biotechnology. Advances in metallomics, the study of metal ions in biological systems, pave the way for developing metal-based drugs and diagnostic tools [5]. Concurrently, advancements in protein engineering have enabled the construction of artificial metalloproteins as minimal catalysts that mimic the functionalities of complex natural enzymes [6,7]. In summary, the diversity of metals in life underscores their indispensable role in maintaining biological systems' structure, function, and regulation [8].

Although the metals covered in this chapter emphasize their prominence and indispensability, yet on the contrary, CHNOPS, a mnemonic term used to describe the six essential elements, carbon (C), hydrogen (H), nitrogen (N), oxygen (O), phosphorus (P), and sulfur (S), is often regarded as the so-called "building blocks of life", as they contribute >90% of the mass of the biological systems [9]. Although valid, this perspective underestimates the significance of metals in sustaining life. Even the simplest prokaryotic microorganisms are dependent on these metals for survival. Several studies have emphasized that transition metals have played a prominent role during prebiotic catalysis [10]. The nascent earth was abundant with transition metals, minerals, and organic matter, especially in clay pores [11,12] and hydrothermal vents [13,14]. These niches provided ambient conditions for these molecules to interact, potentially triggering a cascade of redox reactions that synergistically drove prebiotic electron transfer reactions and energy transduction [15]. Those catalytically active prebiotic molecules likely evolved into the first biologically relevant macromolecules [16,17] through various metabolic events, and the extant enzymes' functionality is reminiscent of those prebiotic reactions, albeit within more diversified and complex biological systems. To summarize, the interplay between metals and biomolecules is a testament to the origins and complexity of life, reflecting the intricate mechanisms that sustain living organisms. This chapter highlights the myriad ways metals contribute to life, their biochemical roles, mechanisms of action, and their implications for health and disease.

2 ALKALI METALS

The alkali metals are characterized by their outermost electrons in the s-orbital, which include a subset of group 1 elements of the periodic table of elements, excluding hydrogen. The alkali metals comprises lithium (Li), sodium (Na), potassium (K), rubidium (Rb), and cesium (Cs). Among these elements, sodium and potassium are abundantly in biology, ubiquitous to all living forms, and are essential to support life. Both sodium and potassium are some of the most abundant elements in the earth's crust, constituting about 2.3 % and 2.1 %, respectively. Sodium and potassium form monovalent cations in biological systems, as spectator ions. These ions are weakly hydrated and have labile interactions with biomolecules [18]; however, they differ in their coordination geometry. The typical coordination site for sodium is six, whereas potassium is eight. Both sodium and potassium are highly abundant in human cellular fluids (150 mM of K^+ and about 10 mM of Na^+). They are primarily involved in physiological processes involving osmotic stress, pH homeostasis, balancing electrolytes, and signaling pathways [19]. Apart from being abundantly found in various cellular fluids, sodium and potassium are essential for the activity of several enzymes, channels and pumps, as they facilitate the transport of these solutes across the bio-membranes. These transporters are essential to the survival of organisms as they maintain the electrolyte balance of the cells. The following are a few classical examples of biomolecules interacting with alkali metals, especially sodium and potassium.

2.1 Ion Channels

Ion channels are pore-forming transmembrane proteins that facilitate the movement of ions through the channel pores across biological membranes such as plasma membranes and organelle membranes [20,21]. These channels' pores are hydrophilic, allowing small water-soluble hydrophilic ions to pass across the hydrophobic membrane. Ion channels are highly efficient proteins that can transfer several million ions per second across the membrane, and the flow of ions is always downhill (passive transport). These ion channels are not coupled to ATP hydrolysis; hence, they cannot perform active transport. The simple flow of ions through these ion channels across the electrochemical gradient is essential to drive several physiological processes such as maintaining resting potential, shaping action potentials, nerve and muscle excitation, sensory transduction, blood pressure regulation, hormonal secretion, cell volume regulation and maintenance of salt and pH [22]. Most ion channels are gated by voltages, meaning they open and close based on the voltage gradient across the membrane. Sodium channels and potassium channels are classical examples of voltage-gated channels. The eukaryotic voltage-gated sodium channels are large multisubunit transmembrane proteins comprising a pore-forming α-subunit (~ 260 kDa) and one or more auxiliary β-subunits (~ 33–36 kDa) [23]. The transmembrane region of the α-subunit has four equivalent domains, and each domain comprises six α-helical transmembrane segments (S1-S6). In total, the α-subunit has 24 α-helical transmembrane segments [24]. Each domain's first four segments (S1-S4) serve as the voltage sensors, whereas the S5-S6 segment constitutes the channel pore [25]. To facilitate the cation binding and transport, the pore's architecture's opening and interior are composed of negatively charged amino acid side chains [26]. The primary sequence of the pore's architecture is highly conserved across all the orthologs, as it plays an essential role in "selectivity filter" screening for only Na^+ ions concurrently, avoiding other similar cations like the K^+ ions. Similarly, voltage-gated potassium channels (VGKCs) play a crucial role in generating and propagating the action potential by repolarizing the membrane, thereby maintaining cell homeostasis [27]. Potassium channels are homotetrameric transmembrane proteins, and each subunit consists of six transmembrane regions (6TM) from S1 to S6. The S4, with positively charged amino acid residues, acts as a voltage sensor. At the same time the S5 and S6 constitute the pore and the selectivity filter, which are more specific for K^+ ions over other cations [28,29]. Apart from voltage-gated channels, there are also ion channels regulated through the binding of secondary messengers that induce conformational changes, leading to the opening of the pore. For example, calcium-activated potassium channels open the pore upon Ca^{2+} ion binding to facilitate the downhill flow of K^+ across the membrane [30].

2.2 Sodium-Potassium Pumps

The sodium-potassium pump (Na^+/K^+ pump) is a transmembrane protein that facilitates the movement of Na^+ and K^+ ions across the plasma membrane due to an electrochemical gradient generated through adenosine triphosphate (ATP) hydrolysis. Since these pumps use ATP hydrolysis to drive the transport, they are also called Na^+/K^+-ATPases or sodium-potassium ATPases and are found among all animals [31]. For every ATP molecule that is hydrolyzed to adenosine diphosphate (ADP) and inorganic phosphate (Pi), the Na^+/K^+ pump exports three Na^+ ions concomitant with the import of two K^+ ions (Figure 2). This stoichiometric difference is critical to generating electrochemical gradients across the membrane. The Na^+/K^+ pumps are large heterotrimeric proteins belonging to the family of P-type ATPases. The α-subunit hosts the binding sites for Na^+, K^+, and ATP, whereas the β-subunit is required for the occlusion of the K^+ ions, and the smaller γ-subunit regulates the pumping activity [32]. The Na^+/K^+ pumps indulge in several physiological processes, such as the excitation of electrical signals, regulation of cell volume, intracellular pH, and maintenance of homeostasis, which is achieved by transporting three sodium ions out of the membrane and importing two potassium ions into the cell [33].

2.3 Alkali Metals in G-Quadruplex (G4)

G-quadruplexes are non-canonical nucleotide structures that are formed at guanine (G)-rich sequences both in deoxyribonucleic acid (DNA) and in ribonucleic acid (RNA). G-quadruplexes' building blocks are formed from four guanines arranged in a square planar geometry known as G-quartets, which are stabilized by Hoogsten-based "non-Watson-Crick" hydrogen bonding [34]. The center of each G-quartet is occupied by a monovalent cation, preferentially K^+ ion and occasionally Na^+ ion, which interacts with the O6 of the guanine residues [35,36]. Figure 3 shows the systematic arrangement of K^+ ions in a typical quadruplex. These G-quartets are arranged as bundles by stacking on top of one another, which are stabilized by π–π interactions, thus forming a functional G4 [37]. The G4s are involved in several physiological functions, such as epigenetic maintenance of chromosomal integrity, regulation of replication, transcription, cell division, and recombination [38]. Owing to its importance in several cellular functions, the G4s are being targeted by biopharma industries to develop selective drugs to cure human diseases, including cancers [39].

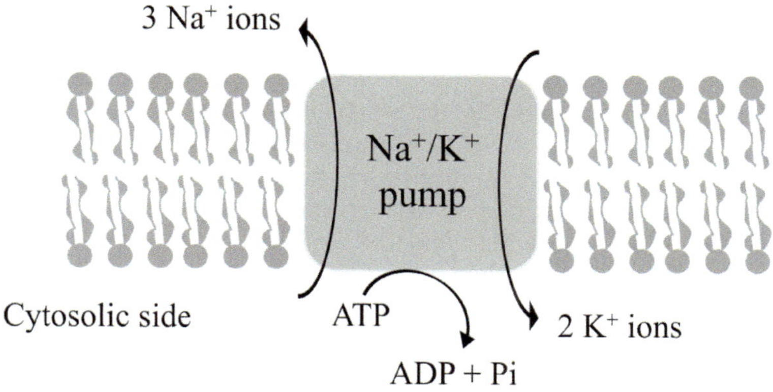

FIGURE 2 Schematic representation of the mode of action of the Na^+/K^+ that is embedded in the plasma membrane primarily composed of a phospholipid bilayer.

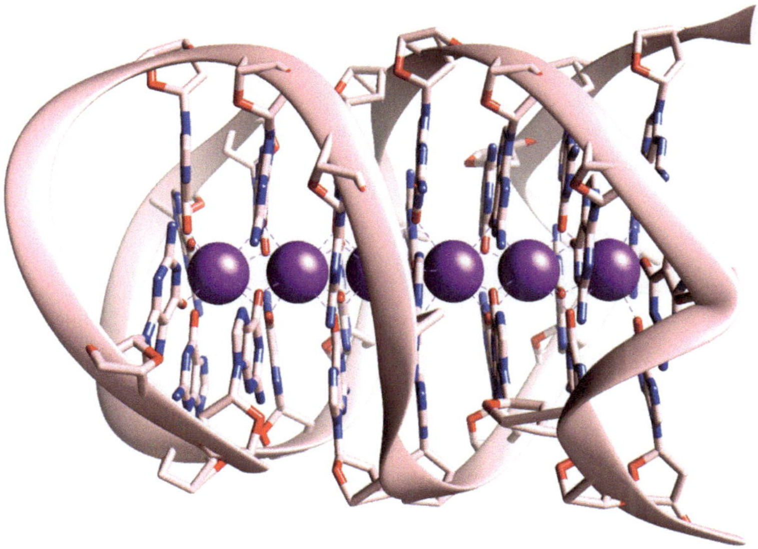

FIGURE 3 The structure of the B-raf dimer DNA quadruplex (PDB: 4H29) shows the systematic arrangement of six potassium ions. The color scheme is as follows: DNA backbone – pink; nitrogen – blue; oxygen – red; potassium – indigo spheres.

3 ALKALINE EARTH METALS

The alkaline earth metals occupy group 2 of the periodic table, consisting of beryllium (Be), magnesium (Mg), calcium (Ca), strontium (Sr), barium (Ba), and the radioactive element radium (Ra) [40]. These metals are characterized by their white and lustrous appearance and complete occupancy with two electrons in their outermost s-orbitals. Only magnesium and calcium occur naturally in biology to support all life forms, although strontium can be found in trace amounts among human bones. Magnesium and calcium are the eighth and fifth most abundant elements in the earth's crust. Since these metals have two electrons in their outermost shell, they can readily donate them to form divalent cations with several biomolecules. The Mg^{2+} ion is typically hexahydrated in solutions, coordinating six oxygens in an octahedral geometry [41]. Unlike Mg^{2+}, the Ca^{2+} ion is usually bound to seven oxygen atoms in a pentagonal bipyramid geometry. Since the ionic radius of Ca^{2+} is slightly larger than Mg^{2+}, the distance of Ca-O is larger than the Mg-O by about 0.3 Å [42]. Since magnesium and calcium bind a myriad of biomolecules and participate in several physiological processes, we have individually described their essential functions in the subsequent subsections of this chapter (3.1 and 3.2).

3.1 Magnesium (Mg)

Magnesium interacts with several biomolecules and is a cofactor in various metalloenzymes. It is critical to several biological processes and is essential to sustain life on Earth. Being a divalent cation, Mg^{2+} in proteins interacts with negatively charged amino acids like aspartic acid (Asp) and glutamic acid (Glu) and binds to the oxygen of the terminal carboxyl group [43]. Apart from interacting with proteins, Mg^{2+} also binds non-proteinaceous biomolecules such as chlorophyll pigments, which coordinates nitrogen instead of oxygen. Although several biomolecules interact with Mg^{2+}, we have scrutinized our compilation to the following significant examples.

3.1.1 Magnesium Ion Transporters

Magnesium is an essential element across all the kingdoms of life. Therefore, Mg^{2+} import is highly critical to all the organisms. Although there are several magnesium transporters, we provide a few well-studied examples here. So far, three distinct families of Mg^{2+} transporters have been identified in prokaryotes: CorA, MgtE, and MgtA/B [44]. CorA is a classic example of Mg^{2+} transporters found in several bacteria and archaea and is well characterized from *Escherichia coli* and *Salmonella typhimurium*. Although CorA is an Mg^{2+} transporter, it also has a moderate affinity toward Co^{2+} and a weaker affinity toward Ni^{2+} [45]. MgtE is a eubacterial transmembrane Mg^{2+} transporter that acts like a Mg^{2+} sensor and plays a key role in maintaining Mg^{2+} homeostasis by regulating intracellular Mg^{2+} levels (Figure 4A) [46]. Unlike CorA a housekeeping gene product, the MgtE expression is controlled by an Mg^{2+} sensing "M-box riboswitch" that responds to cellular Mg^{2+} levels [47,48]. MgtA and MgtB are the only "P-type ATPases" in certain bacteria undertaking Mg^{2+} import [49]. On the other hand, Mrs2 and Lpe10 are two major Mg^{2+} ion-transporting counterparts responsible for maintaining Mg^{2+} homeostasis in eukaryotes [50,51].

3.1.2 Mg^{2+} in Nucleotide-Binding Proteins

Several DNA- and RNA-binding proteins have Mg^{2+} ions at their active site and are essential for their catalytic activities. In this chapter, we have included the following important examples: DNA and RNA polymerases, primases, and guanylate cyclase. DNA polymerases are the key enzymes that play a key role in DNA replication. Several families of DNA polymerases are found among viruses, prokaryotes, and eukaryotes. Although there are differences in the sequences among these families, they share a similar active site composed of negatively charged amino acids, typically aspartic acid (Asp) that coordinates two metal ions, usually with Mg^{2+} or occasionally Mn^{2+} ion [52]. The presence of Mg^{2+} ion at the active site controls the dynamic conformational changes occurring during the dNTP (deoxynucleotide triphosphate) polymerization reaction [53]. Interestingly, the reverse transcriptase (RT), which is an RNA-dependent DNA polymerase found typically in viruses, also has Mg^{2+} ion at its active site [54]. Like DNA polymerases RNA polymerases have two Mg^{2+} ions at their active site coordinated to Asp residues. These Mg^{2+} ions not only play a role in the deportation of the OH at the 3′ end of RNA but also facilitate its nucleophilic attack on the incoming NTP (nucleoside triphosphate), forming a new phosphodiester bond [55]. Hence, Mg^{2+} ions play a key role in the elongation phase of the transcription. On the other hand, DNA primases are a type of ssDNA (single-stranded DNA)-dependent RNA polymerases that catalyze the synthesis of small RNA segments on the ssDNA, known as primers [56]. These RNA primers act like a scaffold for DNA polymerase to initiate replication. Apart from having two Mg^{2+} ions at its active site, the DNA primases also have an N-terminal zinc-binding

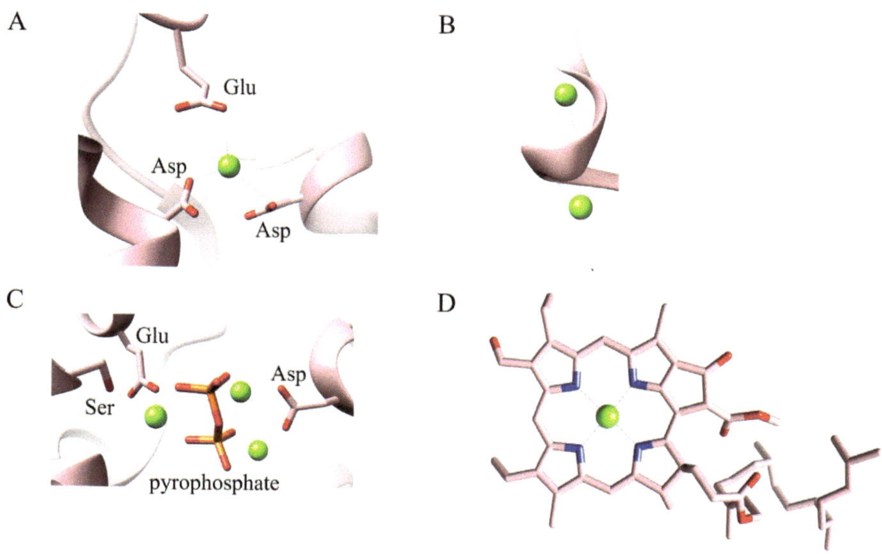

FIGURE 4 The occurrence of Mg in various biomolecules: (A) One of the Mg-binding sites of the magnesium transporter, MgtE (PDB: 2YVX). (B) Magnesium-binding sites in a typical ribozyme RNA loop (PDB: 1GID). (C) The active site of trichodiene synthase harbors a unique trinuclear Mg center (PDB: 1JFG). (D) Stick representation of Chlorophyll B molecule (PDB: 2X20). Color scheme: Protein backbone, RNA backbone and carbon – pink; nitrogen – blue; oxygen – red; phosphorus – vermillion; magnesium – lime green spheres.

domain, with a single divalent Zn^{2+} ion coordinated to four cysteinyl ligands [57,58]. Guanylate cyclase is an Mg^{2+} ion containing metalloenzyme that catalyzes the formation of cyclic guanosine monophosphate (cGMP) from its precursor, guanosine triphosphate (GTP) [59]. cGMP is an important "secondary messenger" that plays a crucial role in signal transduction.

3.1.3 Mg^{2+}-Dependent Ribozymes

Ribozymes are ribonucleic acid metalloenzymes entirely made up of RNA and harbor divalent cations, usually Mg^{2+} ions, at their active site, as depicted in Figure 4B. These ribozymes catalyze the ligation and cleavage of the phosphodiester bonds and play a key role in RNA splicing [60]. Specific ribozymes also mediate the formation of peptide bonds [61]. Artificial Mg^{2+} ion-dependent ribozymes like Diels–Alderase catalyze the formation of C-C bonds and drive the [4 + 2] cycloaddition reaction between anthracene dienes and maleimide dienophiles [62].

3.1.4 Mg^{2+}-Dependent Kinases and Phosphatases

Kinases are transferases involved in the transfer of phosphate moiety to substrates from phosphate, donating high-energy molecules such as NTPs in a process described as phosphorylation [63]. Kinases have been classified into several families, and many of them use Mg^{2+} ions as a cofactor to catalyze the phosphorylation reaction [64]. For example, extracellular signal-regulated kinase 2 (ERK2) of the classical mitogen-activated protein kinase (MAPK) pathway binds one Mg^{2+} ion and is essential for its catalytic activity. MAPK pathway is one of the critical signaling pathways that transduces extracellular signals into cytoplasm and nucleus [65]. Inside the cytoplasm, ERK2, through its immediate downstream effector molecule, Ribosomal s6 kinase (RSK), can trigger the activation of ribosomal protein s6 of the smaller subunit of ribosomes and promote translation. Similarly, inside the nucleus, RSKs alone or in association with ERK2 can bind and activate a variety of transcription factors such as cAMP response element-binding protein (CREB), serum response factor (SRF), estrogen receptor α (ERα), FOS, and JuN and participate in the initiation of transcription [66]. Since kinases have myriad roles in several signaling pathways, mutations in some key enzymes can cause several human diseases, including neurological, hematological, immunological, and cancer [67]. On the other hand, phosphatases are hydrolases that play a key role in the dephosphorylation (removal of a phosphate group) of biomolecules in the presence of a water molecule, and in general, phosphatases have the opposite activity in comparison to kinases [68]. Magnesium-dependent phosphatase-1 (MDP-1) is a eukaryotic phosphatase with two conserved Asp residues that coordinate a single Mg^{2+} ion [69].

3.1.5 Trichodiene Synthase

Trichodiene synthase plays a crucial role in terpenoid biosynthesis by catalyzing the cyclization of farnesyl diphosphate to trichodiene, which is a sesquiterpene that serves as a precursor to over 100 different trichothecene toxins and antibiotics [70]. Trichodiene synthase belongs to the family of lyases, and it is interesting to note that it has a unique trinuclear metallocofactor composed of three Mg^{2+} ions at its active site, as shown in Figure 4C [71]. These enzymes have a unique DDXXD motif at the active site that binds Mg^{2+} ions, which is responsible for driving the cyclization reaction.

3.1.6 Chlorophyll

Chlorophyll is a water-insoluble pigment found in cyanobacteria, algae, and plants' chloroplasts. It allows them to absorb blue and parts of red light of the visible region of electromagnetic radiation (EMR) [72]. This absorbed light is vital for cyanobacteria, algae, and plants to drive photosynthesis, a biological process responsible for converting light energy into carbon-based chemical energy [73]. Photosynthesis is crucial in supporting life on Earth by maintaining the atmospheric oxygen levels and supporting the hierarchical food chain with high-energy biomolecules. Although several types of chlorophyll pigments exist among various photosynthetic organisms, they all fall under the "chlorine" type of pigments, as they have a common central tetrapyrrole segment. Besides having a central tetrapyrrole, chlorophylls have one additional ketonic-based ring. At the center of the tetrapyrrole is a single Mg^{2+} ion coordinated by the nitrogen of each pyrrole ring [74]. Figure 4D shows the structure of Chlorophyll B.

3.2 Calcium (Ca)

Like magnesium, calcium ions bind negatively charged amino acids such as aspartic acid (Asp) and glutamic acid (Glu). However, calcium is more abundant than magnesium, especially among the higher order of eukaryotes. Calcium alone constitutes about ~1.5 % of the total human body weight, most of which resides in bones, cartilage, and teeth. Apart from contributing to the structural integrity of humans and other mammals, calcium plays a significant role in several physiological functions through signal transduction. High levels of Ca^{2+} ions are toxic to the cell. Therefore, the intracellular concentration of Ca^{2+} is usually maintained at 10^{-1} μM, whereas the concentration in the extracellular space is around 1 mM. The cell imports Ca^{2+} ions only during muscle contraction, neurotransmission, or signal transduction to the effector molecule [75]. The following are some key biomolecules whose interactions with calcium are well established.

3.2.1 Calcium ATPase (Ca^{2+}-ATPase)

Ca^{2+}-ATPases are transmembrane proteins belonging to P-type ATPases that are essential for transporting the Ca^{2+} ions across the membrane. Based on their location, they are classified into two major types: plasma membrane Ca^{2+} ATPase (PMCA) and the sarcoplasmic reticulum Ca^{2+} ATPase (SERCA). The PMCA, which is ubiquitous in all eukaryotes, is one of the major proteins responsible for maintaining stable intracellular Ca^{2+} ion levels by exporting excess Ca^{2+} ions into the extracellular space, as excess intracellular Ca^{2+} ion levels can be detrimental to the cells [76]. The SERCA are a special type of Ca^{2+}-ATPases found exclusively in the sarcoplasmic reticulum of the myocytes (muscle cells). Its main function is to export calcium from the cytosol into the lumen of the sarcoplasmic reticulum during muscle relaxation [77].

3.2.2 Calcium as a Secondary Messenger

Secondary messengers are important intracellular signaling molecules that transduce the extracellular signals, triggered by the first messengers' stimulation of the cell membrane receptors. The propagation of this extracellular signal inside the cell is critical to regulating several cellular functions such as replication, transcription, and translation. As a secondary messenger, Ca^{2+} ions play a significant role in muscle contraction, neurotransmission, and in signal transduction of several signaling pathways, especially the phosphoinositide 3-kinase (PI3K) pathway [78,79].

3.2.3 Ca^{2+}/Calmodulin-Dependent Kinases

In the previous section (3.2.2), it was discussed how Ca^{2+} ions act as secondary messengers in communicating cellular signals. However, the signal is not always directly delivered to the effector molecule. Certain downstream effector proteins need another intermediate to access the information the Ca^{2+} ions are carrying. Calmodulin (CaM) is a small Ca^{2+} ion-binding protein that acts as an intermediate sensory molecule and helps in transmitting the Ca^{2+} ion signal to its downstream effector molecule [80]. Figure 5A shows the calcium-binding site in calmodulin from *Paramecium tetraurelia*. The most common examples of this type are the Ca^{2+}/Calmodulin-dependent kinases (CAMKs). Although their function is similar to Mg^{2+} ion-dependent kinases (discussed in Section 3.1.4), the mode of activation is completely different for both of these types as CAMKs can only be activated upon binding to the Ca^{2+}/CAM complex.

3.2.4 Calbindin, Calretinin, and Calsequestrin

Calbindin and calretinin are small vitamin D-dependent calcium-binding proteins that can bind up to four calcium ions. These proteins act like calcium sensors and are important in buffering calcium ions inside mammalian cells. Apart from buffering calcium ions, they are involved in several cellular physiological processes, including signal transduction, extracellular transport, and synapse [81,82]. Figure 5B shows a typical calcium-binding site in calbindin. Calsequestrin is a small calcium-binding protein that buffers calcium ions like calbindin and calretinin. However, it is exclusively found in the sarcoplasmic reticulum (SR) of skeletal and cardiac muscles and plays a vital role in muscle contraction and relaxation. Calsequestrin is to be anchored to membrane proteins inside the lumen of junctional SR [83]. Anchored calsequestrin polymerizes at the junctional SR and forms branches, whose complexity increases with the abundance of calcium ions. Since calbindin, calretinin, and calsequestrin play a significant role in buffering calcium ions and participating in several physiological processes, mutations in these proteins can cause several human diseases.

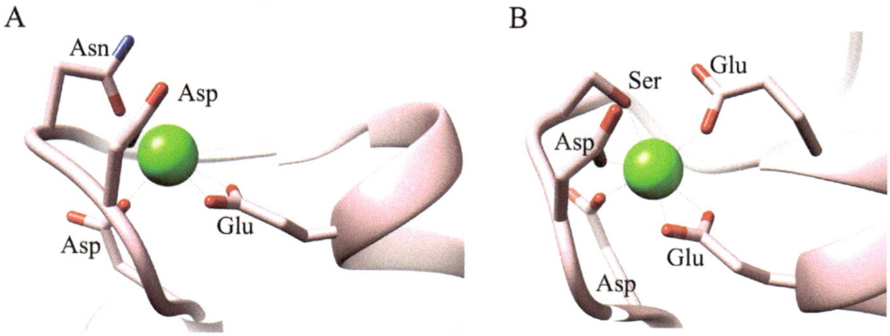

FIGURE 5 The coordination of Ca in various proteins: (A) The calcium-binding site in calmodulin from *Paramecium tetraurelia* (PDB: 1OSA). (B) The calcium-binding pocket in human calbindin (PDB: 6FIE). Color scheme: Protein backbone and carbon – pink; nitrogen – blue; oxygen – red; calcium – green spheres.

3.2.5 Annexin

Annexins or lipocortins are eukaryotic membrane-binding proteins that interact with acidic phospholipids, such as phosphatidic acid and phosphatidylserine phosphatidylinositol in a calcium-dependent manner [84]. Although several annexins have been identified to date, they have a similar architecture in having one annexin core domain and the N-terminal domain [85]. The annexin core domain is highly conserved and comprises four to eight repeats. These annexin repeats are arranged concentrically in a way that appears as a slightly bent disc that encloses a central hydrophilic hole. This hydrophilic hole serves as a voltage-dependent calcium channel and facilitates membrane trafficking. The convex side of the disc has calcium-binding sites, whereas the concave side interacts with the N-terminal domain. Unlike the annexin core domain, the N-terminal domain is not well conserved across organisms. The N-terminal domain has a few post-translational modifications (PTMs) that allow annexins to interact with several molecules and facilitate signal transduction. Annexins are involved in several cellular and extracellular activities, such as membrane trafficking (endocytosis and exocytosis), membrane-cytoskeleton scaffolding, ion channel regulation, and preventing coagulation [86].

3.2.6 Calcium in Bones

Calcium is the fifth most abundant element in human beings, and 99% of it is found in the skeletal system primarily within the bones and teeth. Calcium in human bones occurs in an inorganic mineral form known as hydroxyapatite ($Ca_{10}[PO_4]_6[OH]_2$) [87]. Since it is a mineralized form of calcium, it provides invaluable support and strength to humans and all mammals. Based on the extent of calcification or mineralization, bone tissue is classified into two types: cortical bone and cancellous bone. The cortical bone is a hardened and rigid form of bone tissue with a greater degree of calcification and is mainly found on the surface of the bones. On the other hand, the cancellous bone is a less calcified and dense spongy form of bone, primarily occurring at the junctions of long bones and joints. Although bone is mineralized, it is incredibly dynamic, as it continuously remodels itself to maintain the structural integrity of the skeletal system, bone healing apart from balancing the levels of calcium and phosphorus in the human body. This bone modeling is carried out by several specialized bone cells such as osteoclasts, osteoblasts, and osteocytes [88]. Calcium deficiency can lead to severe consequences to the structural integrity of bone and lead to several human diseases like osteomalacia and osteoporosis [89].

4 TRANSITION METALS

Transition metals are the d-block elements occupying the groups from 3 to 11 of the periodic table. These elements are characterized by their lustrous appearance, with most having good thermal and electrical conductivity properties. Also, most of these metals are strong, with high melting and boiling points. Among the various groups, transition metals from only groups 5 to 11 are biologically relevant and have functional roles. Only these group's first members are found in biology, except group 6, as its first three elements are relevant to biology (Cr, Mo, and W). Since most of these transition metals exist in multiple oxidation states, they can coordinate several different molecules (including both biological and inorganic) and catalyze several biochemical reactions critical to the sustainability of life on Earth.

4.1 Vanadium (V)

Vanadium is the first element in group 5 of the periodic table. Although vanadium is the second most abundant biologically relevant transition metal in the earth's crust, its prominence in biology is related to only a subset of microorganisms. Most known vanadium-binding enzymes have counterparts with iron or molybdenum cofactors that could catalyze the same reaction as the vanadium-binding enzymes. Thus, vanadium is vital only to certain marine microorganisms. A German scientist, M. Henze, first reported the occurrence of vanadium in living microorganisms more than a century ago; since then, several vanadium-containing proteins have been successfully characterized [90]. No known specific transporters are associated with vanadium transport, and it is mainly absorbed in the serum by transferrin (discussed in Section 4.6.1), a type of glycoprotein that usually transports iron. The promiscuity of these proteins enables the cells to bind and transport other closely related cations depending on the cell's requirements.

4.1.1 Vanadocytes and Vanabins

Vanadocytes are a particular type of blood cells exclusively found in ascidians or sea squirts. These ascidians are sedentary marine invertebrates that grow under high saline conditions, naturally occurring in the shallow parts of the oceans. Vanadocytes have an incredible capacity to hold vanadium at a magnitude seven orders higher than the levels found in the oceans [91]. Vanadium in vanadocytes is primarily concentrated in vanadium-binding proteins known as vanabins [92]. Like hemoglobin (discussed in Section 4.6.7) of vertebrates, vanabins are chromogenic proteins, and the presence of vanadium confers their pale green color. Hence, these proteins are also known as hemovanadins. Although vanabins are relatively smaller proteins, they can, however, bind to several vanadium ions individually in a mononuclear state [93].

4.1.2 Vanadium-Dependent Nitrogenase

Perhaps the most studied vanadium-containing enzyme is the vanadium-dependent nitrogenase (Vnf), which converts the inert dinitrogen from the atmosphere into a bioavailable form of ammonia. Vanadium(V)-dependent nitrogenase is common among several species of nitrogen-fixing organisms [94]; however, these organisms also have the molybdenum(Mo)-dependent nitrogenase (Nif) (discussed in Section 4.3.1). The V-dependent nitrogenase is less efficient in catalysis than the Mo-dependent nitrogenase (discussed in Section 4.3.1). Breaking the triple bonds in dinitrogen requires enormous energy, and V-dependent nitrogenases consume up to 36 ATP molecules to drive the reaction. In contrast, Mo-dependent nitrogenase uptakes energy from the hydrolysis of 16 ATP molecules. The V-dependent nitrogenase is a complex two-component system, one being the VFe protein (VnfDKG) and another being the Fe protein (VnfH). The VFe protein is a hexamer of $\alpha_2\beta_2\delta_2$ subunits and holds the active site for reducing dinitrogen utilizing a unique cofactor known as the VFe cofactor (Figure 6A) [95], and the vnfH encoded Fe protein is a $\gamma 2$ homodimer [96]. The VFe protein and Fe protein act as two separable components in nitrogenase.

4.1.3 Vanadium-Dependent Haloperoxidases

Haloperoxidases are peroxidases that utilize hydrogen peroxide to oxidize halides in a two-electron-driven process. Depending on the type of halogen, the haloperoxidases are subdivided into chloroperoxidases, bromoperoxidases, and iodoperoxidases [97]. Although heme-containing haloperoxidases do exist, this chapter focuses only on vanadium-dependent haloperoxidases (VHPOs). Vanadium-dependent haloperoxidases (VHPOs) are mostly found in marine organisms such as brown and red macroalgae and in certain fungi and bacteria. In these enzymes, vanadium is in +5 oxidation state. Figure 6B shows the active site of VHPO from *Ascophyllum nodosum*. The VHPOs adapted a unique bi-bi ping-pong mechanism to trigger the two-electron oxidation of the halides upon hydrogen peroxide binding [98].

4.1.4 Vanadium-Dependent Nitrate Reductases

Nitrate reductases are enzymes that play a key role in reducing nitrate (NO_3^-) to nitrite (NO_2^-), an essential reaction critical to the survival of plants [99]. Although most nitrate reductases contain molybdenum cofactor, enzymes isolated from *Pseudomonas isachenkovii* and *Thioalkalivibrio nitratireducens* were found to coordinate vanadium in their active sites [100,101]. Although molybdenum nitrate reductases additionally require molybdopterin cofactor, the vanadium counterparts do not require additional ligands or cofactors to bind the enzyme, displaying the simplicity of these enzymes. This could also suggest vanadium nitrate reductases may have evolved before the molybdenum enzymes.

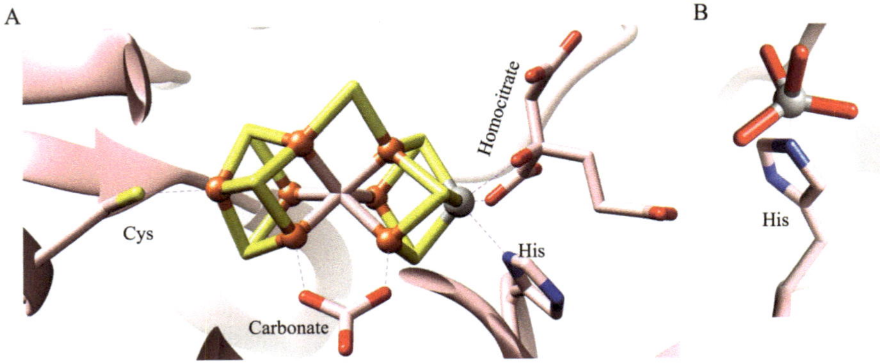

FIGURE 6 (A) The VFe cofactor (7Fe-8S-V-C) of V-dependent nitrogenase. (B) The vanadate at the active site of vanadium-dependent haloperoxidase from *Ascophyllum nodosum* (PDB: 1QI9). Color scheme: Protein backbone and carbon – pink; nitrogen – blue; oxygen – red; sulfur – yellow; iron – orange spheres; vanadium – light gray spheres.

4.2 Chromium (Cr)

Chromium is the first element of group 6 of the periodic table and is the third most abundant transition metal in the earth's crust. Like vanadium, the role of chromium is minimal, and only a few proteins have been characterized so far. Chromium is occasionally considered a micronutrient in mammals. However, its role in biology is poorly understood and highly debated. Recent observations on the chromium flux generated between the bloodstream and the adjacent tissues triggered by insulin led to speculations that chromium might act as a secondary messenger in insulin signaling [102].

4.2.1 Chromodulin

In the 1980s, Yamamoto and co-workers characterized a unique chromium-binding oligopeptide and named it a low-molecular-weight chromium-binding substance (LMWCr), later renamed chromodulin. This was the first reported instance where chromium was found to be associated with biomolecules. This oligopeptide is only about ~1,500 Da in molecular weight, having only four amino acids: glycine, cysteine, glutamate, and aspartate residues. Despite its smaller size, it can still bind up to four chromium ions with high affinity [103]. LMWCr plays a crucial role in mammalian insulin signaling. In insulin-sensitive cells, the LMWCr is in its apo-state, and upon insulin binding to the insulin receptor, the cell membrane creates a flux, facilitating a net flow of chromium ions from the bloodstream into the insulin-sensitive cells. The rise in chromium levels loads the cellular apo-LMWCr with chromium ions. This holo-LMWCr then binds to the transmembrane domain of the insulin receptor and helps prolong its phosphorylated state to facilitate greater absorption of glucose into the cells [104].

4.3 Molybdenum (Mo)

Molydenum is the second element of group 6, with an atomic number of 42. Molybdenum is the only element in the second row of transition metals that has a functional role in biology and is critical to most living organisms [105]. Thus, molybdenum is defined as an essential trace element across all life forms. Although Mo can exist in multiple oxidation states, only Mo(VI), Mo(V), and Mo(IV) are biologically relevant. Unlike most of the other elements, which can occur both as an individual entity or in association with other elemental moieties (cofactors) in biomolecules, Mo's occurrence is however only limited to cofactors i.e. association with other elements. Mo cofactors in biology are unique, and depending on the type of enzyme, they can mediate either the transfer of oxygen or proton/electron.

4.3.1 Mo-Dependent Nitrogenase

As discussed in the section on V-dependent nitrogenase, the Mo-dependent nitrogenase is most energy-efficient at reducing N_2 than the other alternative isoforms (V-dependent nitrogenase and iron (Fe)-only nitrogenase) [106]. Like the V-dependent counterpart, the Mo-dependent nitrogenase is a two-component system consisting of the catalytic MoFe protein and the Fe protein. The MoFe protein is a tetramer of $\alpha_2\beta_2$ subunits where the α subunit harbors the active center for the reduction of dinitrogen to ammonia, known as the FeMo-cofactor or FeMo-co (7Fe-9S-Mo-C-homocitrate) [107] (Figure 7A). Mo in the FeMo-co is the only known example in biology where its occurrence lacks molybdopterin (discussed in Section 4.3.2). In addition to the FeMo-co, the MoFe protein also has a unique iron-sulfur cluster known as the P cluster [8Fe-7S] cluster, which acts as an electron relay in transferring the electrons to the FeMo-co for the ultimate reduction of dinitrogen to ammonia.

4.3.2 Xanthine Oxidase (XO)

XO is one of the major enzymes in the catabolic pathway of purine nucleotides, which catalyzes the oxidation of hypoxanthine to xanthine and, eventually, to uric acid by generating reactive oxygen species (ROS). XO is a Mo-binding enzyme where Mo is bound to molybdopterin, also known as pyranopterin-dithiolate [108]. Molybdopterin is a heterocyclic compound where a pyran is fused to a pterin ring, and the pyran additionally has two essential thiolates that bind either Mo or W. When bound to Mo, the molybdopterin is referred to as Moco. Figure 7B shows the 3D structure of Mo-bound molybdopterin at the active site of XO. Apart from the molybdopterin cofactor, XO also has other cofactors, such as flavin adenine dinucleotide (FAD) and ferredoxin-like [2Fe-2S] clusters (discussed in 4.6.11). Under-excretion of uric acid generated by XO may cause uric acid accumulation in the blood, leading to an inflammatory condition known as gout in humans. Gout is a chronic disease characterized by red and swollen joints. Anti-inflammatory drugs targeting XO, such as allopurinol and febuxostat, are often administered as first-line measures to treat gout patients [109]. A subclass of XO known as aldehyde oxidases (AO) has highly conserved sequence homology, including the cofactor requirements such as Moco, FAD, and [2Fe-2S] clusters. AO, however, plays a key role in the oxidation of aldehydes into carboxylic acids, apart from carrying out the hydroxylation of some heterocyclic compounds [110].

4.3.3 Sulfite Oxidase (SO)

SO is an essential enzyme that catalyzes the oxidation of sulfite (SO_3^{2-}) to sulfate (SO_4^{2-}), the final step in the degradation of "sulfur"-containing amino acids [111]. Similar to XO, SO has a molybdopterin cofactor that is bound to Mo(VI) that serves as the active site for the oxidation of SO_3^{2-} to SO_4^{2-}, during which Mo(VI) accepts two electrons and gets reduced to Mo(IV). Unlike XO, which has additional [2Fe-2S] clusters, SO instead has a heme B (discussed in Section 4.6.7) cofactor. Since oxidation of SO_3^{2-} to SO_4^{2-} is a critical reaction in the sulfur metabolism, any dysregulation of this enzyme can lead to a severe form a condition known as "sulfite oxidase deficiency", characterized by neurological abnormalities, mental retardation, skeletal deformities, and degradation of the brain, often leading to death [112].

4.3.4 DMSO Reductase

DMSO (dimethyl sulfoxide) reductases are Mo-containing enzymes found in bacteria and archaea. They mediate an essential reaction of catalyzing the reduction of DMSO to dimethyl sulfide (DMS), an essential intermediate in the biogeochemical sulfur cycle. This reaction is critical in the anaerobic respiration and the dissimilatory reduction of certain toxic oxoanions, as DMSO serves as a terminal electron

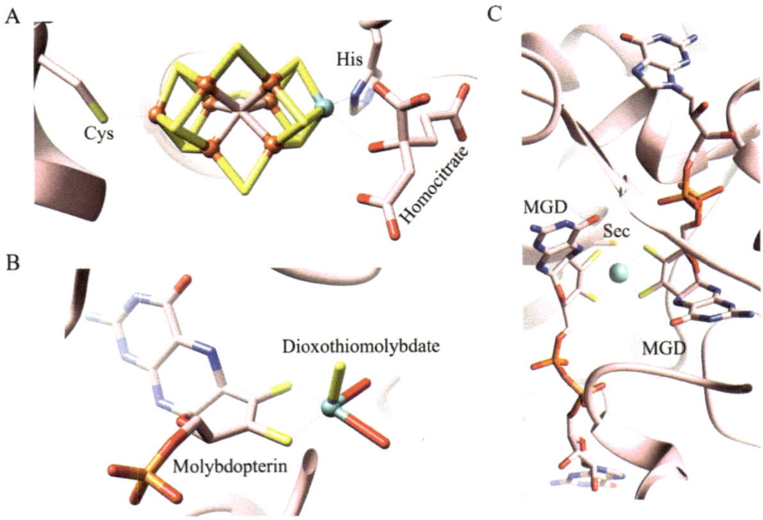

FIGURE 7 (A) The FeMoc-co of Mo-nitrogenase. (B) The molybdopterin-bound dioxothiomolybdate center at the active site of xanthine oxidase from bovine milk (PDB: 1FIQ). (C) The Mo–bis-MGD center at the active site of formate dehydrogenase from *Escherichia coli* (PDB: 1KQF). Color scheme: Protein backbone and carbon – pink; nitrogen – blue; oxygen – red; sulfur – yellow; phosphorus – vermillion; selenium – saffron; iron – orange spheres; molybdenum – turquoise spheres.

acceptor, where the oxygen of the DMSO is transferred to Mo, which is then reduced to water. The molybdopterin found in DMSO reductase is unique and different from XO and SO enzymes, as two molybdopterins (as against one) are required to bind one Mo. These molybdopterins are also modified to bind guanine dinucleotides, forming a novel cofactor known as molybdopterin guanine dinucleotide (MGD). Two such MGDs bind a single Mo center (Mo–bis-MGD) that serves as the active site [113]. Being terminal reductases, they play a key role in anaerobic respiration and lithotrophy, as these enzymes can generate energy solely from inorganic substrates.

4.3.5 Formate Dehydrogenases

Formate dehydrogenases (FDHs) catalyze the oxidation of formate ($HCOO^-$) to carbon dioxide (CO_2), a critical reaction in the geological carbon cycle [114]. Among the FDHs, only the bacterial enzymes have molybdneum, which is held by two molybdopterins moieties additionally bound to guanosine monophosphate, to form a Mo–bis-MGD center similar to the active site observed in DMSO reductases. Figure 7C shows the Mo–bis-MGD center at the active site of FDH from *Escherichia coli.* Bacterial FDHs are complex enzymes, and apart from the Moco, they also have [4Fe-4S] clusters and heme cofactors that participate in the electron transfer pathway. These enzymes under anaerobic conditions (absence of oxygen) and in the presence of $HCOO^-$ and $NO_3{}^-$ play a vital role in the formate:nitrate respiratory chain, in which the oxidation of formate is synergistically coupled to the reduction of nitrate ($NO_3{}^-$) to ($NO_2{}^-$) [115].

4.4 Tungsten (W)

Tungsten is the third element of group 6, with an atomic number of 74. It is the heaviest element and the only member in the third row of transition metals. It has a functional role in biology and plays a key role in the biogeochemical cycles of carbon, nitrogen, and sulfur. Like molybdenum, tungsten does not occur as a free ion and is usually found coordinated to dithiolene groups of two pterin cofactors. The occurrence of tungsten in enzymes is primarily limited to certain microorganisms [116].

4.4.1 Acetylene Hydratases (AHs)

AH is a tungstoenzyme initially discovered in *Pelobactor acetylenicus*, that catalyzes the non-redox hydration (addition of water molecule) of acetylene (C_2H_2) to form acetaldehyde (CH_3CHO) [117]. The catalytic site of AH has a tungsten bound to two pterin cofactors, each of which is bound to guanosine monophosphate [118]. AH additionally contains a [4Fe-4S] cluster that keeps the tungsten in its reduced W(IV) state throughout the catalytic reaction. The [4Fe-4S] cluster is only required to maintain the non-redox state of the tungsten and may not be essential to the catalysis as long as W(IV) state is maintained by other strong reducing agents.

4.4.2 Benzoyl-CoA Reductases (BCRs)

BCRs are bacterial enzymes that mediate the ATP-dependent de-aromatization of benzoyl-CoA to cyclohexa-1,5-diene-1-carbonyl-CoA, analogous to Birch-like reduction [119]. Benzoyl-CoA is the critical intermediate during the biological degradation of various aromatic compounds under anaerobic conditions. BCR was first identified in the denitrifying bacterium, *Thauera aromatica*. The active site of BCRs comprises a mononuclear tungsten center bound to two pterin moieties, as depicted in Figure 8. Apart from the tungstopterin cofactor, the BCRs have a monovalent magnesium ion and three low-potential [4Fe-4S] clusters. The transfer of electrons from the [4Fe-4S] clusters to the active site of BCRs coupled to ATP hydrolysis is like the electron transfer pathway observed in nitrogenases. The redox potential of the BCR substrate/product couple is around −622 mV *vs* SHE (standard hydrogen electrode) and is one of the lowest potentials observed in natural systems, and indeed, an electron transfer from any reduced electron donor to the active site of BCR has to be coupled to an exergonic reaction to drive this uphill reaction [120].

4.5 Manganese (Mn)

Manganese is a transition metal with the atomic number 25 in group 7 of the periodic table. It is an essential trace element for all life forms due to its critical role in various biochemical processes. Manganese typically exists in the (II) and (III) oxidation states (although higher oxidation states occur) in biological systems, where it functions as a cofactor for numerous enzymes [121]. Mn-dependent enzymes are involved in various physiological and metabolic processes, including antioxidant defense, metabolism, and cellular regulation.

4.5.1 MntE and MntP (Mn Transporters)

Mn transporters, such as MntE and MntP, are critical for maintaining manganese homeostasis within cells. MntE and MntP function as efflux transporters, facilitating

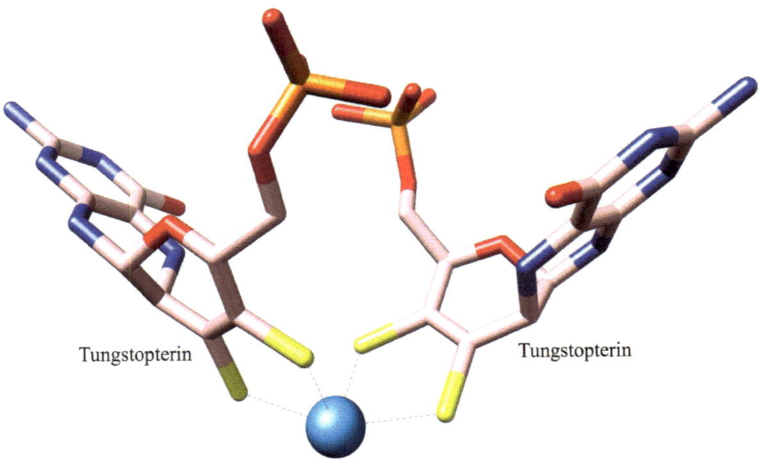

FIGURE 8 The structure of tungstopterin cluster at the active site of Benzoyl-CoA-reductase from *Geobacter metallireducens* (PDB: 4Z3W). Color scheme: Carbon – pink; nitrogen – blue; oxygen – red; sulfur – yellow; phosphorus – vermillion; tungsten – teal sphere.

the export of excess manganese out of the cell to prevent toxicity. This mechanism is vital because excessive intracellular manganese can lead to oxidative stress and cellular damage [122,123]. These transporters work in concert to regulate the intracellular manganese levels, required to maintain optimal concentrations. The precise regulation of manganese is essential for the functioning of numerous Mn-dependent enzymes involved in processes such as antioxidant defense, nucleotide synthesis, and several other metabolic pathways. Disruption in the activity of these transporters can lead to severe health conditions associated with either manganese deficiency or toxicity.

4.5.2 Mn-Dependent Superoxide Dismutase (Mn-SOD)

Mn-SOD is a crucial antioxidant enzyme that protects cells from oxidative damage by catalyzing the dismutation of superoxide radicals ($O_2^{\bullet -}$) into molecular oxygen (O_2) and hydrogen peroxide (H_2O_2). This enzyme plays a vital role in defending against reactive oxygen species (ROS), which can otherwise cause severe oxidative damage to cells. The active site of Mn-SOD contains a mononuclear Mn(III) center (Figure 9A) that undergoes reduction-oxidation between Mn(III) and Mn(II) states

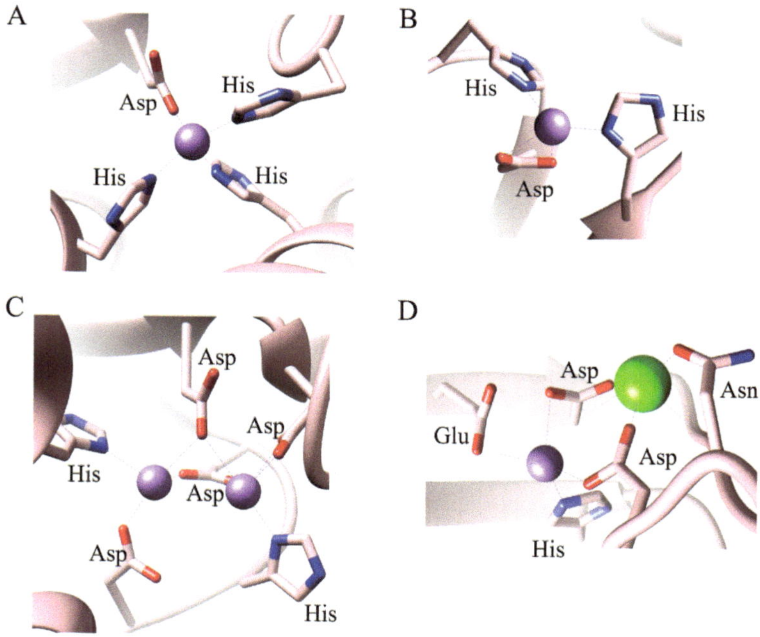

FIGURE 9 (A) The mononuclear Mn(III) center at the active site of the human Mn-SOD (PDB: 1VAR). (B) The active site coordination in Mn-dependent acetone carboxylase from *Xanthobacter autotrophicus* (PDB: 5SVC). (C) The binuclear Mn center at the active site of human arginase (PDB: 2PHA). (D) The hetero binuclear Mn and Ca site in ConA from *Canavalia ensiformis* (PDB: 3CNA). Color scheme: Protein backbone and carbon – pink; nitrogen – blue; oxygen – red; calcium – green sphere; manganese – purple spheres.

during the catalytic process. This cycling is essential for converting two superoxide free radicals into one molecule of oxygen and one molecule of hydrogen peroxide. The H_2O_2 generated is another form of non-radical ROS, which is further detoxified by catalase or glutathione peroxidase to water (H_2O). Mn-SOD is found in mitochondria, which helps to counter the oxidative stress generated during cellular respiration and its activity is crucial for maintaining mitochondrial integrity and function [124]. Deficiency in Mn-SOD activity can lead to increased oxidative stress, contributing to various pathologies, including neurodegenerative diseases, cancer, and inflammatory conditions. Therefore, Mn-SOD is one of the primary targets in treating medical conditions associated with oxidative stress.

4.5.3 Mn-Dependent Acetone Carboxylase

Mn-dependent acetone carboxylase is an enzyme that catalyzes the ATP-dependent carboxylation of acetone to acetoacetate, a key intermediate in the metabolism of acetone. These enzymes belong to the class of ligases, as they are involved in the carbon-carbon bond formation [125]. In certain microorganisms, Mn-dependent acetone carboxylase allows them to utilize acetone as a primary carbon source [126]. Unlike several bicarbonate-incorporating carboxylases that rely on biotin as an essential organic cofactor, Mn-dependent acetone carboxylase utilizes mononuclear manganese ions at the active site, as depicted in Figure 9B. This Mn ion at the active site is critical to the enzyme's catalytic activity, as it plays a crucial role in the positioning of the substrate and in facilitating the transfer of the carboxyl group.

4.5.4 Manganese-Dependent Phosphatase

Manganese-dependent phosphatases are enzymes that catalyze the removal of phosphate groups from various substrates, a key regulatory mechanism in several cellular signaling pathways. These phosphatases are crucial for modulating the activity of proteins involved in numerous cellular processes, including metabolism, growth, and cellular differentiation. Manganese ions play a critical role in the coordination and activation of the substrate, facilitating the nucleophilic attack on the phosphate group of several substrate molecules [127]. This process involves the transfer of phosphate from the substrate to water, resulting in the dephosphorylation of the substrate. By dephosphorylating the substrates, these enzymes regulate the phosphorylation state of key proteins involved in signal transduction, thereby controlling various cellular functions. Dysregulation of phosphatase activity can lead to aberrant cellular signaling and is associated with various diseases, including cancer and metabolic disorders [128].

4.5.5 Arginase

Arginase is an enzyme that catalyzes the hydrolysis of arginine to ornithine and urea, a crucial step in the urea cycle. Arginase exists in two isoforms: arginase I, which is primarily expressed in the liver, and arginase II, which is found in other tissues. The active site of arginase has a unique binuclear Mn center (Figure 9C), which is essential for its activity and proper folding of the enzyme. The manganese ions coordinate with the substrate and facilitate the nucleophilic attack on the guanidinium carbon of arginine, leading to the formation of ornithine and urea. This reaction is critical, as it plays an essential role in converting the toxic ammonia into urea, which can safely be excreted by the kidneys. Arginase is essential for regulating various cellular functions such as senescence, apoptosis, proliferation, inflammation, and autophagy, apart from playing a key role in several diseases such as hypertension, atherosclerosis, stroke, and heart failure [129]. In humans, arginase deficiency leads to a rare form of pathological condition known as hyperargininemia or arginemia, which is an autosomal recessive hereditary disorder. Mutations within the arginase I gene in hepatic cells can lead to arginemia, the rarest of the heritable disorders in ureagenesis, whose symptoms include neurological disorders, dementia, and hyperammonemia. Currently, there are no therapies for this disorder.

4.5.6 Concanavalin A (ConA)

ConA is a lectin (glycoprotein) originally isolated from the jack bean (*Canavalia ensiformis*). As a lectin, it binds to several sugar molecules, such as glucose and mannose, which allows it to agglutinate cells such as erythrocytes and precipitate glycoconjugates. ConA is a heterobimetallic protein with manganese and calcium ions, as depicted in Figure 9D. The presence of these metals does not confer any catalytic role. However, they are required only for structural stability and enhancing their binding affinity for various sugars [130]. ConA's ability to bind sugars with high specificity makes it a valuable tool in cell biology, immunology, and biochemistry and is widely used in studying glycoproteins and cell membranes, as well as for isolating glycoproteins from complex mixtures *via* lectin affinity chromatography.

4.5.7 Glutamine Synthetase (GS)

GS is a metalloenzyme that catalyzes the condensation of glutamate and ammonia to glutamine in an ATP-dependent process [131]. This reaction is critical for nitrogen metabolism and regulating nitrogen levels in the cell and is found across all the domains of life. Glutamine synthetase is one of the significant enzymes responsible for the assimilation of ammonia, which is toxic at high concentrations. GS has a dual role, where it provides nitrogen as a source for anabolic pathways such as amino acid biosynthesis and other metabolic pathways and is also involved in detoxification by removing excess ammonia. GS is a homo-oligomeric enzyme, in which each of the subunits has two Mn(II)-binding sites (although some GS isoforms have Mg) at the active site, where one Mn ion is involved in the transfer of γ -phosphoryl group from the ATP, while the other Mn ion positions the glutamate at the active site, to facilitate the reaction. Several synthetic GS inhibitors, such as methionine sulfoximine (MSO) and phosphinothricin (PPT), bind at the glutamate-binding site. Notably, PPT is widely used as a herbicide that kills unwanted plants and weeds due to toxic effects from the heightened ammonia accumulation [132].

4.5.8 Phosphoenolpyruvate Carboxykinase (PEPCK)

PEPCK is a key enzyme in gluconeogenesis, the metabolic pathway specialized in generating glucose from non-carbohydrate substrates. PEPCK belongs to the family of lyases that catalyze the decarboxylation of oxaloacetate (OAA) to phosphoenolpyruvate (PEP) and carbon dioxide (CO_2) in a GTP-dependent process [133]. PEPCK is at the conjunction between glycolysis and the Krebs cycle and is involved in the first committed step in gluconeogenesis. PEPCK serves as a classic example of an anaplerotic reaction, during which it can reverse its directionality of the reaction leading to the formation of OAA from PEP, even though this reaction is thermodynamically unfavorable. This is possible only when pyruvate kinase, the enzyme involved in the conversion of PEP to pyruvate (the ultimate step in glycolysis) is knocked out or mutated and has been observed only in certain bacteria such as *Bacillus subtilis* [134]. PEPCK has Mn(II) cofactor at its active site, which helps bind and stabilize the enzyme-substrate complex and facilitates the decarboxylation and phosphorylation reactions. As a key player in gluconeogenesis, PEPCK is critical for maintaining blood sugar levels during fasting and intense physical workouts. In mammals, it is found in the liver and kidney, where it plays a crucial role in glucose homeostasis.

4.5.9 Pyruvate Carboxylase (PC)

PC is an enzyme belonging to the class of ligase that catalyzes the ATP-dependent carboxylation of pyruvate to generate oxaloacetate, an important anaplerotic intermediate that is essential in several metabolic pathways, including gluconeogenesis, lipogenesis, biosynthesis of neurotransmitters, and Krebs cycle [135]. Pyruvate carboxylase requires Mn(II) as a cofactor, which is involved in activating the substrate bicarbonate apart from stabilizing the enzyme-substrate complex. PC plays a vital role in maintaining the balance of metabolic intermediates and supporting glucose production and lipid metabolism, particularly in the liver, pancreas, and kidneys. When the PC is defective or deficient, excess pyruvate is converted to lactate instead of OAA, this can lead to lactic acidosis. Deficiency of PC can also lead to other conditions, such as hypoglycemia.

4.5.10 Mn-Dependent Ribonucleotide Reductase (Mn-RNR)

Mn-RNR is an enzyme that catalyzes the reduction of ribonucleotides to deoxyribonucleotides, the building blocks of DNA. Mn-RNR contains a binuclear manganese center that facilitates the removal of the 2′-hydroxyl group from the ribose ring of the ribonucleotides mediated by the formation of a tyrosyl radical at the active site. The manganese ions cycle between different oxidation states, enabling the transfer of electrons required for the reduction of ribonucleotides. Mn-RNR is a key enzyme that maintains the DNA/cell mass ratio constant all the time, especially during DNA replication and repair [136]. Since Mn-RNR is crucial for cell proliferation and maintenance of genomic integrity, it is often targeted to disrupt DNA synthesis in cancer therapies.

4.5.11 Isopropylmalate Synthase (IPMS)

IPMS is an enzyme that catalyzes the "Claisen-like" condensation of acetyl-CoA and 2-oxoisovalerate to yield isopropyl-malate, an important precursor and the first committed step in the leucine biosynthesis [137]. Interestingly leucine, which is the end product of the "leucine biosynthesis pathway" acts as an allosteric inhibitor

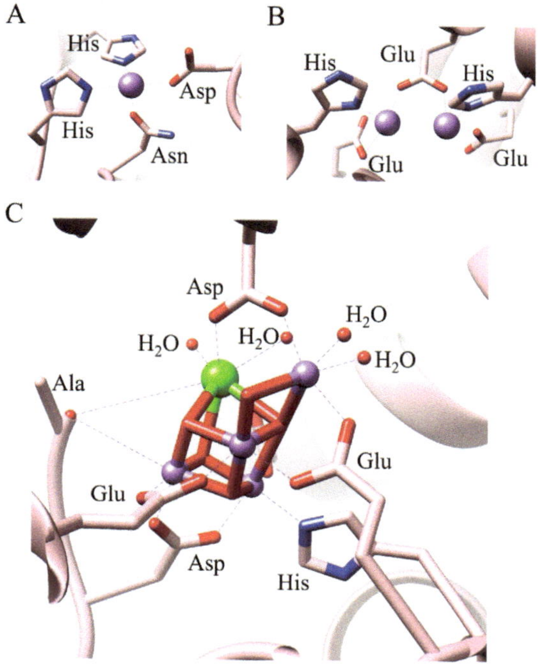

FIGURE 10 (A) The mononuclear Mn center at the active site of isopropylmalate synthase from *Neisseria meningitidis* (PDB: 3RMJ). (B) The binuclear Mn center at the active site of non-heme manganese catalase from *Lactobacillus plantarum* (PDB 1JKU). (C) The 3D structure of the OEC of photosystem II complex from *Thermostichus vulcanus* (PDB: 8IR5). Color scheme: Protein backbone and carbon – pink; nitrogen – blue; oxygen – red; calcium – green sphere; manganese – purple spheres.

to the IPMS, thereby regulating the leucine bioavailability *via* negative feedback mechanism. Apart from leucine biosynthesis, IPMS is also involved in pyruvate metabolism. The active site of IPMS has a mononuclear manganese center, typically coordinated to histidine, aspartic acid and asparagine residues, as depicted in Figure 10A. The Mn center has two major functions: it facilitates the transfer of the acetyl group and plays a critical role in stabilizing the enzyme-substrate complex.

4.5.12 Manganese Peroxidase (MnP)

MnP is a metalloenzyme that has a Mn(II) center at its active site, specialized in the oxidation of various organic substrates using hydrogen peroxide as one of the substrates. This enzyme is primarily found in fungi, where it plays an essential role in the degradation of lignin, a complex polymer found in plant cell walls. Apart from Mn(II) center, MnP also additionally has heme cofactor. During the catalytic reaction, the Mn(II) center of the MnP is oxidized to Mn(III), which then oxidizes lignin and other organic molecules [138]. MnP's activity is responsible for the breakdown of plant biomass, specifically lignin, facilitating nutrient recycling of the soil. MnP is also of a broad interest in bioremediation and other industrial applications for its ability to degrade a wide range of organic pollutants.

4.5.13 Non-Heme Manganese Catalase

Non-heme manganese catalase is an enzyme that breaks down hydrogen peroxide into water and oxygen, thereby protecting cells from oxidative damage. Unlike heme-containing catalases, these enzymes use a binuclear manganese center at the active site, as depicted in Figure 10B. During catalysis, Mn ions undergo redox cycling to facilitate the breakdown of hydrogen peroxide [139]. These types of enzymes are found in certain bacteria and archaea, where they provide a defense mechanism against oxidative stress, thereby contributing to the maintenance of cellular integrity, allowing cells to survive in harsh environments.

4.5.14 Oxygen-Evolving Complex (OEC)

OEC is a critical component of photosystem II in the photosynthetic apparatus of plants, algae, and cyanobacteria. The OEC, also known as the "water splitting complex", contains a cluster of manganese ions that catalyze the splitting of water molecules into oxygen, protons, and electrons [140]. This process is fundamental for oxygenic photosynthesis, providing the oxygen we breathe and contributing to the global carbon cycle. The catalytic site of OEC comprises a complex metal center involving a Mn_4CaO_5 cluster (Figure 10C), in which Mn ions undergo redox changes, facilitating the extraction of electrons from water and the generation of molecular oxygen [141]. The OEC is vital for the conversion of solar energy into chemical energy, driving the synthesis of organic molecules that sustain life on earth [142].

4.6 Iron (Fe)

Iron is a transition metal with the atomic number 26 in group 8 of the periodic table. It is one of the most abundant elements on earth and vital for all life forms. Fe exists in multiple oxidation states; however, Fe(II) (ferrous) and Fe(III) (ferric) states are the primary oxidation states that allow them to participate in a variety of redox reactions. Fe is a crucial component of many proteins and enzymes, playing critical roles in oxygen transport, electron transfer, and a myriad of metabolic reactions. The bioavailability of Fe is very flexible as it can occur independently (either in a mono center or a binuclear center), coordinated to heme cofactor or as iron-sulfur (FeS) clusters, and is one of the most abundant metalloproteins in humans. The genes that make the Fe-bound metalloproteins alone constitute up to 2% of the human genome, and a majority of them are heme-binding proteins (48%), a third are individual Fe-binding proteins (35%), and the FeS proteins constitute the rest (17%) [143]. Since there are several examples of Fe-bound metalloproteins, we have emphasized only some important examples or subclasses based on the occurrence of iron, either individually, in heme, or in the iron-sulfur cluster.

4.6.1 Transferrin

Transferrins are iron-binding glycoproteins found in vertebrates that play a crucial role in iron transport and distribution throughout the body. Typically, transferrins have binding sites for two ferric ions Fe(III), which could be transported across the body through the bloodstream to various tissues and organs, including bone marrow, where iron is utilized for the synthesis of hemoglobin (discussed in Section 4.6.7) stocked, or may be incorporated into other iron-dependent enzymes. Transferrins remarkably have a very high affinity toward binding iron; however, they bind in a reversible manner so that they can deliver the iron cargo to the target sites. Figure 11A shows the typical Fe-binding site in human transferrin. The regulation of transferrin and its receptor-mediated uptake is crucial for maintaining iron homeostasis and preventing iron overload or deficiency [144]. As a serum protein, transferrin serves as one of the major biomarkers of normal health and identifies any abnormalities. The standard reference range of transferrin in a healthy adult blood plasma is 204–360 mg/dL. Levels higher than this range may indicate Fe deficiency anemia.

4.6.2 Ferritin

Ferritin is an Fe-storage protein that helps regulate iron homeostasis within cells and is nearly ubiquitous to all known living organisms. It consists of twenty four subunits that form a hollow globular structure capable of storing up to 4,500 Fe ions in a soluble and non-toxic form and plays a key role in Fe homeostasis by buffering the required amounts to avoid iron deficiency and iron overload. The iron stored in ferritin is released when needed for various cellular processes, such as DNA synthesis, respiration, and detoxification. Ferritin plays a vital role in protecting cells from iron-induced oxidative damage by sequestering free iron and preventing the formation of harmful reactive oxygen species (ROS). Ferritin is essential for maintaining iron balance in the body, and the standard serum reference values in healthy adults are 18–250 ng/mL (males) and 12–160 ng/mL (females). As an essential diagnostic marker, ferritin levels are often monitored during regular blood tests, as any abnormal levels can signify several health conditions such as anemia, hemosiderin, hemochromatosis, hemophagocytic syndrome, and Still's disease [145].

4.6.3 Lactoferrin

Lactoferrin is a Fe-binding glycoprotein found in various secretory fluids, such as milk, saliva, and tears, and in immune cells, such as neutrophils. It has a high affinity toward Fe and plays a key role in Fe homeostasis and immune defense. By sequestering free Fe, lactoferrin limits the availability of Fe to pathogenic bacteria, inhibiting their growth and thereby reducing infections. Additionally, lactoferrin exhibits antimicrobial, anti-inflammatory, and immunomodulatory properties [146]. Its ability to bind and transport Fe makes it essential not only for maintaining Fe homeostasis but also in the innate immune system by boosting the body's defense mechanisms against microbial invasion and oxidative stress.

4.6.4 Rubredoxin

Rubredoxin is a small Fe-binding protein that plays an essential role in electron transfer processes in various microorganisms, particularly in anaerobic bacteria and archaea. Rubredoxins are typically single polypeptide chain proteins with a molecular weight of ~5.5 kDa that contains a single Fe ion coordinated by four cysteine residues, forming a unique Fe $(S\text{-}Cys)_4$ center (Figure 11B) [147]. The iron ion in rubredoxin is tetrahedrally coordinated by four cysteine thiolate ligands, creating a robust and stable structure that facilitates efficient electron transfer. The iron can either accept or donate electrons by cycling between the oxidized (III) and reduced (II) states, and this redox cycling is crucial for participating in various metabolic pathways such as sulfate reduction, nitrogen fixation, and hydrocarbon degradation. Owing to its role in the degradation of alkanes and other hydrocarbons, rubredoxins are considered a valuable tool in the bioremediation efforts aimed at cleaning up oil spills and other environmental pollutants [148].

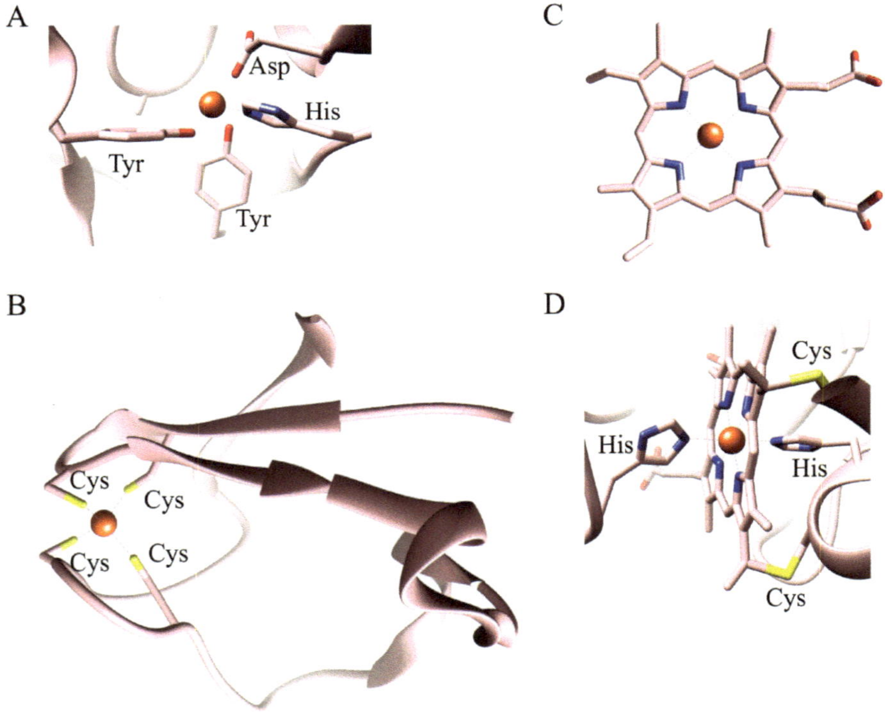

FIGURE 11 (A) The mononuclear Fe-binding site in human transferrin (PDB: 1A8E). (B) Ribbon model of rubredoxin from *Desulfovibrio desulfuricans* with the Fe-binding center coordinated to four cysteine residues (PDB: 6RXN). (C) The stick model of heme B with iron (sphere) at the center of the tetrapyrrole porphyrin ring (PDB: 1GZX). D. The heme C-binding site in cytochrome c-549 from *Limnospira maxima* (PDB: 1F1C). Color scheme: Protein backbone and carbon – pink; nitrogen – blue; oxygen – red; sulfur – yellow; iron – orange spheres.

4.6.5 Fe-Dependent Dioxygenases

Fe-dependent dioxygenases are a group of enzymes that incorporate both atoms of molecular oxygen into their substrates which, in contrast to monooxygenases, incorporate only one atom of molecular oxygen into their respective substrates, while the other atom is reduced to water. The active site of these enzymes has a mononuclear Fe either in II or III oxidation state, which is not coordinated to cysteines as in the case of rubredoxin (discussed in Section 4.6.4), but instead coordinated to tyrosine, histidine ligands, or glutamic acid amino acids depending on the enzyme type [149]. Fe-dependent dioxygenases play a key role in various metabolic pathways, including the biosynthesis of hormones and collagen, degradation of aromatic compounds, regulation of gene expression, and DNA repair.

4.6.6 Siderophores

Siderophores ("iron carriers" in Greek) are small, high-affinity iron-chelating compounds secreted by microorganisms such as bacteria, fungi, and some plants (phytosiderophores). These molecules are crucial in acquiring Fe from the surrounding environments where it is scarce. Fe is essential for various cellular functions, including DNA synthesis, respiration, and a myriad of metabolic pathways, but in aerobic environments, it is predominantly found in insoluble forms like ferric hydroxide. Siderophores help organisms overcome this hurdle by binding Fe with high affinity in a soluble form and facilitating its uptake. Microbial siderophores are diverse in their chemical structures and are broadly classified into four major categories based on their Fe-binding groups: catecholates, phenolates, hydroxamates, and carboxylates [150]. Catecholates, such as enterobactin, contain catechol (benzene-1,2-diol) groups that tightly bind ferric ions. Enterobactin, produced by *Escherichia coli* and other enteric bacteria, is one of the strongest Fe-binding siderophores known in biology. It forms a robust hexadentate complex with iron, effectively solubilizing it for transport. Pyochelin is a classic example of phenolates with a unique heterocyclic thiazoline ring and plays a crucial role in Fe uptake in bacteria such as *Pseudomonas aeruginosa*. Hydroxamates, such as desferrioxamine B and ferrichrome, have hydroxamic acid groups (–C(=O)–N(–OH)–) that bind iron. Carboxylate-type siderophores, such as rhizoferrin, contain carboxyl groups (-COOH) that chelate iron. Despite their structural diversity, all siderophores share the common function of solubilizing and transporting Fe into the cell. When the cells require Fe, they secrete siderophores into the surrounding environment. Once bound to available Fe ions, they form soluble Fe-siderophore complexes, which are then recognized by specific receptors on the cell surface. This interaction allows the transport of these Fe-siderophore complexes into the cell, where they are readily cleaved, releasing the essential Fe to the cell. Siderophores also have several medical and pharmaceutical applications. Sideromycins are siderophore-antibiotic conjugates that have been developed to exploit the Fe acquisition pathways of pathogenic bacteria. These synthetic conjugates utilize siderophores to transport antibiotics into bacterial cells, enhancing their effectiveness. For example, cefiderocol is a siderophore-cephalosporin conjugate used to treat multidrug-resistant bacterial infections [151]. Similarly, siderophores and their synthetic analogs are used in Fe chelation therapy to treat Fe overload conditions, such as thalassemia and hemochromatosis.

4.6.7 Hemoglobin and Myoglobin

Hemoglobin and myoglobin are iron-containing proteins essential for oxygen transport and storage in vertebrates. Hemoglobin is found in red blood cells (RBCs), which constitute up to ~ 96% by dry weight and binds oxygen in the lungs (terrestrial) and gills (aquatic) and transports it to various cells and tissues throughout the body. Hemoglobin also functions as an antioxidant and regulates iron metabolism in certain non-RBC cells such as macrophages and mesangial cells [152]. Hemoglobin is also found in leguminous plants (leghemoglobin) and invertebrates such as nematodes [153,154]. Myoglobin, on the other hand, is exclusively found in muscle tissues (cardiac and skeletal), where it stores oxygen and facilitates its diffusion to mitochondria for cellular respiration. Unlike hemoglobin, myoglobin has a higher affinity toward oxygen and shows no cooperativity toward binding subsequent oxygen molecules. The human hemoglobin is a tetramer composed of two α and two β subunits, and each subunit has one prosthetic tetrapyrrole porphyrin ring known as heme B that coordinates one Fe ion at the center, as shown in Figure 11C [155]. In contrast, myoglobin is a monomer and has a single heme group [156]. Max Perutz and John Kendrew were awarded the Nobel Prize in chemistry for the year 1962 for their pioneering research on hemoglobin and myoglobin, respectively. The Fe ions in both proteins can reversibly bind to oxygen, enabling these molecules to function effectively in oxygen transport and storage. The ability of hemoglobin and myoglobin to bind and release oxygen is critical for maintaining cellular respiration and overall metabolic activities. Deficiencies and malfunction of these proteins lead to many diseases often associated with anemia and hypoxia. As one of the prominent proteins of the blood, hemoglobin serves as one of the most essential diagnostic markers to diagnose several clinical conditions related to human health. The normal hemoglobin values for healthy individuals are 13.2–16.2 gm/dL for males and 12.0–15.2 gm/dL for females. Although myoglobin is in neither blood serum nor plasma, it is still considered an essential blood marker for diagnosing muscle injuries and myocardial infarctions.

4.6.8 Cytochromes

Cytochromes are a family of Fe-containing proteins involved in electron transport and energy production within cells and are ubiquitous to all living forms. These proteins contain heme cofactor, with Fe at the center, which undergoes redox changes to facilitate electron transfer. Cytochromes are key components of the electron transport chain, where they play a crucial role in oxidative phosphorylation. In the eukaryotes, by transferring electrons between different protein complexes, cytochromes help generate a proton gradient across the mitochondrial membrane, ultimately driving ATP synthesis. Depending on the type of bound heme and the nature of coordination to the active site, the cytochromes are classified into four major groups: cytochrome a, cytochrome b, cytochrome c and cytochrome d [157]. Figure 11D shows the heme C cofactor found in cytochrome c. Cytochromes are also found in the chloroplasts of algae and plants. Specifically, they have two c-type cytochromes: a membrane-bound cytochrome *f* (cytochrome c of plants) and a soluble cytochrome, known as c_6 [158]. The versatility of iron in accepting and donating electrons and its ability to undergo redox changes spontaneously, makes it indispensable for the function of cytochromes and in energy metabolism.

4.6.9 Catalase

Catalase is an iron-containing enzyme that catalyzes the decomposition of hydrogen peroxide into water and oxygen, protecting cells from oxidative damage (for non-heme manganese catalase) (discussed in Section 4.5.13). The enzyme contains a heme group with Fe at its active site, which facilitates the breakdown of hydrogen peroxide. This reaction is crucial for cellular defense against oxidative stress, as hydrogen peroxide is a destructive ROS that can damage proteins, lipids, and DNA. Catalase is a highly efficient enzyme with high activity and specificity capable of breaking millions of hydrogen peroxide molecules per second. This turnover number is one of the highest among all known enzymes. Catalase activity in detoxifying hydrogen peroxide highlights the importance of Fe in protecting cells from oxidative damage by ROS [159].

4.6.10 Nitric Oxide Reductase (NOR)

NOR is an enzyme that is involved in the reduction of nitric oxide (NO) to nitrous oxide (N_2O), playing a key role in the greenhouse effect and the nitrogen cycle [160]. This process is one of the key steps in denitrification, which is the conversion of nitrate (NO_3^-) and nitrite (NO_2^-) to nitrogen gas (N_2), thereby recycling the nitrogen from ecosystems and ultimately back to the atmosphere. NORs are predominantly found in various anaerobic bacteria and archaea as membrane-bound enzymes that typically consist of multiple subunits, which can vary depending on the type of organisms. NORs are iron-containing metalloenzymes and in some cases, copper (Cu) is found along with the iron at the active site. Based on the type of metallocofactors and the type of electron donors, the NORs are broadly classified into cNOR, qNOR, and qCuNOR. The cNORs use cytochromes or cupredoxins as electron donors, whereas qNOR uses azurin or quinol, and qCuNOR uses quinol, respectively [161]. The primary function of NOR is to catalyze the reduction of two molecules of nitric oxide (NO) to one molecule of nitrous oxide (N_2O) and water (H_2O). NORs have several industrial and biotechnological applications. For example, in bioremediation and wastewater treatment, engineered microorganisms with enhanced NOR activity are harnessed to remove excess nitrogen from wastewater, thereby reducing nitrate and nitrite pollution and preventing eutrophication.

4.6.11 Iron-Sulfur Proteins

FeS proteins contain FeS clusters that play a key role as single electron carriers mediating electrons within (intramolecular) or between (intermolecular) proteins, contributing to nearly all the metabolic pathways [162]. FeS clusters are widely considered as the earth's primordial catalysts that may have played a key role in the origins of life on earth [163]. FeS proteins are ubiquitous to all forms of life, in which the FeS clusters typically occur as [2Fe-2S], [3Fe-4S], and [4Fe-4S] clusters (Figure 12 A-C) that are coordinated to the sulfur of the cysteine residues, however, occasionally histidine may coordinate FeS clusters. Ferredoxins are well-characterized FeS proteins with either one or two iron-sulfur clusters, majorly involved in transferring electrons. Between the high-potential iron-sulfur proteins (HiPIPs) and the low-potential ferredoxins, they can mediate electron transfer reactions over a wide range of potentials, ranging up to 1 V. Beyond ferredoxins, several enzymes have one or several FeS clusters that mediate the flow of electrons either into or out of the active site of the respective enzyme. Certain metalloenzymes also have modified FeS clusters, such as the P cluster [8Fe-7S] in Mo-dependent nitrogenases (discussed in Section 4.3.1).

4.6.12 [FeFe]-Hydrogenases

[FeFe]-hydrogenases catalyze the reversible oxidation of molecular hydrogen at a unique metallocofactor known as the H cluster, which is composed of a cubane [4Fe-4S] subsite that is bridged to a 2Fe subsite via cysteinyl thiolate [164]. The individual iron ions in the 2Fe subsite have carbon monoxide and cyanide functional groups and are bridged *via* carbon monoxide and an azadithiolate (ADT), which is unique only to [FeFe]-hydrogenases and not seen elsewhere in biology [165]. Figure 12D shows the structure of the H cluster represented in a ball and stick model. Several members of the [FeFe] hydrogenases additionally contain iron-sulfur clusters (discussed in Section 4.6.11), typically [4Fe-4S] and [2Fe-2S] clusters that serve as electron relay to control the flow of the electrons either to be consumed or removed from the H cluster. The reaction mediated by [FeFe]-hydrogenases is fundamental to several metabolic pathways. For example, certain fermentative bacteria employ [FeFe]-hydrogenases to dispose of excess reducing equivalents (electrons) by simply converting them to molecular hydrogen. Electron carriers like ferredoxins (discussed in Section 4.6.11) transport the essential electrons that are used to reduce protons at the active site of the [FeFe]-hydrogenases. Certain microorganisms utilize molecular hydrogen via [FeFe]-hydrogenases as a source of electrons, with the concomitant reduction of downstream electron acceptors such as nitrate, sulfate, carbon dioxide, or fumarate, depending on the type and need of the microorganisms. Some [FeFe]-hydrogenases exhibit significant catalytic bias in catalysis favoring rate acceleration relative in one direction over the other [166]. For example, *Clostridium pasteurianum*, a nitrogen-fixing bacterium under nitrogen-fixing conditions, employs CpII [FeFe]-hydrogenase, which is highly biased to hydrogen oxidation to capture the molecular hydrogen and redirects the electrons to nitrogenase for nitrogen fixation [167].

4.6.13 Iron-Dependent Nitrogenase

Similar to V-dependent nitrogenase (discussed in Section 4.1.2) and molybdenum-dependent nitrogenase (4.3.1), iron-dependent nitrogenase (Anf) plays a key role in reducing dinitrogen to ammonia, Fe-dependent nitrogenase is the least efficient of the nitrogen isozymes consuming up to 40 ATP molecules in reducing dinitrogen to ammonia. The components of Fe-dependent nitrogenase are termed the FeFe

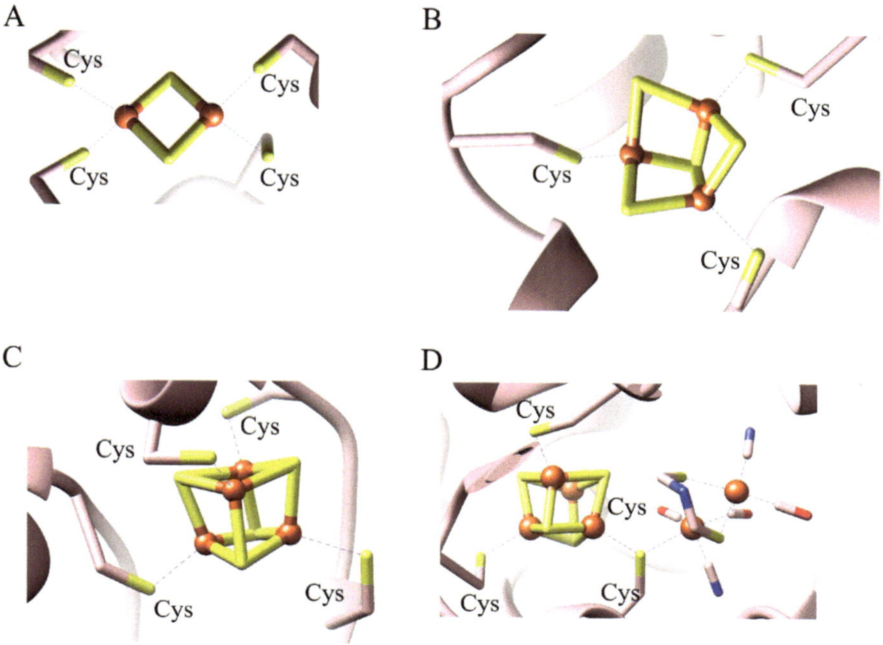

FIGURE 12 The structures of FeS clusters and [FeFe]-hydrogenase active site: The 3D structures of [2Fe-2S] (A) (PDB: 6NAC), [3Fe-4S] (B) (PDB: 8DQV) and [4Fe-4S]. (C) (PDB: 6NAC) clusters represented as "ball and stick" models. (D) 3D structure of the H cluster of CpI [FeFe]-hydrogenase from *Clostridium pasteurianum* coordinated to four cysteine residues (PDB: 6NAC). Color scheme: Protein backbone and carbon – pink; nitrogen– blue; oxygen – red; sulfur – yellow; iron – orange spheres.

protein (AnfDKG) and the Fe protein (AnfH) [168]. Similar to the VFe protein, the FeFe protein is a hexamer of $\alpha_2\beta_2\delta_2$ subunits, in which the α subunit houses the FeFe-co (8Fe-9S-C-homocitrate) [169]. Recent studies have demonstrated that the Fe-dependent nitrogenase could also reduce carbon dioxide to formate and methane [170].

4.7 Cobalt (Co)

Cobalt (Co) is a transition metal with the atomic number 27, positioned in group 9 of the periodic table. It is an essential trace element crucial for various enzymatic functions in biological systems. Co primarily exists in (II) and (III) oxidation states, enabling it to play diverse roles in redox reactions. As a vital component of vitamin B12, cobalt is indispensable for numerous metabolic processes in living organisms. Co^{60} is a synthetic high-intensity γ ray emitting radioactive isotope of cobalt with a half-life ($T_{1/2}$) of about 5.27 years and is generally used in radiotherapy to treat cancer patients [171].

4.7.1 COT1 Cobalt Transporter

COT1 is a transmembrane protein found in various organisms, including bacteria and fungi. This transporter is integral to the regulation of intracellular cobalt concentrations, ensuring that cells maintain adequate levels of cobalt for essential biochemical functions and metal homeostasis while preventing toxic accumulation [172]. COT1 is part of the cation diffusion facilitator (CDF) family of transporters, which are characterized by their ability to transport divalent metal ions across cellular membranes. The COT1 protein typically comprises multiple transmembrane domains that form a channel through which cobalt ions can pass, generating an electrochemical gradient across the membrane.

4.7.2 Cobalamin (Vitamin B12)

Cobalamin, commonly known as vitamin B12, is a cobalt-containing complex that plays a critical role in DNA synthesis, fatty acid metabolism, and amino acid metabolism. It is the only vitamin containing a metal ion, with Co in the center of a corrin ring. The crystal structure of B12 was solved by Dorothy Hodgkin, for which she was awarded the Nobel Prize in chemistry in the year 1964 [173]. Cobalamin exists in several forms, including methylcobalamin and adenosylcobalamin, which are coenzymes in various biochemical reactions. Methylcobalamin (Figure 13A) is essential for converting homocysteine to methionine, a critical reaction in the methylation of several biomolecules, including DNA, phospholipids, and various proteins. Adenosylcobalamin is involved in the isomerization of methylmalonyl-CoA to succinyl-CoA, an essential step in fatty acid and amino acid catabolism. Deficiency in vitamin B12 can lead to pernicious anemia and several neurological disorders [174]. Cobalamin also occurs as a catalytic cofactor in certain enzymes, such as reductive dehalogenases in organohalide respiring bacteria and plays a key role in the dehalogenation of certain environmental persistent organic pollutants (POPs) such as polychlorinated biphenyls and polychlorinated dioxins, thereby contributing to the global halide cycle [175].

4.7.3 Methionine Aminopeptidase (MetAP)

MetAP (Figure 13B) is an enzyme that catalyzes the hydrolytic cleavage of the N-terminal methionine from nascent proteins and polypeptides, a crucial step in post-translational modification. MetAP is a non-corrin type of coenzyme with a binuclear metal center with two cobalt ions [176]. Cobalt ions in the active site facilitate the hydrolysis of the peptide bond, enabling the release of methionine. This enzymatic reaction is vital for the maturation, functioning, and localization of proteins. Inhibition or deficiency of MetAP can lead to aberrant protein processing and various diseases.

4.7.4 Nitrile Hydratase

Nitrile hydratase is a cobalt-containing metalloenzyme enzyme that catalyzes the hydration of nitriles to amides, in a direct nucleophilic attack on the triple bond of the nitrile group [177]. It plays a significant role in the metabolism of nitrile-containing

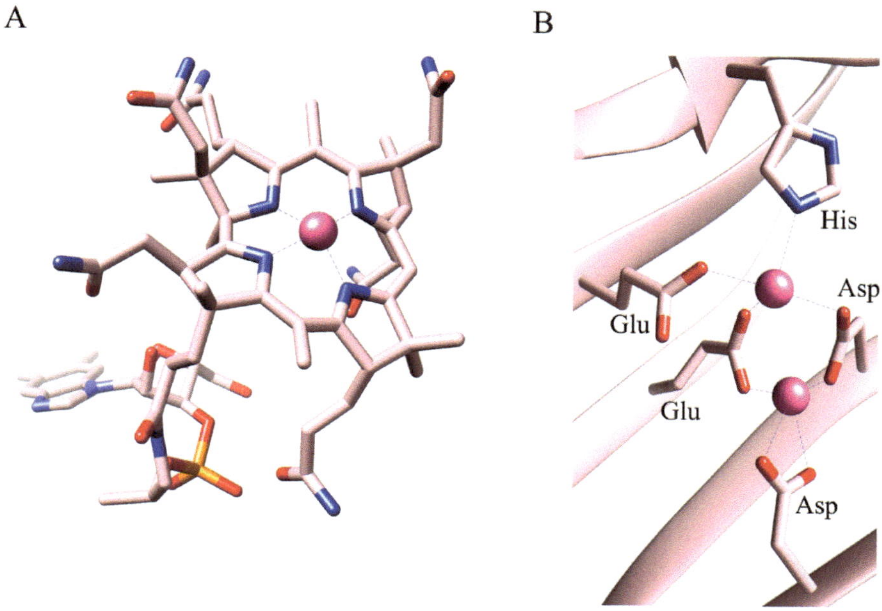

FIGURE 13 The occurrence of cobalt in various biomolecules. (A) The stick model of methylcobalamin with Cu (sphere) at the center of corrin ring (PDB: 8Q4H). (B) The binuclear Co center at the active site of human MetAP (PDB: 1BN5). Color scheme: Protein backbone and carbon – pink; nitrogen – blue; oxygen – red; phosphorus – vermillion; cobalt – hot pink spheres.

compounds and has a broad role in industrial applications in the production of acrylamide, nicotinamide, and other commercially relevant amides. Unlike MetAP, which contains a binuclear cobalt site, nitrile hydratase instead consists of a mononuclear cobalt ion in Co(III) oxidation state without the corrin ring at its active site, coordinated by cysteine and serine residues from a highly conserved CXXCSC motif. This coordination environment facilitates the activation and hydrolysis of the nitriles.

4.7.5 Prolidase

Prolidase, also known as proline dipeptidase, is a cobalt-dependent metalloenzyme that plays a critical role in the final step of protein and peptide degradation. It specifically hydrolyzes dipeptides with a proline residue at the C-terminus, releasing free proline and the N-terminal free amino acid. Prolidase is essential for recycling proline from dietary proteins and endogenous protein turnover, contributing to the overall free proline pool that is required for various metabolic processes. The active site of prolidase has a binuclear metal center composed of two Co ions essential for stabilizing the enzyme structure and facilitating the catalytic reaction. During catalysis, a water molecule is activated, leading to a nucleophilic attack on the carbonyl carbon of the scissile peptide bond. Further, the cobalt ions also stabilize the transition state and the resulting tetrahedral intermediate, allowing efficient hydrolysis of the peptide bond. Prolidase plays a crucial role in collagen metabolism, as collagen is rich in proline and hydroxyproline residues. During collagen degradation, prolidase hydrolyzes the resulting dipeptides, releasing free proline and hydroxyproline, which can be utilized to synthesize new collagen or other proline-rich proteins, thereby contributing to tissue repair and maintenance [178]. In humans, prolidase is encoded by *PEPD* gene, and its activity is highest in tissues with active collagen turnover, such as skin fibroblasts, where it supports wound healing and tissue remodeling. Mutations in the PEPD gene can lead to prolidase deficiency, a rare autosomal recessive disorder characterized by impaired collagen metabolism, skin lesions, photosensitivity, hyperkeratosis and other connective tissue abnormalities [179]. Prolidase has several downstream applications.

4.8 Nickel (Ni)

Nickel is a transition metal with the atomic number 28, located in group 10 of the periodic table. Although Ni is abundant on earth, it occurs as an essential cofactor only in a few enzymes that are involved in various biological processes, such as carbohydrate metabolism and energy metabolism involving hydrogen and carbon [180]. Ni typically exists in the Ni(II) oxidation state in biological systems. It is involved in several redox reactions and is considered an essential trace element in certain bacteria and archaea.

4.8.1 Urease

Urease is a nickel-dependent metalloenzyme that catalyzes the hydrolysis of urea into ammonia and carbon dioxide [181]. Urease was the first enzyme to be crystallized almost a century ago in 1926, for which James Sumner was awarded the Nobel Prize in chemistry in 1946. Interestingly, urease is also the first nickel-containing metalloenzyme identified in which the nickel ions coordinate with histidine residues and water molecules. Urease is crucial for nitrogen metabolism in many bacteria, fungi, and plants. The hydrolysis of urea by urease plays an essential role in nitrogen recycling, as the generated ammonia is critical to soil fertility and agricultural output. Although humans do not have urease, patients infected with *Helicobacter pylori*, the pathogen responsible for gastric cancer, are known to have urease originating from the pathogen itself. *H. pylori* uses urease as a survival and defensive mechanism as the ammonia generated is utilized to neutralize stomach acids, thereby also contributing to its virulence [182]. In humans, the presence of urease is used as a diagnostic tool to detect *H. pylori* infections and gastric cancer.

4.8.2 Methyl-Coenzyme M Reductase (MCR)

MCR, also known as coenzyme B sulfoethylthiotransferase, is a Ni-containing enzyme found in methanogenic archaea that thrives in the guts of ruminants. It catalyzes the final step in methanogenesis; the reduction of methyl-coenzyme M to methane. The active site of MCR contains a Ni corphin cofactor known as F_{430} and is the only natural tetrapyrrole that coordinates to nickel (Figure 14A) [183]. During the catalytic cycle, Ni alternates between different oxidation states, facilitating the

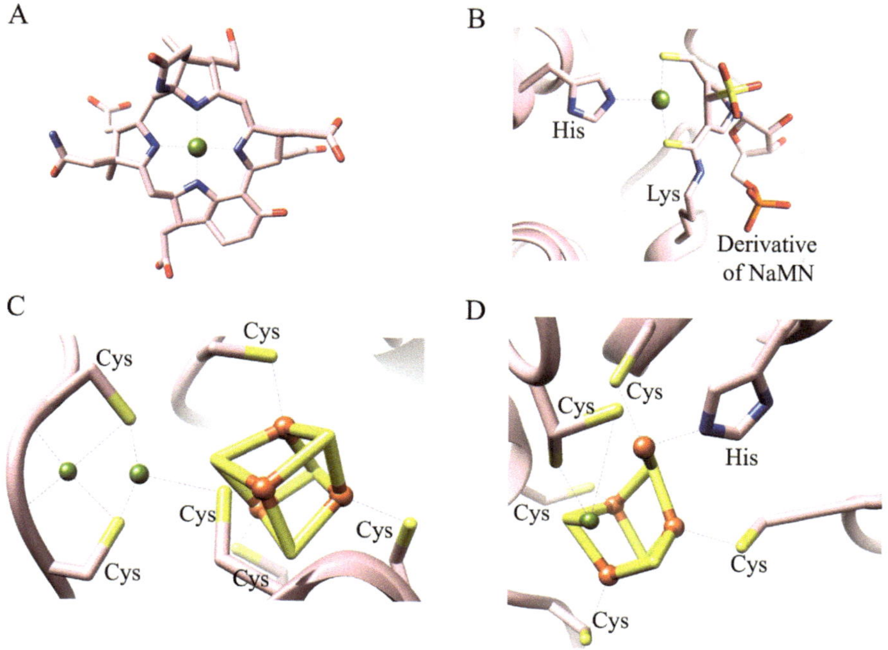

FIGURE 14 (A) The F_{430} with nickel (sphere) at the center of tetrapyrrole (PDB: 7SUC). (B) The active site of lactate racemase from *Lactiplantibacillus plantarum* has a nickel ion coordinated to a unique ligand, a derivative of NaMN, in pincer coordination (PDB: 6C1W). 3D structures of the A cluster (C) and C cluster (D) of ACS/CODH complex from *Moorella thermoacetica* (PDB: 6X5K). Color scheme: Protein backbone and carbon – pink; nitrogen – blue; oxygen – red; sulfur – yellow; phosphorus – vermillion; iron – orange spheres; nickel – olive green spheres.

conversion of methyl-coenzyme M and coenzyme B to methane and a hetero-disulfide [184]. This process is critical for the energy metabolism of methanogens and plays a significant role in the global carbon cycle and biogas production [185].

4.8.3 Lactate Racemase

Lactate racemase is an enzyme that catalyzes the interconversion of L-lactate and D-lactate enantiomers and is found primarily in bacteria. Lactate racemase is essential for lactate metabolism in certain bacteria, enabling them to utilize both enantiomers of lactate for energy production. Lactate is also utilized in the glycosylation of biomolecules; specifically, certain bacteria use D-enantiomer as a component in cell walls. Remarkably, the nickel ion found in lactate racemase has a unique tridentate pincer-like arrangement, which is otherwise uncommon in biology. The mononuclear nickel ion at the active site is coordinated to one histidine residue and a unique derivative of nicotinic acid mononucleotide (NaMN), known as pyridinium-3-thioamide-5-thiocarboxylic acid mononucleotide that is covalently bound to a lysine residue (Figure 14B). This unique ligand binds to the nickel ion in a pincer-like geometry, which is critical to the hydride transfer during catalysis [186].

4.8.4 Acetyl-CoA Synthase (ACS)

ACS belongs to the class of ligases that is involved in the formation of acetyl-CoA from acetate and CoA in an ATP-dependent process. ACS is a key enzyme in the Wood–Ljungdahl pathway (reductive acetyl-CoA pathway), enabling certain bacteria and archaea to utilize hydrogen (H_2) as a source for electrons and carbon dioxide (CO_2) as the ultimate electron acceptor. In certain microorganisms, ACS plays a crucial role in carbon fixation and acetogenesis and the biosynthesis of building blocks of biomolecules. ACS is a nickel-containing metalloenzyme with a unique metallocofactor known as the A cluster composed of a di Ni subsite and a [4Fe-4S] cubane subsite at its active site (Figure 14C). Within the di Ni subsite, the proximal Ni ion is involved in the substrate binding and catalysis [187]. ACS is also widely used in industries in the production of several commercially relevant fatty acids.

4.8.5 Carbon Monoxide Dehydrogenase (CODH)

CODH is a Ni-dependent enzyme that catalyzes the reversible oxidation of carbon (CO) monoxide to carbon dioxide. This enzyme is found in various bacteria and archaea, where it plays a key role in the carbon cycle and energy metabolism. The active site of CODH contains a unique metallocofactor known as the C cluster, which is composed of [Ni–4Fe–4S] center (Figure 14D), where the nickel insertion into the [4Fe-4S] is critical to the activity of the enzyme [188]. Apart from functioning as an independent enzyme involved in the reversible catalysis of CO, CODH can also function as a bifunctional enzyme complex in association with ACS (discussed in Section 4.8.4). This ACS/CODH complex is involved in the Wood–Ljungdahl pathway, which plays a key role in the conversion of carbon dioxide into acetyl-CoA [189]. CODH enables microorganisms to either utilize carbon monoxide as a carbon source, converting it to carbon dioxide under aerobic conditions, or utilize it in anaerobic conditions for acetogenesis or methanogenesis.

4.8.6 [NiFe]-hydrogenase

Ni-containing hydrogenases catalyze the reversible oxidation of molecular hydrogen (H_2) into protons and electrons [190]. These enzymes play a vital role in the energy metabolism of many anaerobic microorganisms, including archaea and bacteria. The active site of [NiFe]-hydrogenases contains both Ni and Fe in which the iron ion is coordinated by non-protein ligands including one CO and two CN^- groups, typical to hydrogenases. On the other hand, the Ni ion is primarily bound to thiolates originating from cysteine residues within the primary coordination sphere. Ni ion at the active site is primarily responsible for the catalytic activity of [NiFe]-hydrogenases, which undergo redox changes between Ni(III), Ni(II), and Ni(I), depending on the oxidation state of the enzyme. In contrast, the Fe ion is redox inactive and remains Fe(II) throughout the catalytic reaction [191]. In addition to the binuclear active site, [NiFe]-hydrogenases also contain accessory [4Fe-4S] clusters that facilitate the transport of electrons into and out of the active site.

4.9 Copper (Cu)

Copper is a transition metal with the atomic number 29 and is located in the first row of group 11 of the periodic table. It is an essential trace metal for all aerobic organisms. This is further supported by the extremely poor solubility of Cu in its reduced state, and its bio-inavailability in the primordial oceans indicates, no wonder, Cu has not been incorporated into enzymes, possibly until the great oxidation event (GOE). Cu is the third most abundant d-block metal in humans, followed by Fe and zinc (Zn). Cu typically occurs as Cu(I) and Cu(II) oxidation states, and its ability to cycle between these redox states makes it a key metallocofactor that drives several metabolic pathways [192]. Cu is essential to several enzymes, where it facilitates oxygen transport, efficient electron transfer mediating various catalytic reactions, and contributes to structural stabilization.

4.9.1 Cytochrome c Oxidase

Cytochrome c oxidase, also known as Complex IV, is a large transmembrane enzyme complex found in bacteria, archaea, and the inner mitochondrial membrane of eukaryotes. It participates in the electron transport chain by catalyzing the reduction of molecular oxygen to water, a crucial step in cellular respiration and energy production. Cytochrome c oxidase contains two Cu centers, a binuclear Cu A center (Cu_A) and a mononuclear Cu B center (Cu_B), which are essential for its function [193]. Although the Cu_B is homogeneously mononuclear, it interacts with an adjacent $heme_{a3}$, forming a catalytic binuclear center. During the reaction, Cu_A receives electrons from cytochrome c and transfers them to Cu_B via heme $_{a3}$ which then facilitates the reduction of oxygen to water [194]. This reduction process generates a proton gradient across the bacterial and mitochondrial membrane, which is utilized in ATP synthesis, an important energy storing unit of the cell. In humans, mutations in the gene coding for cytochrome c oxidase can be fatal leading to several metabolic disorders [195].

4.9.2 Azurin and Plastocyanin

Azurin is a blue Cu protein found in certain bacterial species of *Pseudomonas* and *Methylomonas*. It functions primarily in the bacterial electron transport chain, participating in redox reactions crucial for cellular respiration. Azurin contains a type I copper center (T1Cu), which facilitates the transfer of electrons to enzymes like cytochrome c oxidase (discussed in Section 4.9.1) and nitrite reductase [196]. The mononuclear T1Cu center cycles between I and II oxidation states, allowing azurin to shuttle electrons efficiently. Azurin also potentially plays a key role in cancer therapy. Azurin, by directly interacting with tumor suppressor p53, stabilizes its longevity and triggers apoptosis in cancer cells [197]. Plastocyanin, on the other hand, is a type of blue copper protein involved in the electron transport chain of plants and phytoplankton, such as algae and cyanobacteria, mediating photosynthesis. In plants, it is localized in the thylakoid lumen of chloroplasts and serves as an electron carrier between cytochrome b_6f complex and photosystem I [198]. Plastocyanin has a type I copper center (T1Cu) that is typically bound to two histidine residues, one cysteine and one methionine. This mononuclear Cu center undergoes reversible oxidation and reduction between I and II states, facilitating robust electron transfers. This process is essential for light-dependent photosynthesis reactions, enabling light energy conversion into chemical energy.

4.9.3 Hemocyanin

Hemocyanin is a Cu-containing metalloprotein majorly involved in the transport of oxygen throughout the body via blood and is typically found in various invertebrates especially in arthropods and mollusks. Unlike hemoglobin (discussed in Section 4.6.7), which uses heme iron to bind oxygen, hemocyanin instead uses Cu ions to bind oxygen molecules [199]. The active site is composed of a binuclear type III Cu center (T3Cu) typically coordinated to histidine residues, as depicted in Figure 15A. The binding and release of oxygen by hemocyanin are facilitated by the reversible oxidation and reduction of the Cu ions at the active site.

4.9.4 Tyrosinase

Tyrosinase is a unique bifunctional metalloenzyme involved in catalyzing two distinct reactions: one is the hydroxylation of monophenols, and the other is the subsequent oxidation of the diphenol product into the corresponding quinone [200].

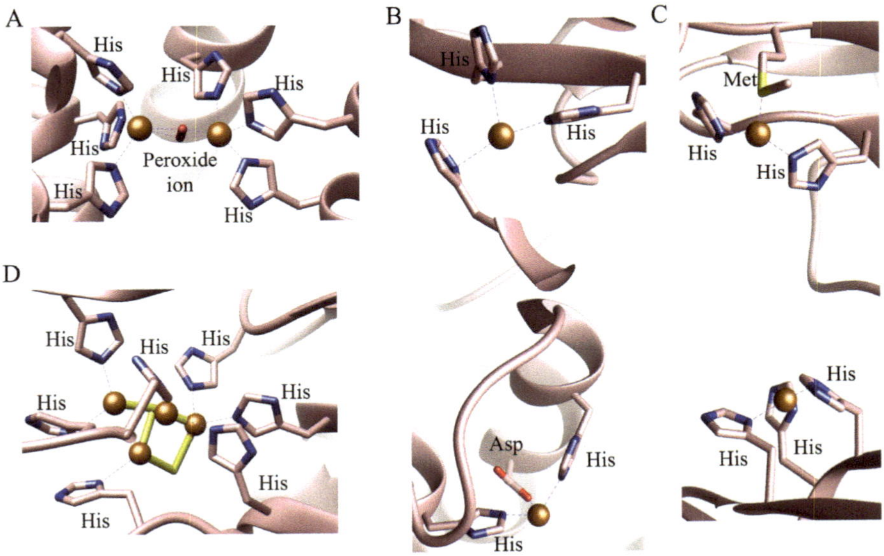

FIGURE 15 (A) Peroxide-bound type III copper center (T3Cu) in hemocyanin from *Limulus polyphemus* (PDB: 1NOL). (B) The two independent mononuclear copper sites in pMMO from Methylocystis (PDB: 4PHZ). (C) The two independent mononuclear type II Cu centers (T2Cu) in PAM from *Rattus norvegicus* (PDB: 8DSJ). (D) 3D structure of CuZ cluster coordinated to 7 His residues in nitrous oxide reductase from *Stutzerimonas stutzeri* (PDB: 3SBP). Color scheme: Protein backbone and carbon – pink; nitrogen – blue; oxygen – red; sulfur – yellow; copper – copper brown spheres.

Tyrosinase controls the rate-limiting step in the classical Raper Mason pathway [201,202], essential for the biosynthesis of melanin, the pigment responsible for coloration in skin, hair, and eyes in humans. The active site of tyrosinase has a type III Cu center (T3Cu) and in mammals, it catalyzes the hydroxylation of tyrosine to DOPA (dihydroxyphenylalanine) and the subsequent oxidation of DOPA to dopaquinone, which further polymerizes to form melanin. Deficiency or mutations in the tyrosinase gene can lead to conditions such as albinism. Polyphenol oxidases (PPOs) are a type of tyrosinases that are typically found in plants and are responsible for the browning of fruits and mushrooms [203].

4.9.5 Dopamine β-Hydroxylase (DBH)

DBH, also known as dopamine β-monooxygenase, is a metalloenzyme that plays a key role in catalyzing the hydroxylation of dopamine to norepinephrine (noradrenaline), a key neurotransmitter in the central and peripheral nervous system. Norepinephrine is also an essential intermediate in the biosynthesis of epinephrine (adrenaline), an important hormone and a neurotransmitter. Both norepinephrine and epinephrine play a crucial role in controlling various behavioral and physiological processes, such as learning, memory, mood, and stress. The DBH is a homodimer, with each subunit having one type II Cu center (T2Cu) [204]. During catalysis, the closed conformation brings these two mononuclear T2Cu sites together, functioning synergistically like a binuclear Cu site. The proper functioning of dopamine β-hydroxylase is essential for maintaining neurological health and function, and any deficiencies can lead to neurological disorders such as dopamine beta-hydroxylase deficiency.

4.9.6 Lysyl Oxidase (LOX)

LOX is an enzyme that catalyzes the oxidative deamination of lysine amino acid into allysine, an aldehyde derivative of lysine. This reaction is fundamental to the health and stability of extracellular matrix (ECM) and connective tissues. The LOX is a Cu enzyme with a mononuclear type II Cu center (T2Cu) that undergoes redox changes and is critical to catalysis. LOX acts explicitly on the lysine residues of the key ECM and connective tissue proteins such as collagen and elastin and converts them into allysine, which makes key covalent cross-linking, strengthening and stabilizing the ECM overall. Therefore, proper functioning of LOX is necessary for maintaining the strength and elasticity of connective tissues. The right balance of LOX is essential to human health, as any deficiencies or defects could lead to connective tissue disorders such as lathyrism and over-expression, can promote metastasis in tumor cells [205].

4.9.7 Galactose Oxidase

Galactose oxidase is a copper-dependent enzyme that catalyzes the two-electron oxidation of primary alcohols, such as D-galactose, to their corresponding aldehydes. The enzyme contains a mononuclear Cu(II) that is coordinated to two tyrosine and two histidine residues and additionally to a tyrosine-derived free radical, which plays a key role during catalysis [206]. This Cu-bound metallo-radical at the active site couples two half-reactions: one is the oxidation of the primary alcohol, and the second is the reduction of molecular oxygen to hydrogen peroxide. Due to its high turnover numbers, galactose oxidase is used in industries to catalyze the conversion of primary alcohols to aldehydes.

4.9.8 Particulate Methane Monooxygenase (pMMO)

Particulate methane monooxygenase (pMMO) is an enzyme found in methanotrophic bacteria that uses methane as their primary source of carbon in catalyzing the oxidation of methane to methanol [207]. This reaction is critical in combating the global greenhouse effect, as methane, one of the major greenhouse gases, contributes significantly to global warming. pMMO is an integral membrane metalloenzyme in which the active site has two Cu ions (Figure 15B). pMMO also has several industrial applications, especially in bioremediation and the production of biofuels.

4.9.9 Peptidylglycine α-Amidating Monooxygenase (PAM)

PAM is a bifunctional metalloenzyme that serves as a monooxygenase and a lyase that plays a key role in the biosynthesis of various neuropeptides and peptide hormones by amidating their C-terminal residues. The PAM mediates two reactions: one is the hydroxylation of the C-terminal glycine residue of the precursor peptides, followed by the cleavage of the N-C bond of the hydroxylated glycine leaving behind the bioactive amidated peptide and glyoxylate. During catalysis, on a "n+1" amino acid long peptide, PAM cleaves away the C-terminal glycine, thereby yielding a "n" amino acid long bioactive peptide with a terminal amide group. PAM has two mononuclear type II Cu centers (T2Cu) (Figure 15C) designated as CuH (H-site) and CuM (M-site), both of which are essential for its catalytic activity [208].

4.9.10 Copper Amine Oxidase (CAO)

CAOs are a group of enzymes that catalyze the oxidative deamination of primary amines to their corresponding aldehydes concomitant with the release of ammonia and hydrogen peroxide. These enzymes are ubiquitous to all aerobic organisms and are primarily involved in various metabolic processes such as cell signaling, cell differentiation, growth, detoxification, and development. In mammals, they also play a key role in the biosynthesis of bioactive amines, such as neurotransmitters and histamines. The active site of CAOs typically contains a mononuclear type II Cu center (T2Cu) along with a unique cofactor known as 2,4,5-trihydroxyphenylalanine quinone (TPQ), a derivative of tyrosine, both of which are required for the catalytic activity [209].

4.9.11 Ceruloplasmin

Ceruloplasmin, also known as ferroxidase enzyme, catalyzes the oxidation of Fe(II) to Fe(III). Ceruloplasmin is a multicopper oxidase found in the blood plasma of vertebrates and plays a key role in iron metabolism by converting the toxic ferrous ion into non-toxic and bioavailable ferric ion, thereby facilitating transferrin (discussed in Section 4.6.1) in binding ferric ions and subsequent transport of iron throughout the body. Ceruloplasmin contains six tightly bound copper ions critical for its oxidase activity, accounting for nearly 95% of total copper in human plasma. Among the six bound Cu ions, there are mononuclear three type I Cu centers (T1Cu), one mononuclear type II Cu center (T2Cu), and a binuclear type III Cu center (T3Cu). The T2Cu and the T3Cu occur in proximity and synergistically work as a trinuclear copper center, which serves as the binding site for oxygen during catalysis [210]. Beyond its role in Fe metabolism, ceruloplasmin also exhibits antioxidant properties by scavenging free radicals and reducing oxidative stress. Deficiency or mutations in ceruloplasmin could disrupt Fe metabolism and cause a Fe overload, further associated with several disorders such as Wilson's disease, aceruloplasminemia, and neurodegeneration [211]. On the contrary, excess ceruloplasmin can lead to copper toxicity and neurodegenerative disorders [212].

4.9.12 Nitrous Oxide Reductase

Nitrous oxide reductase is a metalloenzyme that plays a key role in the final step of bacterial denitrification by catalyzing the reduction of nitrous oxide to dinitrogen. The nitrous oxide reductase has two Cu centers – one is the mixed-valent binuclear copper A center (Cu_A) and a unique, tetranuclear center known as Cu_Z [213]. The composition of Cu_Z is a [4Cu:2S] cluster that is coordinated to seven histidine residues, as shown in Figure 15D. One of the sulfurs, designated as S_{Z1}, is coordinated to all four copper ions, whereas the other sulfur (S_{Z2}) is coordinated to only two Cu ions (Cu_1 and Cu_4), which is oxygen labile, and upon its loss, can lead to an inactive form of the enzyme with a [4Cu:S] center. As a critical enzyme in denitrification, nitrous oxide reductase is vital in regenerating nitrogen gas from the biomass and bringing it back to the atmosphere.

4.9.13 Copper Storage Protein (Csp1)

Cu is essential to nearly all aerobic and facultative anaerobic organisms. Certain microorganisms, such as methane-oxidizing bacteria (methanotrophs), however, require large quantities of Cu to supply enough Cu cofactor to support pMMOs (discussed in Section 4.9.9) and other important cellular Cu metalloenzymes. However, excess copper can lead to severe cytotoxic effects. Hence, excess Cu ions must be properly handled and stored safely to mitigate toxicity. Csp1 is a new class of Cu-storing protein originally identified and characterized from methanotroph *Methylosinus trichosporium*. Csp1 is a small protein of 122 amino acids, arranged in a four-helix bundle pattern that can bind remarkably up to thirteen Cu (I) ions inside the core of the four-helix scaffold (Figure 16) [214]. This kind of metal storage within the core of a helix bundle is exclusive to Csp1 and is not observed elsewhere in nature. The helical bundle's simplicity and inherent properties make it an ideal choice as a scaffold protein to design artificial metalloproteins for various biotechnological applications [215,216].

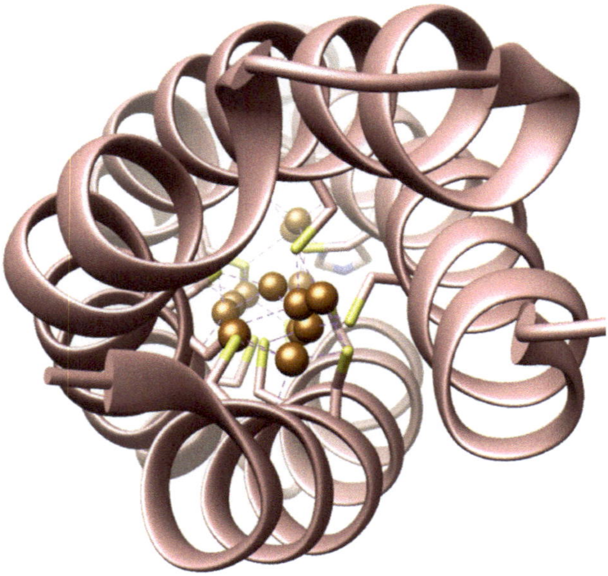

FIGURE 16 The top view of fully loaded Cu binding sites in Csp1 from *Methylosinus trichosporium* (PDB: 5FJE). Color scheme: Protein backbone and carbon – pink; nitrogen – blue; sulfur – yellow; copper – copper brown spheres.

5 NON-TRANSITION D-BLOCK METALS

Group 12 is the last group of d-block metals in the periodic table, which includes three major natural metals: zinc (Zn), cadmium (Cd), and mercury (Hg). Often, these elements are misplaced under the transition group. However, these metals are not transition metals and are therefore described as non-transition d-block metals [217]. Zn, with the atomic number 30, is the only metal from group 12 to have a role in biology. It is an essential trace element and is vital to survival across all the domains of life. In humans, Zn is the second most abundant d-block metal, following iron, and plays a crucial role in numerous biological processes [218]. Zn's ability to form stable complexes with various ligands makes it an indispensable cofactor in many enzymes and proteins, contributing to catalytic activity, structural stability, and regulatory functions. Zn deficiency is described as a type of malnutrition that can lead to diarrhea and several other infections.

5.1 Zinc Finger Proteins (ZNFs)

Zn finger proteins are a large family of proteins that are majorly involved in binding nucleic acids and play a vital role in the regulation of replication and transcription of DNA. These proteins are characterized by the presence of one or more Zn finger motifs, which are composed of a short peptide loop coordinating a mononuclear Zn ion, primarily *via* two cysteines and two histidine residues, known as C2H2 proteins

(Figure 17A); however, other types of coordination exists [219]. These motifs enable ZNFs to bind DNA, RNA, or other proteins, thereby regulating gene expression, DNA double-strand break repair, and other cellular functions such as signal transduction, cell migration, differentiation, and responding to environmental stimuli.

5.2 Alcohol Dehydrogenases (ADHs)

ADHs are ubiquitous in higher organisms that catalyze the reversible oxidation of a wide variety of alcohols into aldehydes or ketones concomitant with the reduction of nicotinamide adenine dinucleotide (NAD^+) to nicotinamide adenine dinucleotide (NADH). These enzymes not only play a key role in the detoxification of alcohols from the body but also in the generation of aldehyde and ketone metabolites required in various metabolic pathways. ADH contains two Zn ions per subunit: of which one is the catalytic Zn center (Figure 17B) that mediates the redox reaction, and the other is the structural Zn ion that stabilizes the enzyme's tertiary structure [220]. In humans, ADH expression levels are highest in the liver; however, basal expression occurs in other tissues and organs, where it helps metabolize ethanol and other alcohols, protecting the body from their toxic effects. Deficiency in ADH leads to

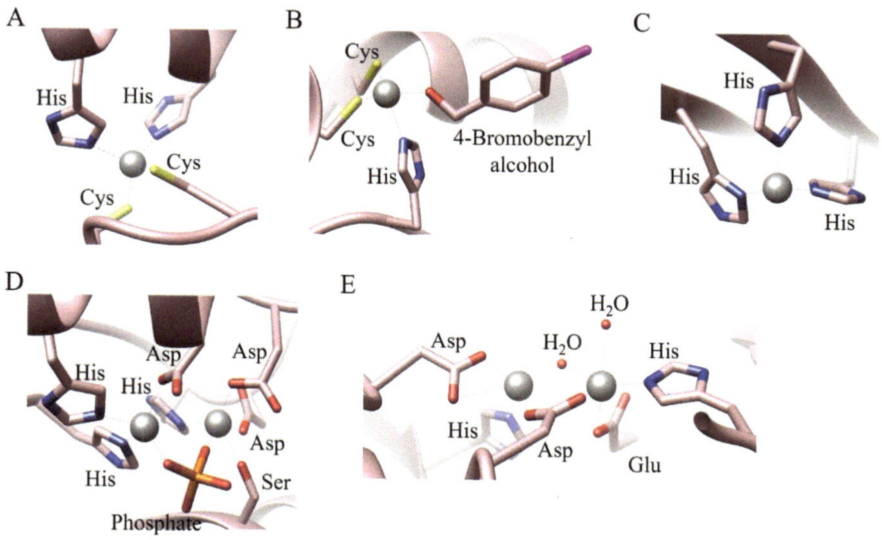

FIGURE 17 The Zn centers in various enzymes. (A) A C2H2-based Zn finger domain from human transcription factor, Spl (PDB: 1SP1). (B) The active site of alcohol dehydrogenase from *Equus caballus* in complex with 4-bromobenzyl alcohol (PDB: 1HLD). (C) Zn coordination at the active site of human carbonic anhydrase (PDB: 6LUX). (D) Phosphate coordinated binuclear Zn center at the active site of alkaline phosphatase from *Escherichia coli* (PDB: 1ALK). (E) Structural arrangement of the binuclear Zn center at the active site of human glutamate carboxypeptidase II (PDB: 2OOT). Color scheme: Protein backbone and carbon – pink; nitrogen – blue; oxygen – red; sulfur – yellow; phosphorus – vermillion; bromine – magenta; zinc – gray spheres.

a genetic disorder known as ALDH2 deficiency or Asian flush syndrome that negatively affects alcohol metabolism [221].

5.3 Carbonic Anhydrase (CA)

CA is a metalloenzyme found across all forms of life that catalyzes the reversible hydration of carbon dioxide to bicarbonate and protons. CA activity is crucial for maintaining acid-base balance in blood and tissues, facilitating carbon dioxide transport, and regulating pH levels. In addition, CA is important in several biosynthetic processes such as urea biosynthesis, bone resorption, lipogenesis, and gluconeogenesis. The active site of CA has a mononuclear zinc ion coordinated by three histidine residues, as shown in Figure 17C. Deficiency in carbonic anhydrase leads to a rare genetic disorder known as carbonic anhydrase VA deficiency or CA-VA deficiency, characterized by metabolic hyperammonemic encephalopathy, feeding intolerance, weight loss, seizures, and coma [222]. CA is also known to be highly elevated in certain human diseases such as glaucoma, obesity, epilepsy, osteoporosis, and even cancer [223]. Clinically approved CA inhibitors (CAIs) are primarily used to lower intraocular pressure in treating patients with glaucoma, apart from various other CAIs as anticancer agents are in the clinical phase [224].

5.4 Alkaline Phosphatase (ALP)

ALP is an enzyme that catalyzes the dephosphorylative hydrolysis of phosphate esters, concomitant with the release of inorganic phosphate [225]. These enzymes are active and can mediate the catalytic reaction only under alkaline conditions at high (basic) pH values. The enzyme requires two Zn ions at its active site (Figure 17D), essential for its catalytic activity and substrate binding. In humans, ALP is found in various tissues, including the liver, bone, kidney, and intestine, where it plays a key role in phosphate metabolism, bone mineralization, and dephosphorylation of biomolecules. Elevated levels of alkaline phosphatase in the blood can indicate liver or bone diseases; therefore, ALP is routinely used as a diagnostic biomarker in identifying disease conditions such as hepatitis or osteomalacia [226]. A low level of ALP is a rare condition; however, it may indicate malnutrition or Zn deficiency, both of which could be managed *via* diet.

5.5 Polymerases

Polymerases are enzymes involved in synthesizing the long chain of nucleic acids. These enzymes are essential for nucleic acid processing, including replication, transcription, recombination, and repair. There are two major types of polymerases based on the template used during the synthesis: DNA polymerase and RNA polymerase [227]. DNA polymerases catalyze the synthesis of DNA molecules during the process of replication. These enzymes have highly conserved structures with Zn ions that stabilize the enzyme's structure and are involved in the catalytic process of dNTP addition sequentially. DNA polymerases are found in all living organisms, ensuring the accurate replication and repair of genetic material. On the other hand,

RNA polymerases are enzymes responsible for the synthesis of RNA molecules from single-stranded DNA templates during the process of transcription [228]. These enzymes contain a Zn ion at their active sites, which coordinates with conserved acidic residues and helps stabilize the enzyme-DNA-RNA complex during transcriptional elongation. Zn-dependent RNA polymerases are found across all domains of life and play essential roles in gene expression and regulation. Depending on the organism, RNA polymerase can be a protein complex (bacteria, archaea, eukarya) or only consist of a single subunit (phages, eukaryotic chloroplasts, and mitochondria). Inhibition of RNA polymerase activity by Zn-binding compounds is a therapeutic strategy for controlling several viral infections and other pathogenic diseases [229]. Any errors due to a mismatch by either of the polymerases during the elongation phase may incur deleterious effects, therefore these enzymes display a very high-fidelity during catalysis.

5.6 Histone Deacetylases (HDACs)

HDACs are enzymes that catalyze the removal of acetyl groups from lysine residues, especially in histone proteins. This deacetylation reaction enables histones to wrap DNA more tightly, thereby leading to repression of transcription and promoting chromatin condensation. Although HDACs activity in the deacetylation of histones is considered an epigenetic modification, it however indirectly plays a crucial role in regulating gene expression, cell cycle progression, and development. The active site of HDACs contains a zinc ion which is essential for the catalytic activity and facilitates the hydrolysis of the acetyl group. Inhibitors of HDACs are currently being explored as therapeutic agents for cancer and other diseases, given their role in modulating epigenetic regulation. Vorinostat is a well-known HDAC inhibitor, approved by the United States Food and Drug Administration (US FDA) and is used in the treatment of cutaneous and peripheral T-cell lymphoma (CTCL) [230].

5.7 Carboxypeptidases

Carboxypeptidases are exopeptidases which specialize in the hydrolysis of peptide bonds at the carboxy terminus of various peptides and proteins [231]. Carboxypeptidases substrate specificity is striking in contrast to that of aminopeptidases such as MetAP (discussed in Section 4.7.3). The proteolytic cleavage by carboxypeptidases is part of post-translational modifications essential in activating several precursor neuroendocrine peptides, including insulin, thereby facilitating several biophysical processes such as blood clotting, wound healing, and signal transduction. The active site of carboxypeptidases primarily has a binuclear zinc center (Figure 17E) that facilitates nucleophilic attack on substrate molecules and helps stabilize the reaction intermediates generated during catalysis. Carboxypeptidases are generally produced as inactive precursors, known as zymogens, to mitigate non-specific proteolytic cleavage on unwanted targets. After successful translocation to a specific target site, the zymogen forms are then converted to active carboxypeptidases by proteolytic cleavage by other proteases. Several carboxypeptidase inhibitors are being developed to treat several human diseases, including cancer.

5.8 Angiotensin-Converting Enzyme (ACE)

ACE is a metallopeptidase that catalyzes the conversion of angiotensin I (proangiotensin) to the active angiotensin II by cleaving off the last two amino acids at the C-terminal end of the angiotensin I. Angiotensin is an important hormone of the renin-angiotensin system that plays a crucial role in vasoconstriction leading to high blood pressure. ACE has a wide substrate specificity, including bradykinin, kallidin, hemopressin, and amyloid β-protein. Interestingly, ACE also participates in the kinin-kallikrein system where it plays a key role in degrading bradykinin, a potent vasodilator. This opposing role of ACE, wherein it generates vasoconstrictors in one pathway and degrades vasodilators in another, synergistically regulates the cardiac sympathetic activity, blood pressure, and electrolyte balance [232]. The active site of the ACE has a mononuclear zinc ion (Figure 18A), which participates in a promoted-water pathway, in which the reaction is initiated via nucleophilic attack by the zinc-bound water molecule, targeting the scissile carbonyl carbon. During this step, deportation of the Zn-bound water occurs resulting in a tetrahedral Zn-OH intermediate, concomitant with the protonation of the proximal glutamic acid residue (E384 in humans) [233]. This protonated glutamic acid facilitates the cleavage of the targeted amide C–N bond, resulting in proteolytic cleavage. Inhibitors of ACE are commonly used as therapeutic agents in the treatment of several human conditions, such as hypertension, cardiac arrest, diabetic nephropathy, and type 2 diabetes mellitus [234].

5.9 Thermolysin

Thermolysin is a thermostable enzyme (stable up to ~ 70°C) that catalyzes the hydrolysis of peptide bonds specifically at the hydrophobic amino acid sites. Thermolysin is a native metalloenzyme produced by certain species of the Gram-positive bacteria

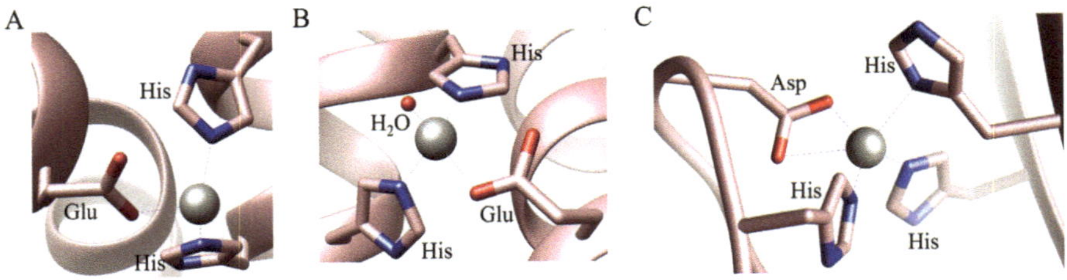

FIGURE 18 (A) Mononuclear Zn center at the active site of human angiotensin-converting enzyme (PDB: 1O86). (B) The active site of thermolysin from *Bacillus thermoproteolyticus* (PDB: 3TMN). (C) Mononuclear Zn center at the active site of human stromelysin-1 (MMP-3) (PDB: 1UEA). Color scheme: Protein backbone and carbon – pink; nitrogen – blue; oxygen – red; zinc – gray spheres.

Bacillus, which has two different types of metals. The active site of thermolysin has a mononuclear Zn ion (Figure 18B), which is essential for the catalytic activity, whereas four calcium ions are required for its structural stability [235]. Thermolysin is synthesized as an inactive precursor form, which undergoes two post-translational proteolytic cleavages by other proteases, before being secreted out of the cell as a secretory active enzyme. This active form of secreted thermolysin breaks down extracellular proteins and peptides into smaller peptides to be absorbed by the bacteria as nutrition, thereby conferring its virulence and pathogenesis. For this reason, thermolysin is often targeted in treating various bacterial infections.

5.10 Matrix Metalloproteinases (MMPs)

MMPs are a family of endopeptidases that catalyzes the proteolytic degradation of extracellular matrix proteins and peptides, thereby playing a crucial role in remodeling extracellular matrix (ECM) apart from contributing to several metabolic activities such as cell signaling, proliferation, migration, differentiation, angiogenesis, apoptosis, tissue remodeling, wound healing, development, and host defense [236]. MMPs contain a zinc ion at the active site within the catalytic domain typically coordinated by His and Asp residues, as shown in Figure 18C. Like thermolysin (discussed in Section 5.9), certain MMPs also contain calcium ions required for their structural stability. Since MMPs have a broad spectrum of substrates, their regulation is critical to mitigate unwanted proteolytic damage of specific essential proteins/peptides. Tissue inhibitors of metalloproteinases (TIMPs) are natural inhibitors of MMPs that play a crucial role in their controlled inactivation. Any uncontrolled dysregulation of MMPs can lead to several human diseases, such as cancer, arthritis, inflammation, cardiovascular diseases, and neurodegenerative disorders [237].

5.11 Metallothioneins (MTs)

MTs (such as MT1, MT2, and SmtA) belong to a family of cysteine-rich, low-molecular-weight metalloproteins (< 7 kDa) that typically bind metals such as Zn and Cu, although Zn is the preferred cation. MTs typically have up to 18–23 cysteines (~ 30%) strikingly with no aromatic amino acids and histidine residues and bind the divalent metal ions through the thiol groups of cysteine residues [238]. MTs can bind up to 4–7 metal ions, typically coordinated in a tetrahedral geometry. Generally, MTs sequester excess Zn ions and release them when required, maintaining Zn homeostasis. MTs can also bind several divalent cations of xenobiotic heavy metals such as cadmium, mercury, silver, arsenic, lead, and bismuth, thereby playing a key role in protecting the cells from metal toxicity and relieving cells from oxidative stress. Since MTs play a key role in metal homeostasis and oxidative stress, any dysregulation in their expression and functioning can lead to several diseases, such as cancer and cardiac abnormalities [239].

5.12 Zinc Transporters

Zn transporters are membrane proteins that regulate the influx and efflux of zinc ions across cellular membranes, maintaining Zn homeostasis within cells and tissues. These transporters play crucial roles in Zn uptake from the extracellular environment, intracellular Zn distribution, and disposal of excess Zn from the cytoplasm. On average, an adult human has about 2.3 g of Zn, of which only 0.1% is found circulating in the plasma. The remaining Zn is found inside the cell, either used as a cofactor in metalloenzymes or stored as metalloproteins. Zrt-/Irt-like proteins (ZIPs) are transmembrane proteins found in mammalian cell membranes that facilitate zinc uptake (influx) either from the extracellular medium or from the lumen into the cytoplasm [240]. In contrast, Zn transporters (ZnTs) are proteins that pump excess Zn out (efflux) of the cytoplasm or into intracellular organelles. Dysregulation of Zn transporters is implicated in various diseases, including neurodegenerative disorders, diabetes, and cancer [241].

6 METALS WITH RARE OCCURRENCE

Several recent studies have demonstrated that lanthanides have metabolic functions in certain microorganisms [242,243]. We report these examples as rare occurrences in biology, as there are lanthanide-free counterparts to most of these examples and they are explicitly inclusive to microorganisms that thrive in environments rich in rare earth metals. Lanmodulin (LanM) is a natural lanthanide-binding protein found in certain methylotrophs that typically coordinates to mononuclear lanthanum (Figure 19A). Lanthanum, with an atomic number of 57, is possibly the heaviest metal in biology that can occur freely without the requirement of another cofactor, as seen in LanM. Depending on the growth conditions and the availability of rare earths in the microenvironment, LanM can also bind neodymium (atomic number 60) and dysprosium (atomic number 66) [244]. The role of LanM is not thoroughly

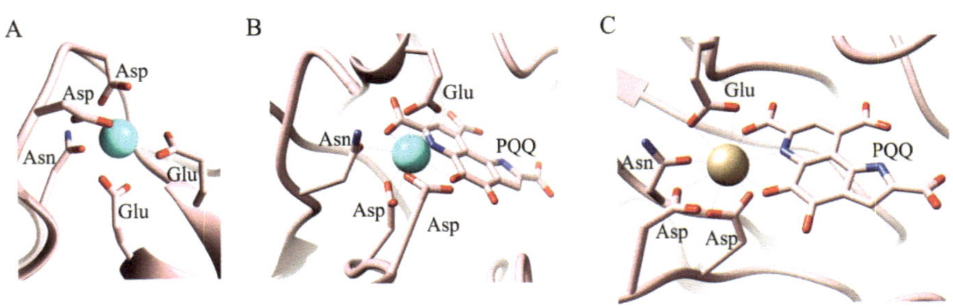

FIGURE 19 (A) A mononuclear La center at the active site of LanM from *Hansschlegelia quercus* (PDB: 8DQ2). (B) La in coordination with PQQ within the active site of lanthanide-dependent MDH from *Methylorubrum extorquens* AM1 (PDB: 6OC6). (C) Ce and PQQ at the active site of MDH from *Methylacidiphilum fumariolicum* SolV (PDB: 4MAE). Color scheme: Protein backbone and carbon – pink; nitrogen – blue; oxygen – red; lanthanum – cyan spheres; cerium – tan sphere.

understood, however it is believed to store lanthanides in methylotrophs, which could utilize these rare metals in enzymes that require lanthanides as cofactors, such as methanol dehydrogenases (MDHs). MDHs play a crucial role in oxidizing methanol to formaldehyde. Typically, these enzymes have calcium as the cofactor, however, certain methanotrophic and methylotrophic organisms additionally have a gene called *xoxF* that encodes for a lanthanide-dependent MDH, which coordinates to lanthanides such as lanthanum (Figure 19B) and cerium (Figure 19C) [245,246]. In contrast to LanM, lanthanide-dependent MDHs require an additional cofactor known as pyrroloquinoline quinone (PQQ), as shown in Figure 19B and Figure 19C. The PQQ is the actual catalytic site that performs the 2 electron oxidization of methanol to formaldehyde, whereas the lanthanides only play a role in stabilizing the PQQ. Recent *in vitro* studies on lanthanide-dependent MDHs revealed that the lanthanides could potentially be replaced with actinides such as americium (atomic number 95) and curium (atomic number 96), with comparable activities to the lanthanide counterparts [247]. These rare occurrences of f-block metals in biology exemplify the diverse role of metals in biology.

7 CONCLUSION

The diversity of metals across biological kingdoms, from prokaryotes to eukaryotes, underscores their essential and multifaceted roles in sustaining life. Transition metals, in particular, are integral to nearly all metabolic processes, influencing critical pathways such as the global cycling of carbon, nitrogen, and sulfur. From their role in prebiotic chemistry to their crucial roles in extant biological systems, metals have been and remain central to the origins, evolution, and sustainability of life on Earth, a role that has often been underrated. This chapter aimed to emphasize the significance of metals, challenging the traditional focus on CHNOPS elements by highlighting the indispensable nature of metals in mediating metabolic reactions, conferring structural stability, and maintaining cellular homeostasis. Their abundance and indispensability reflect the complexity and diversity of biological mechanisms, with far-reaching implications from fundamental biology to practical applications in health, disease, and biotechnology.

ACKNOWLEDGMENTS

This work was supported by the National Institutes of Health (1R01GM138592-01) and the US DOE, Office of Science, Office of Basic Energy Sciences, under award DE-SC0018143 to J.W.P. Bhanu P. Jagilinki researched the subject and wrote the manuscript for this chapter followed by editing by Florence Mus and John W. Peters.

ABBREVIATIONS

ADP adenosine diphosphate
ATP adenosine triphosphate
CDF cation diffusion facilitator
cGMP cyclic guanosine monophosphate

CTCL	cutaneous and peripheral T-cell lymphoma
DMS	dimethyl sulfide
dNTP	deoxynucleotide triphosphate
ECM	extracellular matrix
EMR	electromagnetic radiation
FAD	flavin adenine dinucleotide
GOE	great oxidation event
GTP	guanosine triphosphate
HiPIPs	high-potential iron-sulfur proteins
LMWCr	low-molecular-weight chromium-binding substance
MAPK	mitogen-activated protein kinase
MGD	molybdopterin guanine dinucleotide
NAD^+	nicotinamide adenine dinucleotide
NADH	nicotinamide adenine dinucleotide
NaMN	nicotinic acid mononucleotide
OAA	oxaloacetate
PDB	Protein Data Bank
PEP	phosphoenolpyruvate
Pi	inorganic phosphate
POPs	persistent organic pollutants
PQQ	pyrroloquinoline quinone
PTMs	post-translational modifications
RBCs	red blood cells
ROS	reactive oxygen species
RT	reverse transcriptase
SHE	standard hydrogen electrode
SR	sarcoplasmic reticulum
$T_{1/2}$	half-life
TIMPs	tissue inhibitor of metalloproteinases
TPQ	2,4,5-trihydroxyphenylalanine quinone
US FDA	United States Food and Drug Administration
VGKCs	voltage-gated potassium channels

REFERENCES

1. M.A. Zoroddu, J. Aaseth, G. Crisponi, S. Medici, M. Peana, V.M. Nurchi, *J. Inorg. Biochem.* **2019**, *195*, 120–129.
2. J.W. Peters, *Curr. Opin. Struct. Biol.* **1999**, *9* (6), 670–676.
3. R.A. Festa, D.J. Thiele, *Curr. Biol.* **2011**, *21* (21), R877–883.
4. K. Jomova, M. Makova, S.Y. Alomar, S.H. Alwasel, E. Nepovimova, K. Kuca, C.J. Rhodes, M. Valko, *Chem. Biol. Interact.* **2022**, *367*, 110173.
5. G. Scalese, K. Kostenkova, D.C. Crans, D. Gambino, *Curr. Opin. Chem. Biol.* **2022**, *67*, 102127.
6. Y. Lu, N. Yeung, N. Sieracki, N.M. Marshall, *Nature.* **2009**, *460* (7257), 855–862.
7. H.J. Davis, T.R. Ward, *ACS Cent. Sci.* **2019**, *5* (7), 1120–1136.
8. D. Rossetto, S.S. Mansy, *Front. Cell Dev. Biol.* **2022**, *10*, 864830.

9. B. Kacar, A.K. Garcia, A.D. Anbar, *ChemBioChem.* **2021**, *22* (1), 114–119.
10. W. Martin, M.J. Russell, *Philos. Trans. R. Soc. Lond. B.* **2003**, *358* (1429), 59–83.
11. C. Ponnamperuma, A. Shimoyama, E. Friebele, *Orig. Life.* **1982**, *12* (1), 9–40.
12. P. Laszlo, *Science.* **1987**, *235* (4795), 1473–1477.
13. W. Martin, M.J. Russell, *Philos. Trans. R. Soc. Lond. B.* **2007**, *362* (1486), 1887–1925.
14. W. Martin, J. Baross, D. Kelley, M.J. Russell, *Nat. Rev. Microbiol.* **2008**, *6* (11), 805–814.
15. C. Huber, G. Wächtershäuser, *Science.* **1998**, *281* (5377), 670–672.
16. S.L. Miller, *Science.* **1953**, *117* (3046), 528–529.
17. S.L. Miller, H.C. Urey, *Science.* **1959**, *130* (3370), 245–251.
18. D.W. Smith, *J. Chem. Educ.* **1977**, *54* (9), 540.
19. *The Alkali Metal Ions: Their Role for Life, Metal Ions in Life Sciences*, Volume 16, A. Sigel, H. Sigel, R. K. O. Sigel, Eds, Springer International Publishing, Cham, Switzerland, **2016**, pp. 1–628.
20. S.P.H. Alexander, A. Mathie, J.A. Peters, *Br. J. Pharmacol.* **2011**, *164*(Suppl 1), S137–174.
21. B. Hille, *Biophys. J.* **1978**, *22* (2), 283–294.
22. G. V. Petkov, *Pharmacology*, Academic Press, San Diego, **2009**, pp. 387–427.
23. F.H. Yu, W.A. Catterall, *Genome Biol.* **2003**, *4* (3), 207.
24. C.A. Ahern, J. Payandeh, F. Bosmans, B. Chanda, *J. Gen. Physiol.* **2016**, *147* (1), 1–24.
25. D. Jiang, R. Banh, T.M. Gamal El-Din, L. Tonggu, M.J. Lenaeus, R. Pomès, N. Zheng, W.A. Catterall, *Cell.* **2021**, *184* (20), 5151–5162 e11.
26. Y.M. Sun, I. Favre, L. Schild, E. Moczydlowski, *J. Gen. Physiol.* **1997**, *110* (6), 693–715.
27. D.M. Kim, C.M. Nimigean, *Cold Spring Harb. Perspect. Biol.* **2016**, *8* (5).
28. G. Yellen, *Nature.* **2002**, *419* (6902), 35–42.
29. S.B. Long, E.B. Campbell, R. Mackinnon, *Science.* **2005**, *309* (5736), 897–903.
30. C. Vergara, R. Latorre, N.V. Marrion, J.P. Adelman, *Curr. Opin. Neurobiol.* **1998**, *8* (3), 321–329.
31. J.P. Castillo, H. Rui, D. Basilio, A. Das, B. Roux, R. Latorre, F. Bezanilla, M. Holmgren/, *Nat. Commun.* **2015**, *6*, 7622.
32. J.P. Morth, B.P. Pedersen, M.S. Toustrup-Jensen, T.L.M. Sørensen, J. Petersen, J.P. Andersen, B. Vilsen, P. Nissen, *Nature.* **2007**, *450* (7172), 1043–1049.
33. M.V. Clausen, F. Hilbers, H. Poulsen, *Front. Physiol.* **2017**, *8*, 371.
34. D. Rhodes, H.J. Lipps, *Nucleic Acids Res.* **2015**, *43* (18), 8627–8637.
35. D. Bhattacharyya, G. Mirihana Arachchilage, S. Basu, *Front. Chem.* **2016**, *4*, 38.
36. M.L. Zhang, Y.P. Xu, A. Kumar, Y. Zhang, W.Q. Wu, *Biochemistry.* **2019**, *58* (38), 3955–3959.
37. G.N. Parkinson, M.P. Lee, S. Neidle, *Nature.* **2002**, *417* (6891), 876–880.
38. M.L. Bochman, K. Paeschke, V.A. Zakian, *Nat. Rev. Genet.* **2012**, *13* (11), 770–780.
39. S. Balasubramanian, S. Neidle, *Curr. Opin. Chem. Biol.* **2009**, *13* (3), 345–353.
40. N. N. Greenwood, A. Earnshaw, *Beryllium*, in *Chemistry of the Elements*, 2nd ed., N. N. Greenwood, A. Earnshaw, Butterworth Heinemann, Oxford, **1997**.
41. T. Dudev, C. Lim, *Chem. Rev.* **2003**, *103* (3), 773–788.
42. R. H. Kretsinger, in *Encyclopedia of Metalloproteins*, R. H. Kretsinger, V. N. Uversky, E. A. Permyakov, Eds, Springer, New York, **2013**, pp. 521–536.
43. T. Dudev, J.A. Cowan, C. Lim, *J. Am. Chem. Soc.* **1999**, *121* (33), 7665–7673.
44. D.G. Kehres, M.E. Maguire, *Biometals.* **2002**, *15* (3), 261–270.
45. M.D. Snavely, J.B. Florer, C.G. Miller, M.E. Maguire, *J. Bacteriol.* **1989**, *171* (9), 4761–4766.
46. M. Hattori, N. Iwase, N. Furuya, Y. Tanaka, T. Tsukazaki, R. Ishitani, M.E. Maguire, K. Ito, A. Maturana, O. Nureki, *EMBO J.* **2009**, *28* (22), 3602–3612.

47. C.E. Dann, 3rd, C.A. Wakeman, C.L. Sieling, S.C. Baker, I. Irnov, W.C. Winkler, *Cell.* **2007**, *130* (5), 878–892.
48. A. Ramesh, W.C. Winkler, *RNA Biol.* **2010**, *7* (1), 77–83.
49. D.L. Smith, T. Tao, M.E. Maguire, *J. Biol. Chem.* **1993**, *268* (30), 22469–22479.
50. R. Schindl, J. Weghuber, C. Romanin, R.J. Schweyen, *Biophys. J.* **2007**, *93* (11), 3872–3883.
51. J. Gregan, D.M. Bui, R. Pillich, M. Fink, G. Zsurka, R.J. Schweyen, *Mol. Gen. Genet.* **2001**, *264* (6), 773–781.
52. V.V. Khrustalev, E.V. Barkovsky, T.A. Khrustaleva, *J. Theor. Biol.* **2016**, *395*, 174–185.
53. V.K. Batra, W.A. Beard, D.D. Shock, J.M. Krahn, L.C. Pedersen, S.H. Wilson, *Structure.* **2006**, *14* (4), 757–766.
54. K. Pari, G.A. Mueller, E.F. DeRose, T.W. Kirby, R.E. London, *Biochemistry.* **2003**, *42* (3), 639–650.
55. V. Svetlov, E. Nudler, *Biochim. Biophys. Acta.* **2013**, **1829** (1), 20–28.
56. M.A. Griep, *Indian J. Biochem. Biophys.* **1995**, *32* (4), 171–178.
57. J.L. Keck, D.D. Roche, A.S. Lynch, J.M. Berger, *Science.* **2000**, *287* (5462), 2482–2486.
58. M. Kato, T. Ito, G. Wagner, C.C. Richardson, T. Ellenberger, *Mol. Cell.* **2003**, *11* (5), 1349–1360.
59. Y. Kang, R. Liu, J.X. Wu, L. Chen, *Nature.* **2019**, *574* (7777), 206–210.
60. A.M. Pyle, *Science.* **1993**, *261* (5122), 709–714.
61. M.J. Fedor, J.R. Williamson, *Nat. Rev. Mol. Cell Biol.* **2005**, *6* (5), 399–412.
62. T. Berezniak, M. Zahran, P. Imhof, A. Jäschke, J.C. Smith, *J. Am. Chem. Soc.* **2010**, *132* (36), 12587–12596.
63. S.S. Taylor, A.P. Kornev, *Trends Biochem. Sci.* **2011**, *36* (2), 65–77.
64. A.L. Buchachenko, D.A. Kouznetsov, N.N. Breslavskaya, M.A. Orlova, *J. Phys. Chem. B.* **2008**, *112* (8), 2548–2556.
65. B.P. Jagilinki, N. Gadewal, H. Mehta, H. Mahadik, V. Pandey, U. Sawant, P. A Wadegaonkar, P. Goyal, S. Kumar, A. K Varma, *J. Biomol. Struct. Dyn.* **2015**, *33* (4), 852–860.
66. B.P. Jagilinki, R.K. Choudhary, P.S. Thapa, N. Gadewal, M.V. Hosur, S. Kumar, A.K. Varma, *Cell Biochem. Biophys.* **2016**, *74* (3), 317–325.
67. P. Lahiry, A. Torkamani, N.J. Schork, R.A. Hegele, *Nat. Rev. Genet.* **2010**, *11* (1), 60–74.
68. M.J. Chen, J.E. Dixon, G. Manning, *Sci. Signal.* **2017**, *10* (474).
69. J.D. Selengut, R.L. Levine, *Biochemistry.* **2000**, *39* (28), 8315–8324.
70. D.E. Cane, Q. Xue, B.C. Fitzsimons, *Biochemistry.* **1996**, *35* (38), 12369–12376.
71. M.J. Rynkiewicz, D.E. Cane, D.W. Christianson, *Proc. Natl Acad. Sci. U. S. A.* **2001**, *98* (24), 13543–13548.
72. P. Falkowski, D.A. Kiefer, *J. Plankton Res.* **1985**, *7* (5), 715–731.
73. A. Stirbet, D. Lazár, Y. Guo, G. Govindjee, *Ann. Bot.* **2020**, *126* (4), 511–537.
74. R.D. Willows, *Nat. Prod. Rep.* **2003**, *20* (3), 327–341.
75. A.M. Hofer, *J. Cell Sci.* **2005**, *118* (5), 855–862.
76. M. Brini, E. Carafoli, *Cold Spring Harb. Perspect. Biol.* **2011**, *3* (2).
77. C. Toyoshima, M. Nakasako, H. Nomura, H. Ogawa, *Nature.* **2000**, *405* (6787), 647–655.
78. M. Endo, *J. Pharmacol. Sci.* **2006**, *100* (5), 519–524.
79. D.A. Fruman, H. Chiu, B.D. Hopkins, S. Bagrodia, L.C. Cantley, R.T. Abraham, *Cell.* **2017**, *170* (4), 605–635.
80. M. Zhang, C. Abrams, L. Wang, A. Gizzi, L. He, R. Lin, Y. Chen, P.J. Loll, J.M. Pascal, J.F. Zhang, *Structure.* **2012**, *20* (5), 911–923.

81. J.W. Noble, R. Almalki, S.M. Roe, A. Wagner, R. Duman, J.R. Atack, *Acta Crystallogr. D Struct. Biol.* **2018**, *74* (10), 1008–1014.
82. A.J. Camp, R. Wijesinghe, *Int. J. Biochem. Cell Biol.* **2009**, *41* (11), 2118–2121.
83. Q. Wang, M. Michalak, Calsequestrin, *Cell Calcium.* **2020**, *90*, 102242.
84. V. Gerke, S.E. Moss, *Physiol. Rev.* **2002**, *82* (2), 331–371.
85. M.A. Lizarbe, J.I. Barrasa, N. Olmo, F. Gavilanes, J. Turnay, *Int. J. Mol. Sci.* **2013**, *14* (2), 2652–2683.
86. P. Raynal, H.B. Pollard, *Biochim. Biophys. Acta.* **1994**, *1197* (1), 63–93.
87. *Overview of Calcium*, in *Dietary Reference Intakes for Calcium and Vitamin D*, A. C. Ross, C. L. Taylor, A. L. Yaktine et al., Eds, National Academies Press, Washington (DC), **2011**, 35–75.
88. P. Rowe, A. Koller, S. Sharma. *Physiology, Bone Remodeling*, in *StatPearls*, StatPearls Publishing, Treasure Island, FL, **2023**.
89. L. Song, *Adv. Clin. Chem.* **2017**, *82*, 1–46.
90. M. Henze, Untersuchungen über das Blut der Ascidien. I. Mitteilung, *Biol. Chem.* **1911**, *72* (5–6), 494–501.
91. H. Michibata, T. Uyama, T. Ueki, K. Kanamori, *Microsc. Res. Tech.* **2002**, *56* (6), 421–434.
92. T. Ueki, T. Adachi, S. Kawano, M. Aoshima, N. Yamaguchi, K. Kanamori, H. Michibata, *Biochim. Biophys. Acta.* **2003**, *1626* (1–3), 43–50.
93. T. Hamada, M. Asanuma, T. Ueki, F. Hayashi, N. Kobayashi, S. Yokoyama, H. Michibata, H. Hirota, *J. Am. Chem. Soc.* **2005**, *127* (12), 4216–4222.
94. D. Rehder, *J. Inorg. Biochem.* **2000**, *80* (1–2), 133–136.
95. D. Sippel, O. Einsle, *Nat. Chem. Biol.* **2017**, *13* (9), 956–960.
96. M. Rohde, C. Trncik, D. Sippel, S. Gerhardt, O. Einsle, *J. Biol. Inorg. Chem.* **2018**, *23* (7), 1049–1056.
97. J.M. Winter, B.S. Moore, *J. Biol. Chem.* **2009**, *284* (28), 18577–18581.
98. A. Butler, J.N. Carter-Franklin, *Nat. Prod. Rep.* **2004**, *21* (1), 180–188.
99. C. Sparacino-Watkins, J.F. Stolz, P. Basu, *Chem. Soc. Rev.* **2014**, *43* (2), 676–706.
100. A.N. Antipov, N.N. Lyalikova, T.V. Khijniak, N.P. L'vov, *IUBMB Life.* **2000**, *50* (1), 39–42.
101. A.N. Antipov, D.Y. Sorokin, N.P. L'Vov, J.G. Kuenen, *Biochem. J.* **2003**, *369* (1), 185–189.
102. J.B. Vincent, *Biol. Trace Elem. Res.* **2015**, *166* (1), 7–12.
103. J.B. Vincent, *J. Nutr.* **2000**, *130* (4), 715–718.
104. C.M. Davis, J.B. Vincent, *Biochemistry.* **1997**, *36* (15), 4382–4385.
105. R. Hille, J. Hall, P. Basu, *Chem. Rev.* **2014**, *114* (7), 3963–4038.
106. L.C. Seefeldt, B.M. Hoffman, J.W. Peters, S. Raugei, D.N. Beratan, E. Antony, D.R. Dean, *Acc. Chem. Res.* **2018**, *51* (9), 2179–2186.
107. L.C. Seefeldt, Z.Y. Yang, D.A. Lukoyanov, D.F. Harris, D.R. Dean, S. Raugei, B.M. Hoffman, *Chem. Rev.* **2020**, *120* (12), 5082–5106.
108. C. Enroth, B.T. Eger, K. Okamoto, T. Nishino, T. Nishino, E.F. Pai, *Proc. Natl Acad. Sci.* USA **2000**, *97* (20), 10723–10728.
109. N. Dalbeth, T.R. Merriman, L.K. Stamp, Gout, *Lancet.* **2016**, *388* (10055), 2039–2052.
110. M. Montefiori, F.S. Jorgensen, L. Olsen, *ACS Omega.* **2017**, *2* (8), 4237–4244.
111. C. Kisker, H. Schindelin, A. Pacheco, W.A. Wehbi, R.M. Garrett, K.V. Rajagopalan, J.H. Enemark, D.C. Rees, *Cell.* **1997**, *91* (7), 973–983.
112. E. Karakas, H.L. Wilson, T.N. Graf, S. Xiang, S. Jaramillo-Busquets, K.V. Rajagopalan, C. Kisker, *J. Biol. Chem.* **2005**, *280* (39), 33506–33515.
113. F. Schneider, J. Löwe, R. Huber, H. Schindelin, C. Kisker, J. Knäblein, *J. Mol. Biol.* **1996**, *263* (1), 53–69.

114. A.M. Appel, J.E. Bercaw, A.B. Bocarsly, H. Dobbek, D.L. DuBois, M. Dupuis, J.G. Ferry, E. Fujita, R. Hille, P.J.A. Kenis, C.A. Kerfeld, R.H. Morris, C.H.F. Peden, A.R. Portis, S.W. Ragsdale, T.B. Rauchfuss, J.N.H. Reek, L.C. Seefeldt, R.K. Thauer, G.L. Waldrop, *Chem. Rev.* **2013**, *113* (8), 6621–6658.
115. M. Jormakka, B. Byrne, S. Iwata, *Curr. Opin. Struct. Biol.* **2003**, *13* (4), 418–423.
116. L.E. Bevers, P.-L. Hagedoorn, W.R. Hagen, *Coord. Chem. Rev.* **2009**, *253* (3), 269–290.
117. P.M. Kroneck, *J. Biol. Inorg. Chem.* **2016**, *21* (1), 29–38.
118. G.B. Seiffert, G.M. Ullmann, A. Messerschmidt, B. Schink, P.M.H. Kroneck, O. Einsle, *Proc. Natl Acad. Sci.* USA **2007**, *104* (9), 3073–3077.
119. M. Boll, G. Fuchs, *Eur. J. Biochem.* **1995**, *234* (3), 921–933.
120. O. Tiedt, J. Fuchs, W. Eisenreich, M. Boll, *J. Biol. Chem.* **2018**, *293* (26), 10264–10274.
121. W. Zhu, N.G.J. Richards, *Essays Biochem.* **2017**, *61* (2), 259–270.
122. C.M. Grunenwald, J.E. Choby, L.J. Juttukonda, W.N. Beavers, A. Weiss, V.J. Torres, E.P. Skaar, *mBio.* **2019**, *10* (1).
123. R. Zeinert, E. Martinez, J. Schmitz, K. Senn, B. Usman, V. Anantharaman, L. Aravind, L.S. Waters, *J. Biol. Chem.* **2018**, *293* (15), 5715–5730.
124. A.K. Holley, V. Bakthavatchalu, J.M. Velez-Roman, D.K. St Clair, *Int. J. Mol. Sci.* **2011**, *12* (10), 7114–7162.
125. F. Mus, B.J. Eilers, A.B. Alleman, B.V. Kabasakal, J.N. Wells, J.W. Murray, B.P. Nocek, J.L. DuBois, J.W. Peters, *Sci. Rep.* **2017**, *7* (1), 7234.
126. K.A. Shisler, W.M. Kincannon, J.R. Mattice, J. Larson, A. Valaydon-Pillay, F. Mus, T. Flusche, A. Kumar Nath, S.A. Stoian, S. Raugei, B. Bothner, J.L. DuBois, J.W. Peters , *mBio.* **2024**, *15* (2), e0298723.
127. L. Shi, *Front. Biosci.* **2004**, *9*, 1382–1397.
128. H. Johnson, S. Narayan, A.K. Sharma, *Cancer Cell Int.* **2024**, *24* (1), 11.
129. Z. Li, L. Wang, Y. Ren, Y. Huang, W. Liu, Z. Lv, L. Qian, Y. Yu, Y. Xiong, *Cell Death Discov.* **2022**, *8* (1), 413.
130. S.G. Nonis, J. Haywood, J.W. Schmidberger, E.R.R. Mackie, T.P. Soares da Costa, C.S. Bond, J.S. Mylne, *Plant Cell.* **2021**, *33* (8), 2794–2811.
131. D. Eisenberg, H.S. Gill, G.M. Pfluegl, S.H. Rotstein, *Biochim. Biophys. Acta.* **2000**, **1477** (1–2), 122–145.
132. W.W. Krajewski, R. Collins, L. Holmberg-Schiavone, T.A. Jones, T. Karlberg, S.L. Mowbray, *J. Mol. Biol.* **2008**, *375* (1), 217–228.
133. S. Yu, S. Meng, M. Xiang, H. Ma, *Mol. Metab.* **2021**, *53*, 101257.
134. N. Zamboni, H. Maaheimo, T. Szyperski, H.P. Hohmann, U. Sauer, *Metab. Eng.* **2004**, *6* (4), 277–284.
135. S. Jitrapakdee, M. St Maurice, I. Rayment, W.W. Cleland, J.C. Wallace, P.V. Attwood, *Biochem. J.* **2008**, *413* (3), 369–387.
136. J. Herrick, B. Sclavi, *Mol. Microbiol.* **2007**, *63* (1), 22–34.
137. J.W. de Kraker, K. Luck, S. Textor, J.G. Tokuhisa, J. Gershenzon, *Plant Physiol.* **2007**, *143* (2), 970–986.
138. M. Hofrichter, *Enzyme Microb. Technol.* **2002**, *30* (4), 454–466.
139. J.W. Whittaker, *Arch. Biochem. Biophys.* **2012**, *525* (2), 111–120.
140. N. Cox, D.A. Pantazis, F. Neese, W. Lubitz, *Acc. Chem. Res.* **2013**, *46* (7), 1588–1596.
141. Y. Umena, K. Kawakami, J.R. Shen, N. Kamiya, *Nature.* **2011**, *473* (7345), 55–60.
142. R. Gupta, *Plant Signal. Behav.* **2020**, *15* (12), 1824721.
143. C. Andreini, V. Putignano, A. Rosato, L. Banci, *Metallomics.* **2018**, *10* (9), 1223–1231.
144. H. Kawabata, *Free Radic. Biol. Med.* **2019**, *133*, 46–54.
145. W. Wang, M.A. Knovich, L.G. Coffman, F.M. Torti, S.V. Torti, *Biochim. Biophys. Acta.* **2010**, *1800* (8), 760–769.
146. P.F. Levay, M. Viljoen, *Haematol.* **1995**, *80* (3), 252–267.

147. J.B. van Beilen, M. Neuenschwander, T.H.M. Smits, C. Roth, S.B. Balada, B. Witholt, *J. Bacteriol.* **2002**, *184* (6), 1722–1732.
148. R.S. Peixoto, A.B. Vermelho, A.S. Rosado, *Enzyme Res.* **2011**, *2011*, 475193.
149. T.D. Bugg, S. Ramaswamy, *Curr. Opin. Chem. Biol.* **2008**, *12* (2), 134–140.
150. J. Kramer, O. Ozkaya, R. Kummerli, *Nat. Rev. Microbiol.* **2020**, *18* (3), 152–163.
151. A. Ito, T. Nishikawa, S. Matsumoto, H. Yoshizawa, T. Sato, R. Nakamura, M. Tsuji, Y. Yamano, *Antimicrob. Agents Chemother.* **2016**, *60* (12), 7396–7401.
152. H. Nishi, R. Inagi, H. Kato, M. Tanemoto, I. Kojima, D. Son, T. Fujita, M. Nangaku, *J. Am. Soc. Nephrol.* **2008**, *19* (8), 1500–1508.
153. M.I. Ahmad, S. Farooq, Y. Alhamoud, C. Li, H. Zhang, *Trends Food Sci. Technol.* **2023**, *141*, 104199.
154. A.H. Burr, R. Schiefke, G. Bollerup, *Biochim. Biophys. Acta.* **1975**, *405* (2), 404–411.
155. M.F. Perutz, M.G. Rossmann, A.F. Cullis, H. Muirhead, G. Will, A.C. North, *Nature.* **1960**, *185* (4711), 416–422.
156. G.A. Ordway, D.J. Garry, *J. Exp. Biol.* **2004**, *207* (20), 3441–3446.
157. C.J. Reedy, B.R. Gibney, *Chem. Rev.* **2004**, *104* (2), 617–649.
158. Z. Xie, D. Culler, B.W. Dreyfuss, R. Kuras, F.A. Wollman, J. Girard-Bascou, S. Merchant, *Genetics.* **1998**, *148* (2), 681–692.
159. A. Nandi, L.J. Yan, C.K. Jana, N. Das, *Oxid. Med. Cell. Longev.* **2019**, 9613090.
160. T. Hino, Y. Matsumoto, S. Nagano, H. Sugimoto, Y. Fukumori, T. Murata, S. Iwata, Y. Shiro, *Science.* **2010**, *330* (6011), 1666–1670.
161. C.M. Cordas, A.G. Duarte, J.J.G. Moura, I. Moura, *Biochim. Biophys. Acta.* **2013**, **1827** (3), 233–238.
162. H. Beinert, R.H. Holm, E. Munck, *Science.* **1997**, *277* (5326), 653–659.
163. E. Camprubi, S.F. Jordan, R. Vasiliadou, N. Lane, *IUBMB Life.* **2017**, *69* (6), 373–381.
164. J.W. Peters, W.N. Lanzilotta, B.J. Lemon, L.C. Seefeldt, *Science.* **1998**, *282* (5395), 1853–1858.
165. J.H. Artz, O.A. Zadvornyy, D.W. Mulder, S.M. Keable, A.E. Cohen, M.W. Ratzloff, S.G. Williams, B. Ginovska, N. Kumar, J. Song, S.E. McPhillips, C.M. Davidson, A.Y. Lyubimov, N. Pence, G.J. Schut, A.K. Jones, S.M. Soltis, M.W.W. Adams, S. Raugei, P.W. King, J.W. Peters, *J. Am. Chem. Soc.* **2020**, *142* (3), 1227–1235.
166. D.W. Mulder, J.W. Peters, S. Raugei, *Chem. Commun. (Camb.).* **2021**, *57* (6), 713–720.
167. J.B. Therien, J.H. Artz, S. Poudel, T.L. Hamilton, Z. Liu, S.M. Noone, M.W.W. Adams, P.W. King, D.A. Bryant, E.S. Boyd, J.W. Peters, *Front. Microbiol.* **2017**, *8*, 1305.
168. F.V. Schmidt, L. Schulz, J. Zarzycki, S. Prinz, N.N. Oehlmann, T.J. Erb, J.G. Rebelein, *Nat. Struct. Mol. Biol.* **2024**, *31* (1), 150–158.
169. C. Trncik, F. Detemple, O. Einsle, *Nature Catalysis* **2023**, *6* (5), 415–424.
170. N.N. Oehlmann, F.V. Schmidt, M. Herzog, A.L. Goldman, J.G. Rebelein, *Sci. Adv.* **2024**, *10* (33): p, eado7729.
171. M. Kobayashi, S. Shimizu, *Eur. J. Biochem.* **1999**, *261* (1), 1–9.
172. D.S. Conklin, J.A. McMaster, M.R. Culbertson, C. Kung, *Mol. Cell. Biol.* **1992**, *12* (9), 3678–3688.
173. D.C. Hodgkin, J. Kamper, M. Mackay, J. Pickworth, K.N. Trueblood, J.G. White, *Nature.* **1956**, *178* (4524), 64–66.
174. R. Oh, D.L. Brown, *Am. Fam. Physician.* **2003**, *67* (5), 979–986.
175. K.A.P. Payne, C.P. Quezada, K. Fisher, M.S. Dunstan, F.A. Collins, H. Sjuts, C. Levy, S. Hay, S.E. Rigby, D. Leys, *Nature.* **2015**, *517* (7535), 513–516.
176. W.T. Lowther, D.A. McMillen, A.M. Orville, B.W. Matthews, *Proc. Natl Acad. Sci.* USA **1998**, *95* (21), 12153–12157.
177. Z. Cheng, Y. Xia, Z. Zhou, *Front. Bioeng. Biotechnol.* **2020**, *8*, 352.
178. P. Wilk, E. Wator, M.S. Weiss, *Biochimie.* **2021**, *183*, 3–12.

179. A. Lupi, R. Tenni, A. Rossi, G. Cetta, A. Forlino, *Amino Acids.* **2008**, *35* (4), 739–752.
180. M. Alfano, C. Cavazza, *Protein Sci.* **2020**, *29* (5), 1071–1089.
181. E. Jabri, M.B. Carr, R.P. Hausinger, P.A. Karplus, *Science.* **1995**, *268* (5213), 998–1004.
182. D. Strugatsky, R. McNulty, K. Munson, C.K. Chen, S.M. Soltis, G. Sachs, H. Luecke, *Nature.* **2013**, *493* (7431), 255–258.
183. H. Chen, Q. Gan, C. Fan, *Front. Microbiol.* **2020**, *11*, 578356.
184. A. Gendron, K.D. Allen, *Front. Microbiol.* **2022**, *13*, 867342.
185. T.A. Dinh, K.D. Allen, *Acc. Chem. Res.* **2024**.
186. B. Desguin, T. Zhang, P. Soumillion, P. Hols, J. Hu, R.P. Hausinger, *Science.* **2015**, *349* (6243), 66–69.
187. S.L. Chen, P.E.M. Siegbahn, *Inorg. Chem.* **2020**, *59* (20), 15167–15179.
188. M. Merrouch, M. Benvenuti, M. Lorenzi, C. Léger, V. Fourmond, S. Dementin, *J. Biol. Inorg. Chem. JBIC Publ. Soc. Biol. Inorg. Chem.* **2018**, *23* (4), 613–620.
189. P.S. Adam, G. Borrel, S. Gribaldo, *Proc. Natl Acad. Sci.* USA **2018**, *115* (6), E1166–1173.
190. M.E. Pandelia, W. Nitschke, P. Infossi, M.T. Giudici-Orticoni, E. Bill, W. Lubitz, *Proc. Natl Acad. Sci.* USA **2011**, *108* (15), 6097–6102.
191. H. Ogata, W. Lubitz, Y. Higuchi, *J. Biochem.* **2016**, *160* (5), 251–258.
192. E.T. Adman, *Adv. Protein Chem.* **1991**, *42*, 145–197.
193. T. Tsukihara, H. Aoyama, E. Yamashita, T. Tomizaki, H. Yamaguchi, K. Shinzawa-Itoh, R. Nakashima, R. Yaono, S. Yoshikawa, *Science.* **1995**, *269* (5227), 1069–1074.
194. I.C. Soto, F. Fontanesi, J. Liu, A. Barrientos, *Biochim. Biophys. Acta.* **2012**, *1817* (6), 883–897.
195. M. Brischigliaro, M. Zeviani, *Biochim. Biophys. Acta Bioenerg.* **2021**, **1862** (1), 148335.
196. F. De Rienzo, R.R. Gabdoulline, M.C. Menziani, R.C. Wade, *Protein Sci.* **2000**, *9* (8), 1439–1454.
197. J. Hu, W. Jiang, J. Zuo, D. Shi, X. Chen, X. Yang, W. Zhang, L. Ma, Z. Liu, Q. Xing, *Commun. Biol.* **2023**, *6* (1), 59.
198. A.G. Sykes. In *Long-Range Electron Transfer in Biology.* Springer, Berlin, Heidelberg, **1991**.
199. C.J. Coates, E.M. Costa-Paiva, *Subcell. Biochem.* **2020**, *94*, 233–250.
200. Y. Matoba, T. Kumagai, A. Yamamoto, H. Yoshitsu, M. Sugiyama, *J. Biol. Chem.* **2006**, *281* (13), 8981–8990.
201. H.S. Raper, *Physiol. Rev.* **1928**, *8* (2), 245–282.
202. H.S. Mason, *J. Biol. Chem.* **1948**, *172* (1), 83–99.
203. A.M. Mayer, *Phytochemistry.* **2006**, *67* (21), 2318–2331.
204. T.V. Vendelboe, P. Harris, Y. Zhao, T.S. Walter, K. Harlos, K. El Omari, H.E.M. Christensen, *Sci. Adv.* **2016**, *2* (4), e1500980.
205. S. Kumari, T.K. Panda, T. Pradhan, *Indian J. Clin. Biochem.* **2017**, *32* (2), 134–141.
206. J.W. Whittaker, *Chem. Rev.* **2003**, *103* (6), 2347–2363.
207. M.O. Ross, F. MacMillan, J. Wang, A. Nisthal, T.J. Lawton, B.D. Olafson, S.L. Mayo, A.C. Rosenzweig, B.M. Hoffman, *Science.* **2019**, *364* (6440), 566–570.
208. R.J. Arias, E.F. Welch, N.J. Blackburn, *Protein Sci.* **2023**, *32* (4), e4615.
209. C.M. Chang, V.J. Klema, B.J. Johnson, M. Mure, J.P. Klinman, C.M. Wilmot, *Biochemistry.* **2010**, *49* (11), 2540–2550.
210. N.E. Hellman, J.D. Gitlin, *Annu. Rev. Nutr.* **2002**, *22*, 439–458.
211. Y. Yang, W. Hao, T. Wei, L. Tang, N. Qian, Y. Yang, H. Xi, S. Zhang, W. Yang, *Front. Neurol.* **2022**, *13*, 1058642.
212. M. Arredondo, M. González, M. Olivares, F. Pizarro, M. Araya, *Biol. Trace Elem. Res.* **2008**, *123* (1–3), 261–269.

213. A. Pomowski, W.G. Zumft, P.M.H. Kroneck, O. Einsle, *Nature.* **2011**, *477* (7363), 234–237.
214. N. Vita, S. Platsaki, A. Baslé, S.J. Allen, N.G. Paterson, A.T. Crombie, J.C. Murrell, K.J. Waldron, C. Dennison, *Nature.* **2015**, *525* (7567), 140–143.
215. A.G. Tebo, V.L. Pecoraro, *Curr. Opin. Chem. Biol.* **2015**, *25*, 65–70.
216. B.P. Jagilinki, S. Ilic, C. Trncik, A.M. Tyryshkin, D.H. Pike, W. Lubitz, E. Bill, O. Einsle, J.A. Birrell, B. Akabayov, D. Noy, V. Nanda, *ACS Synth. Biol.* **2020**.
217. W.B. Jensen, *J. Chem. Edu.* **2003**, *80* (8), 952.
218. J.E. Coleman, *Annu. Rev. Biochem.* **1992**, *61*, 897–946.
219. M. Cassandri, A. Smirnov, F. Novelli, C. Pitolli, M. Agostini, M. Malewicz, G. Melino, G. Raschellà, *Cell Death Discov.* **2017**, *3*, 17071.
220. K. Kim, B.V. Plapp, *Biochemistry.* **2020**, *59* (7), 862–879.
221. Y. Matsumura, K.M. Stiles, J. Reid, E.Z. Frenk, S. Cronin, O.E. Pagovich, R.G. Crystal, *Mol. Ther. Methods Clin. Dev.* **2019**, *15*, 72–82.
222. *Carbonic Anhydrase VA Deficiency.* **1993**. University of Washington, Seattle (WA), (USA).
223. C.T. Supuran, *Nat. Rev. Drug Discov.* **2008**, *7* (2), 168–181.
224. A. Garcia-Llorca, F. Carta, C.T. Supuran, T. Eysteinsson, *Front. Mol. Biosci.* **2024**, *11*, 1338528.
225. U. Sharma, D. Pal, R. Prasad, *Indian J. Clin. Biochem.* **2014**, *29* (3), 269–278.
226. J. Verma, D.A. Gorard, *BMJ Case Rep.* **2012**, *2012*, bcr2012006768.
227. C.M. Joyce, T.A. Steitz, *Annu. Rev. Biochem.* **1994**, *63*, 777–822.
228. F. Werner, D. Grohmann, *Nat. Rev. Microbiol.* **2011**, *9* (2), 85–98.
229. A.J. te Velthuis, S.H.E. van den Worm, A.C. Sims, R.S. Baric, E.J. Snijder, M.J. van Hemert, *PLOS Pathog.* **2010**, *6* (11), e1001176.
230. E. Seto, M. Yoshida, *Cold Spring Harb. Perspect. Biol.* **2014**, *6* (4), a018713.
231. F.A. Quiocho, W.N. Lipscomb, *Adv. Protein Chem.* **1971**, *25*, 1–78.
232. I. Fleming, *Circ. Res.* **2006**, *98* (7), 887–896.
233. C. Zhang, S. Wu, D. Xu, *J. Phys. Chem. B.* **2013**, *117* (22), 6635–6645.
234. A.A. Patchett, E. Harris, E.W. Tristram, M.J. Wyvratt, M.T. Wu, D. Taub, E.R. Peterson, T.J. Ikeler, J. ten Broeke, L.G. Payne, D.L. Ondeyka, E.D. Thorsett, W.J. Greenlee, N.S. Lohr, R.D. Hoffsommer, H. Joshua, W.V. Ruyle, J.W. Rothrock, S.D. Aster, A.L. Maycock, F.M. Robinson, R. Hirschmann, C.S. Sweet, E.H. Ulm, D.M. Gross, T.C. Vassil, C.A. Stone, *Nature* **1980**, *288* (5788), 280–283.
235. M. Tajima, I. Urabe, K. Yutani, H. Okada, *Eur. J. Biochem.* **1976**, *64* (1), 243–247.
236. R.P. Verma, C. Hansch, *Bioorg. Med. Chem.* **2007**, *15* (6), 2223–2268.
237. S. Loffek, O. Schilling, C.W. Franzke, Series, *Eur. Respir. J.* **2011**, *38* (1), 191–208.
238. M. Si, J. Lang, *J. Hematol. Oncol.* **2018**, *11* (1), 107.
239. M.O. Pedersen, A. Larsen, M. Stoltenberg, M. Penkowa, *Prog. Histochem. Cytochem.* **2009**, *44* (1), 29–64.
240. B.H. Bin, J. Seo, S.T. Kim, *J. Immunol. Res.* **2018**, *2018*, 9365747.
241. E. Bafaro, Y. Liu, Y. Xu, R.E. Dempski, *Signal Transduct. Target. Ther.* **2017**, *2*, 17029.
242. L.J. Daumann, A. Pol, H.J.M. Op den Camp, N.C. Martinez-Gomez, *Adv. Microb. Physiol.* **2022**, *81*, 1–24.
243. S.M. Krause, T. Johnson, Y. Samadhi Karunaratne, Y. Fu, D.A.C. Beck, L. Chistoserdova, M.E. Lidstrom, *Proc. Natl Acad. Sci.* USA **2017**, *114* (2), 358–363.

244. J.A. Mattocks, J.J. Jung, C.Y. Lin, Z. Dong, N.H. Yennawar, E.R. Featherston, C.S. Kang-Yun, T.A. Hamilton, D.M. Park, A.K. Boal, J.A. Cotruvo, *Nature.* **2023**, *618* (7963), 87–93.
245. N.M. Good, M. Fellner, K. Demirer, J. Hu, R.P. Hausinger, N.C. Martinez-Gomez, *J. Biol. Chem.* **2020**, *295* (24), 8272–8284.
246. A. Pol, T.R.M. Barends, A. Dietl, A.F. Khadem, J. Eygensteyn, M.S.M. Jetten, H.J.M. Op den Camp, *Environ. Microbiol.* **2014**, *16* (1), 255–264.
247. H. Singer, R. Steudtner, A.S. Klein, C. Rulofs, C. Zeymer, B. Drobot, A. Pol, N. Cecilia Martinez-Gomez, H.J.M. Op den Camp, L.J. Daumann, *Angew. Chem. Int. Ed. Engl.* **2023**, *62* (31), e202303669.

4 Trace Metals as Redox Cofactors and Their Role in Controlling Microbial Diversity and Evolution

*Donato Giovannelli**
Department of Biology, University of Naples Federico II, Naples, Italy
Marine Chemistry & Geochemistry Department - Woods Hole Oceanographic Institution, MA, USA
Institute for Marine Biological and Biotechnological Resources, National Research Council of Italy (CNR-IRBIM), Ancona, Italy
Department of Marine and Coastal Science, Rutgers University, New Brunswick, NJ, USA
Earth-Life Science Institute, ELSI, Tokyo Institute of Technology, Tokyo, Japan

Martina Cascone, Gabriella Gallo, Flavia Migliaccio, Bernardo Barosa and Deborah Bastoni
Department of Biology, University of Naples Federico II, Naples, Italy

Jacopo Brusca
Department of Biology, University of Naples Federico II, Naples, Italy
Department of Environmental Science, Informatics and Statistics, University of Venice Ca' Foscari, Venice, Italy

Antonio Longo
Department of Biology, University of Naples Federico II, Naples, Italy

* Corresponding author

DOI: 10.1201/9781003459910-4

Department of Integrative Marine Ecology, Stazione Zoologica Anton Dohrn, Villa Comunale, 80121, Naples, Italy

Luca Tonietti
Department of Biology, University of Naples Federico II, Naples, Italy

Department of Science and Technology, University of Naples Parthenope, Naples, Italy

INAF-OAC, Osservatorio Astronomico di Capodimonte, Naples, Italy

Angelina Cordone
Department of Biology, University of Naples Federico II, Naples, Italy

CONTENTS

ABSTRACT

Over the past ~4 billion years of Earth's history, the geosphere and biosphere have coevolved, influencing each other's evolutionary trajectories and maintaining our planet's habitability. Central to this interaction are the biogeochemical cycles, linking long-term geological reservoirs with the rapid evolution of Earth's biologically dominated outer layers. At the core of these cycles are microbial-encoded proteins that employ redox-sensitive transition metals, such as Fe, Co, Ni, Zn, Mo, W, V, and Cu, as catalytic centers. These proteins allow life to exploit thermodynamic disequilibria in the form of redox disequilibria, with metal cofactors playing a crucial role in precisely mediating redox reactions. Despite the significance of these processes, the relationship between metal availability and the evolution of metabolic diversity has not been thoroughly explored. This chapter reviews the influence of organometallic-enzymatic complexes on metabolism and microbial diversity, emphasizing their critical role in biogeochemical cycles. While metals and inorganic cations are known to be fundamental in biology, interacting with a large number of enzymes, we focus here on a subset of metal-containing enzymes directly interoperating with small inorganic molecules exchangeable with the environment, and discuss their importance in extant ecosystems and their possible role in the emergence and evolution of life. Understanding the impact of metals availability on microbial functional diversity is key to unraveling the coevolution of life and Earth.

KEYWORDS

trace metals, oxidoreductases, cofactors, microbial metabolism, microbial diversity, biogeochemistry, evolution, origin of life

1 INTRODUCTION

Life, at the most fundamental level, is electric [1]. The thermodynamic disequilibrium needed by life to survive and thrive is redox disequilibrium, represented by the transfer of electrons between an electron donor and an electron acceptor. The decoupling of the flow of electrons and protons obtained from redox reactions allows living systems to create a chemiosmotic gradient, control reducing equivalents, and ultimately carry out chemical and mechanical work [1]. By keeping electrons separate from protons, living systems effectively convert a scalar process (a redox reaction) into a vectorial process that has both directionality and magnitude (Figure 1). In essence, the basic energetics of life works as a Volta electrical battery that naturally evolved nearly 4 billion years ago [2]. The foundational work of Amend and Shock et al. [3–5] has shown the availability of hundreds of both demonstrated and theoretical spontaneous redox reactions (with negative Gibbs free energy values under in situ conditions) that can be utilized by living systems to derive energy. In fact, life, especially microbial life, is able to colonize the vast majority of our planet due to the plethora of redox pairs, having a fundamental role in the geochemical cycles of major and several minor elements [6].

Life's ability to mediate electron flow from a donor to a terminal acceptor through a long series of intermediate carriers is fundamental for metabolism and life existence. Throughout the course of evolution, life has evolved complex enzymatic machines that are able to efficiently and precisely control and mediate redox reactions [2,6,7]. These life engines (*sensu* [6]) are enzymes, named oxidoreductases, that often contain a metal cofactor at their catalytic center [1] (Figure 2 and Table 2).

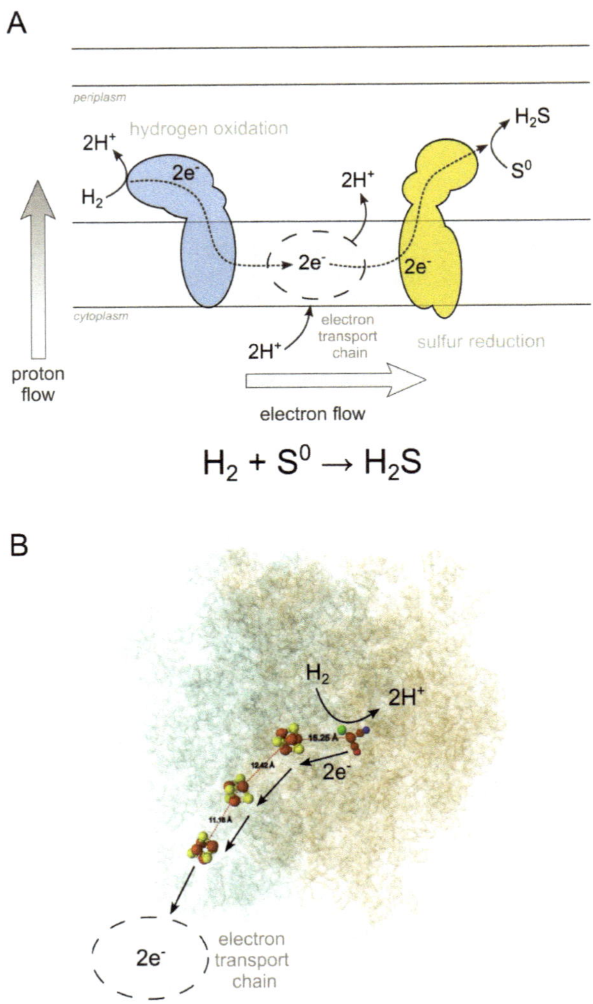

FIGURE 1 Life depends on trace metals for controlling energy conserving reactions. A. Life's energy conserving reactions decouple redox half-reactions diversifying the flux of electrons from protons to perform work. B. Examples of an oxidoreductase protein (NiFe Hydrogenase from *Nitratidesulfovibrio vulgaris* str. Miyazaki F, PDB accession number 4U9H) and its metal centers used to transport electrons and effectively control redox chemistry.

Given the crucial role of electron transfer reactions, the relevance of metal ions for the function of living systems becomes apparent. For instance, trace metals, such as Fe, Co, Ni, Mo, W, V, and Cu, are consistently used by enzymes to mediate electron transfer as catalytic cofactors, whereas metals such as Zn and Mg, are associated with structural roles [1] (Figure 3). In addition to oxidoreductases, a multitude of other enzymes use metals as primary cofactors in their catalytic centers, to stabilize either the substrate or reaction intermediates, or the enzyme itself.

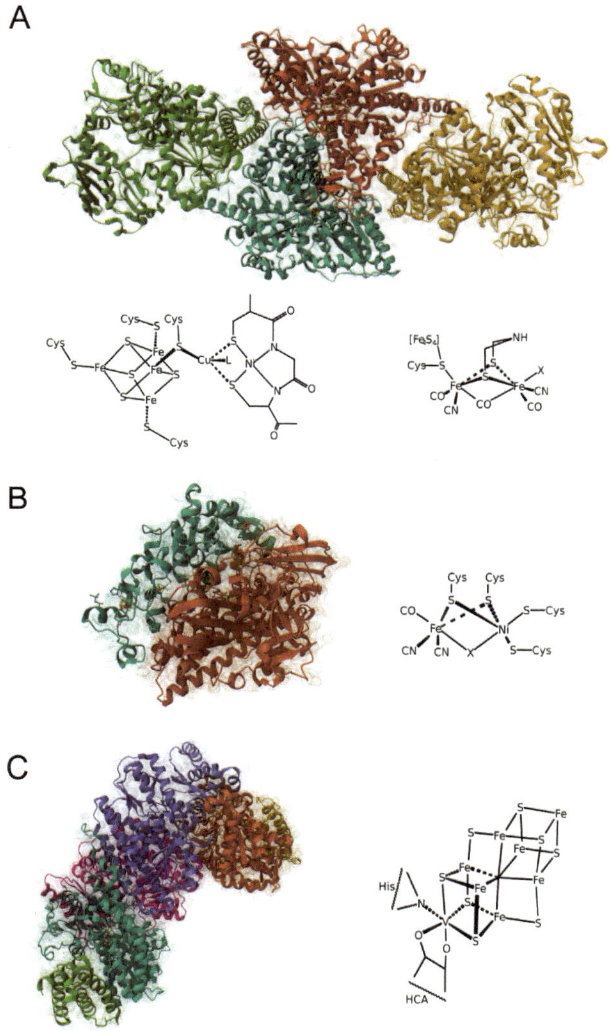

FIGURE 2 Examples of metal-containing oxidoreductases and their metal cofactors. A. The bifunctional anaerobic carbon monoxide dehydrogenase/acetyl-CoA synthase CodH/Acs (PDB 1MJG), containing the Ni-Fe-Cu center (cluster A) and the Fe-[NiFe3S4] cofactor (cluster C) and responsible for the last step of the Wood-Ljungdahl carbon fixation pathway. B. The hydrogen oxidizing hydrogenase, containing the NiFe cluster (PDB 5XLF). C. The V-containing nitrogenase (PDB 5N6Y), containing the FeVco cofactor responsible for nitrogen fixation. All these oxidoreductases represent key enzymes in metabolic reactions that are considered ancestral and important in the early stages of life emergence and evolution.

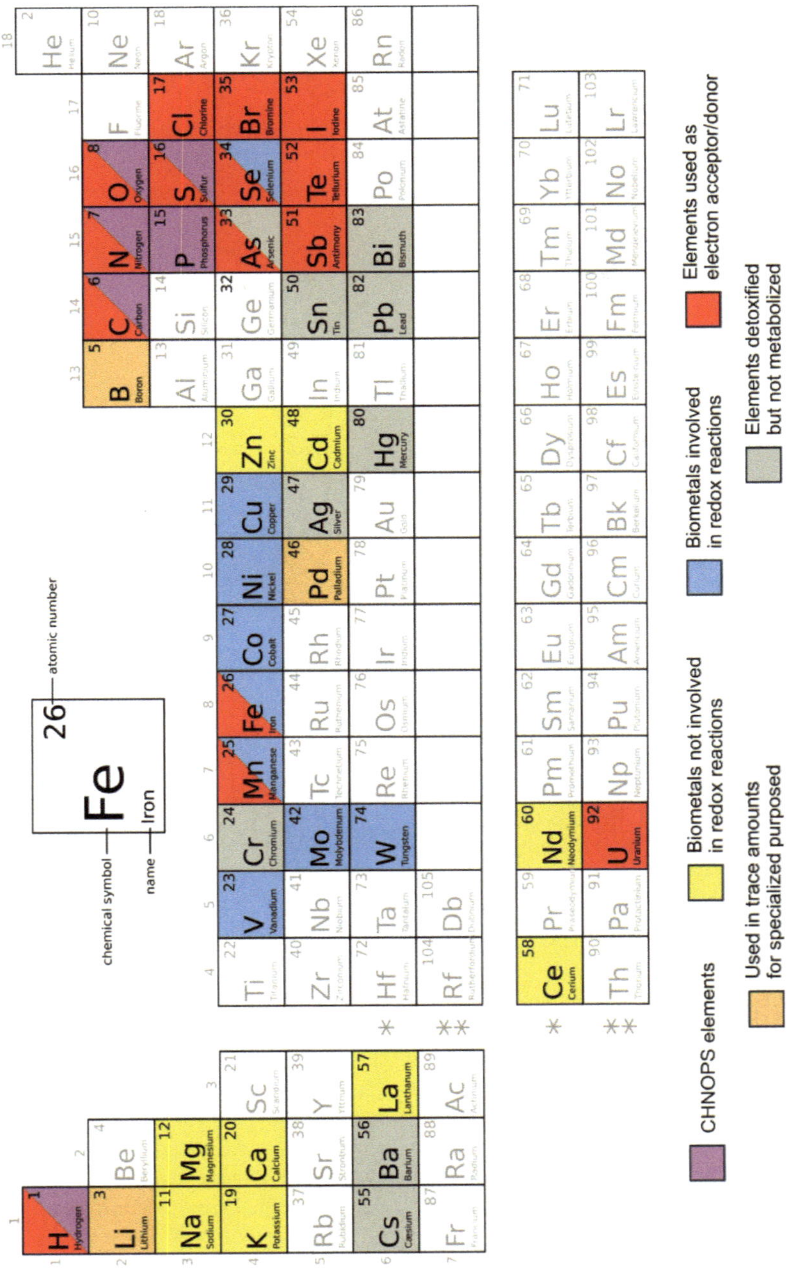

FIGURE 3 Periodic table of biometals colored according to their utilization in biology. Adapted from [116].

TABLE 1
List of Elements Interacting with Life Divided by Their Primary Role in Biology

Symbol	Name	Atomic Number	Oxidation States in Biology	Modern Ocean (%)	Biological Role
CHNOPS elements					
H	Hydrogen	1	+1	11	CHNOPS element, Electron donor
C	Carbon	6	−4, +2, +4	2.8×10^{-3}	CHNOPS element, Electron donor, Electron acceptor
N	Nitrogen	7	−3, +3, +5	5.0×10^{-5}	CHNOPS element, Electron donor, Electron acceptor
O	Oxygen	8	−2	86	CHNOPS element, Electron acceptor
P	Phosphorus	15	+3, +5	7.0×10^{-6}	CHNOPS element, Electron donor
S	Sulfur	16	−2, 0, +2, +4, +6	9.3×10^{-2}	CHNOPS element, Electron donor, Electron acceptor
Essential biometal					
Na	Sodium	11	+1	1.10	Structural
Mg	Magnesium	12	+2	0.13	Catalyst, Structural
K	Potassium	19	+1	4.2×10^{-2}	Structural
Ca	Calcium	20	+2	4.2×10^{-4}	Structural, Biomineralization
V	Vanadium	23	+3, +4, +5	1.5×10^{-7}	Catalyst, RedOx, Electron acceptor
Mn	Manganese	25	+2, +3, +4	2.0×10^{-7}	Structural, Catalyst, RedOx, Electron donor, Electron acceptor
Fe	Iron	26	+2, +3	3.0×10^{-7}	Catalyst, RedOx, Electron donor, Electron acceptor
Co	Cobalt	27	+1, +2, +3	8.0×10^{-9}	Catalyst
Ni	Nickel	28	+1, +2, +3	2.0×10^{-7}	Catalyst, RedOx
Cu	Copper	29	+1, +2	3.0×10^{-7}	Catalyst, RedOx

(*Continued*)

TABLE 1 (CONTINUED)
List of Elements Interacting with Life Divided by Their Primary Role in Biology

Symbol	Name	Atomic Number	Oxidation States in Biology	Modern Ocean (%)	Biological Role
Zn	Zinc	30	+2	5.0×10^{-7}	Structural
Se	Selenium	34	−2, +4, +6	4.5×10^{-8}	Electron acceptor
Mo	Molybdenum	42	+4, +6	1.0×10^{-6}	Catalyst
Cd	Cadmium	48	+2	5.0×10^{-9}	Structural
La	Lanthanum	57	+3	3.4×10^{-10}	Structural
Ce	Cerium	58	+3, +4	1.2×10^{-10}	Structural
Nd	Neodymium	60	+3	2.8×10^{-10}	Structural
W	Tungsten	74	+4, +6	1.2×10^{-8}	Catalyst
Other trace elements					
Li	Lithium	3	+1	1.8×10^{-5}	Used in trace amounts, bioaccumulated
B	Boron	5	+3	4.4×10^{-4}	Used in trace amounts
Other elements					
Cr	Chromium	24	+3	6.1×10^{-8}	Detoxified, involved in electron transport reaction
Br	Bromine	35	−1, +1, +5	6.7×10^{-3}	Electron acceptor
Ag	Silver	47	+1	1.0×10^{-8}	Antimicrobial properties
Sn	Tin	50	+2, +4	1.0×10^{-9}	Bioaccumulated in organotin compounds
Sb	Antimony	51	+3, +5	2.0×10^{-8}	Electron donor, Electron acceptor
Te	Tellurium	52	+4, +6	NA	Electron acceptor

(Continued)

TABLE 1 (CONTINUED)
List of Elements Interacting with Life Divided by Their Primary Role in Biology

Symbol	Name	Atomic Number	Oxidation States in Biology	Modern Ocean (%)	Biological Role
I	Iodine	53	−1, +5	6.0×10^{-6}	Electron donor, Electron acceptor
Cs	Cesium	55	+1	5.0×10^{-8}	Detoxified, Bioaccumulated
Ba	Barium	56	+2	3.0×10^{-6}	Detoxified, Bioaccumulated, biomineralization
Cl	Chlorine	17	−1, +1, +7	2.0	Electron acceptor
As	Arsenic	33	−3,+3, +5	2.0×10^{-7}	Electron donor, Electron acceptor
Hg	Mercury	80	+2	5.0×10^{-9}	Detoxified, Methylated
Pb	Lead	82	+2	3.0×10^{-9}	Detoxified, Bioaccumulated
Bi	Bismuth	83	+3	2.0×10^{-9}	Detoxified, Methylated
U	Uranium	92	+4, +6	3.3×10^{-7}	Electron acceptor

NA = not available

TABLE 2

Biogeochemically Relevant Metal-containing Oxidoreductases and the Catalytic Metal Present in Their Active Center

Cycle	Pathway	Step	Protein Name	Gene	Metal in the Active Center	EC
Carbon	Aerobic methane oxidation	Oxygenation of methane to methanol	Particulate methane monooxygenase alpha subunit	*pmoB1, pmoB2*	Cu	1.14.18.3
			(soluble) Methane monooxygenase component A beta chain	*mmoX*	Fe	1.14.13.25
	Carbon fixation	**CO_2 reduction to formate**	**Formate dehydrogenase H**	***fdhF***	**Mo**	**1.17.98.4**
		CO_2 reduction to formylmethanofuran	**Tungsten formylmethanofuran dehydrogenase subunit fwdA**	***fwdA***	**W**	**1.2.7.12**
		Oxidation of CO to CO_2	(aerobic) Carbon monoxide dehydrogenase medium chain	*coxL*	CuMo	1.2.5.3
		Reduction of CO_2 to CO	**(anaerobic) Carbon monoxide dehydrogenase/acetyl-CoA synthase subunit alpha**	***codh***	**CuNi/Ni**	**1.2.7.4**
Hydrogen	**Hydrogen oxidation**	F420 Reduction	(NiFe) F420-reducing hydrogenase, subunit alpha	*frhA*	NiFe	1.12.98.1
		H_2-respiration	**5,10-methenyltetrahydromethanopterin hydrogenase**	***hmd***	**Fe**	**1.12.98.2**
			Periplasmic [NiFe] hydrogenase large subunit	***hydB***	**NiFe / NiFeSe**	**1.12.2.1**
	Hydrogen production	H_2-production	(FeFe) Iron hydrogenase 1	*hydA*	Fe	1.12.7.2
		H_2-production/Mrp Antiporter	(NiFe) Membrane-bound hydrogenase subunit alpha	*mbhL*	NiFe	1.12.7.2

(Continued)

TABLE 2 (CONTINUED)
Biogeochemically Relevant Metal-containing Oxidoreductases and the Catalytic Metal Present in Their Active Center

Cycle	Pathway	Step	Protein Name	Gene	Metal in the Active Center	EC
Nitrogen	Anammox	From Ammonia to Hydrazine	Hydrazine synthase subunit gamma	*hzsA*	Fe	1.7.2.7
		From Hydrazine to Nitrogen	Hydrazine dehydrogenase	*hdh*	Fe	1.7.2.8
	Assimilatory nitrate reduction	From Nitrate to Nitrite	Nitrate reductase	*narB*	Mo	1.7.7.2
		From Ammonia to Nitrite	Ferredoxin-nitrite reductase	*nirA*	Fe	1.7.7.1
	Dissimilatory Nitrite Reduction	From Ammonia to nitrite	Cytochrome-c nitrite reductase subunit NrfA	*nrfA*	Fe	1.7.2.2
	Denitrification	From Nitrate to Nitrite	Respiratory nitrate reductase 1 alpha chain	*narG*	Mo	1.7.5.1
			Periplasmic nitrate reductase	*napA*	Mo	1.9.6.1
		From Nitric oxide to Nitrite	Copper-containing nitrite reductase	*nirK*	Cu	1.7.2.1
			Nitrite reductase	*nirS*	Fe	1.7.2.1
		From Nitrite to Nitrate	Nitrate reductase [NADPH]	*nasA*	Mo	1.7.7.2
		From Nitrogen to Nitrous oxide	Nitrous-oxide reductase	*nosZ*	Cu	1.7.2.4
		From Nitrous oxide to Nitric Oxide	Nitric oxide reductase	*norB*	Fe	1.7.2.5
	Nitrification	From Ammonia to Hydroxylamine	(Cupredoxin) Ammonia monxygenase beta subunit	*amoB1; amoB2*	Cu	1.14.99.39
			Hydroxylamine reductase	*hcp*	Fe	1.7.99.1
		From Ammonia to Nitrite	Nitrite reductase (NADH) large subunit	*nirB*	Fe	1.7.1.15
		From Hydroxilamine to Nitrite	Hydroxylamine oxidereductase	*hao*	Fe	1.7.2.6
		From Nitrite to Nitrate	nitrite oxidoreductase subunit A	*nrxA*	Mo	1.7.99.-
	Nitrogen fixation	**From Nitrogen to Ammonia**	**Nitrogenase iron-iron protein**	***anfD***	**Fe**	**1.18.6.1**
			Nitrogenase molybdenum-iron protein alpha chain	***nifD***	**Mo**	**1.18.6.1**
			Nitrogenase vanadium-iron protein alpha chain	***vnfD***	**V**	**1.18.6.1**

(Continued)

TABLE 2 (CONTINUED)
Biogeochemically Relevant Metal-containing Oxidoreductases and the Catalytic Metal Present in Their Active Center

Cycle	Pathway	Step	Protein Name	Gene	Metal in the Active Center	EC
Oxygen	Oxygen radicals detoxification	Hydrogen peroxide detoxification	Catalase-peroxidase	*katG*	Fe	1.11.1.21
		Oxygen detoxification	Superoxide oxidase CybB	*cybB*	Fe	1.10.3.17
		Superoxide detoxification	Superoxide dismutase [Ni]	*sodN*	Ni	1.15.1.1
			Superoxide dismutase [Cu-Zn]	*sodC*	Cu	1.15.1.1
			Superoxide dismutase [Mn]	*sodA*	Fe/Mn	1.15.1.1
			Superoxide reductase	*sorA*	Fe	1.15.1.2
	Oxygen respiration	Oxidative phosphorylation	*Cbb3-type cytochrome-c oxidase (subunit II)*	*ccoN*	Cu	7.1.1.9
			Cytochrome ba quinol oxidase subunit 1	*qoxB*	Cu	7.1.1.5
			Cytochrome bd-I ubiquinol oxidase subunit 1	*cydA*	Fe	7.1.1.7
			Putative cytochrome bd-II ubiquinol oxidase subunit AppX	*appC*	Fe	7.1.1.7
			Cytochrome bo(3) ubiquinol oxidase subunit 4	*cyoB*	Cu	7.1.1.3
			Cytochrome-*c* oxidase polypeptide I + III	*ctaD*	Cu	7.1.1.9
	Oxygenic photosynthesis	Water oxidation to oxygen	Photosystem II protein D1 1	*psbA1*	Mn	1.10.3.9
Aerobic sulfur disproportionation	From S-sulfanyl glutathione to glutathione + sulfite	Sulfur dioxygenases	*sdoA*	Fe	1.13.11.18	

(*Continued*)

TABLE 2 (CONTINUED)
Biogeochemically Relevant Metal-containing Oxidoreductases and the Catalytic Metal Present in Their Active Center

Cycle	Pathway	Step	Protein Name	Gene	Metal in the Active Center	EC
	Catalyzes the simultaneous oxidation and reduction of elemental sulfur in the presence of oxygen	Sulfur oxygenase/reductase	*sor*	Fe	1.13.11.55	
Sulfur	Aerobic sulfur disproportionation	From S-sulfanyl glutathione to glutathione + sulfite	Sulfur dioxygenases	*sdoA*	Fe	1.13.11.18
		Catalyzes the simultaneous oxidation and reduction of elemental sulfur in the presence of oxygen	Sulfur oxygenase/reductase	*sor*	Fe	1.13.11.55
	Assimilatory sulfate reduction	Reduction of sulfite to sulfide	Assimilatory sulfite reductase (ferredoxin)	*sir*	Fe	1.8.7.1
			Sulfite reductase [NADPH] hemoprotein beta-component	*cysL*	Fe	1.8.1.2
	DMSO reduction	Catalyzes the conversion of DMSO to dimethyl sulfide	Dimethyl sulfoxide/trimethylamine N-oxide reductase	*dmsA*	Mo	1.8.5.3
		DMSO reduction	Dimethylsulfide dehydrogenase subunit alpha	*ddhA*	Mo	1.8.2.4
	Sulfate reduction	Catalyzes the reduction of sulfite to sulfide	Sulfite reductase, dissimilatory-type subunit alpha	*dsrA*	Fe	1.8.99.5
	Sulfite reduction	Reduces sulfite to sulfide	Coenzyme F420-dependent sulfite reductase	*fsr*	Fe	1.8.98.3
	Sulfite oxidation	Sulfite oxidation to sulfate	Sulfite dehydrogenase subunit A	*soeA*	Mo	1.8.5.6

(Continued)

TABLE 2 (CONTINUED)
Biogeochemically Relevant Metal-containing Oxidoreductases and the Catalytic Metal Present in Their Active Center

Cycle	Pathway	Step	Protein Name	Gene	Metal in the Active Center	EC
	Sulfur disproportionation	From sulfite to sulfate	Sulfite:cytochrome-c oxidoreductase subunit A	*sorA*	Mo	1.8.2.1
	Sulfur reduction	**Catalyzes the cytoplasmic production of hydrogen sulfide in the presence of elemental sulfur**	**Sulfhydrogenase**	***shyB***	**Fe**	**1.12.98.4**
		Sulfur reduction	**Sulfur reductase molybdopterin subunit**	***sreA***	**Mo**	**1.97.1.3/1.12.98.4**
	Thiosulfate oxidation	From thiosulfate to sulfate	Sulfur-oxidation complex	*soxCD*	MoCo	1.8.2.6
		From thiosulfate to tetrathionate	Thiosulfate dehydrogenase	*tsdA*	Fe	1.8.2.2
	Thiosulfate reduction	From thiosulfate to hydrogen sulfide	Polysulfide reductase chain A	*phsA/ psrA*	Mo	1.8.5.5

The element cycle, the pathway, and the metabolic step in which each enzyme is involved are reported, together with the common enzyme name and the Enzyme Commission number (EC number). Enzymes in bold have been considered crucial in the emergence of life. Adapted from [1].

In order to incorporate the necessary metals as cofactors, organisms are required to import them from the environment, making metal availability fundamental in the regulation of metabolism and biological diversity. Interestingly, metal availability has fluctuated during the evolutionary history of our planet in response to changing redox conditions and the great oxygenation events [7]. These changes might have played a key role in the emergence and evolution of biogeochemistry. For instance, the oxygenation of the planet resulted in the appearance of more oxidized forms of nitrogen, such as nitrate, allowing the expansion of anaerobic metabolisms to higher energy levels [8]. Moreover, oxygenation of the planet influenced the availability of the metal that catalyzes the reduction of nitrate, molybdenum [1,7]. Despite the pivotal role of metals in biology, the relationship between their availability and the evolution of metabolism has not been investigated in detail. This chapter reviews the current knowledge about the role of metals in microbial metabolism, drawing attention to the function of these enzymes in biogeochemical cycles and speculating on their role in the origin and emergence of life. While all enzymes are ultimately involved in biogeochemical cycles, we will largely focus on enzymes that are of direct biogeochemical interest, defined as using as substrate or releasing as a product a small inorganic molecule that can then be exchanged with the environment. This definition brings into focus enzymes that directly affect the environmental concentrations and oxidation state or molecules in the surrounding ecosystems. The majority of the enzymes matching this definition are oxidoreductases, but a few additional non-redox enzymes are also discussed.

1.1 Trace Metals in Biological Systems

Metals are elements that easily form cations, behave as good conductors of heat and electricity, and throughout life's history have interacted with biology in a variety of ways. Even though all interactions are relevant for the biogeochemical cycle of the elements, the type of interaction determines to a large extent the role that a metal might have played in the emergence and evolution of life. In the broadest sense, metals can interact with life in four major ways (Figure 4):

1. They serve as catalytic cofactors in redox reactions-catalyzing enzymes [1]. Metals play a central role in controlling the electron flow from and to a redox substrate to derive energy, by finely tuning the midpoint potential of the redox protein to the exact level it requires to interact with a specific substrate. Examples of this are the key enzyme for the Wood-Ljungdahl carbon fixation pathway containing a Ni-Fe-Cu center and a Fe-[$NiFe_3S_4$] cofactor, the NiFe hydrogenases, and the V-containing nitrogenase, responsible for nitrogen fixation (Figure 4A);
2. They function in a non-catalytic way to stabilize protein structures and as quorum sensing signaling molecules for various physiological processes in living systems [9,10]. Examples are the role of Mg in controlling DNA

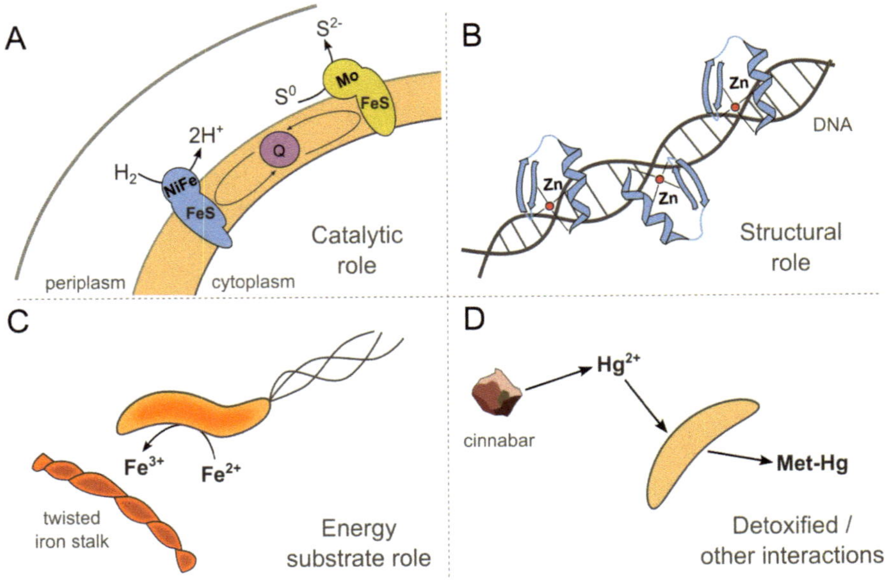

FIGURE 4 The different interactions of life with trace metals. A. Metals used as catalytic cofactors in enzymes and proteins. Depiction of the catalytic roles of nickel and iron in hydrogenases and of molybdenum in sulfur reductases; B. Metals used as structural cofactors or non-catalytic ions in cell communication. Depiction of the structural role of zinc in zinc-finger proteins; C. Metals used as substrates in energy conserving reactions, such as iron(II) as an energy source in iron oxidizing bacteria; D. Metals metabolized or incorporated into biological systems for other reasons, such as the methylation of mercury(II) by sulfate reducers and other microbes.

polymerase activity, the role of Ca as an intracellular signaling molecule, and the stabilizing role of Zn in zinc-finger nucleases (Figure 4B);

3. They can directly be used as redox substrates in energy conserving reactions. In this case, metals play the role of substrate rather than cofactor and can become limiting very quickly. For example, *Acidithiobacillus ferrooxidans* is well known for its ability to catalyze the dissimilatory oxidation of Fe to obtain energy [11,12]. Numerous other examples are described in the literature, such as Mn and Se used both as electron acceptors and electron donors (Figure 4C);
4. They can be incorporated accidentally into biological systems due to similar affinities with respect to the primary target, or they can be detoxified and metabolized without representing an energy source for the cell. Examples are the incorporation of Pb in place of Ca and the methylation of Hg playing a key role in Hg toxicity and biogeochemistry [13] (Figure 4D). The ability of biomolecules to incorporate heavy metals through different uptake mechanisms offers the potential for removing contaminants from polluted environments, a process known as bioremediation [14].

Even though our current understanding of the role that trace metals play in life is ever changing, we argue that they play an essential role in shaping microbial diversity and evolution on our planet. The inventory of elements used by life is incomplete (Figure 3), and additional discoveries are changing our perception regarding the interaction between life and the periodic table. For example, in the past decade, several variants of the well-known enzyme methanol dehydrogenase, which catalyzes the conversion of methanol to formaldehyde in aerobic methanotrophs, have been discovered. These variants incorporate La, Ce, or Nd as cofactors, replacing the traditional Ca [15,16]. This discovery broadens the range of elements used as biological cofactors and reveals a previously unrecognized role for rare earth elements in biological processes. Our understanding of the roles that various metals and elements play in biological systems remains incomplete (as summarized in Table 1), and it is likely that further investigation into uncultured organisms will reveal additional unexpected interactions between elements and life.

1.2 Trace Metals at the Interface between the Geosphere and the Biosphere

Trace metals represent a critical nexus between the geosphere and the biosphere. The elements necessary for sustaining life were likely selected on the basis of their abundance in Earth's crust, their readiness to be mobilized into aqueous environments, and their utility and compatibility with life's needs [17–19]. The availability of these biologically relevant metals is closely tied to the Earth's changing surface redox conditions, which is and was influenced by biology. For instance, the Great Oxygenation Event, triggered by the evolution of oxygenic photosynthesis and a series of geological and biogeochemical feedbacks, led to the oxidation of Earth's surface and hence shifted the availability of biologically relevant trace metals such as Fe, Zn, and Cu, among others (see below). This prompted the evolution of alternative enzymes and microbial metabolic pathways, thus becoming crucial for the evolution of life [19]. These types of complex, cybernetic, feedback mechanisms [20] have played a key role during the coevolution of life and our planet.

Among the building blocks of life, metals occupy an interesting niche at the interface between the geosphere and the biosphere. Several of the metals used by life are part of the major rock-forming elements (Na, K, Ca, Mg, Fe), and are abundant on Earth and expected to be abundant on other rocky planets in the universe [21]. This suggests that overall these elements should be easy to acquire for life, at least in general terms. Other metals are present in trace amounts in the majority of rocks, and their planetary distribution is more patchy, providing a bigger challenge to biological use. However, in spite of their bulk planetary abundance, local conditions, and ecosystem, and planetary hysteresis might sequester specific elements, creating a local scarcity. We discuss this in detail in the following sections with specific examples related to the availability of Fe in modern oceans (see Section 2.1, "Primary Productivity in the Oceans"). Despite all this, life has evolved a variety of processes and mechanisms to scavenge and concentrate metals present in the environment in trace amounts, often investing a large quantity of energy in their uptake.

Metal cofactors' key role in mediating redox chemistry places them at the interface between the geosphere and biosphere [19,22]. In oxidoreductases and other enzymes of biogeochemical interest (as defined in Table 2), metals control the flow of electrons from and to small molecules that are exchanged with the environment and therefore directly participate in biogeochemical cycles and geological processes.

2 BIOMETALS AVAILABILITY AND MICROBIAL (FUNCTIONAL) DIVERSITY

While trace metals are regarded as crucial for microbial metabolism, they have not traditionally been considered major drivers of microbial functional diversity. The distribution of diverse functional profiles is often believed to be a function of either key physico-chemical variables, such as temperature and pH, or the availability of substrates such as organic carbon, electron donors, and acceptors [23–28]. Additionally, the thermodynamic energy landscape available for specific reactions is often invoked to explain microbial functional patterns [29,30]. Although these key factors should always be taken into account in ecological analyses, they alone cannot fully account for the vast diversity of microbial metabolic functions, which often show a high degree of apparent redundancy in diverse environments [19].

Traditionally, the influence of metals on biodiversity has mainly been explored from the perspective of their toxic effects on different organisms and communities [31–34]. On the other hand, only a little information about metabolic requirements for specific trace elements in defined microbial pathways is available, often confined to studies of host-pathogen interactions [35–41]. While the relevant role played by trace metals in biological systems in general was recognized a long time ago, only recent evidence has highlighted their significance in shaping microbial metabolism, for instance in the role of Fe in controlling marine primary productivity and biodiversity (see for pre-1950s' research and [37,42,43], and Co and Cd, through their key roles in vitamin-B12 and the conversion of HCO_3^- to CO_2, respectively [44]).

This section will review the current knowledge about the impact of trace metals on microbial functional diversity across various environments, both natural and artificial.

2.1 Primary Productivity in the Oceans

Oceans cover about 70% of Earth's surface, and they host half of the global net primary production (~105 Pg yr^{-1}), mainly carried out by cyanobacteria and diatoms, which contribute 25 and 20%, respectively [45,46]. The main metabolic pathway adopted by primary producers, namely photosynthesis, not only relies on sunlight as an energy source, CO_2 as a carbon source, and building block nutrient sources (i.e., nitrate, phosphate, and others), but also requires several metalloenzymes, like

RuBisCO and carbonic anhydrase, found in the carbon fixation pathways of phytoplankton [47,48].

The availability of metal ions represents a crucial factor for the health, functions, and dynamics of marine ecosystems [49]. The main inputs of trace metals to the oceans are water-rock interactions and atmospheric transport (dust deposition, river flow, etc.), as well as geothermal emissions (epithermal metals, etc.) and crust recycling throughout geological time [50]. Once dissolved, metals in surface water are subjected to precipitation or can be taken up by living organisms in the euphotic zone via specific or promiscuous transport systems, leading to their biological assimilation [36]. Among the many metals used by phytoplankton and marine organisms, iron plays a crucial role in controlling primary productivity.

The idea that Fe may limit phytoplankton growth in the ocean has been considered for decades. As early as the 1930s, studies have started to investigate the possible role of Fe concentrations in co-controlling primary productivity in areas of the sea where Fe was not supplied from land [51]. Confirmation that iron was a key controller of planktonic primary productivity, with land inputs serving as the primary source of iron, came only in the 1970s–1980s, when new trace-metal clean sampling procedures and analytical methods allowed detailed observations of the distribution of iron in the ocean together with an appreciation of its role in photosynthesis [43]. Given its wide demand, Fe is considered to be a limiting nutrient in oceans (Figure 5), particularly in high nutrient/low chlorophyll regions such as the Southern Ocean, North Pacific, and part of the equatorial Pacific [43]. Primary producers have particularly high Fe requirements due to its functional role in different molecular machines that convert photon-condensed energy into sugars and are involved in nitrogen fixation [43]. For instance, the Calvin cycle, the primary carbon fixation pathway in the extant Earth found in prokaryotic and eukaryotic phototrophs, relies indirectly on Fe for optimal function [43], while the photosynthetic apparatus requires high amounts of iron to ensure the efficient functioning of the electron transport machinery [52]. Fe is also essential for the enzymes that activate RuBisCO, particularly in processes involving ATP and NADPH production during the light-dependent reactions. In addition, cytochromes contain heme groups with Fe cations that can alternate between the two oxidation states (Fe^{2+} and Fe^{3+}), enabling electron transfer during the light-dependent reaction of photosynthesis, fundamental for ATP synthesis [53]. In areas depleted in Fe, several adaptations have evolved to cope with the scarcity of this key metal. For instance, a large number of marine organisms developed efficient Fe transport and accumulation systems, like siderophores, low molecular weight compounds capable of binding to Fe at very low concentrations. According to mathematical modeling, these Fe chelators are more effective in the presence of solid-phase Fe minerals, increasingly widespread after the GOE [54]. Besides, several studies have revealed the expression of different Fe transporters is enhanced in Fe-limiting conditions [55,56]. Prokaryotes are more efficient in the surface ocean at producing and acquiring siderophores and therefore directly compete for Fe with phytoplankton, which have a preference for free Fe. Luckily, siderophores produced by bacterioplankton, contrary to the siderophores produced by soil microorganisms, are photosensitive and quickly degrade in seawater, releasing the

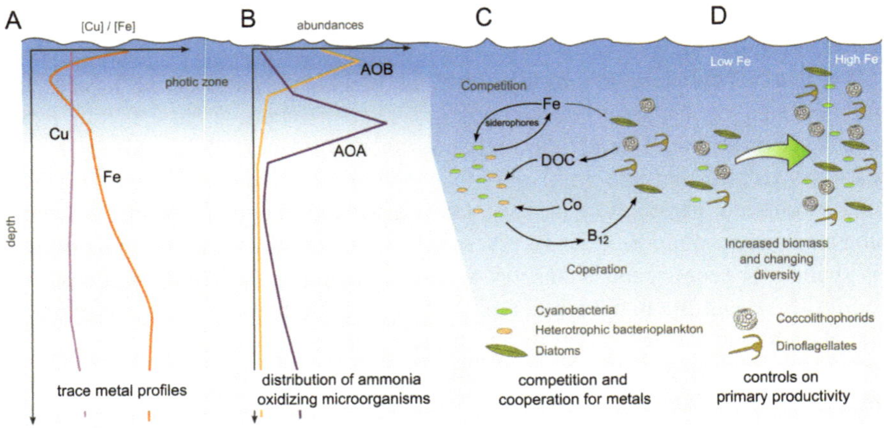

FIGURE 5 Examples of the controls that biometals introduce on microbial metabolism and diversity in the ocean. A. Generalized profile of iron and copper concentrations in the water column of open oceans. B. The distribution of ammonia-oxidizing bacteria and archaea in the water column can be explained by their preference for iron or copper [67]. C. Competition and collaboration for metal cofactors in the oceanic photic zone. Prokaryotes directly compete with phytoplankton for Fe through the use of siderophores. Complex eco-evolutionary feedbacks are present (see main text). Additionally, prokaryotes, also sustained by phytoplankton released DOC convert Co into vitamin B_{12}, a necessary cofactor that phytoplankton obtain directly from their interaction with prokaryotes [57]. D. Effect of Iron availability on primary productivity and diversity.

Fe for phytoplankton use (Figure 5C) [57]. These mechanisms play a key role in survival and in the structuring of marine microbial communities (Figure 5D), and are complemented by the evolution of variants of enzymes and pathways that can substitute Fe with more abundant metals. For example, the Fe-bearing cytochrome CytC6 is a soluble electron carrier that transfers redox equivalents to both photosystem I in photosynthesis and cytochrome-c oxidase in respiration. However, it can functionally be replaced by plastocyanin, a type I blue-copper protein, in most cyanobacteria [46–50]. In surface waters, where Fe is extremely limiting, the switch from CytC6 to plastocyanin has been proved to depend on Cu availability [58]. The replacement of an Fe-dependent soluble electron transporter in phototrophs with a Cu-bearing one might be an early adaptation evolved following changes in ocean redox during the Great Oxidation Event (GEO) both to cope with the sudden reduction in Fe supply and to exploit the increased Cu solubility, likely relieving Fe competition in surface waters [58].

Recently, complex interactions and co-dependencies among diverse trace metals in the ocean have emerged. Fe, together with light, is now recognized as a regulator of the Southern Ocean biological carbon pump, but observations suggest Mn also plays a role. Mn is fundamental for marine photosynthetic activity, as it is involved in the oxygen-evolving complex (OEC), a crucial component of photosystem II [59]. At the catalytic center of OEC there is typically a cluster of four Mn ions, Mn_4CaO_3, where water is oxidized and split into oxygen, protons, and electrons. Recent work has shown a new global biogeochemical model revealing that phytoplankton in over half of the Southern Ocean cannot reach maximum growth rates due to Mn deficiency, negatively affecting the carbon cycle [60]. This limitation is evident in the austral spring and relates to phytoplankton traits, such as the size of photosynthetic antennae and Mn uptake inhibition by high Zn concentrations. Zn is pervasively used by surface phytoplankton, and its concentration follows a nutrient-like trend, such as phosphate and nitrate, being utilized by phytoplankton in the euphotic zone [57] with concentrations in the picomolar range in surface waters.

Similar profiles can be observed for Cd and Co when Zn and other trace metals are depleted [49]. Zn is widely used as an enzyme cofactor, e.g., in alkaline phosphatase, on which primary producers rely for phosphate uptake from organic phosphorus compounds [61], and in the catalytic site of carbonic anhydrase, an ancient and ubiquitous enzyme family, recognized as one of the fastest on Earth in the hydration/dehydration of CO_2 [62]. Carbonic anhydrase usually requires Zn cations to be active, but some subclasses of eukaryotic diatoms can switch between different divalent cations (i.e., Cd or Co), despite their intrinsic toxicity at high concentrations, thus adding another level of intricacies between the role of metals and their usage by living organisms [63]. The overall capability of switching among different metals provides an advantage in oligotrophic water, and the cambialistic nature of some carbonic anhydrase enzymes seems to have evolved in response to very low and changing metal concentrations in the oceans, in what is known as biochemical substitution [49]. Co is currently emerging as a new metal involved in structuring marine microbial communities. Co plays a crucial role as a cofactor in metalloenzymes, such as AroB (3-dehydroquinate synthase) and DapE (Succinyl-diaminopimelate desuccinylase), which are essential for catabolic reactions and amino acid biosynthesis. Even more interestingly, cobalt is the key metal in cobalamin (vitamin B_{12}), and central metabolism enzymes, such as methionine synthase, are vitamin B_{12}-dependent and require Co to function properly [64]. Cobalamin is globally depleted in surface water and largely required by living systems and phytoplankton; however, it is only produced by a few archaeal and bacterial groups [65] (Figure 5C). While these organisms are crucial for cobalamin production, many other microbial taxa rely on cobalamin from these producers, highlighting a complex interdependence in microbial ecosystems [66].

2.2 Other Metabolisms

Besides the role that trace metals play in the ocean influencing primary productivity, additional pieces of evidence of the effect that trace metal availability have on functional diversity have piled up over time with studies from other ecosystems and through trace metal effects on other metabolisms.

As previously mentioned, the differential Fe and Cu availability has been proven to shape phytoplankton metabolism and primary productivity in the surface ocean. However, the GOE-driven shift in marine Fe and Cu availability seems to have also affected the diversity of chemolithotrophic ammonia-oxidizing communities [58,67]

(Figure 5). In fact, the compared physiological responses of the ammonia-oxidizing bacteria (AOB) *Nitrosococcus oceani* and the ammonia-oxidizing archaea (AOA) *Nitrosopumilus maritimus* to unchelated Fe and Cu reveal differential affinities and toxicity threshold towards these trace metals. Genome-wide bioinformatic analysis of metal binding sites and metal transporter systems suggests this pattern may be reflective of broader AOB and AOA communities [67]. Based on this, it has been suggested that the niche separation among AOB and AOA in the oceanic water column may be a consequence of the differential Fe and Cu distribution, and ultimately linked to the evolution of efficient competition mechanisms for trace element acquisition, utilization, and detoxification [67].

Similar competition-driven evolution could have led bacteria and archaea to adopt diverse metabolic strategies to perform methane oxidation and to adapt to the different niche availability of O_2. In fact, while the first step of aerobic methane oxidation in bacteria mainly involves the Cu metalloenzyme methane monooxygenase, anaerobic methanotrophic archaea (ANME) perform reverse methanogenesis, a metabolic pathway mostly based on Fe-metalloenzymes [68,69]. Besides, recent evidence from tectonically active zones of the planet supports the hypothesis that trace metals' environmental availability drives the distribution of specific proteins and microbial groups. In particular, the abundance of Cu-containing and -binding proteins has been shown to correlate with Cu concentrations in the hot springs of Tengchong (South China), while the distribution of prokaryotic groups related to Fe-oxidizing bacteria has been linked to the bioavailable Fe in the subsurface of Costa Rica's convergent margin [70,71].

Additional evidence suggests the emergence of O_2 shaped microbial functional diversity not only by directly influencing the availability of trace metals, but also by exerting selective pressure on the structural and functional properties of different metal cofactors. For instance, glutamate fermentation, carried out by anaerobic heterotrophic bacteria via two alternative metabolic pathways, the 2-hydroxyglutarate and the 3-methylaspartate pathways [72], has remarkably different steps and cofactors, even though they show the same stoichiometry and amount of ATP generated. While the 2-hydroxyglutarate pathway is widespread among strictly anaerobic bacteria as its reactions are catalyzed by O_2-sensitive FeS enzymes, the 3-methylaspartate pathway has been identified in environmental anaerobes subjected to potential O_2 fluctuations. In fact, the latter pathway relies on the O_2-tolerant Co-containing Vitamin B_{12} – one of the most complex in biochemistry, whose biosynthesis requires the expression of 30 additional genes [73]. Adaptation to rising O_2 concentrations could have forced the later evolution of this cofactor [73,74] in spite of the demanding higher energy amounts and complex machinery to synthesize. In extant Earth, oxygen is one of the major components of the atmosphere and it is transiently present at low concentrations even in normally anoxic ecosystems. In these environments, the vitamin B_{12} dependent 3-methylaspartate pathway, despite its high energy demands to synthesize the necessary cofactor, is preferred over the 2-hydroxyglutarate when Co is available, likely because it provides fermenters with greater adaptability to wider O_2 ranges [73,75].

Manipulation experiments on complex communities have provided coherent observations on the effect of trace metal supplementation on several metabolic pathways and thus on microbial diversity. Heterotrophic microbial communities have been studied by supplying coastal and open water with either individual trace metals (namely Fe, Co, Mn, Ni, Zn, Cu) or their combination and then incubated in the dark [76]. The treatment effect highly depended on the in situ trace metals availability: an enhanced prokaryotic growth was observed in response to the addition of all trace metals in shallow open waters only, where these essential nutrients are expected to be scarce; the supplementation of either Fe or Zn alone or the mix of trace metals always stimulated an increase in heterotrophic production rate, but in the case of waters with high trace metal inputs, e.g., deep coastal waters [76]. In fact, Fe-bearing cytochromes and aconitase and Zn-containing alkaline phosphatase count among the metalloproteins essential for heterotrophic metabolism. The same study pointed out that trace metals induced a shift in the marine microbial community towards hydrocarbon-degrading genera, as several of their metabolic pathways involve metalloenzymes [76,77].

Additional evidence of the crucial role that trace metals play in shaping microbial functional diversity and metabolism comes from pure culture studies and attempts to isolate new, previously uncultured microbial groups. For example, nitrogen fixation, a key step in the global nitrogen biogeochemical cycle converting atmospheric N_2 to NH_4^+, is catalyzed by a multimeric, likely ancestral enzyme known as nitrogenase. Three variants of the enzyme are known: a Mo-dependent, a V-dependent, and a Fe-dependent nitrogenase [78–80]. Cultivation attempts in medium lacking fixed nitrogen revealed that *Azotobacter vinelandii* exploits its genomic and metabolic plasticity to adapt to trace metal availability: the Mo-dependent nitrogenase is used to fix nitrogen in the presence of molybdate, while switching to the V-dependent or heterometal-independent (Fe only) nitrogen fixation when Mo is absent and the other trace metals are provided in excess [79]. In attempts to isolate new microbial groups, trace metals might provide a powerful tool to selectively discriminate among competing members of microbial communities. Recent work has demonstrated that the amount of W provided to a microbial mixed culture was crucial for the selective enrichment of a previously uncultured member of the archaeal order *Caldarchaeales*, *Wolframiiraptor gerlachensis* [81]. The metagenome assembled genome of this organism demonstrated a genomic enrichment of W-dependent enzymes, including six W-dependent ferredoxin oxidoreductases and a WO_4^{2-} ABC transport system.

2.3 Evidences from the Built Environment and Biotechnology

Besides the above-mentioned examples, the largest amount of information on the role of trace metals in controlling microbial functional diversity comes from empirical evidence in the field of environmental engineering, especially in the operation of anaerobic bioreactors and biogas production facilities [82–85]. The interest in biotechnological approaches, combining anaerobic digestion with renewable energy recovery, and the consolidated knowledge of the impact of trace elements on microbial activity, has led to the vigorous investigation into the microbiological aspects directly associated with mechanical, chemical, and thermal pre-treatments applied to degrade the initial feedstocks.

While microbes play a crucial role in every stage of anaerobic digestion, our understanding of the microbial consortia involved remains limited due to the scarcity of phylogenetic and metabolic data on predominantly unculturable microorganisms [86,87]. Still, the isolation of strict anaerobic microorganisms is highly challenging because of the requirement for low oxygen concentrations and the high degree of commensalism and mutualism in the communities, making their isolation and characterization difficult. Under such stringent laboratory conditions, the availability of trace metals has been identified as one of the key factors allowing for the selective stimulation of methanogens, as they serve the role of enzymatic cofactors and/or structural constituents, thus influencing biogas production [88]. Nonetheless, culture-independent techniques, such as PCR-based ribotyping, have been successfully utilized to gather insights into these microbial communities. The identified microbes are predominantly from the bacterial phyla Proteobacteria, Firmicutes, Bacteroidetes, Actinobacteria, Acidobacteria, Spirochetes, and Chloroflexi [87,89]. Earlier studies carried out on anaerobic digesters demonstrated that the archaeal population is mainly represented by methanogens, many of which are unculturable, resulting to be very similar to already identified methanogens such as *Methanosarcina barkeri*, *Methanosarcina frisius*, *Methanobacterium formicicum*, with the remainder being thermophilic organisms such as Crenarchaeota or *Thermoplasma* sp. [90]. Several studies have shown that the relative contributions of the different metabolic pathways in the digester can be controlled by different ratios of the trace metals Co, Ni, Fe, Zn, Mo, and W [82,85,91–94]. These elements are typically introduced through the feedstock, and their absence can result in suboptimal anaerobic digestion performance [95] and shifts in the expected products. Observed shifts in the methanogenic archaea populations as a consequence of different trace element combinations and concentrations have been reported, resulting in higher methane production yields [82]. Specifically, alterations in the availability of Ni, Co, Mo, Mn, and W seem to favor hydrogenotrophic methanogens, as proved by the shift from the acetoclastic to hydrogenotrophic pathway for *Methanosarcina* and by the increase of hydrogenotrophic activity for *Methanoculleus* [96].

In some conditions, the addition of a specific concentration of trace elements has been observed to enhance process stability in terms of methane production, thus limiting the accumulation of intermediates [93,97,98]. For instance, the addition of Fe can precipitate the H_2S, reducing its amount in the bioreactor and, at the same time, increasing the bioavailability of other trace metals that do not precipitate in the form of other metal sulfides [99]. A similar function seems to be accomplished by Mo, whose addition into the digester appears to inhibit sulfate reducers and the consequent production of sulfides, thus increasing the bioavailability of trace elements for methanogens and hence methane yields. All of the above supports the role of trace elements supplementation as a potential driver of microbial competition dynamics [100].

Specifically, the availability of both Ni and Co is demonstrated to increase methanogenic growth efficiency, being positively correlated with the concentration of the coenzyme F_{430} and corrinoids detected in the biomass [82,101,102] and necessary cofactors for the methyl coenzyme M reductase, the key enzyme catalyzing the formation of methane, and for the synthesis of Vitamin B_{12}, respectively [103,104].

3 BIOMETALS AVAILABILITY IN SPACE AND TIME

Not all metals are created equal, quite literally. And metals are not distributed uniformly across space and time due to a variety of processes that control their synthesis, delivery, sequestration, and availability at both global and local levels. Metals are created during nucleosynthetic processes in stars and in other star related processes. Diverse metals are created by different processes and are then dispersed throughout the cosmos [105]. After their cosmic generation, metals are incorporated during planetary accretion and differentiation, determining their concentration and distribution within a planet's crust and mantle [106]. Ultimately, a number of geological and geochemical processes, including local redox and microbial competition for the use of the metals, control the local bioavailability of these elements and together they play a huge role in providing life with a minimum set of diverse potential cofactors during its emergence and evolution [22].

3.1 Cosmic Distribution of Biometals

Stars of the main sequence (like the Sun) are made up of H_2 and He in a 3:1 ratio, with a small portion composed of other elements. Both astronomy-defined metals (in astronomy, all the elements heavier than helium are referred to as metals) and proper metals (in chemical terms used throughout this review) are forged from cosmic high energy processes. Nucleosynthetic processes in the extremely hot cores of stars lead to the formation of all elements up to Fe [107]. The relative presence of these elements (up to Fe) in stars is traditionally investigated using measurements of star metallicity, i.e., the relationship between iron and hydrogen [Fe/H] [108]. In astronomy, biometals can be classified into one of two macro-groups: those lighter or heavier than Fe. Elements lighter than Fe are produced in nuclear reactions in the core of stars with a mass larger than two solar masses, while the heavier ones, e.g., Ni, Cu, and Mo, etc., are produced in other kinds of energetic processes such as supernovae rapid neutron capture (r-process), slow neutron capture (s-process), and collision of neutron stars [109]. All these processes require high temperatures and high energy to make the fusion of He possible and to create heavy elements.

On the basis of the Oddo-Harkins rule, in our solar system, elements with even atomic numbers are generally more abundant than those with odd atomic numbers, as particularly evident in the case of rare earth elements [110]. Elements in the interstellar medium (ISM) are incorporated into grains and mainly consist of C, Mg, Si, and Fe, with all other periodic table elements present in smaller amounts. The distribution of these elements across the cosmos is dependent on the volatility of the elements themselves: highly volatile elements such as C, Ar, Kr, N, and O are easily dispersed by stars; moderately volatile elements such as Ti, Se, Zn, Pb, Cl, Na, Ga, K, Cu, and Mn have moderate distribution throughout the universe [111].

The condensation of rock-forming elements such as Mg, Si, and Fe, which includes also the formation of Mg silicates, e.g., forsterite – Mg_2SiO_4 and enstatite – $MgSiO_3$/ Mg,Fe)SiO_3, occurs at high temperatures close to the Sun, concentrating them in the inner solar system where these conditions can be found. The high-temperature condensation explains why these elements are mainly found in terrestrial planets and asteroids [112]. Refractory elements such as V, Ca, and Ti follow a similar path to Mg, Si, and Fe due to the fact that they condense at higher temperatures and thus tend to remain concentrated rather than distributed in the cosmos. This is evident from the comparison between solar and terrestrial compositions, noting that these elements maintain a similar abundance, which implies their localized formation and retention rather than a widespread cosmic presence [113]. On our planet and on other rocky worlds, the distribution of biometals is significantly influenced by their nucleosynthesis, abundance in the galaxy, the processes involved in the planet's formation, and their physical and chemical properties [114–116]. For instance, volatility affects the distribution of these elements to various degrees in the solar system and the whole cosmos, explaining why transition metals are more common on rocky planets close to their host star [117]. Supernovae explosions can also play a major role in the distribution and diffusion of elements in the interstellar medium [118]. As a consequence, the global cosmic enrichment and the mixing of trace metals in the ISM can be very stable over time, with heavy elements concentration decreasing outward from the galactic center. Nonetheless, metal abundance is not always expected to be homogeneous on local or small scales, and some large variations can be observed among nearby stars. Metal distribution in host stars orbiting in distant planetary systems can be investigated by measuring stellar metallicity [108,116]. Thanks to this system, it is possible to obtain the chemical composition of about 146,000 stars [119]. Host star composition could be an indicator of bulk planetary composition and biometal abundance, which is controlled by the metallicity and metal composition of the parent star [116]. On a planetary scale, metal crustal abundance depends on several variables, such as core formation, crustal differentiation, active geology, and redox conditions of the planetary surface and subsurface [120,121]. All these parameters influence the distribution and availability of CHNOPS elements and key biometal cofactors, thus constraining their availability for life and paving the way to the understanding of different habitability conditions of outer planets and distant galaxies [116].

3.2 Planetary Availability of Biometals

On our planet, a large number of geological processes control the inventory and availability of biometals at the surface and near the surface. The solid Earth's metal cycle strongly influences the global distribution of biometals, culminating in their release through volcanic manifestations and hydrothermal circulation, or sequestration as ore deposits in the shallow crust [122]. Crustal differentiation and core formation have had a profound impact on the near-surface availability of biometals. Since most biometals behave as chalcophiles or siderophiles because of their high affinity for sulfur or iron, their distribution is affected by the processes that partition these elements and by their geochemical affinities. Siderophilic biometals such as Ni and Co, which have been largely sequestered by core formation [123], are present in the upper mantle only in trace amounts, where they are partitioned into melts and are sequestered by the crystallization of high Fe minerals such as olivines. The crustal distribution of chalcophile elements such as Cu and Zn, which have a strong affinity for sulfur, will be impacted by processes such as sulfide saturation and degassing [124].

A large number of biometals are redox-sensitive, so their behavior during crustal differentiation and magmatic evolution is strongly influenced by their mobility across different magmatic redox states (oxygen fugacity, fO_2). This implies that different tectonic settings, characterized by different magmatic redox states, affect biometals' surface availability. Thus, the onset of plate tectonics (with current estimates of its onset between 3.8 and 4.0 billion years ago [125], although some authors have suggested a Neoproterozoic origin [126]), might have been a significant turning point in the availability of biometals essential for life. For example, in Cl-rich oxidized magmas like those typical of volcanic arc settings, biometals such as Cu, Mo, V, and Co show increased mobility [127,128], thus facilitating their partitioning into saline aqueous fluids and the prevalent fractionation of the metal-bearing minerals, like magnetite ($Fe^{2+}Fe^{3+}{}_2O_4$), found in these oxidized melts.

Today's surface distribution of biometals can be mainly attributed to magmatic processes. Edmonds et al. [122] have shown how volcanic emissions from different tectonic settings, for instance hotspots and arc-related volcanism, have distinctive metal fingerprints. Subduction zone volcanoes have oxidized and water-rich magmas, and vent acidic HCl-rich fluids enriched in volatile metals that are strongly fluid-mobile and bind to Cl (e.g., Bi, Sb, Cd, Tl) [122]. Magmas at oceanic spreading centers and hotspots are typically more reduced than the ones in subduction zones [127], resulting in extensive sulfide saturation and metal sequestration [124].

Once they reach surface and near-surface environments, the availability of biometals for life is controlled by local processes, such as leaching and weathering events, and direct interactions with biology [129]. Weathering induces mineral decomposition and releases a number of elements that eventually enrich groundwaters, rivers, and ultimately the ocean. The lower the crystallization temperature of a mineral, the higher its resistance to weathering. For example, olivine ($(Mg, Fe)_2SiO_4$) and anorthite ($CaAl_2Si_2O_8$), among the first formed minerals in the discontinuous and continuous reaction series, are very resistant to weathering [130]. Minerals are exposed to direct leaching by water-rock interactions. These interactions take place, for instance, in hydrothermal vents, whose fluids display a broad spectrum of concentrations of different trace metals, leached from the surrounding rocks and carried in saline fluids derived from deep magma bodies [122]. Among these local processes, microbial activity plays a role in metal distribution, causing dissolution, transformation, and precipitation of minerals, as well as contributing to the formation of ore deposits [131,132]. Due to these interactions, microbes colonize nearly every mineral surface, exploiting minerals for physical-chemical protection (from UV radiation, temperature fluctuation, and abrasion), nutrition, and energy [131]. These small ecological niches provided by mineral fractures, cracks, and pores are known as "mineralo-spheres", and supply essential nutrients and elements crucial for microbial life, as well as electron sinks or sources, essential for energy yielding reactions. Microbes can also interact with minerals in processes such as biologically induced mineralization (BIM) and biologically controlled mineralization (BCM). During BIM, microbial metabolism changes the local environment (for instance by modifying the pH or by releasing metabolic byproducts) and, as a consequence, indirectly affects minerals [131,133–135]. BCM is controlled by a set of genes that regulate the import of elements and ions, intracellular biochemical processes, and the formation of crystals within the cell. In this process, microbes have a direct and regulated approach, exerting precise control over the nucleation, growth, and morphology of minerals.

3.3 Changes through Time

The availability of biometals has changed throughout Earth's history, mainly due to changes in our planet's redox chemistry [7]. Since Earth's formation, its surface redox state has evolved, mainly due to events of global oxygenation. [7,136]. Over the past 4.5 billion years, the Earth's interior has undergone a progressive cooling [137]. Potential mantle temperatures have changed from a maximum temperature of ~1500–1600°C to 1350°C over the last 2.5–3 billion years. In this context, even though its total volume has been increasing through time, the average net growth of continental crust has reduced from ca. 3 km^3/year at 3 Ga to 0.8 km^3/year at present [138], suggesting that the loss of isotopic thermal energy and heat energy of the Earth has reduced the interaction between mantle and crust over time [137]. These processes have decreased mantle melt fractionation, influencing the long-term distribution of metals on our planet and impacting differentiation processes within the crust [139]. In this context, the temperature of the convecting mantle has had a primary control on the onset of plate tectonics, and on the composition and tectonic behavior of Earth's lithosphere.

As these deep processes have influenced metal distribution in the Earth's interior, more superficial processes have influenced metal distribution and availability on the planet's surface, profoundly impacting life's evolution. Between 3.8 and 3.4 billion years ago, the archaean oceans were characterized by reducing, anoxic conditions. During this period, redox-sensitive elements such as Cu, Mo, U, Cr, and Re were immobile and insoluble, with limited availability, while other elements such as Fe and Mn were much more soluble than today [140]. The Great Oxidation Event (GOE) profoundly impacted the mobility of elements, their weathering, and delivery to the oceans (Figure 6). For instance, it changed the chemical weathering processes on land, as oxygen reacted with exposed minerals, breaking them down and releasing elements into the environment. The switch to oxic and oxidizing oceans caused an enormous redox change, limiting Fe, Mn, and Co availability, which became less soluble as they oxidized and precipitated out of solution, and increasing Mo, Zn, and Cu solubility and concentrations [141].

4 BIOMETALS AND THE EMERGENCE AND EVOLUTION OF LIFE

4.1 Biometals at the Origin of Life

Whether life originated within deep sea hydrothermal systems, shallow-waters, or inland hot springs, the interactions between water and rocks played a fundamental role in providing the necessary ingredients for the development of life. For instance, serpentinization, the process where Fe(II) containing minerals react with water, resulting in the production of a suite of minerals and the release of molecular hydrogen and alkalinity, has been suggested to have played a fundamental role in the origin of life scenarios [142]. The obtained minerals can in turn be reduced, giving rise to native metals and minerals, which contain Fe and Ni, fundamental biometals

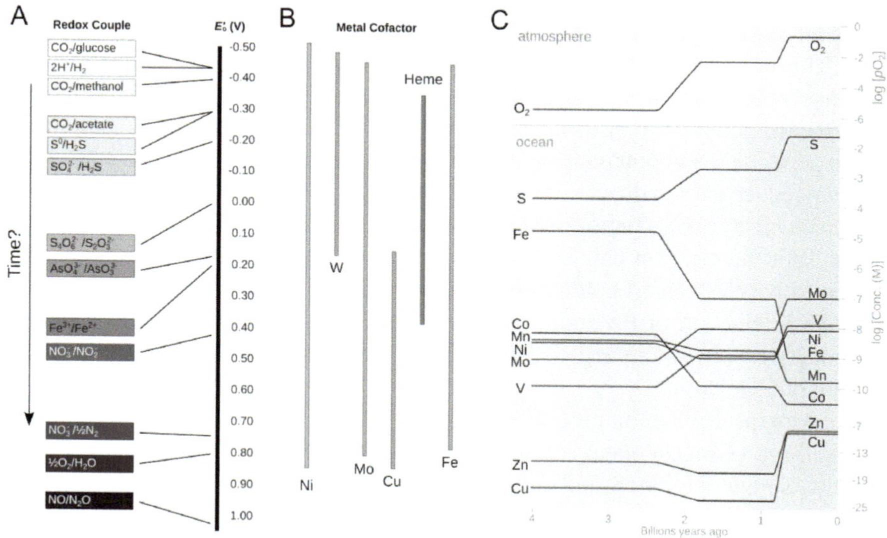

FIGURE 6 Energy conserving redox couples, metal catalytic centers, and changes in trace metal availability over time. Microbial metabolism, transition metals' midpoint potential, and their abundance through time. A. A simplified redox tower used in microbiology to calculate the redox metabolic couples available to life. The order in which more oxidized redox couples appear in the tower coincidentally recapitulates their appearance during the planet's evolution (see [2]); B. Examples of metal cofactors used in metalloproteins and the extent of their midpoint potential, aligned on the E'^0 of panel A. Variations are the result of changes in the position of the metal within the active site, coordinating amino acids, and the structure of the protein; C. Changes in biologically relevant metal concentrations (adapted from [176]) through time as a consequence of planetary surface redox changes. Changing surface redox conditions modified both the availability of redox couples and the availability of transition metals used to access them. Figure adapted from [19].

for living systems. Some of the obtained minerals, such as awaruite (Ni_3Fe), for example, are able to catalyze fundamental prebiotic reactions, such as the reduction of CO_2 to CH_4 and carbides, mainly using an electronic gradient provided by molecular hydrogen [143]. The individual components of this mineral are known for their ability to allow the abiotic formation of organic products such as formate, methanol, acetate, and pyruvate – essential intermediates in the acetyl-CoA cycle [144]. Consequently, on the early Earth, minerals alone can account for the presence of biometals, the establishment of proton gradients, and the synthesis of small organics produced from inorganic compounds such as CO_2 and molecular nitrogen (N_2) in hydrothermal systems. Many other minerals are able to participate in these processes, e.g., olivines and pyroxenes, which are able to oxidize Fe(II) to Fe(III) [145].

Notably, structures resembling minerals are present in enzymes that catalyze metabolic reactions that are considered ancient, such as in the case of acetogenic and methanogenic organisms. In the pathway shared by these groups (acetyl-CoA pathway), the key enzyme carbon monoxide dehydrogenases (CODH) can reduce CO_2 to CO through a catalytic process mediated by an active site containing Ni^0 atoms bonded to a cubane FeS cluster [146]. Since this process has been shown to occur abiotically, this pathway is thought to have appeared before the evolution of CODH [147]. This enzymatic complex is able to generate a transmembrane ion flux in a process involving a flavin-based electron bifurcation mechanism to fix CO_2 [148]. Since the standard potential of H_2 is approximately −414 mV, this is not sufficiently negative to generate a low potential capable of reducing ferredoxin (–500 mV), which in turn can reduce carbon dioxide. To achieve this, the evolution of the electron bifurcation mechanism allows the transfer of an electron pair from H_2 to flavoproteins; the flavins split the electron pair and send one electron to ferredoxin and another to a more positive electron acceptor such as NAD^+ (–320 mV) or an heterodisulfide (–140 mV). The reduction of the more positive electron acceptor energetically supports the reduction of the more negative potential acceptor (ferredoxin), thus enabling the fixation of CO_2 [149].

Ferredoxins are metalloproteins containing one or several FeS clusters involved in a plethora of biochemical reactions as electron carriers [150–152]. Due to their ubiquity, simple structure, and similarity to known minerals, they have been suggested to be ancient and key to the emergence of life [153,154]. Given their low redox potential, of −450 to −600 mV within cells, it suggests their origins in very reducing conditions, such as those present on the early Earth [152,153]. Additionally, a study of the protein folds containing metal cofactors has demonstrated that ferredoxin is one of the two folds occurring at the center of the reconstructed fold network, supporting its ancestral nature [155]. Interestingly, FeS clusters in ferredoxins have been shown to have structures similar to those of FeS minerals, such as greigite [147], and have been shown to assemble spontaneously in solutions containing L-cysteine, $FeCl_3$, and Na_2S, which can be encountered in submarine alkaline hydrothermal systems [154]. FeS dependent proteins, such as ferredoxins, have been shown to be enriched in the proteome of all living organisms, after accounting for ATP- and phosphate-dependent reactions, suggesting an ancient origin of iron sulfur in biology [156,157]. Carbon fixation pathways considered to be ancestral, such as the acetyl-CoA pathway and the rTCA cycle, are also heavily dependent on reduced ferredoxins [158–160]. Together with the presence of metals in all other key enzymes linked to metabolism, and their key role in mediating redox reactions, metals are believed to be crucial for the transition from a prebiotic to a biotic world [161–163]. Recently, a series of papers has demonstrated the possibility of abiotically performing the key reactions of the rTCA cycle through the use of metal ions, supporting a plausible abiotic metal-mediated origin of this carbon fixation in the prebiotic world [164].

4.2 Biometals and the Evolution of Biogeochemistry

Diverse metals are currently used in different steps of all major biogeochemical cycles [1] (see Figure 8 in Section 5). The utilization of metals by prokaryotes generally follows the Irving–Williams series: $Ca^{2+} < Mg^{2+} < Mn^{2+} < Fe^{2+} < Co^{2+} < Ni^{2+} < Cu^{2+} > Zn^{2+}$. This hierarchy may have played a crucial role in promoting the diversification of lineages, shaping the genetic emergence of archaea and bacteria through the formation of biogeochemical cycles [165]. Although the principles underlying metallation and mismetallation of enzymes have yet to be elucidated [166], two main mechanisms are at play: direct insertion of metals in the protein folds mediated by the metal concentration in the protein surrounding and specific physico-chemical properties of the metal and the fold [167], or by the help of specialized chaperones that transfer the metal (or the synthesized organometallic cofactor) to the apoprotein [168]. Given the changes in metal availability due to changing planetary redox conditions, it is possible that the availability of specific metals and their suitability for specific reactions have played a key role in driving the emergence and evolution of biogeochemistry [19].

Similar evolutionary pressures have been invoked in the evolution of specific enzymes that possess variants utilizing different metals as their catalytic center. The ubiquitous superoxide dismutase (SOD) enzyme involved in reactive oxygen species detoxification is an example of the geochemically driven evolution of biomolecules. Following the increase of atmospheric oxygen during the Great Oxidation Event (GOE), SOD enzymes played a key role in the evolution of life within the oxidized Earth's surface [169]. Primitive SODs have been shown to rely on Fe and/or Mn as cofactors, in line with their higher availability when Earth's atmosphere was reductive. Fe was in fact abundant in its most soluble and hence bioavailable form, Fe(II), in the archaean ocean before oxygenic photosynthesis emerged [141]. On the other hand, a more recent SOD family evolved when O_2 diffusion caused Fe precipitation and made Cu comparatively more bioavailable [1,141,169,170]. While it might be tempting to apply similar reasoning to other enzymes that present multiple variants whose metal cofactors have followed clear availability trajectories over time [19], a few well-studied examples suggest that caution is needed.

Nitrogenases, the enzymes responsible for the conversion of dinitrogen (N_2) to ammonia (NH_3), a process known as nitrogen fixation, have probably been one of the key evolutionary innovations which significantly increased global productivity and eventually supported the emergence of higher life forms. At the same time, these enzymes represent a telltale story of the intricacies of the relationship between metal utilization and evolution on a changing planet. Nitrogenases come in three variants, containing Mo, V, or Fe as the catalytic metal at the center of their organometallic cofactor [171] (Figure 7). Given the high availability of Fe in early Earth conditions and the limited solubility of Mo under low oxygen conditions, it is plausible that the Fe form of nitrogenase represents the ancestral enzyme. However, detailed phylogenetic analyses suggest that a Mo containing variant might be the ancestral form of the enzyme [172] and not the Fe containing one. It might be due to a late cambialism of the enzyme that has substituted Fe for Mo during the evolution of its lineage, and therefore the ancestral nitrogenase was indeed Fe containing, or it might be the result of specific pressures selecting for Mo even in a low Mo environment.

Each of these examples suggests a complex relationship between the abundance distribution of chemical elements in the environment and the expression of related biochemistry by living organisms over geological time periods. Defining this relationship will allow us to answer key questions related to the emergence of life and the coevolution of life and our planet [19].

5 CONCLUSION AND FUTURE PERSPECTIVES

Life sustains itself through redox reactions that exploit environmental thermodynamic disequilibria. The complex interactions between metal bioavailability and the evolution of biogeochemistry were shaped by various factors, including stellar processes, protoplanetary disk accretion, planetary differentiation, redox condition changes, plate tectonics, supercontinent assembly, and shifts in dominant volcanism. Understanding the roles of trace metal distribution and availability in influencing microbial functional diversity is key to comprehending the coevolution of life and our planet [19]. This insight has the potential to unlock significant discoveries across fields such as earth sciences, astrobiology, microbial ecology, bioremediation, and biotechnology. Addressing this research area requires an innovative and interdisciplinary approach that combines advances in microbial ecology, biochemistry, and earth sciences.

Despite the critical nature of this topic, our understanding of the role of metals in controlling microbial metabolism and biogeochemistry remains limited. Essential questions about the selective pressures imposed by the redox potentials of substrates and reaction products in selecting specific metals, and the effect of environmental metal availability, remain open. Furthermore, a complete catalog of the elements utilized in protein structures is still lacking. The diversity of organometallic structures has been poorly examined, especially in environmental microbes that are mostly unculturable, making it challenging to investigate protein structures and cofactors using traditional approaches. Recent discoveries, such as the use of lanthanides by enzymes catalyzing key steps in the aerobic respiration of methane [15,16,173], highlight the potential for expanding our knowledge of organometallic cofactors from uncultured microbial groups. This could revolutionize our understanding of how

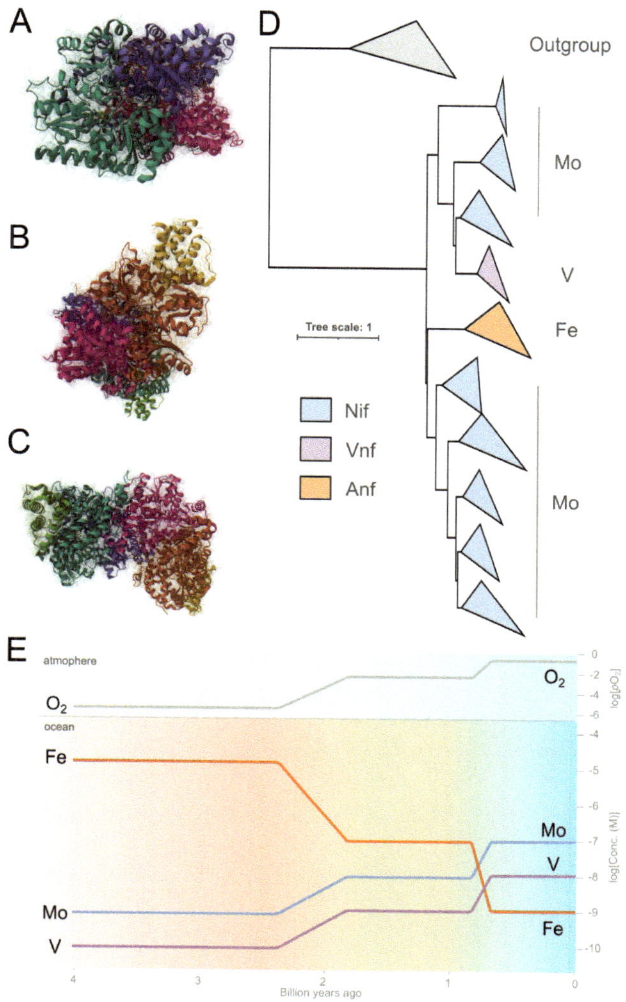

FIGURE 7 Nitrogenase types, their evolutionary history, and the concentrations of relevant trace elements over time. A. The molybdenum-containing nitrogenase from *Azotobacter vinelandii* (PDB 6OP2); B. the vanadium-containing nitrogenase from *A. vinelandii* (PDB 6FEA); C. the iron only nitrogenase from *A. vinelandii* (PDB 8BOQ); D. Phylogenetic tree of the three types of nitrogenase and their respective metals (maximum likelihood tree based on 319 sequences obtained from [172]); E. Concentrations of molybdenum, vanadium, and iron over Earth's history in relationship to atmospheric oxygen concentrations (adapted from [176]).

redox chemistry mediates the interaction between life and our planet, offering promising possibilities in green chemistry and sustainable industry.

Currently, the limited number of metallomic studies and detailed cell metal inventories restrict our ability to investigate the role of trace elements in controlling

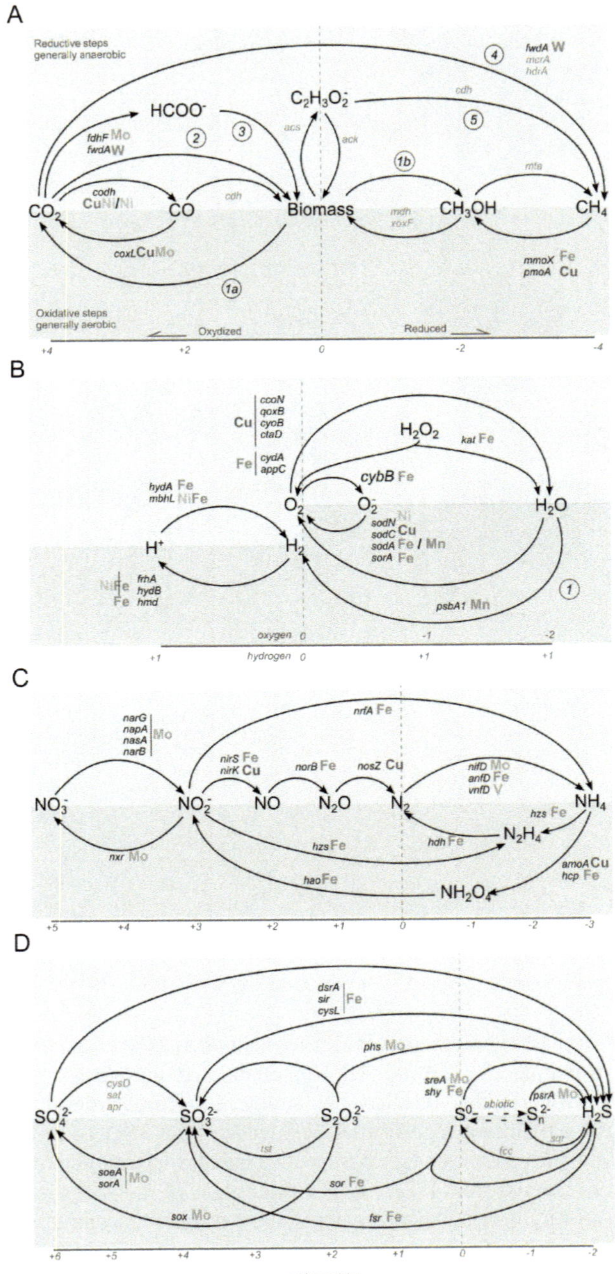

FIGURE 8 Biometals and their role in the carbon (A), hydrogen and oxygen (B), nitrogen (C), and sulfur (D) biogeochemical cycles. For each cycle the reductive side is on the upper section of the cycle highlighted in light gray, while the oxidative side is on the bottom section

metabolism across diverse organisms [174,175], and trace elements are often overlooked as potential explanatory variables in microbial diversity studies [19]. Focusing on the role of trace metals in controlling taxonomic and functional microbial diversity has profound implications for our understanding of life and its evolution on Earth, providing a new framework for interpreting planetary evolution and a guide to the search for life in the universe.

ACKNOWLEDGMENTS

This project has received funding from the European Research Council (ERC) under the European Union's Horizon 2020 research and innovation program (grant agreement No. 948972), project acronym CoEvolve to DG.

ABBREVIATIONS AND DEFINITIONS

ABC transporters	ABC transporters are transmembrane transporters often consisting of multiple subunits, one or two of which are transmembrane proteins and one or two of which are membrane-associated, hence the acronym ABC.
ANME	anaerobic methanotrophic archaea
AOA	ammonia-oxidizing archaea
AOB	ammonia-oxidizing bacteria
AroB	3-dehydroquinate synthase enzyme
ATP	adenosine triphosphate
BCM	biologically controlled mineralization
BIM	biologically induced mineralization
Biogeochemical cycles	The series of abiotic (geological and geochemical) and biotic processes through which elements are transformed and moved across geological and biological reservoirs, maintaining planetary habitability

(CONTINUED)

and highlighted in light blue. Each cycle is ordered left to right based on the oxidation state of the molecules, starting with the most oxidized form. Key enzymes for each step are reported in black if they are key metal-containing oxidoreductases, with the catalytic metal colored according to the Jmol color scheme; other enzymes are in light gray. The numbers in a circle represent complex pathways/processes. (A) Carbon cycle: 1a and 1b, heterotrophy and fermentation; 2, carbon fixation; 3, formate assimilation; 4, methanogenesis pathway: the reported enzyme catalyzes key steps in this multi-enzyme pathway; 5, acetoclastic methanogenesis. (B) Hydrogen and oxygen cycle: 1, abiotic and biotic hydrogen formation; note that the oxidation scale for oxygen and hydrogen are distinct, and hydrogen is reported with the same oxidation state on both sides. (C) Nitrogen cycle. (D) Sulfur cycle: 1, sulfur/thiosulfate oxidation is accomplished by a complex group of enzymes (Sox), of which SoxCD is a metal-containing oxidoreductase. Adapted from [1].

Biometals	Metals that play essential roles in biological systems, often as cofactors in enzymes and proteins, where they are involved in processes such as electron transfer, structural stabilization, and catalysis. Examples include iron, zinc, copper, and manganese.
Chalcophile	In the Goldschmidt classification, an element that preferentially forms sulfide minerals if sulfur is available.
CHNOPS	The abbreviation of the life building blocks carbon, hydrogen, nitrogen, oxygen, phosphorus, and sulfur.
Coevolution	The process by which two or more interacting species, or biological systems, influence each other's evolutionary trajectory over time. In the context of biology, coevolution often involves reciprocal adaptations, where changes in one species or system drive changes in another, leading to closely linked evolutionary paths. This concept can also apply to the mutual influence between life and Earth's environment, such as the coevolution of microbial metabolism and planetary geochemistry.
Cofactor cambialism	The ability of some enzymes to switch metal cofactors depending on the availability of different trace metals in the environment. For example, some carbonic anhydrases can use zinc, cadmium, or cobalt as cofactors.
Cofactor	A non-protein chemical compound that is required for the biological activity of a protein, often an enzyme. Cofactors can be metal ions, organic or organometallic molecules (coenzymes) that assist in catalysis by stabilizing intermediate states or facilitating electron transfer.
CytC6	cytochrome C6
DapE	succinyl-diaminopimelate desuccinylase enzyme
Energy conserving reactions	Redox reactions that are thermodynamically favorable (exergonic) under the in situ conditions and provide energy used by biology in the form of chemical bonds, such as ATP, or as transmembrane gradients, which can later be utilized for cellular functions.
Enzyme Commission (EC) classification	A system for categorizing enzymes based on the reactions they catalyze. Oxidoreductases fall under EC classes 1 and 7.1, playing a crucial role in redox reactions involving electron transfer.

fO_2	Oxygen fugacity is a measure of the partial pressure of gaseous oxygen that is available to react in a particular environment.
Geochemical gradient	A variation in the concentration of chemical elements or compounds in the Earth's crust or environment, often resulting from natural processes such as weathering, volcanic activity, or biological activity. These gradients drive various biogeochemical cycles and provide energy sources for microbial metabolism.
Great Oxidation Event (GOE)	A period around 2.4 billion years ago when atmospheric oxygen levels rose significantly due to the evolution of oxygenic photosynthesis, altering Earth's redox conditions and metal availability, thus influencing biological evolution.
H_2	molecular hydrogen gas
ISM	interstellar medium
Metal	In chemistry, a metal is an element characterized by its ability to lose electrons to form positive ions and its high electrical and thermal conductivity. In astronomy, all elements heavier than He are termed metals. In biological systems, metals often serve as essential cofactors in enzymes, facilitating redox reactions and other cellular processes.
NADPH and NAD^+	nicotinamide adenine dinucleotide phosphate and nicotinamide adenine dinucleotide
OEC	oxygen-evolving complex
Oxidoreductases	A class of enzymes that catalyze redox reactions, facilitating the transfer of electrons from a donor molecule (reductant) to an acceptor molecule (oxidant). These enzymes often utilize metal cofactors to mediate electron transfer.
Redox reaction	A chemical reaction involving the transfer of electrons between two species, where one species undergoes oxidation (loses electrons) and the other undergoes reduction (gains electrons). Redox reactions are fundamental for energy conservation in biological systems.
rTCA	reductive tri-carboxylic acid cycle
RuBisCO	ribulose-1,5-bisphosphate carboxylase/oxygenase enzyme
Siderophile	In the Goldschmidt classification, an element that has a strong affinity for partitioning into liquid metallic Fe.

SOD superoxide dismutase

Trace element A chemical element present in small concentrations (trace amounts) or required in minute quantities for the proper functioning of biological processes. Metals are generally considered trace elements, although the exact meaning of "trace" depends on the system under investigation. For example, Fe is a trace element in the ocean and in biological processes, but it is considered a major element in geology, as a rock-forming elements.

WO_4^{2-} tungstate ion

REFERENCES

1. B. Hay Mele, M. Monticelli, S. Leone, D. Bastoni, B. Barosa, M. Cascone, F. Migliaccio, F. Montemagno, A. Ricciardelli, L. Tonietti, A. Rotundi, A. Cordone, D. Giovannelli, *Essays Biochem.* **2023**, *67*, 653–670.
2. B. I. Jelen, D. Giovannelli, P. G. Falkowski, *Annu. Rev. Microbiol.* **2016**, *70*, 45–62.
3. J. P. Amend, E. L. Shock, *FEMS Microbiol. Rev.* **2001**, *25*, 175–243.
4. E. L. Shock, M. E. Holland, *Geochemical Energy Sources that Support the Subsurface Biosphere*, in *Geophysical Monograph Series*, W.S.D. Wilcock, E.F. DeLong, D.S. Kelley, J.A. Baross, S. Craig Cary, Eds, American Geophysical Union, Washington, DC, 2004, pp. 153–165.
5. J. P. Amend, T. M. McCollom, M. Hentscher, W. Bach, *Geochim. Cosmochim. Acta* **2011**, *75*, 5736–5748.
6. P. G. Falkowski, T. Fenchel, E. F. Delong, *Science* **2008**, *320*, 1034–1039.
7. E. K. Moore, B. I. Jelen, D. Giovannelli, H. Raanan, P. G. Falkowski, *Nature Geosci* **2017**, *10*, 629–636.
8. T. W. Lyons, C. J. Tino, G. P. Fournier, R. E. Anderson, W. D. Leavitt, K. O. Konhauser, E. E. Stüeken, *Nat. Rev. Microbiol.* **2024**, *22*, 572–586.
9. C. Coclet, C. Garnier, S. D'Onofrio, G. Durrieu, E. Pasero, C. Le Poupon, D. Omanović, J.-U. Mullot, B. Misson, J.-F. Briand, *Front. Microbiol.* **2021**, *12*.
10. S. A. Loutet, A. C. K. Chan, M. J. Kobylarz, M. M. Verstraete, S. Pfaffen, B. Ye, A. L. Arrieta, M. E. P. Murphy, *The Fate of Intracellular Metal Ions in Microbes*, in *Trace Metals and Infectious Diseases*, J.O. Nriagu, E.P. Skaar, Eds, MIT Press, Cambridge (MA), 2015.
11. L. Tonietti, M. Esposito, M. Cascone, B. Barosa, S. Fiscale, M. T. M. Tomajoli, T. Sbaffi, R. Santomartino, G. Covone, A. Cordone, D. Giovannelli, A. Rotundi, *Microorganisms* **2024**, *12*, 2407.
12. R. Quatrini, C. Appia-Ayme, Y. Denis, E. Jedlicki, D. S. Holmes, V. Bonnefoy, *BMC Genomics* **2009**, *10*, 394.
13. A. J. Poulain, T. Barkay, *Science* **2013**, *339*, 1280–1281.
14. A. Sulaymon, *J. Chemc. Sci. Technol.* **2014**, *3*, 74–102.
15. N. A. Fitriyanto, M. Fushimi, M. Matsunaga, A. Pertiwiningrum, T. Iwama, K. Kawai, *J. Biosci. Bioeng.* **2011**, *111*, 613–617.
16. R. A. Schmitz, N. Picone, H. Singer, A. Dietl, K.-A. Seifert, A. Pol, M. S. M. Jetten, T. R. M. Barends, L. J. Daumann, H. J. M. Op den Camp, *mBio* **2021**, *12*, 10.1128/mbio.01708-21.

17. R. E. M. Rickaby, *Phys. Eng. Sci.* **2015**, *373*, 20140188.
18. K. A. Remick, J. D. Helmann, *Adv. Microb. Physiol.* **2023**, *82*, 1–127.
19. D. Giovannelli, *Nat. Rev. Earth. Environ.* **2023**, *4*, 597–598.
20. V. I. Vernadsky, *The Biosphere*, Springer, New York, NY, 1998.
21. K. D. Putirka, J. C. Rarick, *American Mineralogist.* **2019**, *104*, 817–829.
22. D. Bastoni, P. H. Barry, D. V. Bekaert, A. Bastianoni, M. Cascone, A. Cordone, G. Covone, M. Edmonds, P. Falkowski, K. G. Lloyd, G. L. Jessen, R. M. Hazen, F. Montemagno, J. M. De Moor, S. Morrison, R. E. M. Rickaby, M. O. Schrenk, L. Tonietti, S. Zahirovic, D. Giovannelli, *Submitted* **2024**.
23. C. E. Nelson, E. K. Wear, *Proc. Natl. Acad. Sci.* **2014**, *111*, 7166–7167.
24. M. Delgado-Baquerizo, A. M. Oliverio, T. E. Brewer, A. Benavent-González, D. J. Eldridge, R. D. Bardgett, F. T. Maestre, B. K. Singh, N. Fierer, *Science* **2018**, *359*, 320–325.
25. S. Chen, F. Wang, Y. Zhang, S. Qin, S. Wei, S. Wang, C. Hu, B. Liu, *Environ. Microbiol.* **2018**, *20*, 980–992.
26. C. Méndez-García, A. I. Peláez, V. Mesa, J. Sánchez, O. V. Golyshina, M. Ferrer, *Front. Microbiol.* **2015**, *6*, 1–17.
27. C. Magnabosco, P. H. A. Timmers, M. C. Y. Lau, G. Borgonie, B. Linage-Alvarez, O. Kuloyo, R. Alleva, T. L. Kieft, G. F. Slater, E. van Heerden, B. Sherwood Lollar, T. C. Onstott, *FEMS Microbiol. Ecol.* **2018**, *94*, fiy089.
28. G. Ren, A. Ma, Y. Zhang, Y. Deng, G. Zheng, X. Zhuang, G. Zhuang, D. Fortin, *Environ. Microbiol.* **2018**, *20*, 2370–2385.
29. N. H. Akerman, R. E. Price, T. Pichler, J. P. Amend, *Geobiology* **2011**, *9*, 436–445.
30. T. Großkopf, O. S. Soyer, *ISME J.* **2016**, *10*, 2725–2733.
31. X. Liu, M. M. I. Chowdhury, M. Zaman, M. Kim, G. Nakhla, *J. Environ. Sci.* **2019**, *82*, 169–178.
32. J. T. Trevors, C. M. Cotter, *J. Indl. Microbiol.* **1990**, *6*, 77–84.
33. K. E. Giller, E. Witter, S. P. Mcgrath, *Soil. Biol. Biochem.* **1998**, *30*, 1389–1414.
34. V. P. Edgcomb, S. J. Molyneaux, M. A. Saito, K. Lloyd, S. Böer, C. O. Wirsen, M. S. Atkins, A. Teske, *Appl. Environ. Microbiol.* **2004**, *70*, 2551–2555.
35. E. M. Bertrand, M. A. Saito, J. M. Rose, C. R. Riesselman, M. C. Lohan, A. E. Noble, P. A. Lee, G. R. DiTullio, *Limnol. Oceanogr.* **2007**, *52*, 1079–1093.
36. F. M. M. Morel, N. M. Price, *Science* **2003**, *300*, 944–947.
37. S. H. Hutner, L. Provosoli, A. Schatz, C. P. Haskins, S. H. Hutner, L. Provosoli, A. Schatz, C. P. Haskins, *Proc. Am. Phil. Soc.* **1950**, *94*, 152–170.
38. A. Pandey, F. Bringel, J.-M. Meyer, *Appl. Microbiol. Biotechnol.* **1994**, *40*, 735–739.
39. M. Cerasi, S. Ammendola, A. Battistoni, *Front. Cell. Infect. Microbiol.* **2013**, *3*.
40. J. E. Jacquot, R. E. A. Horak, S. A. Amin, A. H. Devol, A. E. Ingalls, E. V. Armbrust, D. A. Stahl, J. W. Moffett, *Mar. Chem.* **2014**, *162*, 37–49.
41. L. Ma, A. Terwilliger, A. W. Maresso, *Metallomics* **2015**, *7*, 1541–1554.
42. J. H. Martin, S. E. Fitzwater, *Nature* **1988**, *331*, 341–343.
43. A. Tagliabue, A. R. Bowie, P. W. Boyd, K. N. Buck, K. S. Johnson, M. A. Saito, *Nature* **2017**, *543*, 51–59.
44. E. V. Kupriyanova, N. A. Pronina, *Russ. J. Plant. Physiol.* **2011**, *58*, 197–209.
45. P. Flombaum, J. L. Gallegos, R. A. Gordillo, J. Rincón, L. L. Zabala, N. Jiao, D. M. Karl, W. K. W. Li, M. W. Lomas, D. Veneziano, C. S. Vera, J. A. Vrugt, A. C. Martiny, *Proc. Natl. Acad. Sci.* USA **2013**, *110*, 9824–9829.
46. C. B. Field, M. J. Behrenfeld, J. T. Randerson, P. Falkowski, *Science* **1998**, *281*, 237–240.
47. G. S. Espie, M. S. Kimber, *Photosynth. Res.* **2011**, *109*, 7–20.

48. Y. Matsuda, P. G. Kroth, *Carbon Fixation in Diatoms*, in *The Structural Basis of Biological Energy Generation*, M.F. Hohmann-Marriott, Ed., Springer, Dordrecht, NL, 2014, pp. 335–362.
49. F. Morel, P. Lam, M. Saito, *Ann. Rev. Earth Planet Sci.* **2020**, *48*.
50. T. D. Jickells, A. R. Baker, R. Chance, *Philos. Trans. A Math. Phys. Eng. Sci.* **2016**, *374*, 20150286.
51. J. T. Ruud, *ICES J. Mar. Sci.* **1930**, *5*, 347–360.
52. J. H. Street, A. Paytan, *Met. Ions Biol. Syst.* **2005**, *43*, 153–93.
53. H. Schoffman, H. Lis, Y. Shaked, N. Keren, *Front. Plant. Sci.* **2016**, *7*.
54. J. Wade, D. J. Byrne, C. J. Ballentine, H. Drakesmith, *Pro. Natl. Acad. Sci.* USA, **2021**, *118*, e2109865118.
55. S. M. Payne, A. R. Mey, E. E. Wyckoff, *Microbiol. Mol. Biol. Rev.* **2015**, *80*, 69–90.
56. R. H. Lampe, T. H. Coale, K. O. Forsch, L. J. Jabre, S. Kekuewa, E. M. Bertrand, A. Horák, M. Oborník, A. J. Rabines, E. Rowland, H. Zheng, A. J. Andersson, K. A. Barbeau, A. E. Allen, *Nat. Commun.* **2023**, *14*, 7215.
57. W. Sunda, *Front. Microbiol.* **2012**, *3*.
58. R. García-Cañas, J. Giner-Lamia, F. J. Florencio, L. López-Maury, *PNAS* **2021**, *118*.
59. J. Raymond, R. Blankenship, *Coord. Chem. Rev.* **2008**, *252*, 377–383.
60. N. J. Hawco, A. Tagliabue, B. S. Twining, *Global Biogeochem. Cycles* **2022**, *36*, e2022GB007382.
61. S. González-Gil, B. A. Keafer, R. V. M. Jovine, A. Aguilera, S. Lu, D. M. Anderson, *Mar. EcolProg. Ser.* **1998**, *164*, 21–35.
62. E. L. Jensen, R. Clement, A. Kosta, S. C. Maberly, B. Gontero, *ISME J.* **2019**, *13*, 2094–2106.
63. N. M. Price, F. M. M. Morel, *Nature* **1990**, *344*, 658–660.
64. D. Osman, A. Cooke, T. R. Young, E. Deery, N. J. Robinson, M. J. Warren, *Biochim. Biophys. Acta-Mol. Cell. Res.* **2021**, *1868*, 118896.
65. K. R. Heal, W. Qin, F. Ribalet, A. D. Bertagnolli, W. Coyote-Maestas, L. R. Hmelo, J. W. Moffett, A. H. Devol, E. V. Armbrust, D. A. Stahl, A. E. Ingalls, *Proc. Nat. Acad. Sci.* USA **2017**, *114*, 364–369.
66. M. Beauvais, P. Schatt, L. Montiel, R. Logares, P. E. Galand, F.-Y. Bouget, *Environ. Microbiol.* **2023**, *25*, 3753–3770.
67. R. T. Shafiee, P. J. Diver, J. T. Snow, Q. Zhang, R. E. M. Rickaby, *ISME Commun.* **2021**, *1*, 1–12.
68. S. Berger, A. Cabrera-Orefice, M. S. M. Jetten, U. Brandt, C. U. Welte, *Biochim. Biophys. Acta Bioenerg.* **2021**, *1862*, 148308.
69. J. Glass, V. J. Orphan, *Front. Microbiol.* **2012**, *3*.
70. K. M. Fullerton, M. O. Schrenk, M. Yücel, E. Manini, M. Basili, T. J. Rogers, D. Fattorini, M. Di Carlo, G. d'Errico, F. Regoli, M. Nakagawa, C. Vetriani, F. Smedile, C. Ramírez, H. Miller, S. M. Morrison, J. Buongiorno, G. L. Jessen, A. D. Steen, M. Martínez, J. M. de Moor, P. H. Barry, D. Giovannelli, K. G. Lloyd, *Nat. Geosci.* **2021**, *14*, 301–306.
71. S. Srivastava, H. Dong, B. R. Briggs, *Appl. Environ. Microbiol.* **2020**, *86*, e00581–20.
72. W. Buckel, H. A. Barker, *J. Bacteriol.* **1974**, *117*, 1248–1260.
73. W. Buckel, *Front. Microbiol.* **2021**, *12*.
74. M. Khademian, J. A. Imlay, *Trends Microbiol.* **2021**, *29*, 428–440.
75. M. Ramezani, K. L. Resmer, R. L. White, *FEBS J.* **2011**, *278*, 2540–2551.
76. F. Baltar, A. Gutiérrez-Rodríguez, M. Meyer, I. Skudelny, S. Sander, B. Thomson, S. Nodder, R. Middag, S. E. Morales, *Front. Microbiol.* **2018**, *9*.

77. R. Denaro, F. Crisafi, D. Russo, M. Genovese, E. Messina, L. Genovese, M. Carbone, M. L. Ciavatta, M. Ferrer, P. Golyshin, M. M. Yakimov, *Mar. Genomics* **2014**, *17*, 43–52.
78. R. R. Eady, *Chem. Rev.* **1996**, *96*, 3013–3030.
79. T. L. Hamilton, M. Ludwig, R. Dixon, E. S. Boyd, P. C. Dos Santos, J. C. Setubal, D. A. Bryant, D. R. Dean, J. W. Peters, *J. Bacteriol.* **2011**, *193*, 4477–4486.
80. X. Zhang, D. M. Sigman, F. M. M. Morel, A. M. L. Kraepiel, *Pro. Natl. Acad. Sci.* USA **2014**, *111*, 4782–4787.
81. S. Buessecker, M. Palmer, D. Lai, J. Dimapilis, X. Mayali, D. Mosier, J.-Y. Jiao, D. R. Colman, L. M. Keller, E. St. John, M. Miranda, C. Gonzalez, L. Gonzalez, C. Sam, C. Villa, M. Zhuo, N. Bodman, F. Robles, E. S. Boyd, A. D. Cox, B. St. Clair, Z.-S. Hua, W.-J. Li, A.-L. Reysenbach, M. B. Stott, P. K. Weber, J. Pett-Ridge, A. E. Dekas, B. P. Hedlund, J. A. Dodsworth, *Nat. Commun.* **2022**, *13*, 3773.
82. X. M. Feng, A. Karlsson, B. H. Svensson, S. Bertilsson, *FEMS Microbiol. Ecol.* **2010**, *74*, 226–240.
83. C. J. Banks, Y. Zhang, Y. Jiang, S. Heaven, *Bioresour. Technol.* **2012**, *104*, 127–135.
84. T. Schmidt, M. Nelles, F. Scholwin, J. Pröter, *Bioresour. Technol.* **2014**, *168*, 80–85.
85. M. Carballa, L. Regueiro, J. M. Lema, *Curr. Opin. Biotechno.* **2015**, *33*, 103–111.
86. R. Wirth, E. Kovács, G. Maróti, Z. Bagi, G. Rákhely, K. L. Kovács, *Biotechnol. Biofuels.* **2012**, *5*, 41.
87. A. Chojnacka, P. Szczęsny, M. K. Błaszczyk, U. Zielenkiewicz, A. Detman, A. Salamon, A. Sikora, *PLoS One* **2015**, *10*, e0128008.
88. E. Abdelsalam, M. Samer, Y. A. Attia, M. A. Abdel-Hadi, H. E. Hassan, Y. Badr, *Energy Conv. Manag.* **2017**, *141*, 108–119.
89. R. Chouari, D. Le Paslier, P. Daegelen, P. Ginestet, J. Weissenbach, A. Sghir, *Environ. Microbiol.* **2005**, *7*, 1104–1115.
90. R. Goswami, P. Chattopadhyay, A. Shome, S. N. Banerjee, A. K. Chakraborty, A. K. Mathew, S. Chaudhury, *3 Biotech.* **2016**, *6*, 72.
91. M. B. Osuna, J. Iza, M. Zandvoort, P. N. L. Lens, *Water Sci. Technol.* **2003**, *48*, 1–8.
92. B. Demirel, P. Scherer, *Biomass Bioenerg.* **2011**, *35*, 992–998.
93. A. Karlsson, P. Einarsson, A. Schnürer, C. Sundberg, J. Ejlertsson, B. H. Svensson, *J. Biosci. Bioeng.* **2012**, *114*, 446–452.
94. H. Qiang, Q. Niu, Y. Chi, Y. Li, *Chem. Engg. J.* **2013**, *222*, 330–336.
95. L. Zhang, W. Ouyang, A. Lia, *Procedia Environl. Sci.* **2012**, *16*, 102–111.
96. B. Wintsche, N. Jehmlich, D. Popp, H. Harms, S. Kleinsteuber, *Front. Microbiol.* **2018**, *9*.
97. Y. Jiang, Y. Zhang, C. Banks, S. Heaven, P. Longhurst, *Water Res.* **2017**, *125*, 458–465.
98. B. Shamurad, P. Sallis, E. Petropoulos, S. Tabraiz, C. Ospina, P. Leary, J. Dolfing, N. Gray, *Appl. Energy* **2020**, *263*, 114609.
99. A. Schnürer, *Biogas Production: Microbiology and Technology*, in *Anaerobes in Biotechnology*, R. Hatti-Kaul, G. Mamo, B. Mattiasson, Eds, Springer International Publishing, Cham, **2016**, pp. 195–234.
100. V. Facchin, C. Cavinato, F. Fatone, P. Pavan, F. Cecchi, D. Bolzonella, *Biochem. Eng. J.* **2013**, *70*, 71–77.
101. P. Schönheit, J. Moll, R. K. Thauer, *Arch. Microbiol.* **1979**, *123*, 105–107.
102. K. Kida, T. Shigematsu, J. Kijima, M. Numaguchi, Y. Mochinaga, N. Abe, S. Morimura, *J. Biosci. Bioeng.* **2001**, *91*, 590–595.
103. C. J. Banks, S. Heaven, *Optimisation of biogas yields from anaerobic digestion by feedstock type*, in *The Biogas Handbook*, A. Wellinger, J. Murphy, D. Baxter, Eds, Woodhead Publishing, **2013**, pp. 131–165.

104. L. Šafarič, S. S. Yekta, B. H. Svensson, A. Schnürer, D. Bastviken, A. Björn, *Microorganisms* **2020**, *8*, 598.
105. J. A. Johnson, *Science* **2019**, *363*, 474–478.
106. K. Mezger, A. Maltese, H. Vollstaedt, *Icarus* **2021**, *365*, 114497.
107. S. Ekström, *Front. Astron. Space Sci.* **2021**, *8*.
108. T. J. Looser, F. D'Eugenio, J. M. Piotrowska, F. Belfiore, R. Maiolino, M. Cappellari, W. M. Baker, S. Tacchella, *Mon. Not. R. Astron. Soc.* **2024**, *532*, 2832–2841.
109. A. Frebel, T. C. Beers, *Phys. Today* **2018**, *71*, 30–37.
110. A. M. Nikanorov, *Geochem. Int.* **2016**, *54*, 464–469.
111. D. Schramm, R. Wagoner, *Ann. Rev. Nucl. Sci.* **2003**, *27*, 37–74.
112. K. Lodders, *ApJ* **2003**, *591*, 1220.
113. C. H. Lineweaver, J. A. Robles, in *Bioastronomy 2007: Molecules, Microbes and Extraterrestrial Life*, Volume 420 of the ASP Conference Series, K. J. Meech, J. V. Keane, M. J. Mumma, J. L. Siefert, D. J. Werthimer, Eds, Astronomical Society of the Pacific, San Francisco, **2009**, pp. 83–86.
114. A. E. Ringwood, *Geochim. Cosmochim. Acta* **1966**, *30*, 41–104.
115. J. L. Johnson, H. Li, *Ap. J.* **2012**, *751*, 81.
116. G. Covone, D. Giovannelli, preprint, arXiv **2022**. https://doi.org/10.48550/arXiv.2207.03748.
117. V. Adibekyan, *Geosci.* **2019**, *9*, 105.
118. K. Nomoto, S. Wanajo, Y. Kamiya, N. Tominaga, H. Umeda, *Proc. IAU* **2008**, *4*, 355–368.
119. D. H. Weinberg, J. A. Holtzman, S. Hasselquist, J. C. Bird, J. A. Johnson, M. Shetrone, J. Sobeck, C. Allende Prieto, D. Bizyaev, R. Carrera, R. E. Cohen, K. Cunha, G. Ebelke, J. G. Fernandez-Trincado, D. A. García-Hernández, C. R. Hayes, H. Jönsson, R. R. Lane, S. R. Majewski, V. Malanushenko, S. Mészáros, D. L. Nidever, C. Nitschelm, K. Pan, H.-W. Rix, J. Rybizki, R. P. Schiavon, D. P. Schneider, J. C. Wilson, O. Zamora, *Astrophys. J.* **2019**, *874*, 102.
120. N. Moskovitz, E. Gaidos, *Meteorit. Planet. Sci.* **2011**, *46*, 903–918.
121. T.-A. Suer, C. Jackson, D. S. Grewal, C. Dalou, T. Lichtenberg, *Front. Earth. Sci.* **2023**, *11*.
122. M. Edmonds, E. Mason, O. Hogg, *Annu. Rev. Earth Planet. Sci.* **2022**, *50*, 79–98.
123. N. L. Chabot, D. S. Draper, C. B. Agee, *Geochim. Cosmochim. Acta* **2005**, *69*, 2141–2151.
124. F. Jenner, *Nat. Geosci.* **2017**, *10*.
125. B. F. Windley, T. Kusky, A. Polat, *Precambrian Res.* **2021**, *352*, 105980.
126. W. B. Hamilton, *Lithos* **2011**, *123*, 1–20.
127. S.-Y. Zhao, A. Y. Yang, C. H. Langmuir, T.-P. Zhao, *Sci. Adv.* **2022**, *8*, eabk0718.
128. P. Tollan, J. Hermann, *Nat. Geosci.* **2019**, *12*, 667–671.
129. L. A. Warren, E. A. Haack, *Earth. Sci. Rev.* **2001**, *54*, 261–320.
130. M. C. Davis, W. J. Brouwer, D. J. Wesolowski, L. M. Anovitz, A. S. Lipton, K. T. Mueller, *Phys. Chem. Phys.* **2009**, *11*, 7013–7021.
131. G. M. Gadd, *Microbiol.* **2010**, *156*, 609–643.
132. G. Haferburg, E. Kothe, *J. Basic Microbiol.* **2007**, *47*, 453–467.
133. D. A. Bazylinski, R. B. Frankel, *Rev. Mineral. Geochem.* **2003**, *54*, 217–247.
134. T. D. Hoffmann, B. J. Reeksting, S. Gebhard, *Microbiol.* **2021**, *167*, 001049.
135. H. Dong, L. Huang, L. Zhao, Q. Zeng, X. Liu, Y. Sheng, L. Shi, G. Wu, H. Jiang, F. Li, L. Zhang, D. Guo, G. Li, W. Hou, H. Chen, *Nat. Sci. Rev.* **2022**, *9*, nwac128.
136. A. D. Anbar, A. H. Knoll, *Science* **2002**, *297*, 1137–1142.

137. A. Gando, Y. Gando, K. Ichimura, H. Ikeda, K. Inoue, Y. Kibe, Y. Kishimoto, M. Koga, Y. Minekawa, T. Mitsui, T. Morikawa, N. Nagai, K. Nakajima, K. Nakamura, K. Narita, I. Shimizu, Y. Shimizu, J. Shirai, F. Suekane, A. Suzuki, H. Takahashi, N. Takahashi, Y. Takemoto, K. Tamae, H. Watanabe, B. D. Xu, H. Yabumoto, H. Yoshida, S. Yoshida, S. Enomoto, A. Kozlov, H. Murayama, C. Grant, G. Keefer, A. Piepke, T. I. Banks, T. Bloxham, J. A. Detwiler, S. J. Freedman, B. K. Fujikawa, K. Han, R. Kadel, T. O'Donnell, H. M. Steiner, D. A. Dwyer, R. D. McKeown, C. Zhang, B. E. Berger, C. E. Lane, J. Maricic, T. Miletic, M. Batygov, J. G. Learned, S. Matsuno, M. Sakai, G. A. Horton-Smith, K. E. Downum, G. Gratta, K. Tolich, Y. Efremenko, O. Perevozchikov, H. J. Karwowski, D. M. Markoff, W. Tornow, K. M. Heeger, M. P. Decowski, The KamLAND Collaboration, *Nature Geosci.* **2011**, *4*, 647–651.
138. B. Dhuime, C. J. Hawkesworth, P. A. Cawood, C. D. Storey, *Science* **2012**, *335*, 1334–1336.
139. C. B. Keller, B. Schoene, *Nature* **2012**, *485*, 490–493.
140. J. Hao, D. A. Sverjensky, R. M. Hazen, *Earth Planet. Sci. Lett.* **2017**, *471*, 148–159.
141. M. A. Saito, D. M. Sigman, F. M. M. Morel, *Inorg. Chim. Acta* **2003**, *356*, 308–318.
142. L. Schwander, M. Brabender, N. Mrnjavac, J. L. E. Wimmer, M. Preiner, W. F. Martin, *Front. Microbiol.* **2023**, *14*.
143. M. Preiner, J. C. Xavier, F. L. Sousa, V. Zimorski, A. Neubeck, S. Q. Lang, H. C. Greenwell, K. Kleinermanns, H. Tüysüz, T. M. McCollom, N. G. Holm, W. F. Martin, *Life (Basel)* **2018**, *8*, 41.
144. L. Shi, B. P. Tu, *Curr. Opin. Cell. Biol.* **2015**, *33*, 125–131.
145. M. C. McCanta, M. D. Dyar, *Icarus* **2020**, *352*, 113978.
146. M. Can, F. A. Armstrong, S. W. Ragsdale, *Chem. Rev.* **2014**, *114*, 4149–4174.
147. M. J. Russell, W. Martin, *Trends Biochem. Sci.* **2004**, *29*, 358–363.
148. W. Buckel, R. K. Thauer, *Chem. Rev.* **2018**, *118*, 3862–3886.
149. M. Brabender, D. P. Henriques Pereira, N. Mrnjavac, M. L. Schlikker, Z.-I. Kimura, J. Sucharitakul, K. Kleinermanns, H. Tüysüz, W. Buckel, M. Preiner, W. F. Martin, *Proc. Natl. Acad. Sci.* USA **2024**, *121*, e2318969121.
150. P. S. Garcia, S. Gribaldo, F. Barras, *Biochim. Biophys. Acta - Mol. Cell. Res.* **2024**, *1871*, 119730.
151. G. Fuchs, *Annu. Rev. Microbiol.* **2011**, *65*, 631–658.
152. R. V. Eck, M. O. Dayhoff, *Science* **1966**, *152*, 363–366.
153. D. O. Hall, R. Cammack, K. K. Rao, *Nature* **1971**, *233*, 136–138.
154. S. F. Jordan, I. Ioannou, H. Rammu, A. Halpern, L. K. Bogart, M. Ahn, R. Vasiliadou, J. Christodoulou, A. Maréchal, N. Lane, *Nat. Commun.* **2021**, *12*.
155. H. Raanan, S. Poudel, D. Pike, V. Nanda, P. Falkowski, *Pro. Natl. Acad. Sci.* USA **2020**, *117*, 201914982.
156. J. E. Goldford, H. Hartman, T. F. Smith, D. Segrè, *Cell* **2017**, *168*, 1126–1134.e9.
157. W. F. Martin, R. K. Thauer, *Cell* **2017**, *168*, 953–955.
158. L. G. Ljungdahl, *Annu. Rev. Microbiol.* **2009**, *63*, 1–25.
159. G. Fuchs, *Annu. Rev. Microbiol.* **2011**, *65*, 631–658.
160. R. Braakman, E. Smith, *PLoS Comput. Biol.* **2012**, *8*, e1002455.
161. K. B. Muchowska, E. Chevallot-Beroux, J. Moran, *Bioorg. Med. Chem.* **2019**, *27*, 2292–2297.
162. K. B. Muchowska, S. J. Varma, J. Moran, *Chem. Rev.* **2020**, *120*, 7708–7744.
163. J. Zimmermann, E. Werner, S. Sodei, J. Moran, *Acc. Chem. Res.* **2024**, *57*, 3032–3043.
164. K. Muchowska, S. J. Varma, E. Chevallot-Beroux, L. Lethuillier, G. Li, J. Moran, *Nat. Ecolo. Evolu.* **2017**, *1*.
165. C. L. Dupont, A. Butcher, R. E. Valas, P. E. Bourne, G. Caetano-Anollés, *Proc. Natl. Acad. Sci.* USA **2010**, *107*, 10567–10572.

166. A. W. Foster, D. Osman, N. J. Robinson, *J. Biolo. Chem.* **2014**, *289*, 28095–28103.
167. D. Osman, M. A. Martini, A. W. Foster, J. Chen, A. J. P. Scott, R. J. Morton, J. W. Steed, E. Lurie-Luke, T. G. Huggins, A. D. Lawrence, E. Deery, M. J. Warren, P. T. Chivers, N. J. Robinson, *Nat. Chem. Biol.* **2019**, *15*, 241–249.
168. A. C. Rosenzweig, *Chem. Biol.* **2002**, *9*, 673–677.
169. A.-F. Miller, *FEBS Lett.* **2011**, *586*, 585.
170. K. A. Frye, K. M. Sendra, K. J. Waldron, T. E. Kehl-Fie, *J. Inorg. Biochem.* **2022**, *230*, 111748.
171. Y. Zhao, S.-M. Bian, H.-N. Zhou, J.-F. Huang, *J. Integr. Plant. Biol.* **2006**, *48*, 745–755.
172. A. K. Garcia, H. McShea, B. Kolaczkowski, B. Kaçar, *Geobiology* **2020**, *18*, 394–411.
173. T. Nakagawa, R. Mitsui, A. Tani, K. Sasa, S. Tashiro, T. Iwama, T. Hayakawa, K. Kawai, *PLoS One* **2012**, *7*, e50480.
174. L. L. Barton, F. Goulhen, M. Bruschi, N. A. Woodards, R. M. Plunkett, F. J. M. Rietmeijer, *Biometals* **2007**, *20*, 291–302.
175. L. Galera-Laporta, C. J. Comerci, J. Garcia-Ojalvo, G. M. Süel, *Cell. Syst.* **2021**, *12*, 497–508.
176. A. L. Zerkle, C. H. House, S. L. Brantley, *Am. J. Sci.* **2005**, *305*, 467–502.

5 Electron Transfer Symphony

Unveiling the Enigmatic Role of Inorganic Iron Oxide and Sulfide Nanoparticles as Inorganic Enzymes in Early Earth Environments

Xiaolan Huang
New York State Center for Clean Water Technology, School of Marine and Atmospheric Sciences, Stony Brook University, Stony Brook, NY 11794-6044, USA
xiaolan.huang@ymail.com

CONTENTS

ABSTRACT

This chapter explores the interplay between inorganic nanoparticles and oxidoreductase activity, emphasizing their critical role in Earth's biochemical history. It focuses

DOI: 10.1201/9781003459910-5

on the electron transfer properties of these nanoparticles, particularly iron oxide and sulfide variants, revealing their oxidoreductase activity akin to that of proteins. These nanoparticles, termed "life fossil oxidoreductases", predate complex biological molecules and influence key processes such as redox reactions, detoxification, oxygen formation, organic carbon transformation, phosphate ester hydrolysis, and inorganic denitrification. These abiotic catalytic processes were essential for early life on Earth, providing the material and energy required for life processes.

The investigation emphasizes the structural configuration and electron conformation of these nanoparticles, establishing a foundation for understanding inorganic enzyme mechanisms. It underscores that the electron transfer mechanisms in these inorganic systems are governed by the same unchanging principles as those in organic oxidoreductases, highlighting the seamless continuity between the inorganic and organic worlds. This physics-based approach, rooted in electron transfer principles aligned with the second law of thermodynamics and chemical equilibrium, reveals a fundamental unity in the processes that drive life and nature.

Today, these inorganic nanoparticles continue to function as vital biocatalysts in natural ecosystems, driving critical reactions that maintain ecosystem health. The study also draws a parallel with the evolution of metal centers within proteins, highlighting a shared narrative driven by electron transfer dynamics. This chapter unveils the significant contributions of inorganic nanoparticles and offers insights into the fundamental principles that governed early life processes and adaptations. These inorganic enzymes, with their electron transfer capabilities, remain key players in sustaining life on Earth, leaving an enduring impact on the planet's complex web of life.

KEYWORDS

biocatalytic nanomaterials, cytochrome c, catalytic efficiency, direct electron transfer, early earth environments, enzyme kinetics, electron transport chain, electron hopping dynamics, electric conductivity, life fossil oxidoreductase, inorganic enzyme, inorganic nanozyme, iron oxide, iron sulfides, metal architecture, metalloenzyme, evolution, nanomaterials, oxidoreductase, origin of life, photoreactivation, photothermal transformation, redox reactions

> Life is a partial, continuous, progressive, multiform and conditionally interactive self-realization of the potentialities of atomic electron states
>
> **– John Desmond Bernal**

1 INTRODUCTION

Oxidoreductases, classified as primary class 1 enzymes by the Enzyme Commission (EC), play a crucial role in orchestrating electron transfer (ET) reactions across diverse organisms [1–4]. Their significance is underscored by their involvement in nearly one-third of all enzymatic activities cataloged in the BRaunschweig ENzyme Database (BRENDA) [5], particularly the subset of metal-containing oxidoreductases (EC1 or EC7.1) [6]. It is believed that oxidoreductases were among the first biocatalysts in the origins of life, providing the spark for metabolism [4,6–9]. To some extent, life can be seen as a narrative of electron descent [10]. This notion is closely related to the electron field [11]. ET is essential for all compounds in life, including proteins and genetic molecules (RNA/DNA), and is not limited to chemical reactions involving solutions [11–13]. In fact, ET also occurs on the surfaces of nanoparticles and even within the nanoparticles themselves as biocatalysts [8,9], broadening the scope of electron transfer in biological and environmental systems. Moreover, the interaction between nanoparticles and enzymes, particularly oxidoreductases, can create synergistic effects that enhance catalytic activity and electron flow [14,15].

At their core, oxidoreductases drive the oxidation of one chemical species, acting as an electron donor or reducing agent, while concurrently reducing another species, functioning as an electron acceptor or oxidizing agent. These enzymes are free energy converters. Such redox transformations are elegantly represented as $A^- + B \rightarrow A + B^-$. Notably, these biological catalysts are central to ET, with some also serving as electron exchange mediators within their microenvironment or engaging in intramolecular charge transport [16,17].

The versatility of oxidoreductases is highlighted by their ability to interact with a wide range of substrates. They proficiently act on diverse organic substrates, including alcohols, amines, and ketones. Moreover, their activity extends to inorganic substrates such as small anions (e.g., sulfite) and metals (e.g., mercury). However, their catalytic activity often leads to the generation of reactive oxygen species (ROS) and hydrogen peroxide (H_2O_2) as byproducts [18–21]. In some cases, they can also produce reactive sulfur species (RSS), depending on the sulfur-containing substrates involved.

Instances of oxidoreductases producing ROS and H_2O_2 encompass peroxidases (POD), catalases (CAT), superoxide dismutase (SOD), and oxidases (OXD). PODs exploit H_2O_2 or organic hydroperoxides (R-OOH) as electron donors, with H_2O_2 serving as the electron acceptor during oxidations [22–24]. OXDs facilitate the oxidation of diverse substrates (electron donors) via molecular oxygen (O_2) as the electron acceptor, utilizing hydrogen atoms to produce water or H_2O_2, as seen in enzymes like sulfite oxidase (SOE), glucose oxidase (GOX), or alcohol oxidase (AOX) [25–28]. CATs accelerate H_2O_2 decomposition into water and O_2 [29,30], while SODs disproportionally divide superoxide radicals ($O_2{}^{*-}$) into H_2O_2 and O_2 [31,32].

Intriguingly, certain oxidoreductases directly transfer electrons or employ mediators like cytochrome c (Cyt c) to solid surfaces such as enzymes, electrodes, microorganisms, and nanomaterials [14,33–35]. This phenomenon, termed direct electron transfer (DET) [14,33–39], was initially documented in 1977 by Eddows and Hill [40] and Yeh and Kuwna [41]. Their studies highlighted Cyt c's reversible electrochemistry on gold and tin-doped indium oxide electrodes, observed through cyclic voltammetry. Notably, horseradish peroxidase (HRP) [42] and laccase (Lc) [43] also adhere to carbon electrodes, showcasing DET capacity. Impressively, more than 100 enzymes, with a predominant presence of oxidoreductases, are known to function under DET conditions [14,44–50].

Beyond their established roles in metabolism – spanning aerobic and anaerobic pathways such as glycolysis, the Krebs cycle, oxidative phosphorylation, and amino acid metabolism – oxidoreductases manifest considerable implications in the realm of life's origin [2,7,51]. Furthermore, their impact extends to key processes like photosynthesis in chloroplasts, drug metabolism, and liver detoxification reactions [52]. Notably, oxygen-mediating enzymes, such as POD, CAT, OXD, and SOD, are believed to have emerged before the advent of photosynthesis, particularly oxygenic photosynthesis [53–59]. The intriguing proposition has been put forth that photosystem II, also known as water-plastoquinone oxidoreductase, originated from a calcium-manganese complex ($CaMn_4O_8$) [60,61], evolving from a manganese-containing variant of CAT [7,62–64].

During the Archean era, the primitive atmosphere was mainly composed of carbon dioxide and methane [65,66], but it also harbored significant concentrations of H_2O_2 [67–73], especially following the snowball earth period [74]. The oceans likely contained iron, transition metal sulfides and oxides, and selenide nanoparticles [75–80]. Iron-sulfur clusters, e.g., [2Fe-2S] and [4Fe-4S], formed through the photooxidation of ferrous ions under UV light in laboratory experiments [81]. In this primordial setting, oxidoreductases may have served as pivotal facilitators, engaging in the activation and manipulation of small organic molecules as electron donors or acceptors. Such enzyme-driven processes could have played a pivotal role in the emergence of primordial energy metabolism. Additionally, these enzymes might have played a crucial part in detoxifying harmful reactive species, maintaining redox equilibrium, and enabling fundamental redox reactions that led to the synthesis of essential biomolecules, including amino acids, nucleotides, and cofactors. The dynamic conditions of early Earth, teeming with reactive molecules and nanoparticles, likely necessitated the emergence of efficient ET processes, catalyzed by oxidoreductases.

To unravel the potential role of oxidoreductases in the origin of life, a comprehensive exploration of their properties and capabilities under prebiotic conditions becomes paramount. Experimental studies and theoretical models offer avenues to comprehend how these enzymes could have emerged and evolved in early Earth environments. By illuminating the mechanisms and functions of oxidoreductases within the context of life's inception, we gain a deeper understanding of the fundamental processes that underlie the development of intricate biological systems.

In 2007, Gao et al. made an intriguing discovery that Fe_3O_4 nanoparticles exhibit inherent peroxidase-like activity [82]. This groundbreaking finding gave rise to the field of nanozymes, which encompasses inorganic nanomaterials exhibiting enzyme-like properties [83–86]. In a separate study, Huang observed that inorganic iron oxide nanoparticles in solution can act as phosphatases, facilitating the hydrolysis of phosphorus esters [87–90] (Figure 1). This observation led to the proposal of the "inorganic enzyme hypothesis", suggesting that the metal architecture of inorganic nanoparticles bears similarities to proteins based on the experimental evidence [8,87,90,91]. Since then, nanozymes have gained significant attention in the fields of

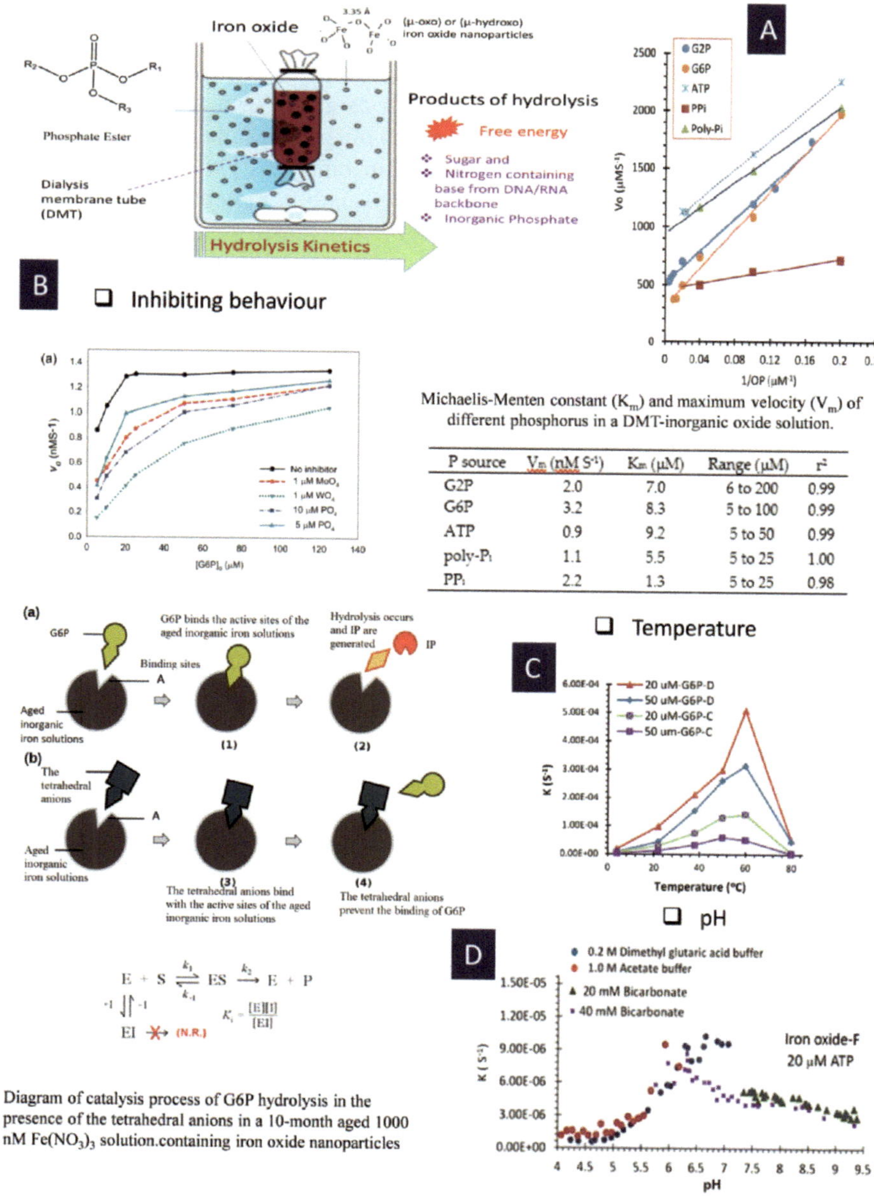

Michaelis-Menten constant (K_m) and maximum velocity (V_m) of different phosphorus in a DMT-inorganic oxide solution.

P source	V_m (nM S^{-1})	K_m (μM)	Range (μM)	r^2
G2P	2.0	7.0	6 to 200	0.99
G6P	3.2	8.3	5 to 100	0.99
ATP	0.9	9.2	5 to 50	0.99
poly-P_i	1.1	5.5	5 to 25	1.00
PP_i	2.2	1.3	5 to 25	0.98

FIGURE 1 Characteristics of inorganic phosphatase [91]. A. The enzyme-like activity of inorganic iron oxide nanoparticles solution using the dialysis membrane tube (DMT), Lineweaver–Burk plot of three organic phosphate compounds (glycerol-2-phosphate G2P, glucose-6-phosphate G6P, and adenosine triphosphate ATP) and two condensed phosphates (polyphosphate Poly-P_i. and pyrophosphate PP_i) [89]. B. The inhibiting behavior of different tetrahedral oxyanions on the hydrolysis of G6P in a 10-month aged, 1000 nM $Fe(NO_3)_3$ solution at room temperature (22°C) [87, 88]. C. The effect of temperature on the hydrolysis velocity of ATP [89]. D. Effect of pH on the hydrolysis velocity of ATP (Ref. [89]). Reprinted from [91].

chemistry and biomedicine [92–96], finding applications in diverse areas, including medicine, agriculture, and environmental protection.

In the medical field, nanozymes have gained significant attention for their potential applications in disease treatment, including cancer [97,98]. They offer advantages such as selective targeting and elimination of specific cells in the body due to their ability to selectively target and eliminate specific cells [99–103]. Ferrihydrite NPs, possessing CAT activity, enhance the effectiveness of radiotherapy [104], while magnetoferritin NPs target and visualize tumor tissues due to the iron oxide core catalyzing the oxidation of peroxidase substrates in the presence of hydrogen peroxide, producing a color reaction used to visualize tumor tissues [105]. Dietary iron oxide NPs with CAT activity mitigate neurodegeneration in a Drosophila-Alzheimer's disease model [106]. They have also been investigated in relation to the emergence of human severe acute respiratory syndrome coronavirus 2 (SARS-CoV-2), the virus responsible for the COVID-19 pandemic [107]. The high POD and CAT activities of inorganic iron polysulfide nanoparticles are responsible for effectively inhibiting *Pseudomonas aeruginosa* and *Staphylococcus aureus*, including drug-resistant strains [108]. These activities enable them to disrupt pathogenic biofilms, making them valuable for disinfecting implant devices like ventilators and blood catheters. Additionally, their enzyme activity accelerates the healing of infected wounds, further extending their potential for preventing or treating biofilm-related infections [108]. Conversely, remarkably active $CeVO_4$, acting as Cyt *c* oxidase (CcO), catalyzes the four-electron reduction of dioxygen to water in the respiratory electron transport chain without releasing any partially reduced oxygen species (PROS) such as superoxide, peroxide, and hydroxyl radicals [109]. These inorganic oxidoreductases also regulate cellular ROS levels, playing a crucial role in maintaining cellular homeostasis [110]. Meanwhile, MoS_2 nanozymes, when administered intravenously, facilitate the incorporation of molybdenum into molybdenum-dependent enzymes in the blood [111]. These findings highlight the potential of iron oxide and other nanoparticles as inorganic enzymes for therapeutic and diagnostic purposes.

The synthesis of inorganic iron oxide nanoparticles is an important technique in agriculture for regulating plant ROS levels in various environments [112–114]. Unlike in biomedicine, where nanoparticles are typically injected into the body, plants can uptake these nanoparticles themselves [112–117]. For example, inorganic iron oxide nanoparticles include those naturally occurring in soils [118–120] and synthesized ones [121–124], which can be absorbed by roots [121–128], leaves [129], or even the surface of seeds [130]. Once inside the plant, these nanoparticles can move to different parts of the plant, affecting various physiological processes [131]. One notable effect of these nanoparticles is their ability to regulate ROS levels in plants, effectively acting as inorganic enzymes to manage oxidative stress [132–134], including various abiotic stresses such as drought [132] and salt stress [135], as well as combating viral [129] and fungal diseases [136], and reducing heavy metal contamination [137–140]. Additionally, inorganic iron oxide nanoparticles have been found to increase the production of photosynthetic pigments in plants and electron transfer, enhancing their photosynthesis and leading to improved growth and development [122,125,127,128,133,137], and can even improve nitrogen fixation in soybeans

[141]. These nanoparticles can also influence plants at the molecular level, impacting gene expression and metabolic pathways [114,116,122,139,142], which can further affect their growth patterns. These applications demonstrate the diverse potential of inorganic nanoparticles as inorganic enzymes in addressing environmental challenges and contributing to sustainable agricultural practices. Inorganic oxidoreductases have also demonstrated effectiveness in detecting, isolating, and neutralizing environmental pollutants and chemical threats [96,143–146]. Their oxidoreductase activity, particularly in advanced oxygen processing, has shown promise in water and wastewater treatment, aiding in the removal of various contaminants, including pharmaceuticals and other organic pollutants [147–150].

However, the mechanisms and nature of these inorganic nanozymes continue to be debated [86,151–157]. Huang further argued that some inorganic nanozymes should be considered as inorganic enzymes rather than artificial enzymes, akin to proteins and RNA, due to their role as biocatalysts [8,9,91]. This perspective highlights the significance of inorganic nanozymes in biocatalytic processes and suggests their potential contribution to the origin and development of life.

This chapter aims to delve deeper into the nature of inorganic nanomaterials as biocatalysts, focusing on their role as examples of inorganic oxidoreductases. It will particularly explore their involvement in ET processes, considering the relationship between the metal structure of nanoparticles and electron hopping, as well as the influence of the electron configuration of the metals involved. By examining these aspects, including the incorporation of light-responsive elements inspired by photoactivity, a better understanding of the ET role of inorganic nanomaterials can be gained. This sheds light on their potential as key players in biocatalysis and their relevance to the origin of life and its earliest evolution.

2 INORGANIC NANOPARTICLES WITH INTRINSIC OXIDOREDUCTASE ACTIVITY

Gao et al. discovered that Fe_3O_4 nanoparticles possess intrinsic POD-like activity in 2007 [82]. It was observed that Fe_3O_4 NPs catalyzed the oxidation of 3,3,5,5-tetramethylbenzidine (TMB) to produce a blue color, di-azo-aminobenzene (DAB) to give a brown color, and *o*-phenylenediamine (OPD) to give an orange color, and their kinetics followed the typical Michaelis-Menten curves. The activity of the Fe_3O_4 nanoparticles was found to be influenced by the size of the particles. The apparent Michaelis-Menten constants (K_m) of HRP and 30 nm Fe_3O_4 particles measured with H_2O_2 as substrate are 3.70 and 154 mM, respectively. In contrast, the respective K_m values of HRP and 30 nm Fe_3O_4 measured with TMB as substrate are 0.434 and 0.098 mM. The catalytic rates (k_{cat}) of HRP and 30 nm Fe_3O_4 particles measured with H_2O_2 or TMB are 3.48 x 10^3 and 8.58 x 10^4 s^{-1} or 4.0 x 10^3 and 3.02 x 10^4 s^{-1}, respectively [82]. Thus, in terms of their catalytic efficiency (k_{cat}/K_m) the nanoparticles (H_2O_2: 560 mM^{-1} s^{-1}; TMB: 3.1 x 10^5 mM^{-1} s^{-1}) are at least as good as their biological counterpart (H_2O_2: 940 mM^{-1} s^{-1}; TMB: 9.2 x 10^3 mM^{-1} s^{-1}). Notably, the activity was related to the particle size. For nanoparticles with a diameter of

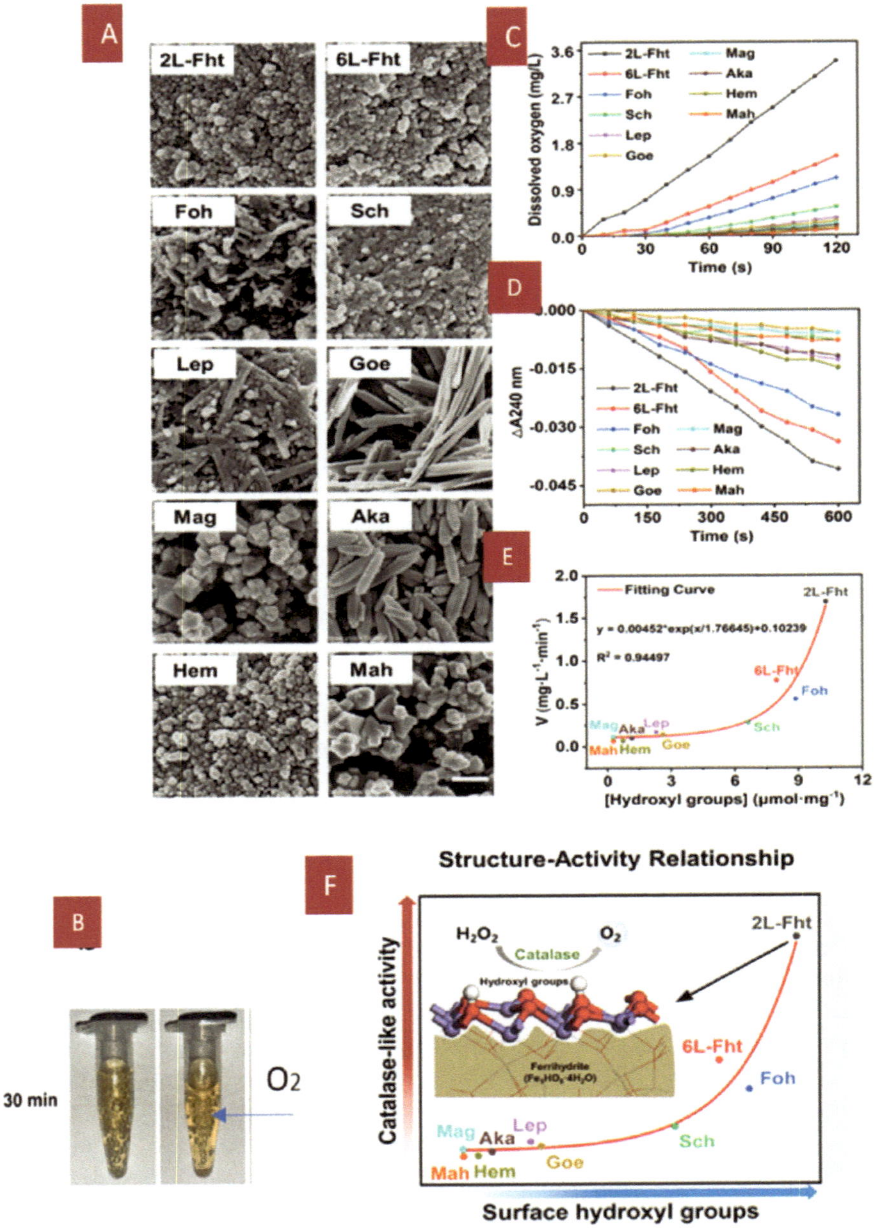

FIGURE 2 Catalase activity of inorganic iron oxide nanoparticles [104]. A. SEM images of 10 different iron oxide nanomaterials synthesized using the method of Cornell and Schwertmann [167], including 2-line ferrihydrite (2L-Fht, $Fe_5HO_8{\cdot}4H_2O$), 6-line ferrihydrite (6L-Fht, $Fe_5HO_8{\cdot}4H_2O$), goethite (Goe, α-FeOOH), akageneite (Aka, β-FeOOH), lepidocrocite

300 nm, the catalytic efficiency of a single nanoparticle was comparable to that of a single HRP molecule [82]. It was also found that many other inorganic iron oxide nanoparticles exhibit intrinsic POD activity [102,158–162]. The specific peroxidase-like activity of pure iron oxide nanoparticles is in the range of several Umg^{-1}, which is much lower than that of HRP with specific activity >150 Umg^{-1} [163].

Additionally, some iron oxide nanoparticles have exhibited multiple activities, including CAT, OXD, and SOD activities associated with H_2O_2 and free radicals [85,99,102,104,161,164–166]. Ferrihydrites, in particular, have been found to possess high intrinsic CAT-like activity and low intrinsic POD-like activity due to the abundance of hydroxyl groups in their nanocrystals (Figure 2) [104]. Ten types of iron oxide NPs were synthesized using the method of Cornell and Schwertmann [167], and their scanning electron microscope (SEM) images are shown in Figure 2A [104]. Each system (50 μg mL^{-1}) was capable of producing molecular oxygen from a 500 mM H_2O_2 solution after incubation for 30 minutes (Figure 2B). A comparison of the formation of dissolved O_2 in 100 mM H_2O_2 solutions containing these iron oxide nanomaterials (each at 10 μg mL^{-1}) is shown in Figure 2C and Figure 2D. The change in absorbance at 240 nm with time indicates the decomposition of H_2O_2. The activity of CAT (expressed as the rate of dissolved oxygen generation) was found to depend on the number of hydroxyl groups on the surface of the iron oxide NPs (Figure 2E) [104]. Iron oxide nanomaterials that have hydroxyl groups in their core structures (2-line ferrihydrite $Fe_5HO_8{\cdot}4H_2O$, 6-line ferrihydrite $Fe_5HO_8{\cdot}4H_2O$, feroxyhyte, δ′-FeOOH) all showed significant CAT activity, while those that did not contain hydroxyl groups (e.g., hematite α-Fe_2O_3, magnetite Fe_3O_4, maghemite, γ-Fe_2O_3) had no CAT activity. This was further confirmed by Fourier-transform

(CONTINUED)

(Lep, γ-FeOOH), feroxyhyte (Foh, δ ′-FeOOH), hematite (Hem, α-Fe_2O_3), maghemite (Mah, γ-Fe_2O_3), magnetite (Mag, Fe_3O_4), and schwertmannite (Sch, $Fe_8O_8(OH)_6SO_4$). Scale bar, 200 nm. B. Photographs of iron oxide nanomaterials (50 μg/mL) catalyzing the formation of oxygen bubbles in 500 mM H_2O_2 solution after incubation for 30 min at 37°C. The bubbles in the vials indicated the oxygen formation. C. The formation of dissolved O_2 in 100 mM H_2O_2 solution containing various iron oxide nanomaterials (10 μg/mL), measured by using a specific oxygen electrode on a multi-parameter analyzer (JPSJ-606L, Leici China). The total volume of the mixture was 5 mL. All reactions were carried out in deoxygenated water at 37°C. D. The decomposition of H_2O_2 (10 mM) added with various iron oxide nanomaterials (10 μg/mL) was detected by the absorbance at 240 nm by using an ultraviolet spectrophotometer (Hitachi U-3900). Iron oxide nanomaterials (10 μg (Fe)/mL) and 3 mL H_2O_2 solution (10 mM) were mixed in a cuvette at 37°C, and then the absorbance change at 240 nm of the reaction system in 10 minutes was continuously measured to determine the ability of iron oxide nanomaterials to decompose H_2O_2. E. The number of hydroxyl groups on the surface of various iron oxide nanomaterials measured by acid-base titration [168] is positively correlated with their corresponding catalase-like activity (expressed by the rate of dissolved oxygen generation). F. Schematic illustration of inorganic catalase activity vs. surface hydroxyl group of iron oxide NPs Reprinted from Ref. [104].

infrared (FTIR) analysis showing a distinct vibration at a wavelength of 3400 cm^{-1} linked to the hydroxyl groups. The amount of hydroxyl groups on the different iron oxide surfaces was quantified by acid-base titrations [168], and it was demonstrated that the surface architecture of iron oxide NPs (i.e., the number of -OH groups per unit mass) and their CAT activity correlate positively and exponentially Figure 2F [104]. Actually, ferrihydrites, functioning as inorganic CAT, have demonstrated in vivo their ability to produce O_2 within the tumor microenvironment, leading to a substantial enhancement of the therapeutic effect of radiotherapy [104,169].

On the other hand, the intrinsic oxidoreductase activity of inorganic nanomaterials extends beyond iron oxide nanoparticles to include iron sulfide nanoparticles. Solid phases of iron sulfides include FeS (mackinawite), $Fe_{1-x}S$ (pyrrhotite), FeS_2 (pyrite), FeS_2 (marcasite), Fe_3S_4 (greigite), and Fe_9S_{11} (smythite) [170]. These nanoparticles have been shown to catalyze the oxidation of peroxidase substrates, resulting in color changes in the reaction solution [171–174]. For instance, suspensions of sheet-like FeS, characterized by their black color, resulted in the production of a blue color due to the oxidation of TMB [173]. The solution exhibited maximum absorbance at 652 nm, corresponding to the oxidation product of TMB. Interestingly, the addition of H_2SO_4 to the TMB and H_2O_2 mixture led to the quenching of the color reaction (Figure 3A). Conversely, the addition of sheet-like FeS alone resulted in an orange solution with a maximum absorbance at 450 nm [173]. The kinetic parameters, such as the K_m, were determined for these nanoparticles, indicating their affinity for the substrate (e.g., TMB) and H_2O_2 [171–174]. The K_m values varied among different nanoparticles, suggesting differences in their catalytic efficiency compared to proteins like HRP.

Fe_7S_8 nanowires (NWs) exhibited POD activity and underwent color changes when exposed to peroxidase substrates and H_2O_2 [171]. The K_m values for Fe_7S_8 NWs with TMB and H_2O_2 indicated their affinity for these molecules. The apparent K_m value for Fe_7S_8 NWs with TMB was determined to be 0.548 mM, significantly lower than that of Fe_3O_4 and even five times lower than that of HRP, indicating that Fe_7S_8 NWs have a higher affinity for TMB [91]. Moreover, the K_m value for Fe_7S_8 NWs with H_2O_2 (0.895 mM) indicates a similar affinity for H_2O_2 compared to HRP and better H_2O_2 affinity than Fe_3O_4 nanoparticles [91]. Additionally, the inorganic POD activity of Fe_7S_8 NWs and HRP was influenced by pH and temperature, with the maximum activity observed at pH 4.0 and 35°C. Notably, Fe_7S_8 NWs exhibited greater stability over a wide pH range compared to HRP. At pH 2 or 10, Fe_7S_8 NWs retained approximately 45% of their POD activity, while HRP showed negligible activity under these extreme pH conditions [171]. This suggests that Fe_7S_8 NWs have a higher tolerance to pH variations compared to HRP. Furthermore, Fe_7S_8 NWs maintained higher POD activity than HRP as the temperature increased [171]. This indicates that Fe_7S_8 NWs are more temperature-stable and can retain their catalytic activity at higher temperatures.

Magnetic Fe_3S_4 nanoparticles also demonstrated POD activity and a high affinity for H_2O_2 [174]. Similar to other iron sulfide nanoparticles, the K_m value of Fe_3S_4 nanoparticles with H_2O_2 (1.16 mM) was approximately 100 times lower than that of Fe_3O_4 nanoparticles (154 mM), highlighting the excellent affinity of Fe_3S_4

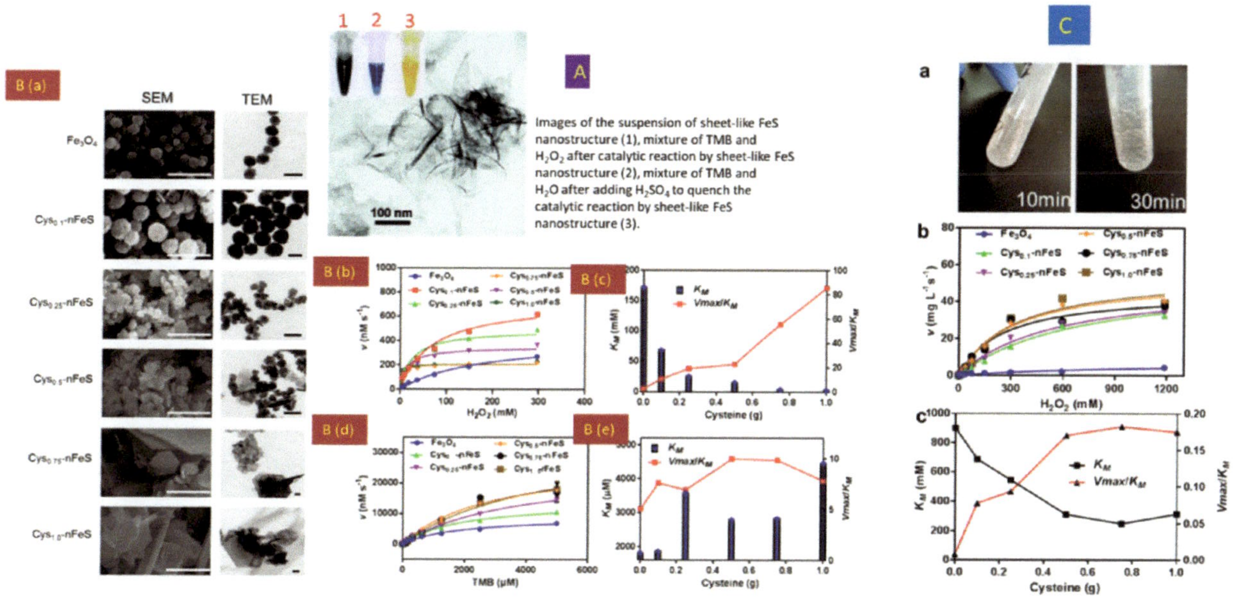

FIGURE 3 Intrinsic oxidoreductase activity of various iron sulfites nanoparticles. A. Color change of suspensions of sheet-like FeS to show the peroxidase activity. (1) Images of the suspension of sheet-like FeS nanostructure, (2) mixture of TMB and H_2O_2 after catalytic reaction by sheet-like FeS nanostructure, and (3), mixture of TMB and H_2O_2 after adding H_2SO_4 to quench the catalytic reaction by sheet-like FeS nanostructure. Reprinted with permission from Ref. [173], Copyright © 2009, John Wiley & Sons. B. Peroxidase activity of various iron sulfides nanoparticles: (a) SEM and TEM images of Cys-nFeS from different amounts of cysteine. Fe_3O_4 represents no cysteine in the reaction. Scale bars in SEM images: 5 μm. Scale bars in TEM images: 200 nm. Representative images are shown. (b). Michaelis-Menten kinetics of peroxidase-like activity of Cys-nFeS (with varied $[H_2O_2]$). (c) The trend of K_m and the ratio of V_{max}/K_m in the kinetics assay with varied $[H_2O_2]$. (d) Michaelis-Menten kinetics of peroxidase-like activity of Cys-nFeS (with varied [TMB]). (e) The trend of K_M and the ratio of V_{max}/K_M in the kinetics assay with varied [TMB], respectively. Reprinted from Ref. [108]. C. Catalase-like activity of Cys-nFeS. (a) The catalase-like activity to decompose H_2O_2 into oxygen (bubble) by Cys-nFeS. Representative images are shown. (b) The trend of K_m and the ratio of V_{max}/K_m in the kinetics assay with varied cysteine. (c) The trend of K_m and the ratio of V_{max}/K_m in the kinetics assay with varied $[H_2O_2]$. Reprinted from Ref. [108].

nanoparticles to H_2O_2 [82]. The K_m value of Fe_3S_4 nanoparticles with TMB as the substrate was determined to be 0.160 mM, lower than that of HRP (0.434 mM) but higher than that of Fe_3O_4 nanoparticles (0.098 mM) [82]. Notably, both the maximum initial velocity (V_m) for TMB and H_2O_2 were lower for Fe_3S_4 nanoparticles compared to HRP and Fe_3O_4 nanoparticles [82].

Additionally, iron polysulfide nanoparticles ($Fe_{1-x}S$ and Fe_3S_4) have been found to exhibit both peroxidase-like and catalase-like activities, allowing them to break down H_2O_2 into free radicals and oxygen [108] (Figure 3B and 3C). Among these nanoparticles, Cys0.5-nFeS demonstrated the highest specific activity, which was nearly 10 times higher than that of Fe_3O_4 nanoparticles. Upon exposure to 50 mM H_2O_2, the color of the nFeS solution (1.0 mg/mL) changed to orange within 30 minutes. This color change resembled the spectrum of NaHS, indicating that the release of free sulfide was facilitated by catalysis [108]. After the release of free sulfides, the precipitate formed by nFeS exhibited low enzyme-like activity, consistent with the catalytic performance of iron oxide [82]. Reactive oxygen species, including hydroxyl (•OH), H_2O_2, O_2*, and singlet oxygen ($^1O_2$), are involved in these biocatalytic reactions, including CAT and OXD activities, which can also contribute to cytotoxicity and therapeutic effects [175–185]. Most of these species also trigger cytotoxicity to fulfill therapeutic effects [186,187].

The significance of the architecture of inorganic nanomaterials, including iron oxide NPs and their catalytic activity, was also investigated using density function theory (DFT) studies [104,166,188,189]. The mechanism for inorganic POD activity involves chemisorption of H_2O_2 onto the nanomaterial surface, generating two hydroxyl adsorbates that undergo two reduction processes to remove the hydroxyl groups from the surface (Figure 4A) [188]. The metal architectures of different iron oxide NPs and other inorganic nanomaterials are shown in Figures 4B and 4C. The chemisorption energy and thus the POD-like activity are dependent on the metal architecture in the nanomaterials (Figure 4D, 4E) [188] as will be discussed in more detail in the following sections.

Iron oxide nanoparticles with intrinsic oxidoreductase activity are not limited to those synthesized in the laboratory. It was confirmed that these magnetic nanoparticles in magnetosomes from magnetotactic bacteria (MTB), after removing the protein membrane, still retained POD activity [190,191]. This observation suggests that the inherent peroxidase activity of these nanoparticles is not solely dependent on the protein components but is an inherent property of the nanoparticles themselves. Furthermore, the peroxidase-like activity of these magnetic nanoparticles has been shown to be associated with light exploration. This implies that these nanoparticles exhibit enhanced catalytic activity in the presence of light [190,191] or photoreactivation. Biogenic iron oxide nanoparticles from *Burkholderia* sp. YN01v [192,193] (Fe_3O_4), *Comamonas testosteroni* ($Fe_{1.44}O_{0.32}(OH)_{3.86}$) [194], and *Acinetobacter* strains ($Fe_{0.96}O_{0.88}(OH)_{1.12}$) [195] also exhibited intrinsic POD-like activity [192–195], SOD-like activity [193], and CAT-like activity [193] (Figure 5A).

Ferritin, an iron storage protein present in various organisms, including archaea, bacteria, and eukaryotes, features an iron core that serves as a natural nano-structure with biogenetic iron minerals [196–199]. In 1999, Belarusian scientists discovered

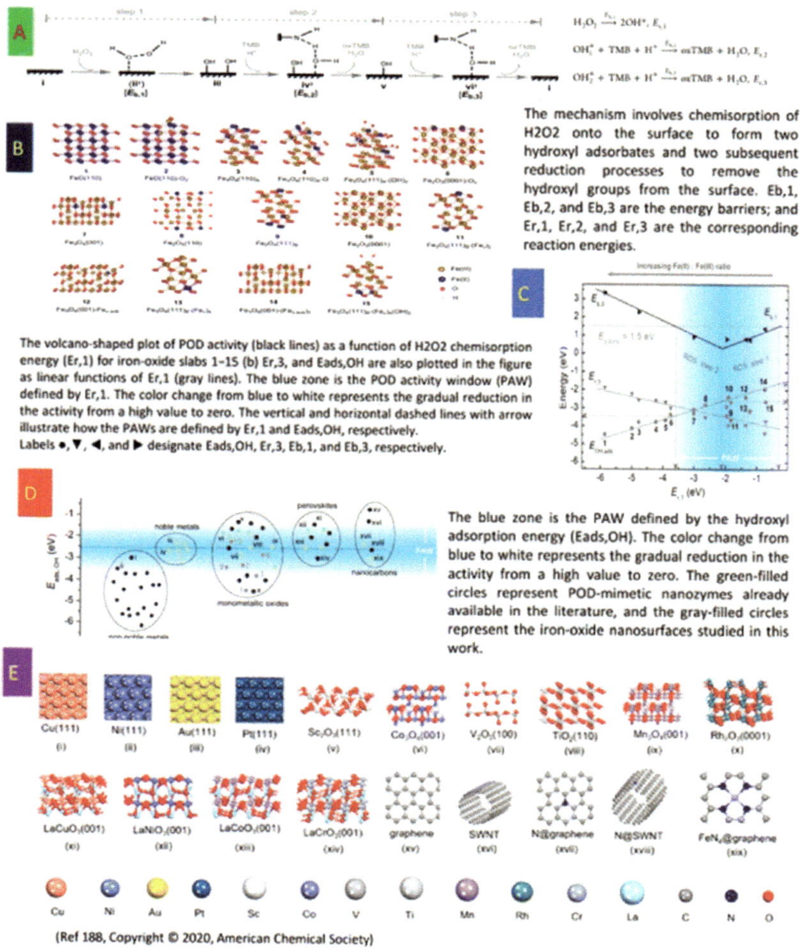

FIGURE 4 Metal architecture of inorganic nanomaterials and POD-like activity. A. Proposed mechanism of the POD-like activity of nanomaterials in three steps. Eb,1, Eb,2, and Eb,3 are the energy barriers; Er,1, Er,2, and Er,3 are the corresponding reaction energies. B. Diagrams of iron-oxide slabs with different crystal structures, exposed facets, defects, and chemical modifications (1–15). The top surfaces were used for the catalysis. FeO, Fe_3O_4, and Fe_2O_3 are ferrous, ferriferous, and ferric oxides. For slab number 2, two O atoms on top; 4, one O atom on top; 5, two hydroxyl groups on top; 6, one O vacancy on top; 11, two Fe vacancies on top; 12, one Fe vacancy on the subsurface; 13, four Fe vacancies on top; 14, two Fe vacancies on the subsurface; 15, two hydroxyl groups and four Fe vacancies on top. C. Structures of the other NMs with different crystal structures, e.g., monometallic oxides (TiO_2, V_2O_5, Mn_3O_4, Fe_3O_4, Co_3O_4, NiO, Cu_2O, ZrO_2, and RuO_2) and perovskites ($LaNiO_3$, $LaCoO_3$, $LaMnO_3$, and $SrFeO_3$). D. Volcano-shaped activity curve. The color change from blue to white represents the gradual reduction in the activity of POD-like activity from a high value to zero. The labels 1–15 represented the corresponding iron oxide structure in (B). E. Overview of POD-like activities of different nanomaterials, corresponding C. The blue zone is the POD activity window defined by the hydroxyl adsorption energy (Eads, OH). Reprinted with permission from [188], Copyright © 2020, American Chemical Society.

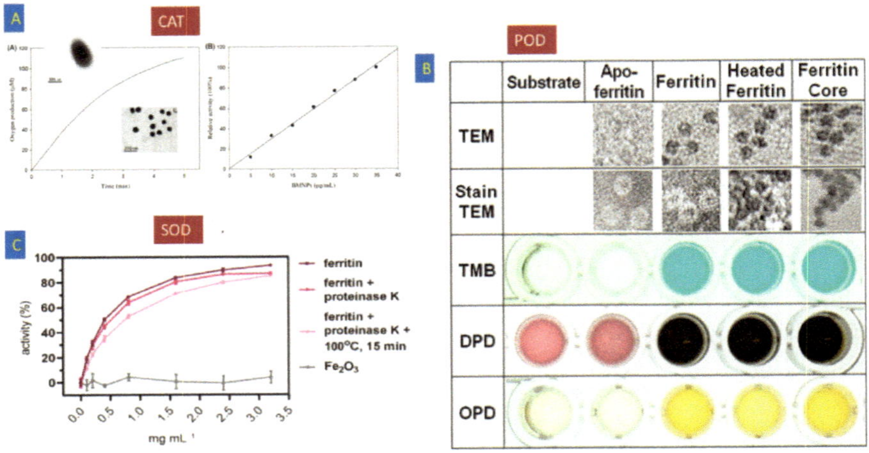

FIGURE 5 Intrinsic oxidoreductase activity of biogenic iron oxide nanoparticles. A. The intrinsic CAT-like activity of BMNPs from *Burkholderia sp.* YN01.Strain YN 01. (a) Time-course curve of H_2O_2 decomposition by BMNPs as a catalase mimic. (b) B. Effect of BMNP concentration on catalytic activity. Reaction conditions: 30 μg ml^{-1} BMNPs or a series of concentrations, 2.5 mmol l^{-1} H_2O_2 and 200 mmol l^{-1} Na_2HPO_4-citric acid buffer (pH 9), or 0.5 U ml^{-1} catalase, 5 mmol l^{-1} H_2O_2 and 200 mmol l^{-1} Na_2HPO_4-citric acid buffer (pH 8). Reprinted with permission from Ref. [193], Copyright © 2019, John Wiley & Sons. B. POD activity from ferritin core, (a) TEM images of ferritin samples and the analysis of its peroxidase-like activity using TMB, DPD, and OPD as substrates at 50 °C. (b) Analysis of peroxidase-like activity of ferritin using p-HPPA (left) and luminol (right) as substrates, (c) the comparison of thermal stability (left) and pH tolerance (right) of peroxidase activity between ferritin and HRP. Reprinted with permission from Ref. [201], Copyright © 2011, American Chemical Society. C. SOD activity from ferritin core, from archaea (*P. furiosus* and *P. yayanosii*, pfFn & pyFn, *S. solfataricus*, ssDps), bacteria (*E. coli*, EcDps), DNA-binding proteins from starved cells, ferritin (FTn) from bacteria (EcFTn), bacterioferritin from bacteria (EcBfr), and eukaryotes (*H. sapiens*) heavy chain ferritin (HFn) and light chain ferritin (LFn). The inhibition rate, indicative of SOD-like activity, of biomineralized ferritins with a final protein concentration of 50 μg/mL, presented as the mean ± SD (n = 3 independent experiments). Reprinted from Ref [202].

that ferritins containing iron crystallites exhibit POD activities, acting as biocatalysts in the oxidation of TMB by hydrogen peroxide in an aqueous acetate buffer solution (pH 4.2) [200]. This activity follows Michaelis-Menten kinetics and has been demonstrated through linear Lineweaver–Burk plots for TMB (up to 3.0 mM) and H_2O_2 (up to 50 mM). Notably, this discovery predates the breakthrough work by Gao and colleagues in 2007 [82]. Subsequently, in 2011, the POD activity of the mineral core of ferritin was further confirmed using TMB, OPD, and N, N-diethyl-1,4-phenylenediamine (DPD) as substrates at 50 °C [201] (Figure 5B). These experiments indicated that the POD activity of horse spleen ferritin originates from its ferric nanocore rather than the ferritin protein itself [201].

Further investigations into the iron cores of various ferritins have revealed that these biogenic iron cores, when isolated from the proteins, exhibit enzyme-like activities such as POD, CAT, OXD, and SOD due to their metal structure, distinct from the organic compounds comprising ferritins, particularly the amino acid sequences in these proteins (Figure 5C) [202]. The study encompassed ferritins from different biological groups potentially spanning different evolutionary lineages. In the archaea domain, ferritins from *Pyrococcus furiosus* (WP_011011871.1, labeled as pfFn), *Pyrococcus yayanosii* (WP_013905435.1, labeled as pyFn), and DNA protection protein from *Sulfolobus solfataricus* (NP_343470.1, labeled as ssDps) were examined. Representing bacteria were bacterioferritin (EcBfr) from *Escherichia coli* str. K-12 substr. MG1655 (NP_417795.1), non-heme ferritin (EcFTn) from enterobacteriaceae (WP_000917208.1), and DNA binding protein (EcDps) from starved *Escherichia coli* str. K-12 substr. MG1655 (NP_415333.1). Eukaryotic ferritins were represented by Heavy chain (HFn) from *Homo sapiens* (NP_002023.2) and Light chain (LFn) from *Homo sapiens* (NP_000137.2). This classification into different groups underscores the diversity of the iron cores from ferritins and their enzyme-like activity across various biological domains [202].

3 PHOTOREACTIVATION OF INORGANIC OXIDOREDUCTASES

Light occupies a pivotal role in governing complex biological processes, particularly within the domain of proteins. This phenomenon, termed photoreactivation, underscores a critical facet in the evolutionary progression of life [203]. Notably, certain enzymes categorized as oxidoreductases exhibit remarkable responsiveness to light, unraveling a persistent enigma that continues to captivate the scientific community [204]. Among these enzymes, PODs and OXDs, characterized by their iron-containing heme groups, emerge as central players in orchestrating ET during redox reactions. An intriguing illustration of this phenomenon is observed in HRP isozymes, specifically HRP A2, which demonstrates sensitivity to both ultraviolet (UV) and blue light [205,206]. These isozymes, featuring heme moieties, govern the reduction of H_2O_2 and the oxidation of a diverse array of organic and inorganic compounds [207–209]. The precise mechanisms that underlie this light-induced transformation remain enigmatic, with implications of electron tunneling facilitated by aromatic amino acids such as tryptophan (Trp) and tyrosine (Tyr) [11]. This electron tunneling, hinged on the aromatic side chains of Trp and Tyr amino acids, potentially facilitates the photoreactivation process in PODs and OXDs. These enzymes, housing heme groups with iron as the central metal center, assume a pivotal role in electron acceptance and donation during redox reactions [210–212].

This influence of light extends beyond protein-based oxidoreductases to encompass inorganic nanozymes. Inorganic nanozymes, encompassing entities such as magnetite within MTB [190,191], α-Fe_2O_3 nanoparticles [213], TiO_2 nanoparticles [214], Cu_2Se hollow nanocubes [215], cobalt sulfide (Co_9S_8) nanodot [216], nanorods of CdS [217] and CuO [218], showcase photoreactivation phenomena. For instance, under light exposure, α-Fe_2O_3 nanoparticles demonstrate a reduced K_m for a substrate TMB compared to the reaction without light. Furthermore, both the V_{max} and K_{cat} of the substrate in the presence of light significantly surpass those observed without light [213]. This suggests heightened substrate affinity and catalytic efficiency under light irradiation. Similarly, the photoreactivation response is evident in magnetosomes and naked magnetosomes, wherein light exposure leads to a notable reduction in K_m for substances like TMB and H_2O_2. Though the effect of light on V_{max} varies, overall, their activities demonstrate greater sensitivity to light conditions [190]. CuO nanorods also exhibit peroxidase-like activity using TMB, OPD, and 2,2′-azinobis (3-ethylbenzothiazoline-6-sulfonic acid) (ABTS) as a substrate [218] (Figure 6A, 6B). Under visible light irradiation, the calculated K_m values for both ABTS and H_2O_2 are lower than those without photoillumination (Figure 6C, 6D). Notably, the K_m value for H_2O_2 experiences a significant decrease from 14.8 to 3.4 mM, a reduction of 4.35-fold, while the K_m value for ABTS shows a modest decrease of 1.5-fold under visible light irradiation. This implies that under photoirradiation conditions, the affinity of CuO nanorods toward H_2O_2 significantly improves, suggesting that fewer H_2O_2 molecules are needed to generate higher quantities of •OH radicals. Additionally, the V_{max} values for both ABTS and H_2O_2 consumption are higher during photoillumination compared to those in the dark, with an improvement of 2.7- and 19.8-fold, respectively. The elevated V_{max} values under photoirradiation indicate that the rate of •OH radical production will be over ~20 times faster than under non-irradiated conditions. The valence-band energy level of CuO nanorods is measured at 4.98 eV, accompanied by an optical band gap of 1.44 eV [219]. This positions CuO nanorods favorably to stimulate •OH radical generation through the enhancement of their catalytic properties upon photoexcitation with visible light (Figure 6E). These investigations provide concrete evidence that visible light functions as a potent trigger to significantly augment the oxidative properties of CuO nanorods, generating larger quantities of •OH radicals (low K_m) at an accelerated rate (high V_{max}).

The combination can further enhance these processes. The realm of composite nanoparticles also contributes to this phenomenon with light-sensitive nanoparticles, such as TiO_2. For example, core–shell structured Fe_2O_3@TiO_2 nanocomposites [220], TiO_2 nanotubes coated with MoS_2 nanoflowers (TiO_2NTs@MoS_2) [221], Fe_3O_4-TiO_2/reduced graphene oxide (Fe_3O_4-TiO_2/rGO) nanocomposite [222], plasmonic gold nanoparticle-modified Mn_3O_4 nanozyme (Mn_3O_4-Au) [223], Fe_3O_4/Ag/Bi_2MoO_6 nanoparticle [224], and quaternized chitosan (QCS)/silver (Ag)/cobalt phosphide (CoP) nanocomposites [225], all exhibit photoresponsive attributes. This intriguing feature endows their catalytic activity with modulation and enhancement through the harnessing of light energy.

Moreover, the phenomenon of photoreactivation extends beyond visible light to encompass the enigmatic domain of near-infrared-II (NIR-II) light (1000–1700 nm). A notable example of an inorganic nanozyme displaying photoreactivation under NIR-II light is the Cu_2MoS_4 nanoparticle (CMS NP) [226]. This nanoparticle showcases both OXD and POD activities upon NIR-II light exposure. In the presence of NIR-II light, CMS NPs demonstrate OXD activity using L-ascorbic acid (AA) as a substrate. The K_m and V_{max} of CMS NPs for AA are determined as 12.06 μM and 0.11 μM s^{-1}, respectively. Furthermore, CMS NPs display OXD activity with glutathione (GSH) and TMB as substrates [140]. Under NIR-II light irradiation, GSH levels

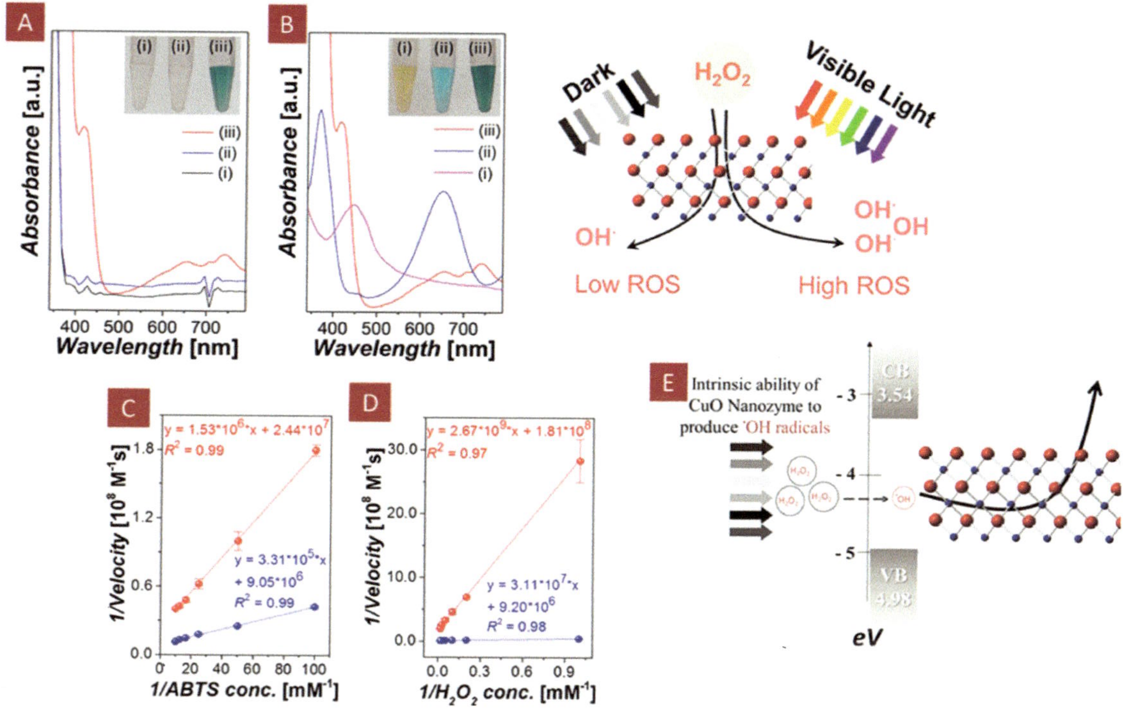

FIGURE 6 POD activity of CuO nanorods (CuO NRs) under UV–visible light. A. UV–visible absorbance spectra of ABTS oxidation under different conditions: (i) ABTS + H_2O_2; (ii) CuO NRs + ABTS; (iii) CuO NRs + ABTS + H_2O_2. B. UV–visible absorbance spectra of different peroxidase substrates using CuO NRs in the presence of H_2O_2: (i) OPD; (ii) TMB; (iii) ABTS. Insets show the color of post-reaction solutions. C. Double-reciprocal Lineweaver–Burk plots of the catalytic activity of CuO NRs at a fixed concentration of H_2O_2 versus varying concentrations of the ABTS. The red and blue fittings represent the experiments conducted without and with visible light illumination, respectively. D. Double-reciprocal Lineweaver–Burk plots of the catalytic activity of CuO NRs at a fixed concentration of ABTS versus varying concentrations of H_2O_2. The red and blue fittings represent the experiments conducted without and with visible light illumination, respectively. E. Suggested mechanism of CuO NRs promoted by UV-visible light. The favorable energy-band structure of CuO NRs allows them to spontaneously produce •OH radicals. Reprinted with permission from Ref. [218]. Copyright © 2018, American Chemical Society.

exhibit a decrease of more than twofold compared to reactions without light irradiation, indicating enhanced OXD activity of CMS NPs. POD activity of Cu_2MoS_4 nanoparticles is also evident under NIR-II light. The K_m of CMS NPs for TMB and H_2O_2 is determined as 1.36 mM and 25.46 mM, respectively. The V_{max} of CMS NPs for TMB and H_2O_2 measures 27.29×10^{-8} M s^{-1} and 42.81×10^{-8} M s^{-1}, respectively. Notably, the absorbance of the oxidized TMB product following treatment with CMS NPs and NIR-II light shows a twofold increase compared to CMS NPs alone, indicating efficient acceleration of POD catalysis by NIR-II light. These findings highlight the dual enzyme-like capabilities of Cu_2MoS_4 nanoparticles, acting as both oxidase and peroxidase, with their catalytic activities significantly amplified by NIR-II light irradiation.

4 ELECTRON TRANSFER MEDIATOR AND LIFE FOSSIL OXIDOREDUCTASE

Cytochrome, a ubiquitous protein distributed across a spectrum of organisms including prokaryotes, archaea, plants, animals, fungi, and unicellular organisms [49], holds significant prominence due to its multifaceted role in mediating ET processes during oxidative reactions [227]. As a member of the c-type Cyt family, it belongs to class I and is distinguished by its distinctive CXXCH amino acid motif [228]. This motif, characterized by two cysteine (C) residues separated by two variable amino acids (any-any), followed by a histidine (H) residue, forms the binding site for the heme group embedded within the protein's structure.

Residing primarily within the inner membranes of mitochondria and chloroplasts, Cyt *c* orchestrates electron transport chains, essential for energy production [49,229]. Within these chains, electrons are initially transferred from the metal center of an oxidoreductase (Complex III or Coenzyme Q—Cyt *c* reductase) to Cyt *c*. Subsequently, Cyt *c* shuttles these electrons to terminal electron acceptor proteins (Complex IV or Cyt *c* oxidase) or to electrodes in experimental contexts. This ET mechanism is pivotal for generating adenosine triphosphate (ATP), the principal cellular energy currency, through mitochondrial ATP synthesis and chloroplast photosynthesis. This orchestrated electron relay by Cyt *c* sustains redox equilibrium and fuels energy generation within cells [230].

Distinguished by a versatile active site in the form of its heme group, Cyt *c* engages in interactions with a diverse array of active-site cofactors and neighboring moieties. These interactions precipitate varied catalytic activities and ET reactions. Active-site and neighboring cofactors such as flavin adenine mononucleotide (FMN), flavin adenine dinucleotide (FAD), pyrroloquinoline quinone (PQQ), molybdopterin cofactors (Moco) bound to molybdenum atoms, and distinct iron-sulfur clusters (Fe_2S_2, Fe_3S_3, Fe_4S_4) confer Cyt *c* with its versatile functional attributes [49,229]. These attributes enable Cyt *c* to partake in oxidative processes by receiving electrons from catalytic cofactors and subsequently transferring these electrons to terminal electron acceptor proteins or electrodes. The heme group itself, known as Haem, contributes intrinsic catalytic capabilities by facilitating redox reactions [231].

Cyt *c*'s catalytic capabilities extend beyond its role in protein environments, encompassing interactions with inorganic nanoparticles that manifest intriguing enzyme-like behaviors. For instance, when paired with WO_3 nanoparticles, Cyt *c* assumes significant peroxidase-like activity, particularly in the presence of water and hydrogen peroxide [232]. Likewise, Cu_2O nanoparticles exhibit Cyt *c* oxidase (CcO) activity, effectively emulating the functionality of the mitochondrial enzyme and proficiently catalyzing the four-electron reduction of oxygen, mirroring the mitochondrial CcO enzyme [233]. Notably, cadmium sulfide exhibits bifunctional NADH oxidase and Cyt *c* reductase activities in the presence of Cyt *c*, facilitating NADH recycling. This catalytic interplay is further influenced by the participation of photo-generated holes and superoxide anions (O_2^{*-}) [217].

A striking discovery in the realm of Cyt *c* electron mediators is exemplified by the catalytic prowess of $CeVO_4$ nanozymes [109] (Figure 7). These nanozymes exhibit the mitochondrial enzyme CcO and utilize Cyt *c* as the biological electron donor in the four-electron reduction of molecular oxygen to water under physiological pH conditions. Remarkably, this catalytic conversion exhibits no release of partially reduced oxygen species (PROS), affirming the efficient and clean nature of the process. Additionally, $CeVO_4$ nanozymes display robust SOD activity, effectively neutralizing detrimental superoxide radicals and safeguarding cells from oxidative harm [109].

Cyc *c* electron mediators are believed to have emerged approximately 3.5 billion years ago during the Archean era, a pivotal epoch marked by the formation of Earth's geological and environmental foundations. During this epoch, MTB displayed exceptional adaptations and biomineralization capabilities [234,235]. These microorganisms harnessed specialized organelles known as magnetosomes to navigate by aligning themselves with Earth's magnetic field [236–238]. Within these magnetosomes, small membrane-bound compartments, resided magnetic minerals such as magnetite (Fe_3O_4) and greigite (Fe_3S_4) [234,235]. Proteins like MamE, MamP, and MamT, housing CXXCH c-type Cyc-binding motifs, were integral components of the magnetosome membrane [235–241], underscoring the early integration of Cyc c electron mediators into cellular functions during this formative era.

The Archean atmosphere harbored significant concentrations of H_2O_2, presenting challenges associated with the presence of ROS to nascent life forms [64,242–246]. ROS, including H_2O_2, are metabolic byproducts capable of inducing oxidative damage to vital cellular components like DNA, proteins, and lipids. This oxidative stress can lead to mutations, cellular dysfunction, and even cell death, posing a potential threat to the survival and evolution of organisms. Defense mechanisms like SOD, CAT, and POD evolved prior to the emergence of oxygen-producing microorganisms and the great oxygenation event, serving as early guardians against ROS-induced damage [57,73,247–251]. The presence of these enzyme systems across diverse domains of life, including anaerobic environments, attests to the imperative need for ROS protection [252–256].

Laboratory investigations suggest that diverse iron oxide nanoparticles, precursors to magnetite, underwent initial precipitation and formation within MTB [234,257–264]. These nanoparticles subsequently became encapsulated within lipid bilayer

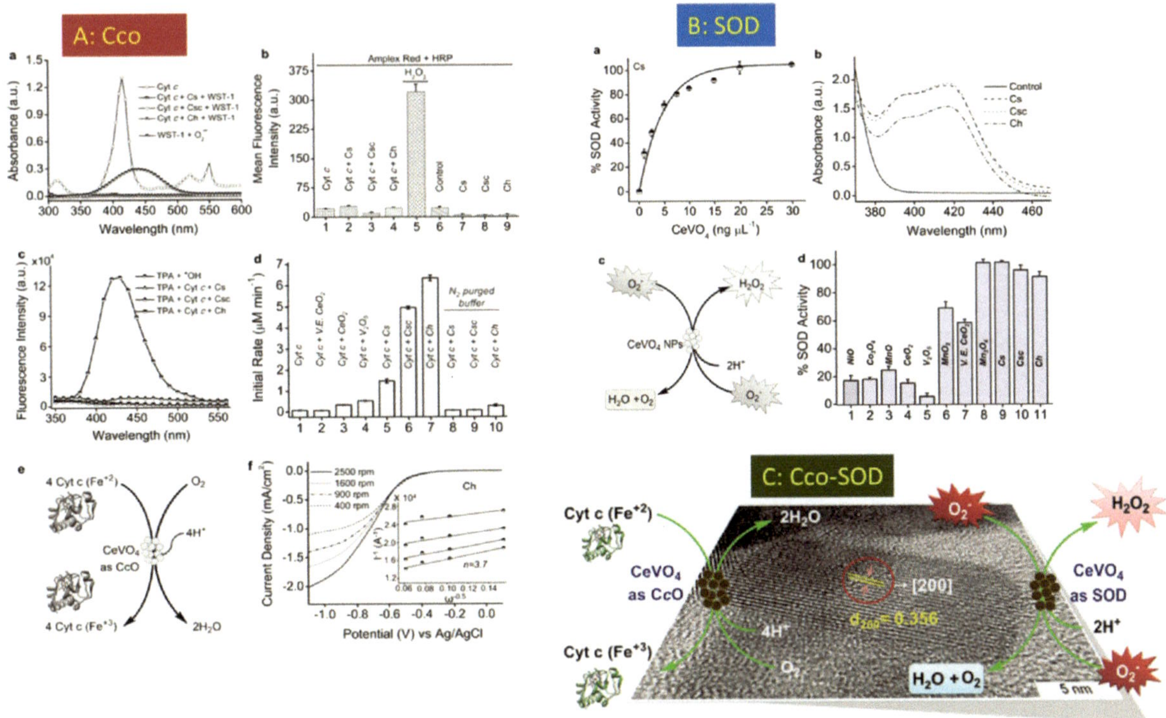

FIGURE 7 Enzymatic activities of Cyt-$CeVO_4$ nanoparticles. A. (a) Superoxide production during CcO-like activity of $CeVO_4$ using WST-1. (b) Determination of H_2O_2 production using Amplex red during SOD-like reaction of $CeVO_4$ nanoparticles. (c) Measurement of hydroxyl radicals via fluorescence spectroscopy using TPA. (d) A comparison between CcO-like activity of $CeVO_4$ and other metal oxide nanoparticles. (e) Schematic representation of the 4-electron reduction of oxygen. (F). Current-potential curve obtained for Ch using rotating ring-disk voltammetry under O_2-saturated conditions at different rotation rates. Inset: Koutecky-Levich plots at various rotation rates for Ch. B. (a) The SOD activity of the inorganic $CeVO_4$. (b) Formation of H_2O_2 during SOD-like reaction of $CeVO_4$ nanoparticles. (c) Scheme depicting SOD-like activity. (d) A comparison of $CeVO_4$ nanozymes' activity with SOD-mimetic nanoparticles and other metal oxides. C. Scheme $CeVO_4$ with Cyc c exhibiting Cyc c oxidase and SOD. Reprinted with permission from Ref. [109]. Copyright © 2019, John Wiley and Sons.

membranes reminiscent of the structures found in mitochondria and chloroplasts [265]. Endocytosis has been defined as "the de novo production of internal membranes from the plasma membrane lipid bilayer" [266] and as the primary pathway for nanoparticles to enter cells, including viruses and bacterial [115,117,267, 268]. Iron oxide nanoparticles like Fe_3O_4 and ferrihydrite could traverse lipid bilayers and enter cellular compartments without disrupting plasma membranes [267,269]. The POD activity exhibited by magnetite within magnetosomes likely conferred a survival advantage in environments rich in hydrogen peroxide [190,191,252–256].

This intriguing concept introduces the notion of a "life fossil oxidoreductase" [91] shedding light on the remarkable role of magnetite nanoparticles enclosed within magnetosomes. These nanoparticles appear to function as enduring enzymes vital for ET in redox reactions [8,9], implying an adaptation to endure high hydrogen peroxide conditions and oxidative stress throughout Earth's history. Notably, synthesized Fe_3S_4 nanoparticles have demonstrated a significant affinity for hydrogen peroxide, accompanied by POD activity [108,174,182,270,271]. These inorganic nanoparticles' activities in environments rich in H_2O_2 could have been a driving force behind the symbiogenesis of MTB. The invasion by MTB may have been instigated by the magnetite or greigite nanoparticles, acting as catalysts in the process of symbiogenesis due to their unique enzyme activity in conditions of high hydrogen peroxide. This interdependence between microorganisms and inorganic magnetite or greigite nanoparticles suggests a profound enzyme activity adaptation in response to elevated hydrogen peroxide conditions. Consequently, this symbiotic relationship possibly played a pivotal role in the microorganisms' ability to thrive and adapt in environments characterized by heightened hydrogen peroxide levels.

Genetic analyses detailing magnetosome assembly reveal intricate steps encompassing membrane biogenesis, protein sorting, iron transport, magnetite nucleation, crystal mineralization, and chain assembly [272,273]. These biomineralization processes offer invaluable insights into the early evolutionary stages that shaped life's complexity. The convergence of Cyc c electron mediators and magnetosomes, coupled with the notion of the "life fossil oxidoreductase" [91], illuminates the strategies adopted by early life during the Archean era to thrive in the face of challenges such as hydrogen peroxide. It underscores the innovative utilization of inorganic nanoparticles for survival and adaptation on Earth, paving the way for the evolution and diversity of life forms we observe today.

5 METAL ARCHITECTURES IN INORGANIC NANOPARTICLES

In the preceding sections, we have delved into the intricate world of inorganic nanomaterials with inherent oxidoreductase activity, exploring the fascinating concept of these materials as biocatalysts. The multifaceted nature of inorganic nanomaterials, with their unique ability to facilitate ET and redox reactions, has unveiled a gateway to understanding their significance in the origin of life and their potential applications across diverse scientific domains.

As we continue our journey deeper into the realm of inorganic nanomaterials, the following sections will focus on a specific and crucial subset, including the metal architecture of inorganic iron oxide and sulfides, ET within nanoparticles, and their inherent oxidoreductase activity. Building upon the foundational knowledge established in the preceding sections, which encompassed the inherent oxidoreductase activity of nanomaterials, we now venture into the realm of ET mechanisms within these nanoparticles. Our exploration will traverse the landscape from traditional electron band theory to the intricate nuances of electron conductivity within nanoparticles, all the while weaving in the theoretical insights from earlier sections.

The following sections (Sections 5 to 8) aim to unravel the dynamic metal architectures of nanoparticles and the intricacies of ET within iron oxide and sulfide nanoparticles, investigating the interplay between their unique properties and the underlying ET mechanisms. We will delve into key factors that govern this behavior, including band gap characteristics, charge ordering phenomena, and structural dynamics. By probing these aspects, we will gain a deeper understanding of how ET processes unfold within these nanomaterials as inorganic enzymes. This journey holds promise not only in expanding our theoretical grasp of ET but also in illuminating implications ranging from the emergence of life, biocatalysts, and evolution of protein. By bridging the gap between theoretical insights and empirical observations, we aim to unlock new dimensions of understanding regarding ET phenomena in the context of inorganic nanomaterials.

Generally speaking, metal architecture in iron oxides is a function of the ferric-ferrous composition Fe^{3+}/Fe_{total} and the hydroxylation ratio OH/Fe_{total} (Figure 8A) [167,274,275]. A Back-Figges δ-Keggin cluster (Fe_{13}), containing 13 iron and 40 oxygen atoms, constitutes the basic structural unit of ferrihydrite [276] or other iron oxide nanoparticles (Figure 8B) [277]. The central tetrahedrally coordinated Fe is connected to 12 peripheral octahedrally coordinated Fe atoms arranged in edge-sharing groups of three by oxo bridges. As a result, iron oxide nanoparticles between 2 and 6 nm can be viewed as a three-dimensional packing of these clusters, with adjacent clusters connected by a common pair of edge or corner or face, even by combinations-shared octahedra, forming oxo bridges in the bare cluster (Figure 8C) [276]. The distance between Fe and Fe depends on the type of connections, with the corner-sharing arrangement having the longest distance (3.39 to 3.70 Å) and the face-sharing arrangement having the shortest distance (2.88 Å) (Figure 8D) [167,278].

Magnetite has an inverse cubic spinel structure, in which the metal ions are shared between tetrahedral and octahedral sites. The tetrahedral sites are occupied by Fe(III) ions, while the octahedral sites contain both Fe(III) and Fe(II) ions in the same amount. The magnetite formula thus can be written as $Fe(III)^{tetra}[Fe(II)Fe(III)]^{octa}(O^{2-})_4$ to represent the position of the ferrous and ferric ions in the different sites of the structure [279]. Similar to magnetite, maghemite also exhibits a spinel crystal structure with all the iron cations in the trivalent state, and the charge neutrality of the cell is guaranteed by the presence of cation vacancies. The maghemite structure can be obtained by creating 8/3 vacancies out of the 24 Fe sites in the cubic unit cell of magnetite. These vacancies are located in the octahedral sites, and therefore the structure of maghemite can be approximated as a cubic unit cell with the composition $\left(Fe_8^{III}\right)_A [Fe_{\frac{40}{3}}^{III} \square_{\frac{8}{3}}]_B O_{32}$, where □ represents a vacancy, A

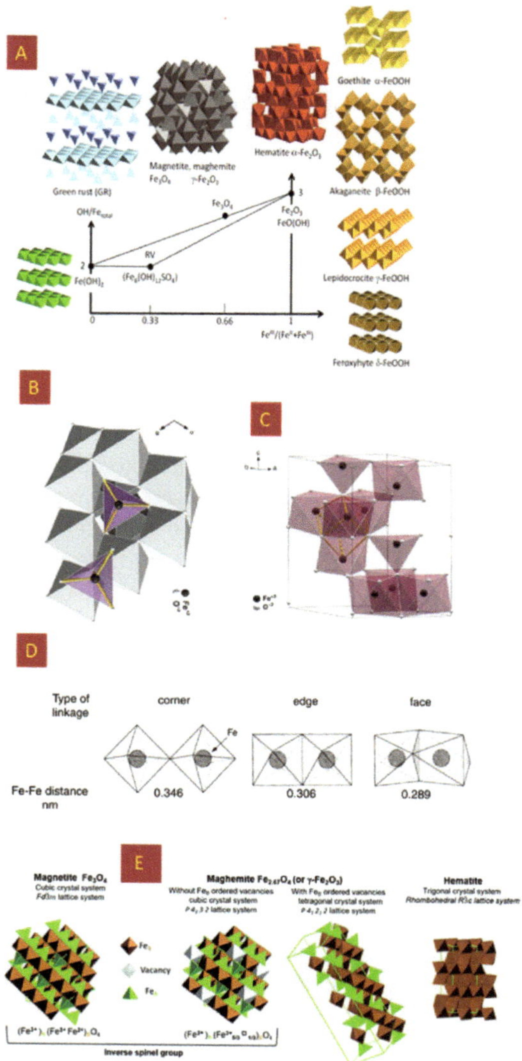

FIGURE 8 Metal architecture of iron oxide. A. Main structure types of iron oxides [274]. Reprinted with permission from Dr Jean-Pierre Jolivet. B. The Baker-Figgis d-Keggin cluster Fe_{13} [276]. Polyhedral representation of the ideal ferrihydrite structure viewed along the c axis. The central FeO_4 tetrahedra are surrounded by 12 FeO_6 octahedra. Reprinted with permission from Ref [276], Copyright © 2007, American Association of the Advancement of Science. C. The basic structural motif consists of a central FeO_4 tetrahedra with 12 FeO_6 octahedra. The hexagonal unit cell for Ferrihydrite, Fe-oxo-Fe bridge at the edge-shared octahedral [276]. The bonded atoms (yellow) define a cubane-like moiety that connects the basic structural motif of the model. Reprinted with permission from Ref [276], Copyright © 2007, American Association of the Advancement of Science. D. The Fe-Fe distance and linkage of octahedra in Fe(III) oxides [278]. The Fe-Fe distance and linkage of octahedra in Fe(III) oxides [167, 278]. Reprinted with permission from Ref [167], Copyright © 2003, John Wiley and Sons. E. Crystal structures of magnetite, maghemite, and hematite. Reproduced from Ref. [287] with permission from the Royal Society of Chemistry.

indicates tetrahedral, and B indicates octahedral coordination sphere [167,280–282]. Hematite, like magnetite and maghemite, contains trivalent iron cations. However, the arrangement of Fe(III) and oxygen atoms in hematite is different, forming a primitive rhombohedral structure with a trigonal-hexagonal unit [167,283–285]. In the crystal structure of trigonal hematite, the oxygen atoms are stacked in approximately closed-packed layers along the threefold axis. These layers are arranged in a hexagonal pattern [286]. The representation of crystal structures and space group data of magnetite, maghemite (with ordered and disordered vacancy distributions), and hematite is presented in Figure 8E [287].

It is worth noting that mixed-valent Fe minerals can be synthesized and occur through both abiotic and biotic pathways in nature. A schematic representation of these pathways is provided in Figure 9A [288]. Factors such as oxygen, ROS, light, nitrate (NO_3^-), different forms of iron (II) and iron (III), as well as phosphorus concentrations can influence the transformation of iron oxide minerals [288,289] or metal architecture of iron oxide nanomaterials [167,275]. These external influences can induce changes in the oxidation state, reactivity, and properties of iron, impacting its behavior in both biological and biogeochemical processing [167,290–293]. The study conducted by Usman et al. demonstrates the transformation of ferric (oxyhydr)oxides, such as ferrihydrite, lepidocrocite, and goethite, into magnetite in the presence of Fe(II) ions [294]. The reactivity and transformation sequence were found to be dependent on several factors, including the initial mineralogy of the oxyhydroxides, aging time, and solution chemistry. The order of reactivity for the transformation into magnetite was observed to be ferrihydrite > lepidocrocite > goethite [294]. Field transmission electron microscopy revealed that magnetite, with variable stoichiometry, particle size, and surface area, was obtained from different ferric oxyhydroxides.

Interestingly, the magnetite formed from all initial ferric oxyhydroxides was non-stoichiometric, regardless of the imposed stoichiometric conditions. In the case of goethite, the transformation into magnetite was particularly striking, as observed by Mössbauer spectroscopy over aging time. The results suggested a solid-state transformation of goethite into magnetite, driven by ET that facilitated the spinel ordering between adsorbed Fe(II) and the ferric oxide [294] (Figure 9B). Furthermore, the reactivity of magnetite could vary depending on the source of the ferric mineral precursor. Magnetite produced from ferrihydrite exhibited closer stoichiometry, smaller particle size, and higher surface area, which could contribute to its higher reactivity [294].

Solar irradiation also plays a role in forming iron oxide nanoparticles. Low levels of ferric ions (Fe(III)) would have been produced at the ocean surface as a result of photooxidation, even in the absence of oxygen in the atmosphere [75]. Solar irradiation promotes ferrihydrite to transform into goethite compared to dark conditions [79,80]. For example, the enhanced pathways of sunlight-driven ferrihydrite transformation in the presence of dissolved oxygen, without the initial addition of dissolved Fe(II), further demonstrated that sunlight significantly facilitated the conversion of ferrihydrite to goethite [80]. At the ferrihydrite interface, photoinduced

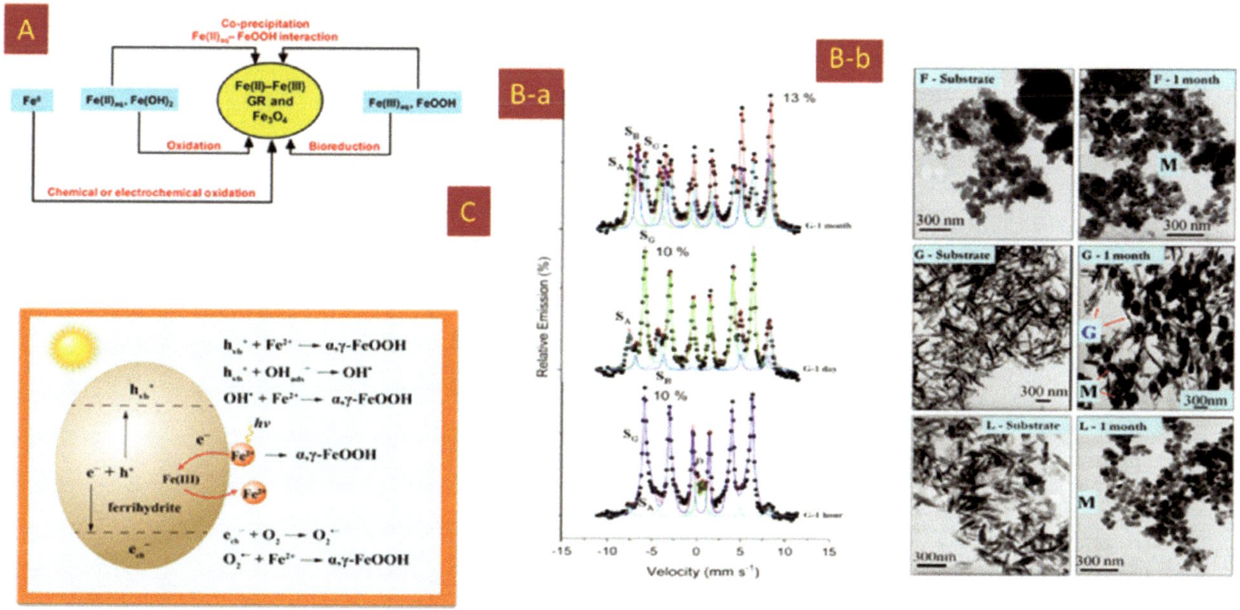

FIGURE 9 Dynamic metal architecture of iron oxide nanoparticles in the environment. A. Main pathways for the formation of magnetite (Fe_3O_4) and green rust (GR) [288], including abiotic and biotic processing. Reprinted with permission from Ref [288], Copyright © 2007, American Chemical Society. B. Magnetite NPs formation from the ferrihydrite (F), lepidocrocite (L), and goethite (G) NPs in the aged solution with the supply of ferrous ions [294]. Reprinted with permission from Ref [294], Copyright © 2012, Elsevier. C. Interaction of solar radiation. Reprinted with permission from Ref [79], Copyright © 2019, Americal Chemical Society.

ET processes generated redox active species, including hole-electron pairs, reactive radicals, and Fe(II). Regarding the production of hydroxyl radicals, three pathways were suggested: water oxidation (5.0%), reduction of dissolved oxygen (40.9%), and photolysis of Fe(III)-hydroxyl complexes (54.1%) [80]. Additionally, the results indicated that under acidic conditions, superoxide radicals played a major role as oxidants for Fe(II) reoxidation, promoting the transformation of ferrihydrite. It is worth noting that the presence of inorganic ions such as chloride, sulfate, and nitrate not only influenced the hydrolysis and precipitation of Fe(III) but also impacted the generation of radicals through photoinduced charge transfer reactions. (Figure 9C) [80].

Magnetite (Fe_3O_4) nanoparticles are capable of converting into maghemite (γ-Fe_2O_3), not only by oxidation with oxygen but also by various ion and/or electron transfers through the solid–solution interface [295]. The elimination of an electron from a surface ferrous ion with the formation of a ferric ion and a cationic vacancy in the octahedral lattice to maintain electrical neutrality. Electron mobility within the solid allows for the renewal of the ferrous sites at the surface. This results in maghemite formation within the particles. At the same time, cationic vacancies are created by the migration of ferric ions to the surface to preserve the electroneutrality of the particles. Oxidation conditions affect processing [295]. Even in the absence of oxygen, 10 nm-sized stoichiometric magnetite particles ($Fe^{2+}/Fe^{3+} = 0.5$) in aqueous solutions over a biologically and environmentally relevant pH range (4–7) are still not stable. Fe^{2+} is released into the solution due to the H^+-promoted dissolution process, which further leads to the partial oxidation of magnetite to a magnetite-maghemite solid solution, demonstrated by the X-ray magnetic circular dichroism (XMCD) spectra and transmission electron microscopy (TEM) observation [296]. Under harsh conditions, for example, 9 nm magnetite completely transferred into the maghemite after 24 h in a 0.07 mol HNO_3 solution. When heated to 60 °C and stirred vigorously in an air atmosphere in DI water, the process was much slower; only complete oxidation of ferrous iron ions was observed, and the crystal phase of maghemite had not yet formed over 24 h [297]. Moreover, recent research revealed that when magnetite 111 is prepared under oxidizing or reducing conditions, the metal architecture at the surface, i.e., iron versus oxygen fractional surface terminations, is also different. A larger fraction of Fe-termination was found in the magnetite preparation under reducing conditions [298].

When iron oxide nanoparticles come into contact with bacteria, their metal architectures can undergo significant changes. An intriguing example of this phenomenon involves *Shewanella oneidensis* MR-1, a strain of bacteria capable of reducing metals. This bacterium has been observed to play a role in the transformation of hematite nanoparticles into magnetite, showcasing a microbial-mediated process [299]. The presence of *Shewanella oneidensis* MR-1 triggers a substantial reduction in hematite, leading to the creation of significant amounts of Fe(II). An astonishing alteration in the crystalline structure of hematite is also evident in this process. The hexagonal arrangement of hematite shifts to a cubic system, as evidenced by advanced techniques such as high-resolution transmission electron microscopy combined with selected area electron diffraction (SAED), Raman spectroscopy, and analyses of magnetic properties [299]. The reductive dissolution of hematite catalyzed by the bacterium results in the formation of intermediate states and the emergence of novel chemical environments involving Fe(II)/Fe(III) complexes characterized by monodentate and bidentate coordination patterns. This transformative process involves the cleavage of iron-oxygen bonds within the hematite structure, leading to the development of microstructures that complex with Fe. These microstructures play a pivotal role in facilitating the creation of biogenic magnetite. Furthermore, observations from electron paramagnetic resonance (EPR) spectra indicate a diminishing EPR intensity over time, suggesting changes in the composition and local structures of both hematite and magnetite throughout the transformation process. The employment of DFT calculations further substantiates the earlier observation regarding the octahedral configuration that arises as a consequence of Fe(II) production [299].

In another study, Fe_3O_4 nanoparticles were exposed to a solution containing *E. coli* for one hour. As a result, approximately 46–48% of γ-Fe_2O_3 was generated, while only 22% of γ-Fe_2O_3 was generated in the control group (pure water, pH 5, 30°C). This was determined by comparing the changes in Fe–K-edge XANES spectra. No structural modifications of maghemite nanoparticles were observed, indicating that maghemite remained highly stable after direct contact with bacteria [300].

A growing body of research has illuminated the pivotal role of metal ion architecture within nanoparticles in governing inorganic oxidoreductase activities, which encompass POD activity of iron oxide nanoparticles, alongside considerations of size, shape, and surface area [104,301–313]. Experimental evidence has underscored that Fe_3O_4 nanoparticles exhibit heightened POD activity when contrasted with γ-Fe_2O_3 and α-Fe_2O_3 nanoparticles [314]. Curiously, alterations in catalytic processes or cycling time can induce changes in the activity of these nanoparticles. A significant revelation is the transformation of magnetite nanoparticles from Fe_3O_4 to γ-Fe_2O_3 during catalyst cycling, which leads to compromised ET rates at the nanoparticle surface, consequently reducing POD activity (Figure 10). The specific POD activity (a_{nano}) of Fe_3O_4, α-Fe_2O_3, and γ-Fe_2O_3 nanoparticles is reported as 1.79, 0.45, and 0.03 $U{\cdot}mg^{-1}$, respectively [253]. Additionally, the specific activity of Fe_3O_4 nanoparticles experiences a notable decline over time, a trend supported by spectroscopic measurements [314]. Importantly, changes in the metal architecture extend beyond the nanoparticle surface. Prolonged reaction times lead to a gradual oxidation of interior Fe(II) within the Fe_3O_4 nanozymes. This implies that both surface Fe(II) and internal Fe(II) contribute to the POD-like activity of Fe_3O_4 nanozymes. The Fe(II) ions housed within Fe_3O_4 nanoparticles facilitate ET to the surface through the Fe(II)-O-Fe(III) chain, aiding in the regeneration of surface Fe(II) ions and thereby enabling sustained POD-like catalytic reactions. However, a limiting factor emerges as excess oxidized Fe(III) ions migrate from the lattice during this process, acting as the rate-limiting step [314]. Consequently, the catalytic activity of the regenerated nanoparticles gradually diminishes as their oxidation state increases [314].

The intricate interplay between minerals and microbiology presents a compelling demonstration of the crucial role played by metal architecture within nanoparticles, a phenomenon that pervades the natural world. This interaction becomes particularly conspicuous at the juncture between actively proliferating fungi and the iron-rich mineral hematite, giving rise to a remarkable process—the substantial generation of

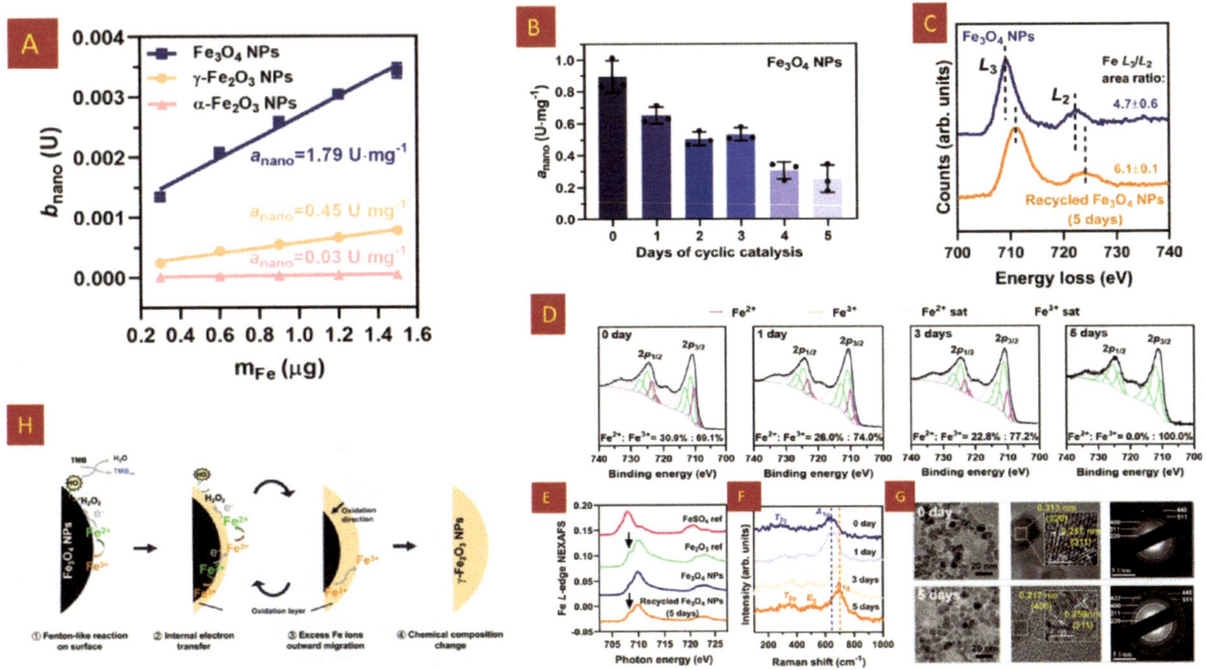

FIGURE 10 Structure change of magnetite nanoparticles in the POD catalytic processing. A. The specific activity (a_{nano}) of these three IONPs with TMB as colorimetric substrates. B. Kinetic study of a_{nano} values of Fe_3O_4 NPs with the days of cyclic catalytic reaction. Error bars represent standard deviation from three independent measurements. C. Comparison of Fe L_2, and L_3 spectra of Fe_3O_4 NPs before and after 5 days of cyclic POD-like reactions. D. The fitted Fe2p XPS spectra of Fe_3O_4 NPs recycled after catalysis on days 0, 1, 3, and 5. E. The Fe L-edge NEXAFS spectra of Fe_3O_4 NPs and recycled Fe_3O_4 NPs after 5 days of catalysis in comparison with the reference spectra of $FeSO_4$ and Fe_2O_3. F. Raman spectra of Fe_3O_4 NPs recycled after catalysis on days 0, 1, 3, and 5. G. TEM, HRTEM images, and SAED pattern of Fe_3O_4 NPs and recycled Fe_3O_4 NPs after 5 days of catalysis. Images were collected at least three times for each type of NPs. H. Schematic diagram of the catalytic mechanism of the POD-like activity for Fe_3O_4 NPs. Reprinted from Ref [314].

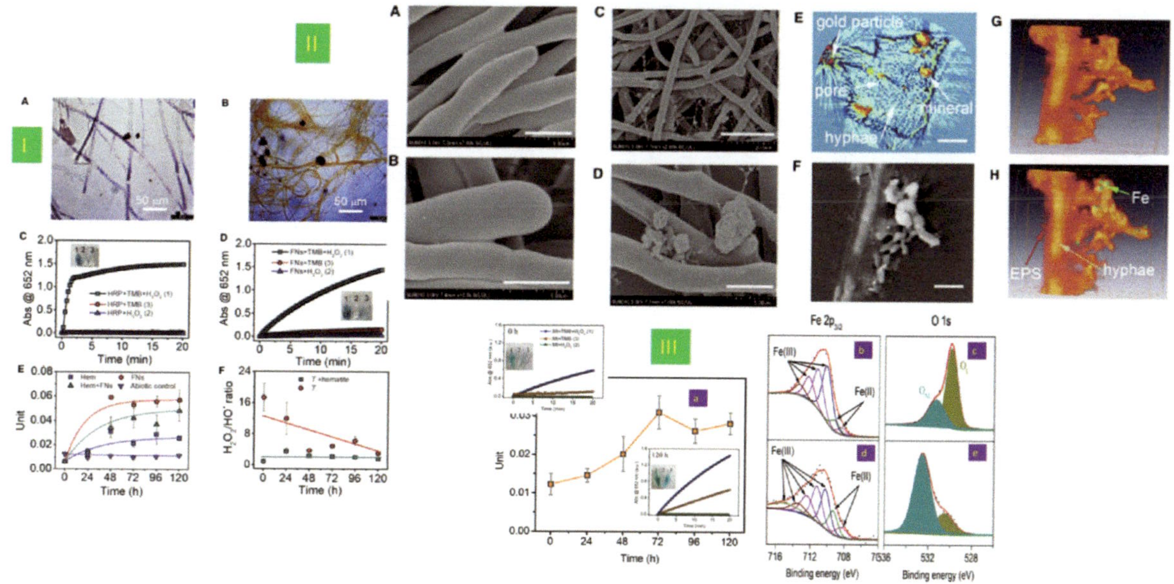

FIGURE 11 Structural modulation of biogenic iron oxide nanoparticles by fungi and implications for POD activity. A. Enzyme-like activities of biogenic ferrihydrite nanoparticles (FN) by fungi. (a and b) Location of H_2O_2 (blue in a) and O_2^{*-} (brown in b) in hyphae shown by staining. (c and d) Progress and yield of the TMB colorimetric reaction catalyzed by 50 ng/mL HRP (c) in comparison to that catalyzed by FNs. (e) Variation of peroxidase-like activity in culture with hematite, hematite + EPS, FNs, and the abiotic control (i.e., hematite exposed to a pH of 3.6; n = 3). (f) Relative abundance of H_2O_2 versus HO* in culture with and without FNs (n = 2 or 3). B. Morphology of fungal hyphae in the absence or presence of iron nanoparticles. (a and b) Cryo-SEM images of fungal biomass in the absence of hematite. Scale bar represents 5 μm. (c and d) Cryo-SEM images of fungal biomass in the presence of hematite (120-h cultivation). Scale bars represent 20 μm (C) or 5 μm. (e) Synchrotron radiation transmission X-ray microscopy (SR-TXM) image showing the presence of iron minerals, pores, and fungal hyphae in fungal-mineral aggregates (120 h). Scale bar represents 200 μm. (f) TEM image of fungal-mineral culture samples (120 h). Scale bar represents 5 μm. (g and h) Element-specific 3D volume renderings for the pre-absorption edge of (g) Fe at E1 = 706 eV and (h) the absorption edge of Fe at E2 = 710 eV. Reprinted with permission from Ref [315]. Copyright 2020, Elsevier C. Effect of time on the interaction of fungi and magnetite nanoparticles on the activity of the peroxidase. The upper left corner shows the zero time and the down right corner shows the 120 h. The corresponding X-ray photoelectron spectroscopy spectra of Fe 2p3/2 (b, d) and O 1s (c, e) at the cultivation times of (b, d) 0 h and (d, e) 120 h respectively. Reprinted with permission from Ref [316], Copyright © 2021, John Wiley and Sons.

biogenic ferrihydrite nanoparticles [315] (Figure 11A). These nanoparticles exhibit inherent POD activity and possess the ability to neutralize hydrogen peroxide, alterations in the non-lattice oxygen within the iron architecture predominantly contribute to the catalytic behavior of these nanomaterials [315]. Interestingly, this catalytic activity deviates from conventional iron redox chemistry and is attributed to the presence of non-lattice oxygen. This discovery expands our understanding of the role of metal architecture in influencing enzyme-like activity. In addition to their structural influence, fungi are actively involved in mediating the catalytic reactions of these biogenic ferrihydrite nanoparticles. Techniques such as cryo-scanning electron microscopy, high-resolution 3D volume rendering (at a scale of 30 nm), and biomass analyses confirm the presence of the organism and its participation in these catalytic reactions [315] (Figure 11B).

Another compelling example involves the incubation of magnetite nanoparticles with *Trichoderma guizhouense* [316]. This interaction results in a substantial enhancement of the nanoparticles' POD activity. Following a 72-hour cultivation period, the average POD activity of the magnetite nanoparticles significantly surpasses that of the original magnetite [316]. Detailed high-resolution X-ray photoelectron spectroscopy analysis reveals changes in the metal architecture of the magnetite nanomaterials, particularly affecting the non-lattice oxygen (Figure 11C). This observation underscores nature's capacity to generate inorganic nanoparticles with oxidoreductase activity through modifications in their metal architecture. The interplay between fungi and iron oxide minerals leads to structural transformations that have a direct impact on the nanoparticles' catalytic properties. This insight not only sheds light on the intricate relationship between microorganisms and nanomaterials but also unveils the potential role of metal architecture in dictating catalytic behaviors, highlighting the intricate complexity of these biological and chemical processes.

Recent study on the biogenic iron cores of ferritin further emphasizes the importance of metal structure, particularly in early life [202], supporting the concept of "life fossil oxidoreductases" [91]. Upon the removal of the protein, the iron core from prokaryotic ferritin, specifically both pfFn and pyFn from archaea *P. furiosus* or *P. yayanosii*, demonstrates higher SOD activity than that of eukaryotic ferritins, such as HFn and LFn from *H. sapiens* [202] (Figure 5C). Additionally, the phosphate-to-iron ratio in the iron core, primarily determined by the ferritin structures, increases, while their corresponding SOD catalytic capability decreases from the inorganic core [202]. For instance, the iron core from ssDps, the DNA protection protein from the archaea *Sulfolobus solfataricus*, also shows significantly lower SOD activity than that of pfFn and pyFn [202]. Meanwhile, with evolutionary processes, eukaryotes have a higher phosphorus content, leading to a decrease in their corresponding SOD activity from the iron core. The presence of phosphate in the iron core transforms it from a single crystalline structure to an amorphous iron phosphate-like structure, leading to a decreased affinity for the hydrogen proton of the ferrihydrite-like core [202]. The study provides valuable insights into the intricate relationship between the metal structure of ferritins, particularly the iron core, and their catalytic activities. To some extent, all of these iron cores in ferritins can also be seen as "life fossil oxidoreductases" [91]. These findings underscore the significance of the composition and evolution of ferritin structures in the context of early life and have potential implications for understanding the role of inorganic materials in biochemical processes.

An investigation focused on the residues of PVC dichlorination and iron chips treated with subcritical water unveiled that the wastewater contained Fe_2O_3 nanoparticles exhibiting inherent peroxidase-like properties due to the distinctive metal architecture of iron oxide nanoparticles [317]. These instances underscore that the metal architectures of iron oxide nanomaterials are far from fixed. Within both ferrous and ferric ions, octahedral and tetrahedral coordinated interstitial sites exist within a closely packed O_2^-/OH^- anionic lattice. The distinctive attributes of these nanoparticles emerge from their electron configuration and coordination with oxygen, including the presence of oxygen vacancies (OVs) [167,280,316,318–323].

This perspective can be extended to encompass various iron sulfide nanoparticles. Similar to iron oxide nanoparticles, diverse iron sulfide particles can also undergo structural modifications due to chemical and microbial interactions [276–285]. Ferrous iron reacts with sulfide to produce FeS under anaerobic conditions, which can then undergo a temperature, redox potential, and pH-dependent maturation process resulting in the formation of oxidized mineral phases such as greigite or pyrite [324]. Tracing oxidative maturation pathways by varying the electrode potential, nanoparticulate $n(Fe^{2+}S^{2-})(s)$ was found to oxidize to a Fe^{3+} containing FeS phase at −0.5 V vs. Ag/AgCl (pH = 7). In a subsequent oxidation, polysulfides are proposed to give a material that is composed of Fe^{2+}, Fe^{3+}, S^{2-}, and polysulfide (S_n^{2-}) species, with its composition described as $Fe_{1-3x}^{2+}Fe_{2x}^{3+}S_{1-y}^{2-}(S_n^{2-})$. These findings suggest the existence of a metastable FeS phase located at the junction of a metal-based oxidation path between FeS and greigite ($Fe^{2+}Fe^{3+}{}_2S_4^{2-}$) and a ligand-based oxidation path between FeS and pyrite ($Fe^{2+}(S_2)^{2-}$) [324].

Additionally, trace metals (Co, Cu, Mo, Ni, Zn) were found to accelerate pyrite formation [325]. These metals can facilitate polysulfide formation or precipitate as nanoparticles, acting as nuclei for pyrite growth. Depending on the initial metal concentration, various morphologies and structures were observed, including tabular precipitates (<150 nm), aggregates resembling microframboids (100–250 nm), octahedral (300–1500 nm), and rose-like particles (1000–3000 nm). This size-shape particle continuum was interpreted as stages of pyrite growth via particle attachments. These interactions can lead to shifts in atomic and ionic arrangements within the nanoparticles, ultimately reshaping their metal architecture and impacting their properties.

Metal architecture plays a pivotal role in determining the functionalities and reactivity of nanomaterials, making them highly versatile, especially in response to biological changes and catalytic behaviors, particularly in the context of nanozymes [326–335]. Defect nanostructures are inherent in natural metal minerals, including iron oxide and iron sulfide minerals, and play a pivotal role in various geological and environmental processes. These defects, which may include vacancies like oxygen, sulfur, or metal vacancies, are not anomalies but rather natural and integral components of these materials [336–340]. They contribute significantly to the functionalities and reactivity of these nanomaterials, rendering them remarkably versatile. The interplay between iron oxide and iron sulfide nanomaterials with environmental factors and biological systems introduces novel avenues for comprehending the inception of life and the evolution of proteins, along with their ecological functions. Studying these interactions provides valuable insights into the fundamental processes that contributed to life's emergence on our planet. Inorganic nanoparticle structures are not rigid and unchanging entities; they dynamically respond to various external influences, including both abiotic and biotic factors, as well as catalytic processing. These factors significantly impact the behavior of biocatalysts and have implications for the importance of metal center stabilization in the evolution of proteins [8,91].

6 ELECTRON CONDUCTION MECHANISMS IN INORGANIC NANOPARTICLES

In the realm of solid-state physics, the concept of energy bands plays a pivotal role in deciphering the electrical properties of materials. The valence band (VB) represents the highest energy band within a material's electron structure, encompassing a range of energy levels typically filled with electrons and engaged in chemical bonding [341]. Conversely, the conduction band (CB) pertains to the range of energy levels wherein electrons can move freely within the crystal lattice. Here, they act as charge carriers, facilitating the flow of electric current. These mobile carriers can arise from intrinsic non-stoichiometric pathways or the introduction of dopants (foreign atoms) into the material's structure [341].

Electron conductivity, synonymous with electrical conductivity, denotes a material's capacity to conduct an electrical current, contingent on the ability of electrons to traverse through the material [342,343]. The ease with which charged particles, such as electrons or protons, navigate through a medium in response to an electric field is quantified as electron mobility—an integral factor influencing conductivity [342]. Amplified electron mobility augments conductivity, as electrons gain greater freedom of movement, enhancing the efficacy of current flow. Conversely, carrier concentration—the abundance of charge carriers, like electrons or holes, present in a material—also shapes conductivity. Heightened carrier concentration translates to elevated conductivity, as a larger pool of particles is available to facilitate electric current [342,343]. As anticipated, the electrical conductivity of iron oxide nanoparticles fluctuates with nanoparticle concentration. Empirical findings showcase that in a Fe_3O_4 nanofluid featuring varying volume fractions, its electrical conductivity escalates alongside rising temperature and weight fraction [344]. For instance, at 25 °C, the electrical conductivity of the base fluid magnifies significantly from 0.39 μS cm^{-1} to 2.419 mS cm^{-1} upon incorporating 4 vol% of iron oxide nanoparticles. This corresponds to an astounding 6000 times enhancement [344]. Evidently, the electrical conductivity of the nanofluid tightly hinges on the volume fraction of iron oxide nanoparticles.

The characteristics of materials as semiconductors or insulators intricately intertwine with the notion of the band gap [341]. Semiconductors and insulators bear a hallmark feature known as the band gap—an energy differential essential for propelling a bounded valence electron into a mobile carrier capable of unfettered movement within the crystal lattice, ultimately facilitating electric current [342,343]. The band gap conventionally denotes the energy disparity (quantified in electron volts) between the apex of the valence band and the nadir of the conduction band. Semiconductors typically harbor a moderately small band gap, empowering them to conduct electricity under specific circumstances [343]. Provided with adequate energy, electrons from the valence band can be excited into the conduction band, engendering liberated charge carriers. This process underpins the selective conduction of current by semiconductors, rendering them indispensable in electronic components like transistors and diodes. By contrast, insulators feature a wider band gap, impeding electrons from vaulting to the conduction band, even with external energy input. As a result, insulators manifest poor electrical conductance, effectively obstructing current flow. Notably, the concept of the band gap extends beyond semiconductors, permeating the realm of general nanomaterials. In the milieu of inorganic nanoparticles, the band gap stands as a pivotal determinant in ET processes, dictating the energy requisite for electron migration between the nanoparticle and other molecules or surfaces.

ET stands as a foundational process, pivotal across a gamut of systems, fostering interactions between inorganic nanomaterials and substantial organic entities, including diverse nanomaterials and large organic compounds (e.g., proteins, DNA) [14,34,35,345], as well as microorganisms [33,346,347]. This process exerts paramount influence over numerous biological mechanisms, particularly those revolving around oxidoreductases' ET. Diverse factors modulate ET efficiency, with the band gap emerging as a key determinant for the inorganic nanomaterials. The band gap intimately links to the nanoparticles' electron configuration and structural attributes, encompassing size, structure, surface effects, composition, doping, and quantum confinement [343,348]. To exemplify, the electron configuration of Fe(II) (iron ion with a +2 charge) and Fe(III) (iron ion with a +3 charge) is [Ar] 3d6 and [Ar] 3d5, respectively. In contrast, a neutral oxygen atom (O) assumes an electron configuration housing 2 unpaired electrons within the 2p subshell. These electron configurations contribute to a multifaceted electronic landscape within nanoparticles, fostering electron delocalization and mobility. Within the crystalline lattice, the transfer of electrons unfolds amid Fe(II) and Fe(III) ions, both inhabiting diverse oxidation states. The fusion of Fe 3d electrons with oxygen affords leeway for electron delocalization and mobility within the nanoparticle's structural matrix. In turn, this underpins a phenomenon christened electron hopping—where localized electrons traverse between Fe(II) and Fe(III) ions. The impetus propelling this electron hopping derives from the oxidation state disparity between ions and the band gap, a metric representing the energy prerequisite for electron transitions. The occurrence of electron hopping is postulated by theoretical models pertaining to myriad iron oxide nanomaterials, encompassing green rust [349], magnetite [350], iron oxyhydroxides [351,352], and hematite [353,354]. Additionally, experimental observations have confirmed electron hopping on the surfaces of iron (oxyhydr)oxide [355], α-Fe_2O_3 [356,357], and γ-Fe_2O_3[358] NPs. The electron small-polaron hopping rate, quantifying the velocity of electron hopping, emerges as a pivotal factor impacting the kinetics of numerous iron redox reactions. This rate hinges on the short-range structural topology [355] or lattice expansion of iron oxide [356], holding precedence over nanoparticle dimensions, suspension pH, or the infusion of multiple electrons per nanoparticle [359]. These nuances underscore the central role played by short-range structural topology and lattice expansion of iron oxide in orchestrating efficient electron hopping, highlighting the elemental role of electron hopping stemming from the nanoparticles' iron and oxygen electron configuration.

Materials characterized by elevated electron conductivity, featuring heightened electron mobility and carrier concentration, exhibit superior capabilities for effective ET, thereby catalyzing ET reactions adeptly. These materials hold paramount significance in shepherding the ET steps integral to biological functions and technological applications alike. Conversely, materials endowed with diminished electron conductivity may confront hindrances in expediting ET processes. In such scenarios, the pace or efficiency of ET reactions could wane, impairing the overall functionality of implicated systems.

The electrical conductivity of iron oxide nanoparticles is subject to an array of influences, including the crystal structure and the spatial disposition of iron ions. Notably, charge ordering (CO) emerges as a pivotal phenomenon underpinning electrical conductivity properties. Charge ordering denotes the organized arrangement of distinct metal oxidation states within a material's crystal lattice. In the context of iron oxide nanoparticles, charge ordering surfaces in magnetite at temperatures below the Verwey transition temperature of 120 K [360]. Within iron oxide nanoparticles, the Fe-Fe distance within octahedral chains wields significant sway over charge ordering and electron mobility. The gap separating iron ions within these chains reverberates through the extent of charge transfer and electron hopping across different oxidation states, thereby wielding sway over the overall electrical conductivity of the material.

Both α-Fe_2O_3 and γ-Fe_2O_3 emerge as n-type semiconductors [167], denoting their capacity to conduct electricity through the orchestrated movement of electrons. Hematite's band gap measures 2.2 eV, whereas maghemite's slightly undercuts it at 2.03 eV [167]. On the converse, wustite (FeO) upholds status as a p-type semiconductor [167], with electricity conduction transpiring through hole movement, characterized by a band gap of 2.3 eV [167]. In the scheme of things, goethite, lepidocrocite, and δ-FeOOH host comparably modest electrical conductivities in contrast to hematite and maghemite. Goethite exhibits a band gap of 2.10 eV, lepidocrocite clocks in at 2.06 eV, and δ-FeOOH boasts a band gap of 1.94 eV [167]. An intriguing case manifests in magnetite, renowned for its multifarious electrical properties, embodying both p-type and n-type semiconductors within its structural confines. It features a minute band gap measuring 0.1 eV [167]. Intriguingly, magnetite's electrical conductivity, boasting a Fe(III)/total Fe ratio of 2/3, stands several multiples higher compared to hematite, characterized by a Fe(III)/total Fe ratio of 1 [361]. Magnetite operates as an electron conductor bearing a distinct spin orientation, while donning the mantle of insulator or semiconductor for electrons aligned with the opposite spin orientation. This unique property categorizes magnetite as a half-metallic material [362]. Rigorous experimental scrutiny and density-functional calculations validate the presence of spin-split band energies, substantiating its half-metallic nature [363–366].

Furthermore, investigations into sphere-shaped magnetite nanoparticles reveal shifts in electron mobility that align with temperature changes across a wide range (173–373 K) [367]. The electrical conductivity of these nanoparticles is closely linked to AC frequency and temperature interactions, as demonstrated by conductivity measurements. Additionally, dielectric properties and AC conductivity at 273 K reveal significant differences related to grain boundaries and grain conductivity, providing insights into long-range and short-range mobility (Figure 12A, 12B) [367]. These findings highlight a continuum between two conduction mechanisms—barrier hopping and non-overlapping small polaron tunneling—that encompass high temperatures and low frequencies (Figure 13C) [367]. The interplay between these mechanisms is further emphasized in the conductivity dispersion, effectively encapsulating both long-range and short-range mobility.

At lower temperatures, electron conductivity primarily occurs through charge carrier hopping between Fe(III) and Fe(II) ions. This hopping process involves two distinct modes: one where the electrons hop between adjacent Fe(III) and Fe(II) ions within a single nanoparticle and another where the electrons hop between different nanoparticles. These modes contribute to a gradual increase in conductivity with rising temperature. As temperatures increase further, electron hopping remains the dominant mechanism, but the specific mode of hopping may change. These shifts in conductivity mechanisms illustrate the intricate interplay between temperature and the structural dynamics of nanoparticles, ultimately impacting electron mobility within the Fe_3O_4 nanoparticles. This electron transport phenomenon is fundamentally governed by the interplay of tunneling of small polarons, electron hopping, and movement between Fe(III) and Fe(II) ions.

Additionally, the electrical conductivity of magnetite evolves in response to pressure alterations. Empirical studies divulge that under pressure ranging from 0 to 20 GPa, the resistivity of Fe_3O_4 plummets by over an order of magnitude, reaching a nadir. However, beyond 20 GPa, the resistivity resurges, ultimately doubling by 48 GPa (Figure 13D) [368]. This propensity for resistivity change underscores the pronounced susceptibility of Fe_3O4's electrical conductivity to shifts in pressure. Notably, the electronic attributes of Fe_2O_3 and FeO also undergo transformation under pressure and structural modifications [369–371]. Temperature and pressure stand out as pivotal factors sculpting iron architecture in nanomaterials, accentuating alterations in Fe-Fe distance. These shifts, in turn, reverberate through the atomic structure of iron oxide nanoparticles, impacting the charge ordering phenomenon (CO) and ET processes. The minimal Fe-Fe distance within the crystalline structure looms large as a critical factor dictating the nature and temperature at which charge ordering transpires (Figure 12E) [372–378]. The electrical conductivity properties of iron oxide nanoparticles are further molded by the transmission of electrons between Fe(III) and Fe(II) ions. Adjustments to the Fe-Fe distance can impede or facilitate

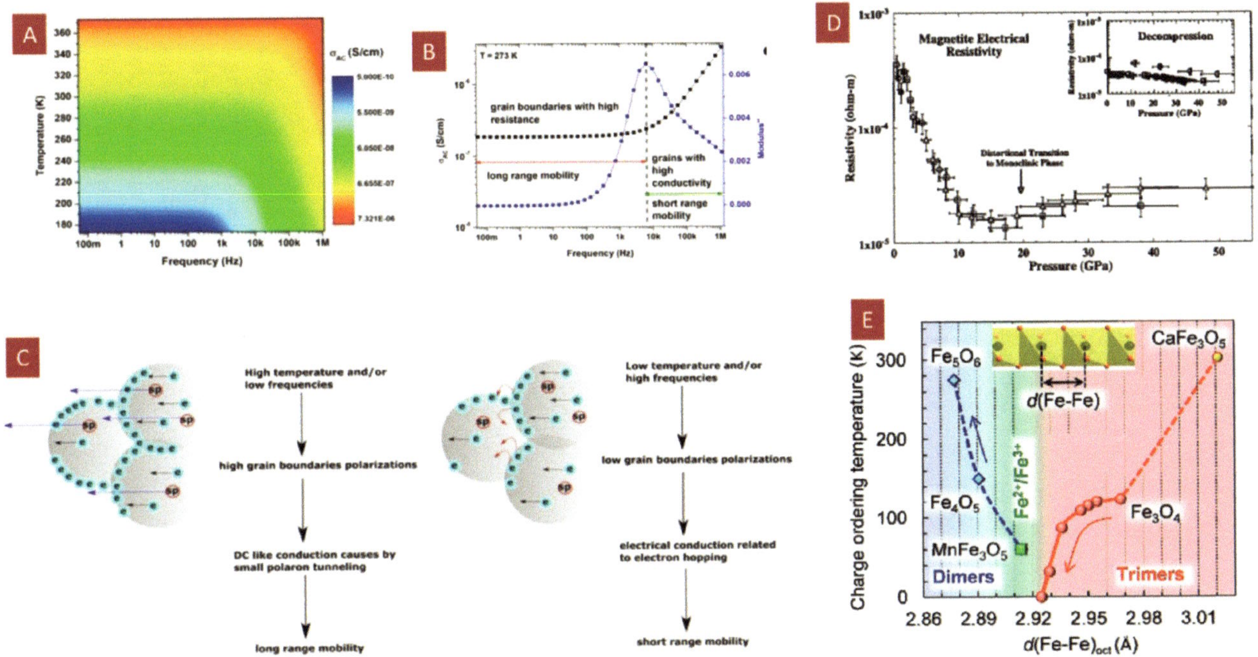

FIGURE 12 Magnetite nanoparticles structure and electrical characteristics. A. Surface plot of AC conductivity in the function of temperature and frequency of Fe_3O_4. Reprinted from Ref [367]. B. Dielectric properties and AC conductivity at 273 K, with marked regions related to the grain boundaries and grain conductivity of Fe_3O_4, and corresponding to them long and short range mobility with the border corresponding to the maximum electric modulus peak *M"*, which is associated with the conduction relaxation processes. Reprinted from Ref [367]. C. Two conduction mechanisms in different temperature and frequency regions, Reprinted from Ref [367]. D. Response of electronic resistivity of Fe_3O_4 under different pressure [368], Reprinted with permission from Ref [368], Copyright © 1997, John Wiley and Sons. E. Dependence of the charge-ordering transition temperature for Fe_5O_6, Fe_4O_5, $MnFe_3O_5$, Fe3O4, and $CaFe_3O_5$ on the minimal Fe–Fe distances in their octahedral iron chains. Reprinted from Ref [378].

electron migration between these ions, invariably precipitating changes in electrical conductivity. Appreciating the nexus between the Fe-Fe distance, charge ordering, and ET in iron oxide nanoparticles assumes pivotal significance in unraveling their electrical conductivity. Notably, shifts in the Fe-Fe distance—stemming from temperature and pressure shifts—unearth invaluable insights into the mechanics of electron transport, such as electron hopping and tunneling of small polarons. These shifts in the Fe-Fe distance stand as an inherent aspect of the metals' atomic structure and electron configuration, engendering repercussions for charge ordering and ET.

The electrochemical traits of iron oxide nanoparticles augur their capacity for charge transfer. Cyclic voltammetry curves and electrochemical impedance spectroscopy (EIS) spectra concerning hematite and magnetite nanoparticles elucidate electrons within these nanoparticles as electroactive species essential for charge transfer [379] (Figure 13). Notably, high-purity iron oxide nanoparticles distinctly exhibit superior charge concentration, mobility, and conductivity vis-à-vis their bulk counterparts, as attested by Hall effect measurements.

Greigite (Fe_3S_4) emerges as a ferrimagnetic compound mirroring the inverse spinel structure akin to oxide magnetite (Fe_3O_4) [380–382]. Within greigite, Fe(II)

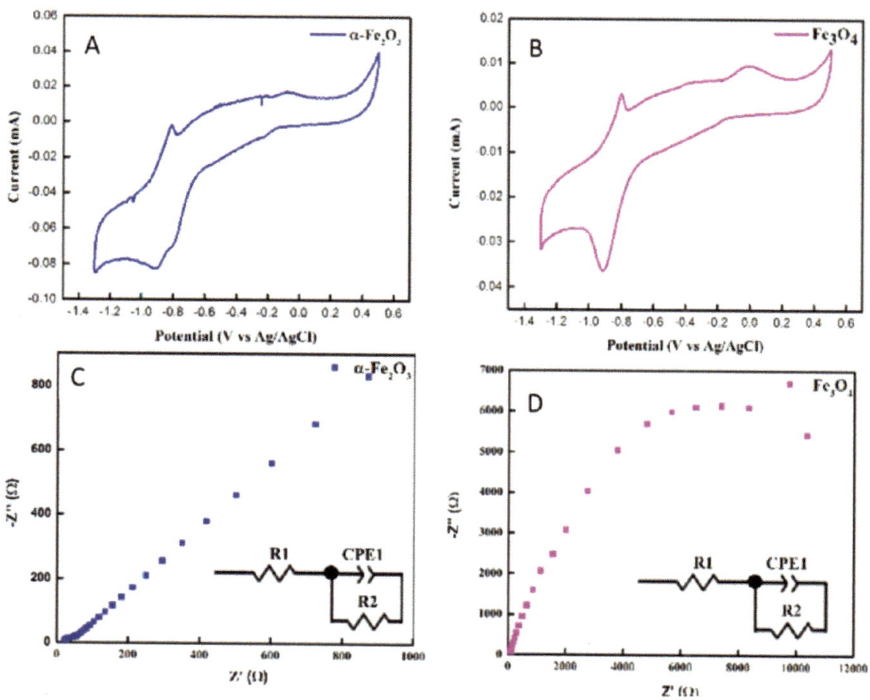

FIGURE 13 Cyclic voltammetry curves of α-Fe_2O_3 nanoparticles (A) and Fe_3O_4 nanoparticles (B), and EIS spectra of α-Fe_2O_3 nanoparticles (C) and Fe_3O_4 nanoparticles (D), Reprinted from Ref [379].

and Fe(III) centers maintain a 1:2 ratio, fostering a mixed-valence compound. Tetrahedral sites harbor Fe cations with magnetic moments oriented in opposition to those in the octahedral sites, contributing to a net magnetization [383,384]. Greigite also boasts high spin quantum numbers on both metal sites, characterizing it as a half metal replete with an S vacancy structure. This configuration gives rise to a magnetic moment below 4.0 μB per formula unit [385] (Figure 14A). Notably, electron hopping between Fe(II) and Fe(III) is manifest at octahedral sites [381,386]. When comparing the traits of Fe_3S_4 and Fe_3O_4, it transpires that mean charges on octahedral Fe rest at 1.0 e^- and 1.7 e^-, respectively, whereas on tetrahedral Fe, they elevate to 1.1 e^- and 1.8 e^- for Fe_3S_4 and Fe_3O_4, respectively. Notably, the saturation magnetization (Ms) in sulfides, greigite inclusive, falls marginally shy of that exhibited by oxides such as magnetite [385]. As a result, the resistivity of sulfides is lower than that of oxides (Figure 14B) [385]. Greigite typically manifests as a precursor to pyrite (FeS_2) within oxygen-depleted sulfate-reducing sedimentary milieus. Here, the presence of sulfide ions expedites its formation. However, greigite's inception is not solely bound to such environments; it can also materialize at various diagenesis stages, with dissolved iron Fe(II) and sulfide (S^{2-}) lurking within the pore fluids of sedimentary rocks. The growth of greigite responds to the availability of these elements and prevailing chemical and environmental conditions characteristic of diagenesis [380].

Iron pyrite (FeS_2) mirrors the crystal structure of sodium chloride (NaCl), assuming a face-centered cubic sublattice. This sublattice is home to diamagnetic Fe (II) ions occupying Na positions and S-S dimers (S_2^{2-}) oriented along 1 1 1 planes, thus monopolizing the anion positions [387]. At the corners and face centers of the cubic unit cell, Fe atoms establish an octahedral coordination, while S_2^{2-} dimers find their residence at the cubic unit cell's center and midpoints, heralding hemihedral symmetry for pyrite [388,389] (Figure 14C). Notably, the coordination of Fe and S in pyrite equates to a charge-compensated FeS_6 molecule. Nonetheless, S vacancies are a common find in both natural and synthetic FeS_2 crystals, endowing them with an S:Fe ratio of 1.9 to 2.08 [390,391]. The advent of S vacancies tweaks local Fe coordination, transitioning from FeS_6 to FeS_5, consequently altering the point group symmetry from octahedral to tetragonal pyramidal. This alteration can entail Fe atoms being coordinated with FeS_4, FeS_3, and so forth, if S vacancies are uniformly distributed [390,392]. The deviation from stoichiometric composition, coupled with impurity elements within solid solution and excitation across the energy gap, fosters the generation of free charge carriers within pyrite [393–397]. Pyrite's band gap amounts to approximately 0.90 eV [395,396] (Figure 14D), catapulting it into the realm of semiconductors, nurturing both p-type and n-type conductivity. This manifests in a broad range of conductivities spanning from 0.02 to 562 $\Omega\ cm^{-1}$, averaging 48 $\Omega\ cm^{-1}$[395]. Recent forays into synthesized pyrite nanowires, boasting diameters ranging from 4 to 10 nm and lengths exceeding 2 mm, affirm their status as highly p-doped semiconductors. This is substantiated by an average resistivity of 0.18 ± 0.09 Ω cm and carrier concentrations hovering around 1021 cm^{-3} [395]. Pertinently, electron hopping materializes between Fe(II) and Fe(III) on octahedral crystal surfaces and between conductive sulfur-deficient grain cores ensconced within nominally stoichiometric FeS_2 [396–399].

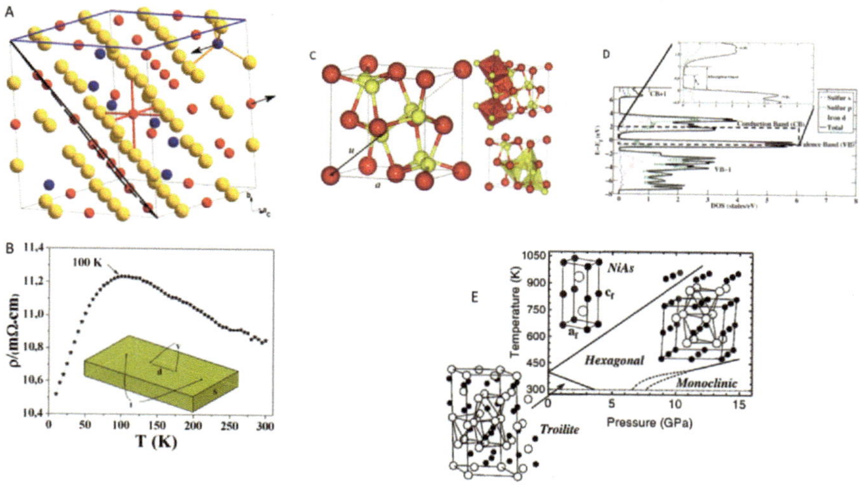

FIGURE 14 Structure and electronic property of three iron sulfide nanoparticles. A. Crystal structure of Fe_3S_4 with the (001) and (111) planes outlined in blue and black, respectively. Sulfur atoms (yellow spheres) form a cubic close-packed lattice: 1/8 of the tetrahedral A sites are occupied by Fe^{3+} (blue spheres) and 1/2 of the octahedral B sites are occupied by Fe^{2+} and Fe^{3+} (red spheres) equally. The magnetic moments on the A and B sites are antiparallel and aligned along the [100] crystallographic axis (indicated by arrows). B. Resistivity of greigite between 5 K and 300 K, and the corresponding contact geometry (inset) Reprinted with permission from Ref [385], Copyright © 2014, American Chemical Society. C. Crystal structures of pyrite FeS_2 with the Fe-S octahedral and tetrahedral coordination. (a) The simple-cubic primitive unit cell of FeS_2. The transition metal atoms occupy the corners and the face centers of this cell. Each transition metal atom is in the center of a sulfur octahedron. (b) The transition metal octahedra are corner sharing in the pyrite structure. In addition, every sulfur is part of a dimer connecting neighboring octahedra, Reprinted with permission from permission from Ref [389], Copyright © 2020, American Physical Society. D. Calculated density of states from FeS_2. The inset shows a zoomed-in region near the Fermi level. Though the band gap may be lower, the onset of optical absorption is consistently measured to be 0.9 eV. Reprinted with permission from Ref [396], Copyright © 2016, American Chemical Society. E. Structural phase diagram of FeS under different pressure and temperature conditions. The dashed phase boundaries (6.5–7.5 Gpa) demarcate a range where two phases coexist. Straight solid lines in the structures show the unit cell as well as the octahedral coordination of Fe (small shaded circles) by S (larger open circles) atoms. Reprinted with permission from Ref [400], Copyright © 1999, American Physical Society.

Iron sulfide nanoparticles, akin to their iron oxide counterparts, yield to structural alterations in the face of pressure shifts [400] or temperature changes [401] (Figure 14E). The crystal structure of iron sulfide (FeS) assumes a tetragonal form, with iron atoms assuming coordination by four equidistant sulfur atoms, thus embracing tetrahedral coordination [402,403]. Conversely, pyrrhotite ($Fe_{1-x}S$) adopts a hexagonal crystal structure patterned after the NiAs arrangement. Here, metal atoms recline in octahedral coordination, and anions nestle within a trigonal prismatic arrangement. Notably, pyrrhotite possesses the capacity to generate iron vacancies, ushering modifications in its crystallographic and magnetic properties. As the incidence of iron deficiency (Fe vacancies) escalates, the hexagonal structure buckles, transitioning into a monoclinic configuration. This transformation dovetails with a shift in magnetic ordering, evolving from antiferromagnetic to ferrimagnetic. This magnetic shift transpires due to the existence of uncompensated magnetic moments within the iron-vacancy-rich strata [400,401,404]. Both structures enthrall octahedral Fe-S coordination.

Natural pyrrhotite is frequently an amalgamation of superstructures, entwining monoclinic and hexagonal constituents, and assumes an overall intermediate composition [170,391,403]. Fe_7S_8, donning a monoclinic structure, witnesses a pronounced upswing in magnetic anisotropy energy, owing to the vacancy framework, when juxtaposed with FeS [404]. Conversely, the symmetrical facet of FeS spells out an overall net zero magnetic moment throughout the unit cell. Meanwhile, the low-symmetry facet of Fe_7S_8 showcases ferrimagnetism, stemming from the uncompensated magnetic moments resident in iron-vacancy-rich layers. Pyrrhotite's electronic attributes prove enthralling; the vacancy-free specimen ($x = 0$), christened troilite, unfurls a metallic state in resistance, plummeting into superconductivity below 4.5 K [405]. Nevertheless, as Fe vacancies snowball ($x \geq 0.05$), superconductivity languishes, giving way to semiconducting behavior [406,407]. Ultrathin Fe_7S_8 nanosheets debut with a surplus of delocalized electrons, primed to facilitate ET during catalytic reactions. With the overlapping d orbitals of Fe(II) and Fe(III) during catalysis, these ultrathin pyrrhotite nanosheets burgeon into highly efficacious Fe-based electrocatalysts for water oxidation [408]. It bears noting that, analogous to iron oxides, iron sulfides are not immune to phase transformations elicited by both chemical and microbial interactions [326–335]. These interactions wield influence over iron sulfides' properties and behavior, rendering them potent actors in the theatre of electron conductivity and catalytic endeavors.

Exploring the world of inorganic nanomaterials unveils a complex interplay of ET mechanisms within iron oxide and sulfide nanoparticles. This journey delves into band gaps, electron mobility, conductivity, charge ordering, electron hopping, proton tunneling phenomena, and structural dynamics. Understanding the intricate relationship between ET and conductivity, particularly at the molecular level within large biomacromolecules (proteins/DNA), remains a key area of investigation [8,9,17,409–412], essential for unraveling the complexities of ET processes within intricate systems. It is noteworthy that both inorganic nanoparticles and large biomacromolecules (proteins/DNA) [413] exhibit electron transfer, including electron hopping and proton tunneling behavior, due to their unique structural and chemical properties from a physics perspective. This shared capability potentially acts as a bridge between inorganic materials and life, reflecting the synergy between solid-state physics, governing the electronic properties of inorganic nanomaterials, and protein chemistry, central to biological systems [8,9]. This interdisciplinary fusion resonates significantly in modern biology and chemistry, where ET processes are crucial components in diverse biological functions and technological advancements.

7 ELECTRON TRANSFER IN INORGANIC OXIDOREDUCTASE

Within the realm of biological systems, it is imperative to bear in mind that the genesis of all electrons invariably stems from biological reduction and oxidation reactions, regardless of whether an active biocatalyst or protein orchestrates the process [414]. Attaining a comprehensive comprehension of chemical equilibrium necessitates a discerning recognition of electrons' pivotal role. Irrespective of the presence or absence of active biocatalysts or proteins, electrons emerge as cardinal protagonists, masterminding the intricacies inherent in redox reactions. While engaging in catalytic undertakings, oxidoreductases, fascinating catalysts in their own right, do not themselves manifest net electron production. The veritable crux of their catalytic prowess resides in facilitating electron-transfer processes, which constitute fundamental underpinnings of both biological and catalytic domains. This resonates with the essential import of electron flux [341].

In this context, oxidoreductases assume the mantle of enigmatic conductors, orchestrating a symphony of electron choreography. Though these enzymatic architects do not engender net electron generation during their catalytic endeavors, their quintessence is profoundly intertwined with the realm of electron-transfer modalities. Functioning as adept mediators, certain oxidoreductases seamlessly shuttle electrons, either extrinsically or within intricate intramolecular passages, a phenomenon denoted as electron transport (ET_p) [16,17]. It is noteworthy that electron mobility predominantly ensues through ionic channels within the biological milieu. Furthermore, the attributes of electron transport within proteins display an intimate nexus with their electron-transfer proficiency and redox potential [415,416].

To grasp the intricacies of the inorganic catalysis process, consider the essential role of electrons in biological processing. In many cases, the electron flux is low, and reaction rates are sluggish. In contrast to metal ions, which directly participate in these biochemical reactions, inorganic nanoparticles act as catalysts due to their unique structure and surface properties. Consider a scenario where a simple chemical equilibrium process involving electron transfer is established with a fixed equilibrium constant (K_{eq}) for the biological reaction. While most of the electrons involved in the reaction are provided by the chemical constituents of the system, nanoparticles with electron hopping behavior can generate additional electrons, disrupting the chemical equilibrium governed by the fixed K_{eq}. This disruption significantly enhances their role as catalysts, leading to increased electron flux and higher reaction rates. This phenomenon aligns with the second law of thermodynamics, which states that the disorder in a closed system always increases over time.

For instance, Fe_3O_4 nanoparticles exhibit a distinctive POD activity by adsorbing H_2O_2 onto their surface, initiating redox reactions [104,188,314]. This process differs from the conventional Fenton reaction, where the generation of hydroxyl radicals takes place in solution with metal ions [417,418]. The pivotal step in hydroxyl radical generation with Fe_3O_4 nanoparticles involves redox reactions between H_2O_2 and the Fe(II) sites on the Fe_3O_4 surface [419,420]. This orchestrated surge re-establishes equilibrium, injecting momentum into biological redox cascades and expediting metabolic processes. This symbolizes the essence of oxidoreductase functionality—a robust sustenance of ET flux, propelling the rhythmic cadence of reduction and oxidation. This interplay fosters a catalytic process fueled by the intrinsic electron hopping propensity of these nanoparticles distinct from conventional protein-based catalysis, as exemplified by HRP, wherein the genesis and retention of hydroxyl radicals transpire within the Fe-porphyrin ring throughout the ET sequence [22,23]. The Fe(III) site nestled proximal to His170 and enveloped by a porphyrin architecture interfaces with Arg38 and His42, harmonizing to reorient H_2O_2 [23,421].

The concept of "intrinsic oxidoreductase-like activity" extends its dominion beyond the precincts of iron oxide or sulfur nanoparticles. A diverse cohort of inorganic nanomaterials, encompassing nanoparticles of transition metal oxides, sulfides, and even selenides, emerges as emblematic exemplars [77,422–430]. Renowned for their heightened electron transfer capacity and electron hopping, these assemblies even exhibit properties reminiscent of superconductivity, similar to their inorganic counterparts such as iron oxides and sulfides. For instance, the mixed-valence vanadium pentoxide (V_2O_5) showcases semiconducting traits [431], underpinned by electron hopping dynamics within V^{4+} and V^{5+} ions [432]. This substrate exemplifies intrinsic POD, GOx, and glutathione peroxidase (GPx) activity [308,433–437]. Moreover, Mn_3O_4 NPs, characterized by polaron hopping of electron holes between Mn^{4+} and Mn^{3+} within octahedral sites [438], manifest intrinsic, CAT, GPx, and SOD activities [310,439,440]. Intriguingly, the direct electron hops within Mn-Mn chains within MnO_2 NPs [441,442] manifest as POD, CAT, OXD, and SOD activities [443,444]; while Mn_2O_3, as electron shuttle [445,446], has the strongest OXD-like activity [447]. Furthermore, Co_3O_4 NPs exhibit semiconducting attributes marked by Co^{3+}-Co^{2+} hopping [358,448,449], concurrently demonstrating intrinsic POD and CAT activities [304,450–452]. In parallel, molybdenum disulfide (MoS_2), distinguished as an exceptional semiconductor [453], is underscored by its capacity for hopping transport [454] bestowing it with inherent POD, CAT, and SOD activity under physiological conditions (pH 7.4, 25 °C) [455,456]. Even α-FeSe and Cu_2O/CuO NPs, renowned for superconductivity [405,457–459], harbor the intrinsic mantle of POD activity [172,309,460–462]. Nanoparticles with lower band gaps and electron hopping, such as titanium dioxide (TiO_2) [463,464], manganese selenide (MnSe) [465], and molybdenum selenide ($MoSe_2$) [466], also display intrinsic POD activity [467–469]. A noteworthy instance is nanocrystalline ceria (CeO_2), renowned for its high electron conductivity and electron hopping attributes [470,471], affording it the unique capacity to directly convert Ce^{4+} to Ce^{3+} due to oxygen vacancies [472]. This renders ceria akin to the functionalities of oxidoreductases (POD, CAT, OXD, SOD), nucleases, urease, photolyases, and phosphatase [307,311,473–478].

The nuanced intricacies of their performance transcend dimensions of mere geometry or surface attributes. The preeminent determinant, the metal architecture of these nanomaterials, stands as the chief choreographer sculpting enzymatic cadence, orchestrated through the artistry of electron hopping, akin to a symphonic ballet rendition of electrons pirouetting within their lattice, supported by the fact that the metal architecture in nanomaterials, rather than their size, shape, and surface area, as well as the defect nanostructures, plays a crucial role in their enzymatic activity associated with ET functions in catalyst processing [104,166,188,189,301–303,306–312,479,

480]. Keep in mind that the metal architecture of nanoparticles is not stable or unchangeable; it can be easily altered by environmental conditions, whether abiotic or biotic, even during the catalytic process. This observation includes iron oxide nanoparticles, as discussed in the previous section, which exhibit intricate ET processes, particularly electron hopping with their metal structures. In comparison to inorganic iron oxide nanoparticles, inorganic iron sulfide has a high ET rate coupled with low resistivity, leading to improved enzyme activity, as previously described [405–407].

A celestial luminary graces the stage—light. A phenomenon eternally etched in the annals of physics, light introduces an exalted facet to the realm of electron contribution. Among the troupe of photoactive inorganic semiconducting nanoparticles, the enchanting NH–MoO_{3-x} nanobelts emerge as a captivating example, lending their voice to the symphony of photons [481] (Figure 15). This elegant spectacle finds a counterpart in the stately MoO_3 nanobelts, a tale of two performers distinguished by their distinct band gaps.

Within the canvas of a DFT model, MoO_3 unveils a band gap of 2.592 eV (Figure 15B)—an electronic symphony anchored between the valence band maximum and the conduction band minimum. In the spotlight, the mesmerizing NH–MoO_{3-x} takes center stage with a bewitching band gap of 1.793 eV, gracefully narrowing the chasm by 0.8 eV (Figure 15C). Amidst this spectral waltz, defect energy levels surface like musical notes between the conduction and valence bands, invoking the potential for an elegant electron exchange.

Much like performers gracing a grand stage, electrons ascend to these defect levels before embarking on their journey to the conduction band, crafting an enchanting symphony of ET. Their structure becomes a torchbearer for POD (Figure 15D, 15E), OXD (Figure 15F), and CAT (Figure 15G, 15H) activities, showcasing the remarkable generation of reactive oxygen species. These musical notes weave a tale of heightened electron conductivity and light-induced enzymatic fervor, immersing the audience in an intricate interplay of temperature, energy bands, and catalytic behavior. The dance of NH–MoO_{3-x} nanobelts gracefully unfurls, narrating a captivating story of light and electrons choreographed with meticulous precision. These inorganic nanomaterials, bathed in the radiance of photons surpassing their band gaps, unveil a mesmerizing performance. The unfolding narrative witnesses the birth of direct electron egress and the emergence of electron hole dyads. Illuminated by photons, electrons embark on an intricate choreography, ascending from valence to conduction tiers. These charged pairs traverse a path culminating at the band's precipice, radiating excess energy. An elegant ritual of thermalization, overseen by the resonance of acoustic phonons, harmoniously channels this surplus energy into an exquisite display of molecular ballet. And so, the saga continues. Enhanced electron conductivity, sculpted by nuanced temperatures, engineers a crescendo, amplifying the resonant enzymatic fervor of these nanoparticles (Figure 15I). Heat, transformed into photonic harmonies, unfurls an eloquent sonnet of photothermal transformation, transmuting these nanoparticles into resonant conduits of both light and electron flux. This harmonious synergy, summoned forth by luminal rays and

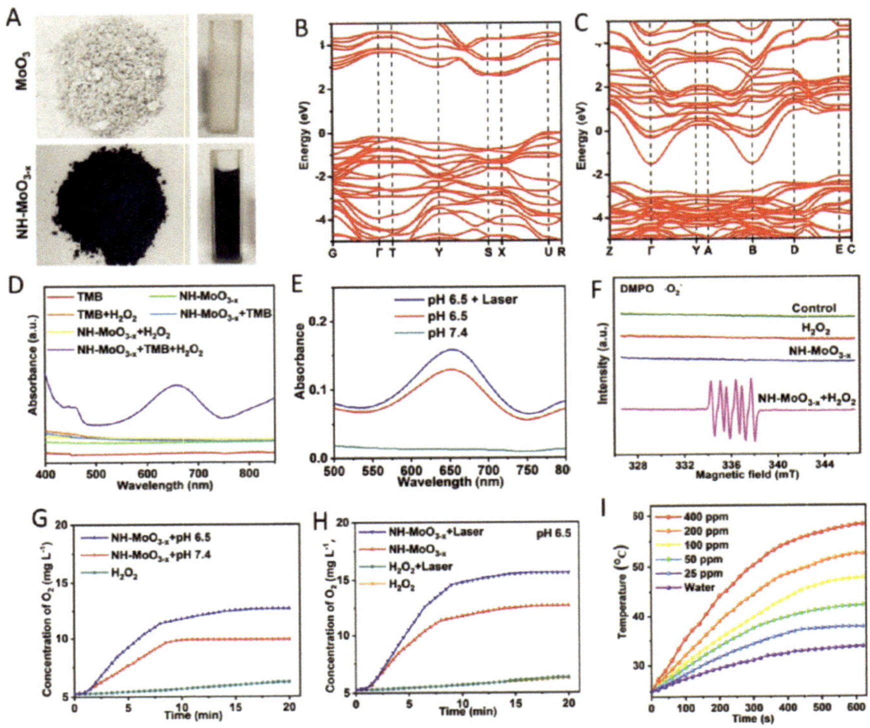

FIGURE 15 Effect of near-infrared-II (NIR-II) light on the oxidoreductases activity of MoO_3 nanobelts. A. Digital photographs in powder and solution of inert MoO_3 and activated NH–MoO_{3-x} nanobelts B. Electronic energy band of MoO_3 based on DFT model. C. Electronic energy band of NH–MoO_{3-x} based on DFT model. D. UV/Vis absorption spectra of ox TMB after different treatments (concentrations of NH–MoO_{3-x}: 50 μg mL^{-1}, H_2O_2: 1.0 mM, TMB: 1.0 mM, pH 6.5). E. NIR-II laser-enhanced POD-like catalytic activity. F. OXD-like catalytic activity testified by ESR spectra of NH–MoO_{3-x} in the presence of H_2O_2 using DMPO as the capture agent. G. CAT-like catalytic activity of oxygen generation test results for NH–MoO_{3-x}@BSA in PBS with different pH values at 25 °C (concentration of NH–MoO_{3-x}: 50 μg mL^{-1}, H_2O_2: 1 mM). H. NIR-II laser effect on CAT-like catalytic activity. I. Heating curves of NH–MoO_{3-x}@BSA with various concentrations by irradiation using a 1064 nm NIR-II laser for 10 min at 1.0 W cm^{-2}. Reprinted with permission from Ref. [481], Copyright © 2022, John Wiley & Sons.

the meandering of electrons, begets an exhilarating crescendo, magnifying enzymatic potency [482–485].

The introduction of Cyc-c electron mediators, seamlessly woven into this intricate dance, stands as a testament to the multifaceted elegance inherent in these nanomaterials—an orchestration of catalytic prowess directed by the skillful hands of light, structure, and the enigmatic Cyc-c electron mediators. This remarkable contribution extends far beyond the domain of enzyme activity within synthesized nanoparticles. For instance, cadmium sulfide coupled with Cyc c serves as a bifunctional NADH oxidase and Cyc c reductase, showcasing the versatile roles that Cyc-c mediators play in orchestrating ET dynamics [217]. Furthermore, $CeVO_4$ nanozymes, performing the functions of the mitochondrial enzyme CcO, utilize Cyt c as a biological electron donor in the four-electron reduction of molecular oxygen, thus integrating SOD-like functions [109]. These instances underscore the vast potential of Cyc-c mediators in influencing catalytic transformations.

This influence resonates deeply within the realm of biochemical transformations, and it even reaches back to the earliest stages of life, exemplified by the intriguing case of MTB [234,235]. Notably, such orchestrations transcend the boundaries of inorganic nanoparticles, finding harmonious alignment with the orchestration of life's ancient redox processes. A striking example lies within the formation of biogenic iron oxide nanoparticles through biomineralization [234,235], referred to as "life fossil oxidoreductases" [91], where Cyc-c mediators play a pivotal role in modulating ET pathways of magnetite within magnetosomes of MTB [235–241], as described in Section 4.

All iron cores in ferritins can also be recognized as another example of "life fossil oxidoreductases" [91]. The observation of superoxide-diminishing activity in various biogenic iron cores of ferritin further emphasizes the significance of the metal structure in early life [202]. After the removal of the protein, the iron core from prokaryotic ferritin, specifically both pfFn and pyFn from archaea *P. furiosus* or *P. yayanosii*, demonstrates higher SOD activity than that of eukaryotic ferritins, such as HFn and LFn from *H. sapiens* [202] (Figure 5C). The iron/phosphate ratio in the iron core, a characteristic mainly determined by the structures of ferritins, also affects their SOD catalytic capability [202]. The presence of phosphate in the iron core transforms it from a single crystalline structure to an amorphous iron phosphate-like structure, leading to a decreased affinity for the hydrogen proton of the ferrihydrite-like core. For example, ssDps, the DNA protection protein from the archaea *Sulfolobus solfataricus*, exhibits this behavior [202]. Additionally, eukaryotes also have a higher phosphorus content due to evolutionary processes [202]. These findings underscore the significance of the composition and evolution of ferritin structures in the context of early life and have potential implications for understanding the role of inorganic materials in biochemical processes.

Within the intricate tapestry of inorganic nanozymes, the introduction of Cyc-c electron mediators emerges as one of the most pivotal modes of ET. Its significance reaches profound depths, extending far beyond the realm of synthetic experiments. In the synesthetic choreography of electron movement, these mediators establish a dynamic bridge connecting light, structure, and catalytic activity. This bridge not only unites nanomaterials and biochemical systems but also transcends temporal boundaries, linking the intricate dance of electrons in contemporary catalysis with the primordial rhythms of life's redox processes.

This intricate interplay extends its influence into unexplored domains, particularly within the context of the origin of life and its nascent evolution. In the rich orchestration of inorganic oxidoreductases, electron transfer modes assume prominence, as elucidated in Figure 16. Here, electron hopping within the inorganic nanoparticles assumes a pivotal role, with the interplay of light and Cyc-c providing a visual

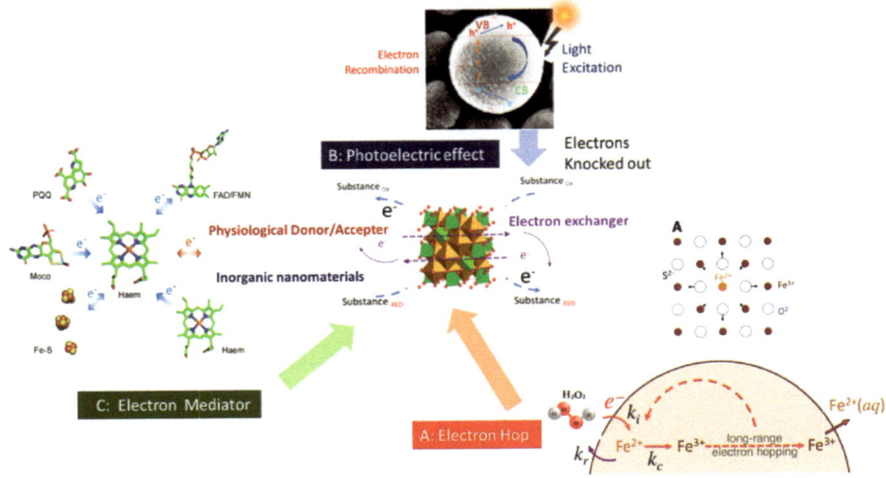

FIGURE 16 Electron transfer modes of the inorganic oxidoreductase. A. Electron hopping in inorganic nanomaterials. B. Photoelectron effect and photothermal effect. C. Electron mediation by cytochrome c. Reprinted from Ref. [9].

demonstration of the intricate pathways that underlie catalytic efficacy. Much like the reverberations of a symphony that persist long after its final note, the involvement of Cyc-c electron mediators extends beyond the confines of the laboratory, resonating through the corridors of scientific enlightenment. This augments our comprehension of the nuanced interplay between inorganic nanozymes and the expansive domain of biochemical processes. In this grand orchestration, electron hopping, in conjunction with the influence of light, finds its integral place as a harmonious melody within the composition of nanoscale catalysis. Furthermore, Cyc-c emerges as a conduit bridging the inorganic and organic realms, particularly salient in the context of the origin of life, connecting the elemental processes of early Earth to the intricate biochemical choreographies of living organisms.

The pivotal link between inorganic nanozymes and their biological counterparts, the proteinaceous oxidoreductases, stands out as a cornerstone within this narrative. Notably, a majority of these inorganic nanozymes, represented by a diverse array of transition metal sulfite or oxide nanoparticles such as iron oxide NPs as iron oxide NPs [82,102,104,161], iron sulfide NPs [108], Mn_3O_4, [310,439], MnO_2 NPs [443,444], V_2O_5 [308,433–437], and MoS_2 [455,456,486], exhibit remarkable parallels with their cognate oxidoreductases (proteinaceous enzymes) [42–50], universally possessing DET capabilities [14]. Even enzymes engaged in the hydrolysis of phosphate esters, known as phosphatases, share profound associations with protein-mediated ET processes [487–489]. These phosphatases mirror the attributes of vanadium haloperoxidase, a specific subclass of oxidoreductases, encapsulating the fundamental essence of ET paradigms [490–493].

As we delve deeper into the interplay between these inorganic enzyme-like activities and the emergence of oxygenic processes, we uncover a captivating narrative that suggests a gradual progression of enzymatic functions in relation to oxygen. This predates the more established enzymatic systems commonly associated with the advent of photosynthesis. This vantage point underscores the intricate and complex timeline of biochemical evolution, shedding new light on the origins of oxygen-related enzymatic processes.

These inorganic enzyme-like activities, particularly the inorganic CAT [494], undoubtedly played a substantial role in the gradual release and accumulation of oxygen—a pivotal event that profoundly influenced Earth's atmosphere and environmental conditions, notably during the Great Oxidation Event [495,496]. The observation that these activities predate the emergence of more familiar proteinous enzymes like POD, CAT, OXD, and SOD adds compelling weight to the proposition that oxygen-related enzymatic processes were active before the era of photosynthesis [53–59]. This insight highlights a fascinating narrative of biochemical evolution, underscoring the essential role of these inorganic enzyme-like activities in the intricate timeline of Earth's oxygenation.

Moreover, the fascinating link between photosystem II, a core component of oxygenic photosynthesis, and the manganese-containing form of catalase proteins suggests an intriguing evolutionary connection [7,62–64]. This further reinforces the hypothesis that these inorganic enzyme-like activities were indeed precursors or precursory processes to the later-evolved proteinous enzymes associated with oxygen metabolism. The activity of inorganic SOD is also necessary to control the superoxide radical level of organisms living in an aerobic environment [32,249,250, 497–499].

The unique electrical attributes of inorganic iron oxide and sulfur nanoparticles stem from their intrinsic metallic architecture, including defects such as oxygen or metal vacancies [336–340,500–505]. Iron oxide and sulfur nanoparticles exhibit distinctive electron configurations, influenced by d-orbital splitting, a result of ligand interactions within coordination complexes. This phenomenon dictates the coloration, reactivity, and catalytic activity of these compounds.

The electron configurations of Fe(III) and Fe(II) ions are distinctly delineated as $1s^2\ 2s^2\ 2p^6\ 3s^2\ 3p^6\ 3d^5$ and $1s^2\ 2s^2\ 2p^6\ 3s^2\ 3p^6\ 3d^6$, respectively. Within the atomic realm, the d orbitals of iron undergo a division into two sets, characterized by differing energy levels—an occurrence known as d-orbital splitting. This phenomenon is intricately influenced by the crystal field effect, an outcome of ligand interactions within coordination complexes. The energy differential between these sets of d orbitals governs the coloration and reactivity displayed by various iron compounds. The electron configuration of the oxide ion (O^{2-}) stands as $1s^2\ 2s^2\ 2p^6$, wherein the incorporation of two electrons instigates occupation within the 2p orbital. Correspondingly, the electron configuration of sulfide ions (S^{2-}) is structured as $1s^2\ 2s^2\ 2p^6\ 3s^2\ 3p^6$, with the additional two electrons also inhabiting the 2p orbital. This intricate interplay of electron arrangement unfolds through the harmonious coordination of iron's electron configuration with oxygen or sulfur moieties. The orchestration is further accentuated by the presence of vacancies, lending a complex yet harmonizing quality to the ensemble.

In this symphonic interplay, a captivating phenomenon emerges—semiconductivity. Characterized by diminutive band gaps, these nanomaterials become adept at facilitating efficient electron translocation. The boundaries of this phenomenon extend even further, venturing into the domain of superconductivity—a remarkable realm akin to the illustrious pyrrhotite. This intrinsic proclivity for spontaneous electron hopping takes form as a symphony of motion, an intricate dance played out across the canvas of octahedral surface locales. This dance is meticulously choreographed by Fe 3d electrons, their harmonious partnership with oxygen or sulfur partners giving rise to a mesmerizing display of intermolecular choreography. In essence, the symphony of electron configurations, orbital interactions, and atomic vacancies composes a harmonious score, producing the symphonic marvel that is the electrical attributes of inorganic iron oxide and sulfur nanoparticles. This resonant interplay serves as a testament to the elegance inherent in the interrelationship between atomic structures and their emergent properties—a dance that shapes the breathtaking panorama of nanoscale reactivity and catalysis.

It is noted that electron hopping, where electrons move between localized sites within a material, is fundamental and observed in both inorganic nanoparticles and biomacromolecules like proteins and DNA [413]. In inorganic nanoparticles, this process occurs within the crystal lattice, influenced by factors such as particle size, shape, and surface chemistry. For example, smaller nanoparticles may exhibit enhanced electron hopping due to pronounced quantum effects. In biomacromolecules, electron hopping is vital in various biological processes [17,413,506–508]. Proteins, with their complex structures, facilitate electron transfer through specific amino acid residues or cofactors. Metal ions or organic cofactors within proteins can act as electron carriers, allowing electrons to move along redox-active sites. Similarly, DNA, despite its role in genetic information storage, exhibits electron hopping behavior under certain conditions, influencing processes like DNA repair and oxidative damage.

The mechanisms underlying electron hopping in both inorganic nanoparticles and biomacromolecules are intricately linked to their structures and characteristics. In inorganic nanoparticles, the arrangement of atoms in the intrinsic metallic architecture and the presence of defects or impurities affect electron hopping. In biomacromolecules, the specific arrangement of amino acids or nucleotides, as well as the presence of metal ions or organic cofactors, modulate this process. Understanding these mechanisms is essential for elucidating the role of ET in biological systems and for designing functional materials with tailored electronic properties. Moreover, comprehending the connection between the metal center structure of proteins and the metal structure in inorganic nanoparticles is crucial for understanding the evolution and emergence of life [8,9]. This dynamic relationship has profound implications for biological functions and technological applications, potentially shaping the course of evolution on Earth.

8 BRIDGE OF INORGANIC NANOPARTICLES IN BIOMOLECULAR EVOLUTION

As we journey back to the conditions of early Earth, a spotlight is cast upon the establishment of fundamental biomolecules that have left indelible marks in the annals of our planet's history. Among these molecules, primitive amino acids, adenine, nucleobases, and thiodepsipeptides stand as the cornerstones of this biochemical narrative [509–512]. In this intricate interplay, amino acids emerge as protagonists, serving as the elemental building blocks of proteins and playing a pivotal role as conduits for ET processes, particularly within the realm of oxidoreductases. These amino acids boast diverse functional groups, ranging from amines to carboxylic acids, each eagerly participating in the choreography of ET reactions.

Amidst this intricate dance, the stage is set for the introduction of key electron relays—tryptophan and tyrosine residues. These entities come forth as maestros of electron flow, orchestrating efficient electron hopping and rapid electron tunneling across substantial distances within the intricate folds of metalloenzymes [11,13, 513–516]. As the dance continues, an additional layer of complexity is woven into the narrative—the interplay of light activation. This interplay introduces a symphonic interlude, where photons harmonize with matter, enriching the catalytic activity of enzymes and contributing to the captivating intricacies of biochemical transformations.

Within this ensemble, certain proteins stand out as virtuosos, housing heme groups with iron at their core. Consider POD [23,517,518], CAT [519,520], Fe-SOD [32], and cytochrome P450 peroxygenase [521,522], as prime examples. Together, they collaborate to construct the iconic Fe-oxo-Fe bridge, a structural marvel crucial to the catalytic procession of metal centers. This structure assumes a pivotal role in the delicate ballet of electron acceptance and donation during redox reactions [210–212]. Through their intricate coordination, they imbue the dance of electrons with grace and purpose, ushering in moments of transformation that define the very essence of life's processes.

The journey of functional groups extends beyond nature's enzymatic boundaries, venturing into the realm of nanomaterials. While the selection of amino acids by humans for nanomaterial modification showcases deliberate innovation, the natural selection of amino acids within proteins through evolution is a result of intricate biological processes. This diverse array of amino acids and their arrangement, shaped over millions of years, represents one of nature's greatest products—driving the complexity and functionality of living organisms. Precise nanomaterial modification with amino acids or other functional groups illustrates human ingenuity, yielding heightened ET rates and catalytic efficiency [523–543]. These adept adjustments amplify surface interactions, culminating in more efficient ET processes. A notable example of this symphony is evident in histidine-modified Fe_3O_4 nanozymes, where a significant increase in apparent affinity K_m for the substrate H_2O_2 leads to a remarkable twentyfold enhancement in catalytic efficiency k_{cat}/K_m [523]. Furthermore, research involving iron polysulfide nanoparticles ($Fe_{1-x}S$ and Fe_3S_4) unveils POD and CAT activities [108].

Even when molybdenum trioxide nanoparticles reveal limited inherent electron conductivity [544], the introduction of triphenylphosphonium ions emerges as a masterful orchestrator, directing the activities of sulfite-oxidizing enzymes with precision and finesse [545]. Similarly, the strategic utilization of near-Infrared-II (NIR-II) light unveils its transformative power, deftly lowering the band gaps of NH–MoO_{3-x} nanoparticles. This nuanced reduction in band gaps acts as a catalyst, triggering the liberation of electrons and catalyzing the remarkable functions of POD, CAT, and OXD [481].

It is worth noting that the metal center of the protein's oxidoreductase is stabilized by their amino acids [8,9,91]. However, their catalytic prowess and judicious selection are not solely due to their structural arrangements; rather, they are amplified by the symphony of hydrogen bonds intricately woven among amino acids within the protein. This symphony of intermolecular interactions acts as a stabilizing force, elevating their catalytic capacities to new heights. These revelations reverberate as echoes throughout the narrative, underscoring the vital role that functional groups, particularly amino acids, play in the intricate choreography of metalloenzymes. Their impact is most profoundly felt at the secondary and tertiary structure levels, where they engage in a seamless dance with the evolving metalloenzyme frameworks. Like adept artisans, these functional groups assume the mantle of stabilizers—a concept elegantly championed by Huang [8,9,91]. Yet, their contribution extends beyond mere stability; they lay the very cornerstone upon which the early path of life was etched. Their presence acts as a catalyst, instigating the emergence of catalytic processes that, across epochs, would weave the intricate tapestry of life as it is known today.

In a role akin to skilled conductors, these functional groups lead the grand orchestra of biochemical transformations, deftly guiding the symphony of evolution. With each delicate movement, they forge the foundations of the intricate dance that is life—a dance that continues to unfold in awe-inspiring complexity and beauty. Yet, despite the richness and depth of this symphony, the evolution of metalloenzymes has often been spotlighted through the lens of primary and secondary protein structures [11,515,516,546–549]. This gaze heavily relies upon gene codes or DNA, casting somewhat limited light on tertiary/quaternary structures and the integral metal centers [550–552]. However, within these tertiary structures, an enigma takes root—an enigma of metal center formation. While many studies have delved into the intricacies of folds, a significant emphasis has been placed on the secondary structure [553,554], often relegating metal centers and tertiary structures [555,556] to the shadows.

The story of ET and its evolution remains partially obscured, despite acknowledgments of the role played by iron-sulfur cluster proteins as electron carriers and transfer agents [557–563]. Amidst these uncertainties, the pivotal question lingers—did metal centers precede primary and secondary protein structures, or did they arise in tandem or even later in the evolutionary journey? Does the formation of these metal centers intertwine with DNA/RNA synthesis, or does it emerge as a distinct process [8,9,91]? The selection of metals within metalloenzymes appears to be intricately woven with the fabric of geochemical processes, rather than solely being dictated by genetic materials like DNA or RNA [418,564–570].

A pivotal study conducted by Eck and Dayhoff in 1966 heralded a paradigm-shifting concept, unveiling the remarkable potential of ferredoxin to harness photon energy even in the absence of a fully established genetic code [571]. This extraordinary attribute emanated from its inorganic active sites—clusters of [FeS]—and its elegantly simple, repetitive amino acid sequences. Subsequent experiments and computational models lent credence to this audacious hypothesis, suggesting the birth of a protoferredoxin, characterized by [2Fe-2S] and [4Fe-4S] motifs. These motifs emerged through the photooxidation of ferrous ions under UV light, facilitated by the duplication of an iron-sulfur tripeptide motif [81,572]. The significance of ferredoxins reverberates across the spectrum of life, from the ancient anaerobic bacteria to the most sophisticated plants and animals, making them invaluable subjects for delving into the evolution of metalloproteins [559,573,574].

Weiss et al.'s groundbreaking research cast an illuminating spotlight on the prevalence of FeS clusters and radical reaction mechanisms within the last universal common ancestor (LUCA)—the origin of all cellular life forms [575]. This revelation offered a tantalizing glimpse into the early utilization of FeS clusters, acting as pioneers that facilitated ET and orchestrated redox reactions within the nascent cells of the primordial world. In parallel, investigations into the enigmatic realm of hydrothermal vents, often hailed as the crucibles of life's origin, unveiled strikingly similar observations [77,423,576]. Notably, the metal centers of Fe-S cluster proteins and diverse iron sulfides exhibited intriguing parallels, reflecting the elegance of nature's design [577–579] (Figure 17). For instance, the [2Fe2S] motifs gracing eukaryotic ferredoxins and Rieske proteins showcased a symphony of coordination between double Fe atoms and two equidistant sulfur atoms [580,581]. In contrast, the quartet of Fe atoms harmoniously coordinated with four equidistant sulfur atoms in high potential iron-sulfur proteins and iron regulatory proteins (IRPs), composed the resonant tune of [4Fe4S] [550,582–584]. Rubredoxin, with its graceful choreography, spotlighted a single iron atom partnered with four equidistant sulfur atoms, a [1Fe4S] motif of eloquent simplicity [585,586]. Moreover, these intricate dancers on the molecular stage were accompanied by their corresponding protein ligands—the harmonious rhythms of [3Fe 3S] in bacterial ferredoxins [587], and the intricate melodies of [6Fe 6S] in *Desulfovibrio vulgaris* (Hildenborough) [588].

The catalytic activity of cubane-type Fe4S4 clusters in metalloproteins, such as biotin synthase [589], aconitase [590], and (E)-4-hydroxy-3-methylbut-2-enyl pyrophosphate reductase (IspH) [591], as well as in synthetic M4S4 clusters, highlights their role in the emergence of life and the formation of organic compounds from inorganic precursors [592]. Recent studies have revealed that Fe–S clusters with low-valent Fe^{1+} centers can adopt various electronic configurations crucial to their catalytic activity. For instance, when CO binds to a synthetic $[Fe_4S_4]^0$ cluster, it triggers the generation of Fe^{1+} centers through intracluster electron transfer, facilitating redox reactions. Similarly, CO binding to an $[Fe_4S_4]^+$ cluster induces electron delocalization, enabling the activation of C–O bonds without highly negative redox states [593]. Metalloproteins containing Fe_4S_4 clusters can catalyze the reduction of CO and CO_2 to hydrocarbons [594–596], which is significant in the context of early Earth life.

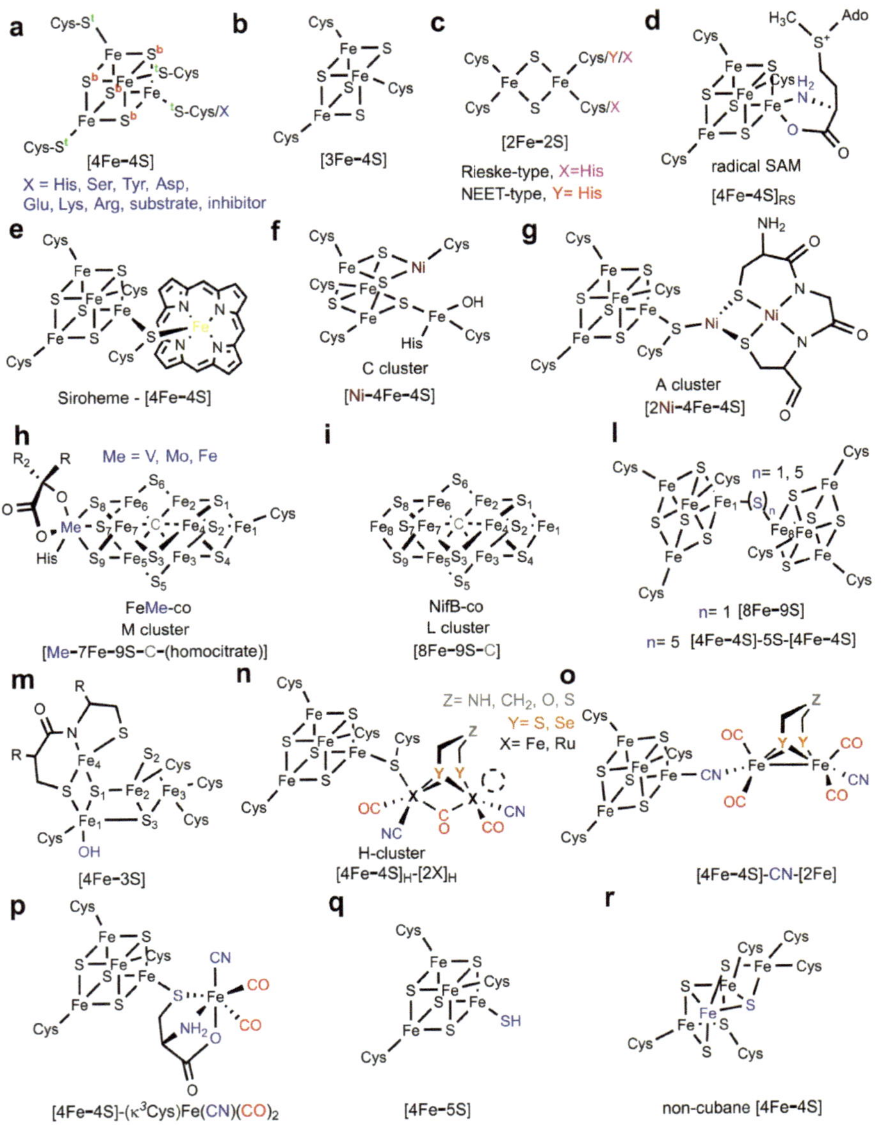

FIGURE 17 Structures of common and non-conventional FeS clusters. (a) [4Fe–4S] cluster, X in blue highlights a different cluster ligation, with terminal (Fe–S^{t}) and bridging (Fe–S^{b}) sulfides indicated in green and red, respectively; (b) [3Fe–4S] cluster; (c) [2Fe–2S] cluster including Rieske-type (X = His, pink) and NEET-type (Y = His, red) clusters; (d) [4Fe–4S] cluster with bound S-adenosyl-L-methionine, O and NH_2 cluster coordination by SAM are marked in blue; (e) siroheme [4Fe–4S] site of nitrite and sulfite reductases, the heme iron is displayed in dark yellow; (f) C cluster [Ni–4Fe–4S] in carbon monoxide dehydrogenase; (g) a cluster [2Ni–4Fe–4S] from acetyl-CoA synthase; the Ni ions in f and g are depicted in brown;

Pyruvate is a central metabolite in archaea, bacteria, and eukarya, where iron-sulfur enzymes play a crucial role in linking pyruvate to carbon fixation pathways and thioester biochemistry [597,598]. The $FeS/S/FeS_2$ system can catalyze the interconversion of hydroxyl acids and keto acids [599]. Recent studies have shown that natural iron sulfide pyrrhotite acts as an "oxidoreductase" catalyst in the conversion of pyruvic acid to lactic acid [600] and in the reduction of CO_2 [601]. However, these studies have not demonstrated this activity with typical enzyme kinetics, in contrast to the previously mentioned examples in this review. In fact, this work is more closely related to the mineral surface hypothesis as a general catalyst, focusing on the iron-sulfur world [422] and its relevance to evolutionary biochemistry [602,603].

The role of iron sulfides extends beyond organic carbon formation to include nitrogen processes such as fixation and transfer [604,605], even though it may not act as a direct biocatalyst. For instance, molybdenum can coprecipitate with iron as sulfides, serving as catalysts and direct reductants using nitrite (NO_2^-) and protons (H^+) as substrates [606,607]. The selectivity of reduction products depends on the amount of molybdenum present, with approximately 10% Mo optimizing NH_4^+/NH_3 production from NO_2^-, minimizing competitive H_2 formation from protons when a secondary reductant is present. This function is similar to that of nitrogenase, where iron sulfides incorporate secondary metals like molybdenum to form the metal center of nitrogenase [562].

The transformation of nitrogen into ammonia is another primarily chemical reduction reaction, but it can also occur catalytically or directly [608,609], similar to the function of Fe-S proteins, which go beyond mere reduction. Iron sulfides have been observed to act as mediators in nitrogen species interactions, serving as both direct reductants and catalysts [610]. Troilite (FeS) exhibits the largest difference between direct and catalytic reduction, while amorphous iron sulfide demonstrates

(CONTINUED)

(h) nitrogenase FeMe-co [Me–7Fe–9S–C–(homocitrate)] M cluster, C (carbon) is represented in gray and Me (metal) in blue; (i) nitrogenase NifB-co [8Fe–9S–C] L cluster, C (carbon) is marked in gray; (l) [4Fe–4S]-S_n-[4Fe–4S] clusters (n = 1, 5) common to RimO, MiaB, and double-cubane cluster proteins, S is depicted in blue; (m) [4Fe–3S] proximal cluster of the oxygen-tolerant membrane-bound [NiFe]-hydrogenase, the bound hydroxy (OH) ligand is depicted in blue; (n) H-cluster of [FeFe]-hydrogenases and its synthetic substitutions marked with X (black, Fe in the native form), Y (orange, S in the native form), and Z (gray, NH in the native form), CO and CN^- are depicted in red and blue, respectively; (o) [4Fe–4S]–CN-[2Fe] cluster in HydF maturase, the bridging CN^- is shown in blue, CO and CN^- ligands are represented as in n; (p) [4Fe–4S]-(κ^3Cys)Fe(CN)$(CO)_2$ cluster in the [FeFe]-hydrogenase maturase HydG, a non-proteinaceous cysteine residue (κ^3-Cys) coordinates a fifth ion via its amino (NH_2, blue), carboxy (COO^-, blue), and thiolate (S^-, blue) functionalities, and CO and CN^- ligands are depicted as in n; (q) [4Fe–5S] cluster common to sulfurating enzymes, the extra SH is depicted in blue; and (r) non-cubane [4Fe–4S] cluster of heterodisulfide reductase HdrB comprising a fused [2Fe–2S] and [3Fe–4S] cluster sharing one S and one Fe ion (depicted in blue). Reprinted with permission from [579], Copyright © 2022, Elsevier.

the greatest selectivity for ammonia/ammonium (NH_4^+/NH_3) production, acting as both a direct reductant and a catalyst [609].

Green rust has been reported to be related to abiotic denitrification (the reduction of nitrite) [611–614], stoichiometrically reduced to ammonium [615], even without ammonium production [616,617]. This reduction occurs in both external and internal reaction sites of minerals, with the latter corresponding to nitrite ions intercalated in the interlayer space of the green rust structure. Green rust's metal structure predates that of the soluble methane monooxygenase (sMMO), which has a specifically di-iron reaction center [618,619]. This metal architecture, found in the green rust's octahedral coordination of reactive $Fe^{2+/3+}$, is geometrically similar to the metal center in sMMO [425,428,430,618–620].

sMMO, found in certain bacteria, enables them to utilize methane as a carbon and energy source [618,619]. The enzyme catalyzes the oxidation of methane to methanol, a reaction difficult to replicate chemically. It consists of three components: a hydroxylase, a reductase, and a regulatory protein. The hydroxylase is responsible for methane oxidation, the reductase provides the necessary electrons, and the regulatory protein controls enzyme activity [618,619]. Methane oxidation has been shown to occur alongside the oxidation of green rust to lepidocrocite and magnetite, producing methanol and halogenated hydrocarbons in the presence of seawater [621]. The similarities in metal ion coordination between minerals and enzymatic cofactors suggest mechanistic and evolutionary links between abiotic chemical mineralogy and biological metabolism. This geochemical pathway for CH_4 oxidation unveils a previously unrecognized carbon cycling mechanism in both ancient and modern environments, shedding light on mineral-mediated reactions critical to the synthesis of organic compounds needed for the origin of life.

While these findings highlight the role of inorganic nanoparticles in carbon and nitrogen transfer on early Earth, even as catalysts, they have not directly demonstrated enzyme-like kinetics. Nevertheless, iron sulfides, in conjunction with anaerobic microbes, play a significant role in modern wastewater treatments, facilitating the conversion of NO_3^- to N_2 and reducing the risk of water eutrophication [622–633].

Of exceptional note is the symphony of iron sulfides, where a solitary iron atom is coordinated with four equidistant sulfur atoms, resonating with the enthralling duet of superconductivity and high oxidoreductase activity. This interplay involves foundational organic carbon formation, nitrogen transfer, and oxygen generation through hydrogen peroxide, predating photosynthesis [8,89,91]. This symphonic revelation hints at the potential roles of iron sulfides within the biochemical processes that shaped the Last Universal Common Ancestor (LUCA) and the evolution of the genetic code, as described above.

The intricate architecture of ferredoxin, as unveiled by Eck and Dayhoff [571], aligns with Huang's conceptual framework, positioning metalloenzymes as sophisticated functionalized nanomaterials [8,9,91]. The parallels between inorganic nanoparticles and metalloenzymes suggest a shared evolutionary lineage. The precise details of metalloenzyme evolution from mineral precursors, including iron-sulfur cluster proteins, remain mysterious. The question of whether the metal active site or the sequence of amino acids and nucleotides within the primary structure evolved first continues to drive active research [9,91,157]. Exploring this evolutionary trajectory will deepen our understanding of life's origins and development on Earth, revealing the fundamental processes that shaped biological complexity.

Within this captivating saga, where the intricate dance of ET bridges the realms of nanomaterials and life's molecules, a mesmerizing interplay of matter and energy unfolds—an ode to catalytic brilliance and the enigmatic origins of life. This symphony of creation weaves together the delicate threads of chemistry, physics, and biology, reverberating through time as a testament to the wondrous dance of the electron and the extraordinary marvels it has bestowed upon us—the very genesis of life itself. As John Desmond Bernal once said, "Life is a partial, continuous, progressive, multiform, and conditionally interactive self-realization of the potentialities of atomic electron states [634]", and as Albert Szent-Györgyi echoed, "Life is nothing but an electron looking for a place to rest [10]".

At the core of this symphonic convergence, ET materializes through the ethereal waltz of electron hopping and the radiant ballet of photons, seamlessly united by the graceful presence of Cyc-c mediators. Together, they form a spellbinding tapestry—a biocatalyst—that unfolds within the grand theater of chemical equilibrium. What's more, this spectacular performance shares its stage with biological counterparts—proteins that embody the essence of oxidoreductase functionality. These inorganic nanozymes, attuned to the intricate choreography of electrons, join in harmony with their protein companions, each playing its part in the elegant dance of ET.

In this breathtaking panorama, whether inorganic or biological, a luminous narrative emerges—an ode to the balance of chemical processes. The irresistible allure of electrons transcends boundaries, inviting an indomitable symphony that merges the worlds of inorganic matter and organic life. It is a symphony that orchestrates a harmonious union, guiding a dance of reactivity and transformation that echoes through the realms of chemistry and biology. Just as musical notes blend and harmonize to compose a masterpiece, the dance of electrons, photons, and Cyc-c mediators creates a resonant composition—a masterpiece of catalytic brilliance that unfurls on the grand stage of nature's theater.

From this view, there is no fundamental difference between the inorganic and organic worlds when it comes to electron transfer—they both follow the same universal physical laws:

- Universal Nature of Electron Transfer: Electron transfer is a process at the heart of both inorganic and organic systems. Whether in living proteins or inorganic nanoparticles, the movement of electrons follows consistent principles governed by quantum mechanics, making ET a unifying process across all matter.
- No Distinction in Physical Laws: The mechanisms that drive ET are governed by the same laws of physics in both inorganic and organic environments, including principles like Coulomb's law and quantum tunneling. Despite differences in the materials' compositions, the physics remains constant.

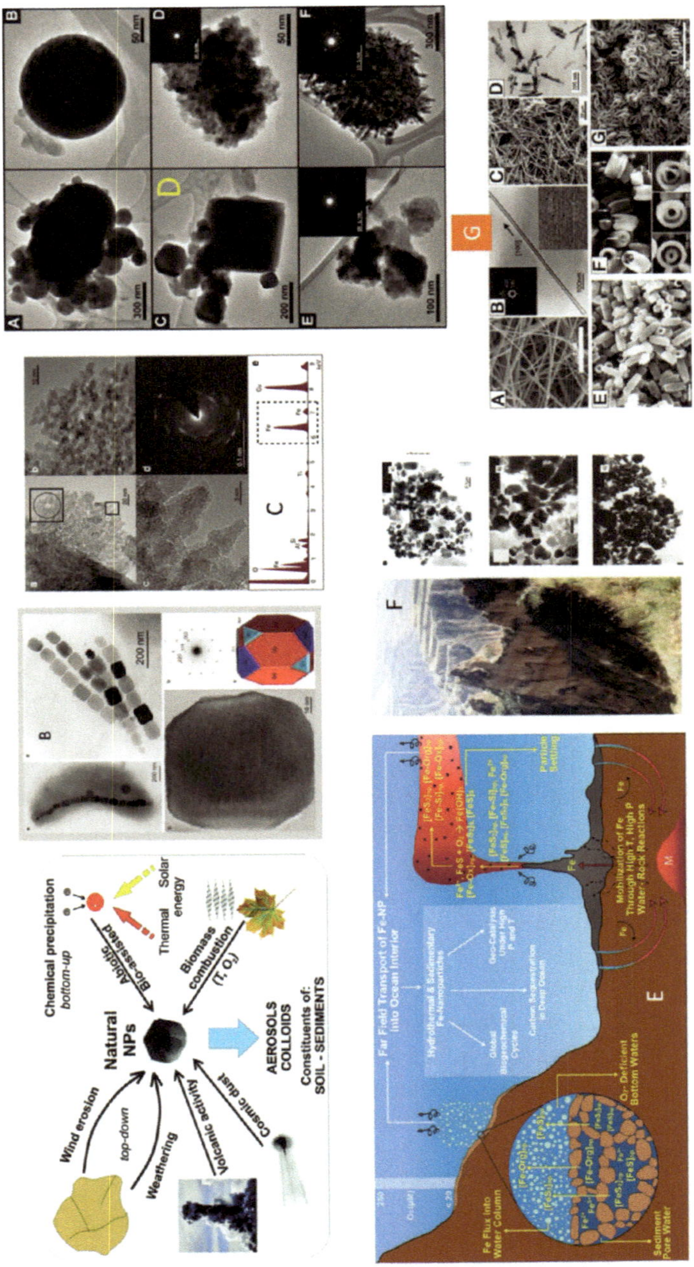
Chemical precipitation
bottom-up
Abiotic
Bio-assisted
Thermal
Solar energy
Biomass combustion
(T, O2)
Natural NPs
AEROSOLS
COLLOIDS
Constituents of:
SOIL - SEDIMENTS
Wind erosion
top-down
Weathering
Volcanic activity
Cosmic dust
Far Field Transport of Fe-NP into Ocean Interior
Hydrothermal & Sedimentary Fe-Nanoparticles
Global Biogeochemical Cycles
Carbon Sequestration in Deep Ocean
Particle Settling
Mobilization of Fe Through High T, High P Water–Rock Reactions
O2- Deficient Bottom Waters
Fe Flux into Water Column
Sediment Pore Water
E
F
G
C

FIGURE 18 Natural iron nanoparticles in the earth environments. A. Natural processes leading to the formation of nanoparticles in the environment, Reprinted with permission from Ref [636], Copyright © 2015, Royal Society of Chemistry. B. Magnetite nanoparticles in magnetosomes with elongated prismatic habits from magnetotactic *Alphaproteobacteria.* (a) TEM image of a cell of a vibrioid MTB from Lake Mead, Nevada, containing a chain of elongated magnetosomes. (b). TEM image of two double chains of elongated magnetosomes from a freshwater coccus. (c) High-resolution TEM image of a magnetosome from a freshwater coccus with (d) its selected-area electron diffraction pattern (in [1–10] orientation) and (e) a morphological model that consists of six large and six small dodecahedral faces, and smaller faces of the cube and octahedron. The elongation direction is [111]. Reprinted from Ref [645]. C. Iron oxide nanoparticles in Greenland and Antarctic Ice Sheets, Nanoparticulate ferrihydrite B5–10 nm in diameter has been identified. Images (b) and (c) are enlargements of (a), as indicated. The diffraction signal (d) shows some crystalline structure owing to possible impact of nano-clay particles, and potentially nano-hematite, but also the characteristic diffuse ferrihydrite rings are identifiable. EDS (e) analysis of the area further confirms Fe-dominated material. Reprinted with permission from Ref [646] Copyright © 2014, Springer Nature. D. Iron oxide nanoparticles in street dust urban environment, (A) aggregate of spherical magnetite particles, (B) Fe-rich fly ash, (C) magnetite particle with an octahedral shape, (D) hematite, (E) ferrihydrite, and (F) goethite. Reprinted with permission from Ref [648] Copyright © 2016, Royal Society of Chemistry E. Iron nanoparticles from the hydrothermal vents, including various iron oxide and sulfide nanoparticles. They are important geochemical actors across naturally occurring marine redox gradients such as high-temperature hydrothermal vents and reducing sediments where metal turnover is orders of magnitude higher compared to other ocean zones. Reprinted from Ref [650] F. Iron oxide nanoparticles are commonly presented in various soils. (Left) A cross-section of loess deposits in Chinese Loess Plateau with loess layers (formed in dry and cold environment) and paleosol layers (formed in warm environment). Paleosol layers display strong susceptibility signals than the loess layers due to the presence of nano-maghemite. (Right) Transmission electron micrographs of magnetically extracted, sub-micrometer pedogenic magnetite (likely partially oxidized toward maghemite), from palaeosol (a) S5, (b) and (c) S1, Luochuan (Qinjiazhai section). Reprinted with permission from Ref [644], Copyright © 2016, Elsevier. G. Iron oxide nanoparticles with various sizes and sharpness. Example morphologies of one-dimensional and three-dimensional NPs: (A) nanowires, (B) long nanotubes, (C) nanoneedles, (D) nanorods, (E) short nanotubes, (F) tube-in-tubes, and (G) nanorings. Reprinted from Ref [652].

- Electron Transfer in Inorganic vs. Organic Systems: Iron-sulfur clusters in proteins, or electron movement in metal oxides, demonstrate similar electron transfer behaviors, highlighting that ET mechanisms transcend the boundary between living and nonliving matter.
- Catalysis Across Boundaries: Catalytic processes facilitated by ET, whether in a biological enzyme or an inorganic nanoparticle, are driven by the same basic principles. Inorganic and organic catalysts both function by reducing activation energies and facilitating electron flow in reactions.
- Bridging the Two Worlds: This continuity of ET mechanisms between the inorganic and organic realms suggests a shared evolutionary legacy, where inorganic processes may have functioned as early biocatalysts, playing a pivotal role in the origin of life.
- Implications for Evolution and Modern Catalysis: Recognizing the universality of ET provides insights not only into the origins of life but also into the development of synthetic catalysts that replicate biological functions, advancing technologies in fields like medical treatments, agriculture application, energy conversion, and environmental protection.

In this way, the dance of electrons serves as a bridge, linking the inorganic and organic worlds through the same unchanging principles, reminding us of the seamless unity that underpins all of nature's complexity. It is worth noting that inorganic oxidoreductase activity continues to function in our current Earth's environment. Recent studies demonstrate that Fe_2O_3 nanoparticles, derived from PVC dichlorination residues and iron chips treated with subcritical water, exhibit inherent peroxidase-like properties [317]. This is mirrored by the biogenic iron oxide nanoparticles, which are formed through interactions between fungi and iron oxide nanoparticles [315,316], or within magnetosomes in MTB [190,191] or the biogenic iron cores of [200–202] due to the unique metal architecture of iron oxide nanoparticles.

These naturally occurring iron oxide/sulfide nanoparticles, distributed across diverse environments, continue to function as biocatalysts, playing critical roles in redox reactions—an insight yet to be fully recognized. While the potential of synthesized nanoparticles remains largely untapped, the naturally occurring ones demonstrate remarkable catalytic capabilities [635].

Natural iron oxide nanoparticles are widely distributed across diverse environments, including soils, water, rocks, and living organisms [636] (Figure 18). Extensive research and documentation on these nanoparticles have been conducted [167,170,186,274,403,637–644]. Examples of environments where iron oxide nanoparticles can be found include biogenic iron minerals [643], such as those from MTB [234,238,645] (Figure 18B), various ice sheets [646] (Figure 18C), as well as in fly ash and street dust [647,648] (Figure 18D), high pH hydrothermal vents [649,650] (Figure 18E), and in different soil types [641,642,644,651] (Figure 18F). These nanoparticles form through various mechanisms [639], resulting in a wide range of sizes, shapes, and structures [652] (Figure 18G).

Similarly, iron sulfide nanoparticles, including some biogenic iron sulfides [653], are found in many marine sediments [170,391,654,655], various soils [170,403,656], and hydrothermal vent plumes [650,657]. They are formed through a combination of geological processes, such as the weathering of sulfide minerals and volcanic activity, as well as biogenic processes involving sulfate-reducing bacteria (SRB) in anaerobic environments [170,391]. These nanoparticles have also been identified on early Earth, predating the existence of life [76,78,237,289]. Some are believed to have played a direct role in the origins of life and the evolution of living organisms [77,422–430].

This continued functionality of iron oxide/sulfide nanoparticles in natural environments highlights their enduring role, both ancient and contemporary. Their biocatalytic activity not only enhances our understanding of the origins of life but also underscores the interconnectedness of inorganic and organic realms through the universal process of electron transfer.

As we stand in awe of this intricate dance, we are reminded of the profound interconnectedness of the universe and the remarkable journey that has led us here. This symphony of electrons and molecules inspires us to explore further, deepening our understanding of life's origins and fueling future discoveries.

9 CONCLUSION

The research presented highlights the remarkable oxidoreductase activity of inorganic iron oxide and sulfide nanoparticles. Their unique metal architectures and electron conductivities play crucial roles in conferring enzyme-like activity. Electron hopping disrupts chemical equilibrium, enhancing electron transfer, a phenomenon that stems from their intricate electron configurations.

These nanoparticles function as semiconductors with low band gaps, facilitating efficient electron transfer. Spontaneous electron hopping occurs at octahedral surface sites, including defective sites, further promoting electron transfer during biological redox reactions. This conductivity and hopping capability are essential for their enzyme-like activity.

Importantly, the electron transfer processes facilitated by these nanoparticles bear no fundamental distinction from those seen in organic oxidoreductases. The flow of electrons across the inorganic and organic worlds follows the same unchanging principles, highlighting the seamless unity in the way nature orchestrates complex biochemical processes. This "dance of electrons" bridges the inorganic and organic, underscoring the interconnectedness of life and non-life on Earth.

Inorganic iron oxide and sulfide nanoparticles, as the earliest known oxidoreductases, acted as biocatalysts before the emergence of proteins and RNA, and continue to function in this role today. They played integral roles in ancient biochemical processes, contributing to redox reactions, detoxification, oxygen formation, organic carbon transformation, phosphate ester hydrolysis, and inorganic denitrification long before life emerged. These processes were essential for the conditions necessary for life to arise, providing both energy and key materials. Even today, these nanoparticles still play a critical, though often unrecognized, role as biocatalysts in natural ecosystems, influencing biogeochemical cycles and ecosystem stability.

Their intrinsic oxidoreductase-like activity is not confined to artificially synthesized materials; it is also found in biogenic and microbe-influenced nanoparticles. These particles have been active participants in Earth's biogeochemical cycles for billions of years, shaping nutrient availability and ecosystem composition.

In conclusion, the oxidoreductase activity of inorganic iron oxide and sulfide nanoparticles—encompassing electron conductivity, hopping, and mediation by Cytochrome c electron mediators—is essential for their enzyme-like behavior and contributions to Earth's history and ecosystems. The seamless flow of electrons across both inorganic and organic worlds offers profound insights into early life processes, adaptations, and the continued influence of these nanoparticles on the delicate balance of our planet's ecosystems. These inorganic nanoparticles serve as captivating "life fossil oxidoreductases", providing a window into the origins and development of life on Earth.

Author Contributions

X.L.H is the sole author of this manuscript. Conceptualization, X-L H.; methodology, X-L H.; investigation, X-L H.; visualization, X-L H.; writing—original draft, X-L H.; writing—review and editing, X-L H.

CONFLICTS OF INTEREST

There are no conflicts to declare.

ACKNOWLEDGMENTS

Many thanks to Dr. Simon Duval and Dr. Wolfgang Nitschke for inviting me to contribute this chapter to the book. It is a great honor to have the opportunity to further develop my hypothesis on inorganic enzymes, initially based on my observations of phosphate ester hydrolysis promoted by inorganic iron oxide nanoparticles.

This work is dedicated to my family, Wei Sun and Jack Jixiang Huang, for their love, endless support, encouragement, and sacrifices. I would also like to express my deep appreciation to Drs. Michael J. Russell Gerhard Schenk and the late Dr. R.J.P. Williams for their invaluable feedback, as well as to Drs. Robert Atlas, Jia-Zhong Zhang, and Peter B. Ortner for their encouragement since my initial observation in 2007. Special thanks to Dr. Jean Pierre Jolivet for his kind permission to use the different solid phases of iron oxide (Figure 9A). The work presented here was developed independently over several decades, drawing on my background in soil science and biogeochemistry. I am sincerely grateful for the support I have received from all of my employers. However, the views and findings expressed are solely those of the author and do not represent the positions of any current or former employer.

ABBREVIATIONS

2L-Fht	2-line ferrihydrite
6L-Fht	6-line ferrihydrite
AA	ascorbic acid
ABTS	2,2′-azinobis (3-ethylbenzothiazoline-6-sulfonic acid)
AKA	akageneite
AOX	alcohol oxidase
ATP	adenosine triphosphate
BRENDA	BRaunschweig ENzyme Database
CAT	catalase
CO	charge ordering
CB	conduction band
CcO	Cyt c oxidase
Cyt c	cytochrome c
DAB	di-azo-aminobenzene
DET	direct electron transfer
DFT	density function theory
DMT	dialysis membrane tubes
DPD	N, N-diethyl-1,4-phenylenediamine
EC	Enzyme Commission
EDX	Energy Dispersive X-Ray Analysis
EPR	electron paramagnetic resonance
ESE	electron spin echo
ET	electron transfer
$\mathbf{ET_p}$	electron transport
FAD	flavin adenine dinucleotide
FMN	flavin adenine mononucleotide
Foh	feroxyhyte
FTIR	Fourier-transform infrared spectroscopy
GOE	goethite
GOX	glucose oxidase
GPx	glutathione peroxidase
GSH	glutathione
G2P	glycerol 2-phosphate
G6P	glucose-6-phosphate
Hem	hematite
HRP	horseradish peroxidase
k_{cat}	catalytic rates
K_m	Michaelis-Menten constants
Lc	laccase
Lep	lepidocrocite
LUCA	last universal common ancestor
Mag	magnetite
Mah	maghemite

Mo_{co}	molybdopterin cofactors
MTB	magnetotactic bacteria
NADPH oxidase	nicotinamide adenine dinucleotide phosphate oxidase
NIR-II	near-infrared-II
NP	nanoparticles
NW	nanowire
OPD	*o*-phenylenediamine dihydrochloride
OVs	oxygen vacancies
OXD	oxidases
PAPs	purple acid phosphatases
POD	peroxidase
Poly-P_i	polyphosphate
PP_i	pyrophosphate
PQQ	pyrroloquinoline quinone
ROS	reactive oxygen species
Sch	schwertmannite
SEM	scan electron microscopy
SOD	superoxide dismutase
SOE	sulfite-oxidizing enzyme
SAED	selected area electron diffraction
SRB	sulfate-reducing bacteria
SR-TXM	synchrotron radiation transmission X-ray microscopy
TEM	transmission electron microscopy
TMB	3,3',5,5'-tetramethylbenzidine
Trp	tryptophan
TRIS	tris(hydroxymethyl)-aminomethane
Tyr	tyrosine
UV	ultraviolet
VB	valence band
VHPO	vanadium-containing haloperoxidases
V_{max}	maximum reaction rate
XANES	X-ray absorption near edge structure
XPS	X-ray photoelectron spectroscopy
XRD	X-ray diffraction

REFERENCES

1. R.J.P. Williams, *CMLS* **1997**, *53* (10), 816–829.
2. P.G. Falkowski, T. Fenchel, E.F. Delong, *Science* **2008**, *320* (5879), 1034–1039.
3. J. Dongun Kim, S. Senn, A. Harel, B.I. Jelen, P.G. Falkowski, *Philos. Trans. R. Soc. Lond. B Biol. Sci.* **2013**, *368* (1622), 20120257.
4. K.N. McGuinness, N. Fehon, R. Feehan, M. Miller, A.C. Mutter, L.A. Rybak, J. Nam, J.E. AbuSalim, J.T. Atkinson, H. Heidari, N. Losada, J.D. Kim, R.L. Koder, Y. Lu, J.J. Silberg, J.S.G. Slusky, P.G. Falkowski, V. Nanda, *Proteins* **2024**, *92* (1), 52–59.

5. L. Sellés Vidal, C.L. Kelly, P.M. Mordaka, J.T. Heap, *Biochim. Biophys. Acta Proteins Proteom.* **2018**, *1866* (2), 327–347.
6. B. Hay Mele, M. Monticelli, S. Leone, D. Bastoni, B. Barosa, M. Cascone, F. Migliaccio, F. Montemagno, A. Ricciardelli, L. Tonietti, A. Rotundi, A. Cordone, D. Giovannelli, *Essays Biochem.* **2023**, *67* (4), 653–670.
7. A. Harel, Y. Bromberg, P.G. Falkowski, D. Bhattacharya, *Proc. Natl. Acad. Sci. USA* **2014**, *111* (19), 7042–7047.
8. X.-L. Huang, J.R. Harmer, G. Schenk, G. Southam, *Front. Chem.* **2024**, *12*, 1349020.
9. X.-L. Huang, *iScience* **2024**, *27* (5), 109555.
10. J. Trefil, H.J. Morowitz, E. Smith, *Am. Sci.* **2009**, *97* (3), 206–213.
11. H.B. Gray, J.R. Winkler, *Chem. Sci.* **2021**, *12* (42), 13988–14003.
12. D. Porath, A. Bezryadin, S. De Vries, C. Dekker, *Nature* **2000**, *403* (6770), 635–638.
13. R.A. Marcus, N. Sutin, *BBA Rev. Bioenerg.* **1985**, *811* (3), 265–322.
14. D. Ratautas, M. Dagys, *Catalysts* **2020**, *10* (1), 9.
15. T. Adachi, Y. Kitazumi, O. Shirai, K. *Catalysts* **2020**, *10*, 236.
16. J.R. Winkler, H.B. Gray, *Chem. Rev.* **2014**, *114* (7), 3369–3380.
17. C.D. Bostick, S. Mukhopadhyay, I. Pecht, M. Sheves, D. Cahen, D. Lederman, *Rep. Prog. Phys.* **2018**, *81* (2), 026601.
18. K. Apel, H. Hirt, *Ann. Rev. Plant Biol.* **2004**, *55* (1), 373–399.
19. H. Bayr, *Crit. Care Med.* **2005**, *33* (12 Suppl), S498–S501.
20. P. Sharma, A.B. Jha, R.S. Dubey, M. *J. Bot.* **2012**, *2012*, 217037.
21. M. Valko, D. Leibfritz, J. Moncol, M.T.D. Cronin, M. Mazur, J. Telser, *Int. J. Biochem. Cell Biol.* **2007**, *39* (1), 44–84.
22. J.N. Rodríguez-López, D.J. Lowe, J. Hernández-Ruiz, A.N. Hiner, F. García-Cánovas, R.N. Thorneley, *J. Am. Chem. Soc.* **2001**, *123* (48) 11838–11847.
23. N.C. Veitch, *Phytochemistry* **2004**, *65* (3), 249–259.
24. F.K. de Oliveira, L.O. Santos, J.G. Buffon, *Food Res. Int.* **2021**, *143*, 110266.
25. K.R. Messner, J.A. Imlay, *J. Biol. Chem.* **2002**, *277* (45), 42563–42571.
26. V. Leskovac, S. Trivić, G. Wohlfahrt, J. Kandrač, D. Peričin, *Int. J. Biochem. Cell Biol.* **2005**, *37* (4), 731–750.
27. D. Jancura, J. Stanicova, G. Palmer, M. Fabian, *Biochemistry* **2014**, *53* (22), 3564–3575.
28. U. Kappler, J.H. Enemark, *J. Biol. Inorg. Chem.* **2015**, *20* (2), 253–264.
29. A. Deisseroth, A.L. Dounce, *Physiol. Rev.* **1970**, *50* (3), 319–375.
30. M. Alfonso-Prieto, X. Biarnés, P. Vidossich, C. Rovira, *J. Am. Chem. Soc.* **2009**, *131* (33), 11751–11761.
31. I. Fridavich, *Annu. Rev. Biochem.* **1995**, *64*, 97–112.
32. Y. Sheng, I.A. Abreu, D.E. Cabelli, M.J. Maroney, A.F. Miller, M. Teixeira, J.S. Valentine, *Chem. Rev.* **2014**, *114* (7), 3854–3918.
33. F. Kracke, I. Vassilev, J.O. Krömer, *Front. Microbiol.* **2015**, 6.
34. R.D. Milton, S.D. Minteer, *J. R. Soc. Interface* **2017**, *14* (131), 20170253.
35. H. Chen, O. Simoska, K. Lim, M. Grattieri, M. Yuan, F. Dong, Y.S. Lee, K. Beaver, S. Weliwatte, E.M. Gaffney, S.D. Minteer, *Chem. Rev.* **2020**, *120* (23), 12903–12993.
36. E.E. Ferapontova, S. Shleev, T. Ruzgas, L. Stoica, A. Christenson, J. Tkac, A.I. Yaropolov, L. Gorton, *Direct electrochemistry of proteins and enzymes*, in: E. Paleček, F. Scheller, J. Wang (Eds.), *Perspectives in Bioanalysis*, Elsevier, 2005, pp. 517–598.
37. X. Dominguez-Benetton, I. Srikanth, Y. Satyawali, K. Vanbroekhoven, D. Pant, *J. Microb. Biochem. Technol.* **2013**, *2013*, 1–20.
38. K. Kano, O. Shirai, Y. Kitazumi, K. Sakai, H.-Q. Xia, *Fundamentals of DET-type bioelectrocatalysis*, in: K. Kano, O. Shirai, Y. Kitazumi, K. Sakai, H.-Q. Xia (Eds.), *Enzymatic Bioelectrocatalysis*, Springer Singapore, Singapore, 2021, pp. 57–78.
39. E.V. Suprun, *Electrochem. Commun.* **2021**, *125*, 106983.

40. M.J. Eddowes, H.A.O. Hill, *J. Chem. Soc. Chem. Commun.* **1977**, *21*, 771b–772.
41. Y. Peter, K. Theodore, *Chem. Lett.* **1977**, *6* (10), 1145–1148.
42. A.I. Yaropolov, V. Malovik, S.D. Varfolomeyev, I.V. Berezin, *Dokl. Akad. Nauk SSSR* **1979**, *249*, 1399.
43. M.R. Tarasevich, A.I. Yaropolov, V.A. Bogdanovskaya, S.D. Varfolomeev, *J. Electroanal. Chem.* **1979**, *104*, 393–403.
44. L. Gorton, A. Lindgren, T. Larsson, F.D. Munteanu, T. Ruzgas, I. Gazaryan, *Anal. Chim. Acta* **1999**, *400* (1), 91–108.
45. E.E. Ferapontova, T. Ruzgas, L. Gorton, *Anal. Chem.* **2003**, *75* (18), 4841–4850.
46. S. Shleev, J. Tkac, A. Christenson, T. Ruzgas, A.I. Yaropolov, J.W. Whittaker, L. Gorton, *Bioelectron.* **2005**, *20* (12), 2517–2554.
47. Y. Liu, R. Yuan, Y. Chai, D. Tang, J. Dai, X. Zhong, *Sens. Actuators B Chem.* **2006**, *115* (1), 109–115.
48. C. Léger, P. Bertrand, *Chem. Rev.* **2008**, *108* (7), 2379–2438.
49. J. Liu, S. Chakraborty, P. Hosseinzadeh, Y. Yu, S. Tian, I. Petrik, A. Bhagi, Y. Lu, *Chem. Rev.* **2014**, *114* (8), 4366–4469.
50. P. Bollella, L. Gorton, R. Antiochia, *Sensors* **2018**, *18* (5), 1319.
51. A.D. Goldman, J.A. Baross, R. Samudrala, *PLoS One* **2012**, *7* (9), e39912.
52. K. Hussein Mahdi, Oxidoreductases: *Significance for humans and microorganism*, in: M. Mahmoud Ahmed (Ed.), *Oxidoreductase*, IntechOpen, Rijeka, 2020, Ch. 8.
53. C.P. McKay, H. Hartman, *Orig. Life Evol. Biosph.* **1991**, *21* (3), 157–163.
54. R.E. Blankenship, H. Hartman, *Trends Biochem. Sci.* **1998**, *23* (3), 94–97.
55. F. Baymann, E. Lebrun, M. Brugna, B. Schoepp-Cothenet, M.T. Giudici-Orticoni, W. Nitschke, *Philos. Trans. R. Soc. Lond. B Biol. Sci.* **2003**, *358* (1429), 267–274.
56. J.L. Kirschvink, R.E. Kopp, *Philos. Trans. R. Soc. Lond. B Biol. Sci.* **2008**, *363* (1504), 2755–2765.
57. M.A. Inupakutika, S. Sengupta, A.R. Devireddy, R.K. Azad, R. Mittler, *J. Exp. Bot.* **2016**, *67* (21), 5933–5943.
58. J. Jabłońska, D.S. Tawfik, *Nat. Ecol. Evol.* **2021**, *5* (4), 442–448.
59. J. Jabłońska, D.S. Tawfik, *Protein Sci.* **2022**, *31* (5), e4310.
60. M.J. Russell, A.J. Hall, A.R. Mellersh, *On the dissipation of thermal and chemical energies on the early earth: The onsets of hydrothermal convection, chemiosmosis, genetically regulated metabolism and oxygenic photosynthesis*, in: R. Ikan (Ed.), *Natural and Laboratory-Simulated Thermal Geochemical Processes*, Dordrecht, Kluwer Academic Publishers, 2003, pp. 325–388.
61. M.J. Russell, A.J. Hall, *News* **2002**, *113* (113), 6–12.
62. N. Lane, *Oxygen: The Molecule That Made the World*, Oxford University Press, 2002, London.
63. J.M. Olson, R.E. Blankenship, *Photosyn. Res.* **2004**, *80* (1–3), 373–386.
64. I. Ślesak, M. Libik, B. Karpinska, S. Karpinski, Z. Miszalski, *Acta Biochim. Pol.* **2007**, *54* (1), 39–50.
65. J.F. Kasting, J.B. Pollack, D. Crisp, *J. Atmos. Chem.* **1984**, *1* (4), 403–428.
66. T.W. Lyons, C.T. Reinhard, N.J. Planavsky, *Nature* **2014**, *506* (7488), 307–315.
67. Y. Zuo, Y. Deng, *Geochim. Cosmochim. Acta* **1999**, *63* (19), 3451–3455.
68. M.J. Borda, A.R. Elsetinow, M.A. Schoonen, D.R. Strongin, *Astrobiology* **2001**, *1* (3), 283–288.
69. J.K. Lee, K.L. Walker, H.S. Han, J. Kang, F.B. Prinz, R.M. Waymouth, H.G. Nam, R.N. Zare, *Proc. Natl. Acad. Sci. USA* **2019**, *116* (39), 19294–19298.
70. H. He, X. Wu, H. Xian, J. Zhu, Y. Yang, Y. Lv, Y. Li, K.O. Konhauser, *Nat. Commun.* **2021**, *12* (1), 6611.

71. J.M. Jenkins, W.H. Brune, D.O. Miller, *J. Geophys. Res. Atmos.* **2021**, *126* (9), e2021JD034557.
72. J. Stone, J.O. Edgar, J.A. Gould, J. Telling, *Nat. Commun.* **2022**, *13* (1), 4529.
73. I. Slesak, H. Slesak, J. Kruk, *Astrobiology* **2012**, *12* (8), 775–784.
74. M.C. Liang, H. Hartman, R.E. Kopp, J.L. Kirschvink, Y.L. Yung, *Proc. Natl. Acad. Sci. USA* **2006**, *103* (50), 18896–18899.
75. P.S. Braterman, A.G. Cairns-Smith, R.W. Sloper, *Nature* **1983**, *303* (5913), 163–164.
76. H.D. Holland, *The Geologic History of Seawater, Treatise on Geochemistry*, Elsevier Inc. 2007, pp. 1–46.
77. W. Nitschke, M.J. Russell, *J. Mol. Evol.* **2009**, *69* (5), 481–496.
78. A. Bekker, N.J. Planavsky, B. Krapež, B. Rasmussen, A. Hofmann, J.F. Slack, O.J. Rouxel, K.O. Konhauser, *Iron formations: Their origins and implications for ancient seawater chemistry*, in: *Treatise on Geochemistry*: 2nd ed., Elsevier Inc. 2013, pp. 561–628.
79. Z. Shu, L. Liu, W. Tan, S.L. Suib, G. Qiu, X. Yang, L. Zheng, F. Liu, *Environ. Sci. Technol.* **2019**, *53* (15), 8854–8861.
80. Z. Shu, Z. Pan, X. Wang, H. He, S. Yan, X. Zhu, W. Song, Z. Wang, *Environ. Sci. Technol.* **2022**, *56* (19), 14188–14197.
81. C. Bonfio, L. Valer, S. Scintilla, S. Shah, D.J. Evans, L. Jin, J.W. Szostak, D.D. Sasselov, J.D. Sutherland, S.S. Mansy, *Nat. Chem.* **2017**, *9* (12), 1229–1234.
82. L. Gao, J. Zhuang, L. Nie, J. Zhang, Y. Zhang, N. Gu, T. Wang, J. Feng, D. Yang, S. Perrett, X. Yan, *Nat. Nanotechnol.* *2* (9) **2007**, 577–583.
83. H. Wei, E. Wang, *Chem. Soc. Rev.* *42* (14) **2013**, 6060–6093.
84. J. Wu, X. Wang, Q. Wang, Z. Lou, S. Li, Y. Zhu, L. Qin, H. Wei, *Chem. Soc. Rev.* *48* (4) **2019**, 1004–1076.
85. L. Gao, *Enzyme-like property (nanozyme) of iron oxide nanoparticles*, in: X.-L. Huang (Ed.), *Iron Oxide Nanoparticles*, https://www.intechopen.com, London UK, 2022.
86. L. Gao, H. Wei, S. Dong, X. Yan, *Adv. Mater.* *36* (10) **2024**, 2305249.
87. X.L. Huang, J.Z. Zhang, *Sediment-water exchange of dissolved organic phosphorus in Florida bay*, in: P.O. Joseph, M. Prospero (Eds.), *CIMAS Seventh Year Annual Report* (NOAA Cooperative Agreement NA17RJ1226), University of Miami, Miami, FL, 2007, pp. 72–73.
88. X.L. Huang, J.Z. Zhang, *RSC Adv.* **2012**, *2* (1), 199–208.
89. X.L. Huang, *Astrobiology* **2018**, *18* (3), 294–310.
90. X.L. Huang, *Iron oxide nanoparticles: An inorganic phosphatase*, in: I. Sinha, M. Shukla (Eds.), *Nanocatalysts*, Interchopen, London, 2019, pp. 97–124.
91. X.-L. Huang, *New J. Chem.* **2022**, *46* (32), 15273–15291.
92. Y. Huang, J. Ren, X. Qu, *Chem. Rev.* **2019**, *119* (6), 4357–4412.
93. M. Liang, X. Yan, *Acc. Chem. Res.* **2019**, *52* (8), 2190–2200.
94. S. Singh, *Front. Chem.* **2019**, *7*, 46.
95. W. Yang, X. Yang, L. Zhu, H. Chu, X. Li, W. Xu, *Coord. Chem. Rev.* **2021**, *448*, 214170.
96. C. Hong, X. Meng, J. He, K. Fan, X. Yan, *Particuology* **2022**, *71*, 90–107.
97. H. Wei, G. Li, J. Li, *Biomedical Nanozymes, from Diagnostics to Therapeutics*, 1st ed., Springer Singapore, Singapore, 2023.
98. Y. Liu, X. Yan, H. Wei, *Medical nanozymes for therapeutics*, in: N. Gu (Ed.), *Nanomedicine*, Springer Nature Singapore, Singapore, 2023, pp. 285–329.
99. L. Gao, X. Yan, *Nanozymes: Biomedical Applications of Enzymatic Fe3O4 Nanoparticles from In Vitro to In Vivo, Advances in Experimental Medicine and Biology*, Springer, 2019, pp. 291–312.
100. X. Zhang, G. Li, G. Chen, D. Wu, Y. Wu, T.D. James, *Adv. Funct. Mater.* **2021**, *31* (50), 2106139.

101. D.O. Lopez-Cantu, R.B. González-González, A. Sharma, M. Bilal, R. Parra-Saldívar, H.M.N. Iqbal, *Coord. Chem. Rev.* **2022**, *469*, 214685.
102. L. Gao, K. Fan, X. Yan, *Theranostics* **2017**, *7* (13), 3207–3227.
103. L. Gao, X. Yan, *Adv. Exp. Med. Biol.* **2019**, *1174*, 291–312.
104. R. Zhang, L. Chen, Q. Liang, J. Xi, H. Zhao, Y. Jin, X. Gao, X. Yan, L. Gao, K. Fan, *Nano Today* **2021**, *41*, 101317.
105. K. Fan, C. Cao, Y. Pan, D. Lu, D. Yang, J. Feng, L. Song, M. Liang, X. Yan, *Nat. Nanotechnol.* **2012**, *7* (7), 459–464.
106. Y. Zhang, Z. Wang, X. Li, L. Wang, M. Yin, L. Wang, N. Chen, C. Fan, H. Song, *Adv. Mater.* **2016**, *28* (7), 1387–1393.
107. J. Wang, Q. Xie, H. Song, X. Chen, X. Zhang, X. Zhao, Y. Hao, Y. Zhang, H. Li, N. Li, K. Fan, X. Wang, *J. Nanobiotechnol.* 21 (1) (**2023**).
108. Z. Xu, Z. Qiu, Q. Liu, Y. Huang, D. Li, X. Shen, K. Fan, J. Xi, Y. Gu, Y. Tang, J. Jiang, J. Xu, J. He, X. Gao, Y. Liu, H. Koo, X. Yan, L. Gao, *Nat. Commun.* **2018**, *9* (1), 3713.
109. N. Singh, G. Mugesh, *Angew. Chem. Int. Ed.* **2019**, *58* (23), 7797–7801.
110. Y. Wang, X. Jia, S. An, W. Yin, J. Huang, X. Jiang, *Adv. Mater.* **2023**, *36*, 2301810.
111. M. Cao, R. Cai, L. Zhao, M. Guo, L. Wang, Y. Wang, L. Zhang, X. Wang, H. Yao, C. Xie, Y. Cong, Y. Guan, X. Tao, Y. Wang, S. Xu, Y. Liu, Y. Zhao, C. Chen, *Nat. Nanotechnol.* **2021**, *16* (6), 708–716.
112. L. Zhao, L. Lu, A. Wang, H. Zhang, M. Huang, H. Wu, B. Xing, Z. Wang, R. Ji, *J. Agric. Food Chem.* **2020**, *68* (7), 1935–1947.
113. Z. Cui, Y. Li, H. Zhang, P. Qin, X. Hu, J. Wang, G. Wei, C. Chen, *J. Agric. Food Chem.* **2022**, *70* (42), 13445–13463.
114. P. Deng, Y. Gao, L. Mu, X. Hu, F. Yu, Y. Jia, Z. Wang, B. Xing, *Proc. Natl. Acad. Sci. USA* **2023**, *120* (25), e2301885120.
115. K. Kettler, K. Veltman, D. van de Meent, A. van Wezel, A.J. Hendriks, *Environ. Toxicol. Chem.* **2014**, *33* (3), 481–492.
116. C. Ma, J.C. White, J. Zhao, Q. Zhao, B. Xing, *Annu. Rev. Food Sci. Technol.* **2018**, *9*, 129–153.
117. A.A. Sembada, I.W. Lenggoro, *Nanomaterials (Basel)* **2024**, *14* (2), 131.
118. M. Gajdardziska-Josifovska, R.G. McClean, M.A. Schofield, C.V. Sommer, W.F. Kean, *Eur. J. Mineral.* **2001**, *13* (5), 863–870.
119. N. Rodríguez, N. Menéndez, J. Tornero, R. Amils, V. de la Fuente, *New Phytol.* **2005**, *165* (3), 781–789.
120. R.I. Khalilov, A.N. Nasibova, V.A. Serezhenkov, M.A. Ramazanov, M.K. Kerimov, A.A. Garibov, A.F. Vanin, *Biophysics* **2011**, *56* (2), 316–322.
121. H. Zhu, J. Han, J.Q. Xiao, Y. Jin, *Environ. Monit. Assess.* **2008**, *10* (6), 713–717.
122. H. Tombuloglu, Y. Slimani, G. Tombuloglu, M. Almessiere, A. Baykal, *Chemosphere* **2019**, *226*, 110–122.
123. H. Tombuloglu, Y. Slimani, T.M. AlShammari, M. Bargouti, M. Ozdemir, G. Tombuloglu, S. Akhtar, H. Sabit, K.R. Hakeem, M. Almessiere, I. Ercan, A. Baykal, *Environ. Pollut.* **2020**, *266*, 115391.
124. G. Chen, S. Taherymoosavi, S. Cheong, Y. Yin, R. Akter, C.E. Marjo, A.M. Rich, D.R.G. Mitchell, X. Fan, J. Chew, G. Pan, L. Li, R. Bian, J. Horvat, M. Mohammed, P. Munroe, S. Joseph, *Sci. Rep.* **2021**, *11* (1), 159.
125. M.H. Ghafariyan, M.J. Malakouti, M.R. Dadpour, P. Stroeve, M. Mahmoudi, *Environ. Sci. Technol.* **2013**, *47* (18), 10645–10652.
126. J.H. Kim, Y. Oh, H. Yoon, I. Hwang, Y.S. Chang, *Environ. Sci. Technol.* **2015**, *49* (2), 1113–1119.
127. J. Li, J. Hu, C. Ma, Y. Wang, C. Wu, J. Huang, B. Xing, *Chemosphere* **2016**, *159*, 326–334.

128. N. Al-Amri, H. Tombuloglu, Y. Slimani, S. Akhtar, M. Barghouthi, M. Almessiere, T. Alshammari, A. Baykal, H. Sabit, I. Ercan, S. Ozcelik, *Ecotoxicol. Environ. Saf.* **2020**, *194*, 110377.
129. L. Cai, L. Cai, H. Jia, C. Liu, D. Wang, X. Sun, *J. Hazard. Mater.* **2020**, *393*, 122415.
130. D.M. Kasote, J.H.J. Lee, G.K. Jayaprakasha, B.S. Patil, *ACS Sustain. Chem. Eng.* **2019**, *7* (5), 5142–5151.
131. T. Zheng, Q. Zhou, Z. Tao, S. Ouyang, *Sci. Total Environ.* **2023**, *904*, 166643.
132. N.G.M. Palmqvist, G.A. Seisenbaeva, P. Svedlindh, V.G. Kessler, *Nanoscale Res. Lett.* **2017**, *12*.
133. L. Yan, P. Li, X. Zhao, R. Ji, L. Zhao, *Sci. Total Environ.* **2020**, *718*, 137400.
134. Y. Liu, Z. Xiao, F. Chen, L. Yue, H. Zou, J. Lyu, Z. Wang, *Sci. Total Environ.* **2021**, *780*, 146578.
135. D. Singh, D. Sillu, A. Kumar, S. Agnihotri, *Environ. Sci. Nano* **2021**, *8* (5), 1308–1325.
136. H. Ashraf, T. Anjum, S. Riaz, T. Batool, S. Naseem, G. Li, *J. Nanobiotechnol.* **2022**, *20* (1), 8.
137. M. Rizwan, S. Noureen, S. Ali, S. Anwar, M.Z. Rehman, M.F. Qayyum, A. Hussain, *J. Soils Sediments* **2019**, *19* (11), 3749–3759.
138. M. Sarraf, K. Vishwakarma, V. Kumar, N. Arif, S. Das, R. Johnson, E. Janeeshma, J.T. Puthur, S. Aliniaeifard, D.K. Chauhan, M. Fujita, M. Hasanuzzaman, *Plants* **2022**, *11* (3), 316.
139. T. Lu, X. Wang, X. Cui, J. Li, J. Xu, P. Xu, J. Wan, *Environ. Pollut.* **2023**, *337*, 122578.
140. N. Manzoor, T. Ahmed, M. Noman, M. Shahid, M.M. Nazir, L. Ali, T.S. Alnusaire, B. Li, R. Schulin, G. Wang, *Sci. Total Environ.* **2021**, *769*, 145221.
141. X. Cao, L. Yue, C. Wang, X. Luo, C. Zhang, X. Zhao, F. Wu, J.C. White, Z. Wang, B. Xing, *ACS Nano* **2022**, *16* (1), 1170–1181.
142. S. Silva, M.C. Dias, D.C.G.A. Pinto, A.M.S. Silva, *Plants* **2023**, *12* (3), 491.
143. Y. Meng, W. Li, X. Pan, G.M. Gadd, *Environ. Sci. Nano* **2020**, *7* (5), 1305–1318.
144. C.A.S. Ballesteros, L.A. Mercante, A.D. Alvarenga, M.H.M. Facure, R. Schneider, D.S. Correa, *Mater. Chem. Front.* **2021**, *5* (20), 7419–7451.
145. V.K. Yata, *Applications of nanozymes in wastewater treatment*, in: H.K. Daima, N. Pn, E. Lichtfouse (Eds.), *Nanozymes for Environmental Engineering*, Springer International Publishing, Cham, 2021, pp. 95–110.
146. E.L.S. Wong, K.Q. Vuong, E. Chow, *Sensors* **2021**, *21* (2), 408.
147. H. Sá, M. Michelin, T. Tavares, B. Silva, *Biomolecules* **2022**, *12* (10), 1489.
148. M. Bilal, A.R. Bagheri, D.S. Vilar, N. Aramesh, K.I.B. Eguiluz, L.F.R. Ferreira, S.S. Ashraf, H.M.N. Iqbal, *J. Chem. Technol. Biotechnol.* **2022**, *97* (2), 420–435.
149. L.Y. Martínez-Zamudio, R.B. González-González, R.G. Araújo, J.A. Rodríguez Hernández, E.A. Flores-Contreras, E.M. Melchor-Martínez, R. Parra-Saldívar, H.M.N. Iqbal, *Curr. Opin. Environ. Sci. Health* **2024**, *37*, 100522.
150. Ç. Kip, G. Çolak, M. Polat, A. Tuncel, *J. Mol. Struct.* **2024**, *1301*, 137375.
151. S. Scott, H.M. Zhao, A. Dey, T.B. Gunnoe, *ACS Catal.* **2020**, *10*, 14315.
152. H. Wei, L. Gao, K. Fan, J. Liu, J. He, X. Qu, S. Dong, E. Wang, X. Yan, *Nano Today* **2021**, *40*, 101269.
153. A. Robert, B. Meunier, *ACS Nano* **2022**, *16* (5), 6956–6959.
154. H. Fan, R. Zhang, K. Fan, L. Gao, X. Yan, *ACS Nano* **2024**, *18* (4), 2533–2540.
155. M. Zandieh, J. Liu, *Adv. Mater.* **2024**, *36* (10), 2211041.
156. C.J. Neal, E. Kolanthai, F. Wei, M. Coathup, S. Seal, *Adv. Mater.* **2024**, *36* (10), 2211261.
157. S. He, L. Ma, Q. Zheng, Z. Wang, W. Chen, Z. Yu, X. Yan, K. Fan, *Bioactive. Mater.* **2024**, *42*, 284–298.
158. X.Q. Zhang, S.W. Gong, Y. Zhang, T. Yang, C.Y. Wang, N. Gu, *J. Mater. Chem.* **2010**, *20* (24), 5110–5116.

159. C. Peng, B. Jiang, Q. Liu, Z. Guo, Z. Xu, Q. Huang, H. Xu, R. Tai, C. Fan, *Energy Environ. Sci.* **2011**, *4* (6), 2035–2040.
160. K.N. Chaudhari, N.K. Chaudhari, J.S. Yu, *Catal. Sci. Technol.* **2012**, *2* (1), 119–124.
161. Z. Chen, J.J. Yin, Y.T. Zhou, Y. Zhang, L. Song, M. Song, S. Hu, N. Gu, *ACS Nano* **2012**, *6* (5), 4001–4012.
162. Q. Liu, L. Zhang, H. Li, Q. Jia, Y. Jiang, Y. Yang, R. Zhu, *Mater. Sci. Eng. C* **2015**, *55*, 193–200.
163. B. Jiang, D. Duan, L. Gao, M. Zhou, K. Fan, Y. Tang, J. Xi, Y. Bi, Z. Tong, G.F. Gao, *Nat. Protoc.* **2018**, *13*, 1506.
164. T. Qin, R. Ma, Y. Yin, X. Miao, S. Chen, K. Fan, J. Xi, Q. Liu, Y. Gu, Y. Yin, J. Hu, X. Liu, D. Peng, L. Gao, *Theranostics* **2019**, *9* (23), 6920–6935.
165. Y. Gu, Y. Huang, Z. Qiu, Z. Xu, D. Li, L. Chen, J. Jiang, L. Gao, *Sci. China Life Sci.* **2020**, *63* (1), 68–79.
166. S. Guo, L. Guo, *J. Phys. Chem. C* **2019**, *123* (50), 30318–30334.
167. R.M. Cornell, U. Schwertmann, *The Iron Oxides: Structure, Properties, Reactions, Occurrence and Uses*, 2nd ed., Wiley-VCH Verlag GmbH & Co. KGaA, 2003.
168. H. Tamura, A. Tanaka, K.-y. Mita, R. Furuichi, *J. Colloid Interface Sci.* **1999**, *209* (1), 225–231.
169. Y. Zu, Z. Wang, H. Yao, L. Yan, *J. Mater. Chem. B* **2023**, *11* (14), 3071–3088.
170. D. Rickard, G.W. Luther, *Chem. Rev.* **2007**, *107* (2), 514–562.
171. W.T. Yao, H.Z. Zhu, W.G. Li, H.B. Yao, Y.C. Wu, S.H. Yu, *Chem. Plus. Chem.* **2013**, *78* (7), 723–727.
172. A.K. Dutta, S.K. Maji, D.N. Srivastava, A. Mondal, P. Biswas, P. Paul, B. Adhikary, *ACS Appl. Mater. Interfaces* **2012**, *4* (4), 1919–1927.
173. Z. Dai, S. Liu, J. Bao, H. Ju, *Chemistry* **2009**, *15* (17), 4321–4326.
174. C. Ding, Y. Yan, D. Xiang, C. Zhang, Y. Xian, *Acta* **2016**, *183* (2), 625–631.
175. X. Nie, L. Xia, H.-L. Wang, G. Chen, B. Wu, T.-Y. Zeng, C.-Y. Hong, L.-H. Wang, Y.-Z. You, *ACS Appl. Mater. Interfaces* **2019**, *11* (35), 31735–31742.
176. C. Kantar, O. Oral, N.A. Oz, *Chemosphere* **2019**, *237*, 124440.
177. J. Gao, Y. Liu, X. Xia, L. Wang, L. Shao, T. Cai, W. Dong, *Appl. Surf. Sci.* **2019**, *463*, 863–871.
178. S. Agnihotri, T. Mohan, D. Jha, H.K. Gautam, I. Roy, *ACS Omega* **2020**, *5* (1), 597–602.
179. W. Ding, H. Liu, W. Zhao, J. Wang, L. Zhang, Y. Yao, C. Yao, C. Song, *ACS Appl. Bio Mater.* **2020**, *3* (9), 5905–5912.
180. C. Song, W. Ding, W. Zhao, H. Liu, J. Wang, Y. Yao, C. Yao, *Biosens. Bioelectron.* **2020**, *151*, 111983.
181. Y. Wang, X. Shen, S. Ma, Q. Guo, W. Zhang, L. Cheng, L. Ding, Z. Xu, J. Jiang, L. Gao, *Biomater. Sci.* **2020**, *8* (9), 2447–2458.
182. X. Meng, D. Li, L. Chen, H. He, Q. Wang, C. Hong, J. He, X. Gao, Y. Yang, B. Jiang, G. Nie, X. Yan, L. Gao, K. Fan, *ACS Nano* **2021**, *15* (3), 5735–5751.
183. H. Ren, J. Yong, Q. Yang, Z. Yang, Z. Liu, Y. Xu, H. Wang, X. Jiang, W. Miao, X. Li, *Acta Pharm. Sin. B* **2021**, *11* (10), 3244–3261.
184. C. Huang, Z. Liu, M. Chen, L. Du, C. Liu, S. Wang, Y. Zheng, W. Liu, *J. Nanobiotechnol.* **2021**, *19* (1), 457.
185. C. Song, H. Liu, L. Zhang, J. Wang, C. Zhao, Q. Xu, C. Yao, *Sens. Actuators B Chem.* **2022**, *353*, 131131.
186. Y. Yuan, L. Wang, L. Gao, *Front. Chem.* **2020**, *8*. https://doi.org/10.3389/fchem.2020.00818.
187. Y. Shan, W. Lu, J. Xi, Y. Qian, *Front. Chem.* **2022**, *10*, 1000709.
188. X. Shen, Z. Wang, X. Gao, Y. Zhao, *ACS Catal.* **2020**, *10* (21), 12657–12665.
189. Z. Wang, X. Shen, X. Gao, *J. Phys. Chem. C* **2021**, *125* (42), 23098–23104.

190. K. Li, C. Chen, C. Chen, Y. Wang, Z. Wei, W. Pan, T. Song, *Enzyme Microb. Technol.* **2015**, *72*, 72–78.
191. L. Hu, T. Song, Q. Ma, C. Chen, W. Pan, C. Xie, L. Nie, W. Yang, *8th International Conference on the Scientific and Clinical Applications of Magnetic Carriers* **2010**, *1311*, 369–374.
192. Y. Pan, N. Li, J. Mu, R. Zhou, Y. Xu, D. Cui, Y. Wang, M. Zhao, *Appl. Microbiol. Biotechnol.* **2015**, *99* (2), 703–715.
193. Y. Pan, Y. Wang, X. Fan, W. Wang, X. Yang, D. Cui, M. Zhao, *Environ. Microbiol. Rep.* **2019**, *11*, 140–146.
194. A. Ahmed, A. Abagana, D. Cui, M. Zhao, *BioMed Res. Int.* **2019**, *2019*, 7127869 .
195. A.Y. Abagana, M. Zhao, M.Y. Alshahrani, K.U. Rehman, S. Andleeb, J. Wang, S.M. Bukhari, *Biomass Convers. Biorefinery* **2022**, *14*, 3453–3462.
196. S.C. Andrews, *Biochim. Biophys. Acta Gen. Subj.* **2010**, *1800* (8), 691–705.
197. N.E. Le Brun, A. Crow, M.E.P. Murphy, A.G. Mauk, G.R. Moore, *Biochim. Biophys. Acta Gen. Subj.* **2010**, *1800* (8), 732–744.
198. S. Narayanan, E. Firlar, M.G. Rasul, T. Foroozan, N. Farajpour, L. Covnot, R. Shahbazian-Yassar, T. Shokuhfar, *Nanoscale* **2019**, *11* (36), 16868–16878.
199. M. Plays, S. Müller, R. Rodriguez, *Metallomics* **2021**, *13* (5), mfab021.
200. G.S. Arapova, A.N. Eryomin, D.I. Metelitza, *Russ. J. Bioorg. Chem.* **1999**, *25*, 369–376.
201. Z. Tang, H. Wu, Y. Zhang, Z. Li, Y. Lin, *Anal. Chem.* **2011**, *83* (22), 8611–8616.
202. L. Ma, J.-J. Zheng, N. Zhou, R. Zhang, L. Fang, Y. Yang, X. Gao, C. Chen, X. Yan, K. Fan, *Nat. Commun.* **2024**, *15* (1), 233.
203. L. Schmermund, V. Jurkaš, F.F. Özgen, G.D. Barone, H.C. Büchsenschütz, C.K. Winkler, S. Schmidt, R. Kourist, W. Kroutil, *ACS Catal.* **2019**, *9* (5), 4115–4144.
204. S.H. Lee, D.S. Choi, S.K. Kuk, C.B. Park, *Angew. Chem. Int. Ed.* **2018**, *57* (27), 7958–7985.
205. M.T. Neves-Petersen, S. Klitgaard, A.S.L. Carvalho, S.B. Petersen, M.R.A. De Barros, E.P. E Meloz, *Biophys. J.* **2007**, *92* (6), 2016–2027.
206. L.M. Shannon, E. Kay, J.Y. Lew, *J. Biol. Chem.* **1966**, *241* (9), 2166–2172.
207. G. Herrmann, E. Jayamani, G. Mai, W. Buckel, *J. Bacteriol.* **2008**, *190* (3), 784–791.
208. J. Raymond, J.L. Siefert, C.R. Staples, R.E. Blankenship, *Mol. Biol. Evol.* **2004**, *21* (3), 541–554.
209. L.M. Rubio, P.W. Ludden, *Ann. Rev. Microbiol.* **2008**, *62*, 93–111.
210. D.V. Bent, E. Hayon, *J. Am. Chem. Soc.* **1975**, *97* (10), 2612–2619.
211. M. Sivaraja, D.B. Goodin, M. Smith, B.M. Hoffman, *Science* **1989**, *245* (4919), 738–740.
212. D.V. Bent, E. Hayon, *J. Am. Chem. Soc.* **1975**, *97* (10), 2599–2606.
213. M. Zhu, Y. Dai, Y. Wu, K. Liu, X. Qi, Y. Sun, *Nanotechnology* **2018**, *29* (46), 465704.
214. S. Lin, W. Tan, P. Han, X. Li, J. Li, Z. Nie, K. Li, *Nano Res.* **2022**, *15*, 9073–9081.
215. X. Wang, X. Zhong, H. Lei, Y. Geng, Q. Zhao, F. Gong, Z. Yang, Z. Dong, Z. Liu, L. Cheng, *Mater.* **2019**, *31* (16), 6174–6186.
216. S. Lin, Y. Wang, Z. Chen, L. Li, J. Zeng, Q. Dong, Y. Wang, Z. Chai, *ACS Sustain. Chem. Eng.* **2018**, *6* (9), 12061–12069.
217. H. Wang, J. Chen, Q. Dong, X. Jia, D. Li, J. Wang, E. Wang, *Nano Res.* **2022**, *15* (6), 5256–5261.
218. M.N. Karim, M. Singh, P. Weerathunge, P. Bian, R. Zheng, C. Dekiwadia, T. Ahmed, S. Walia, E. Della Gaspera, S. Singh, R. Ramanathan, V. Bansal, *ACS Appl. Nano Mat.* **2018**, *1* (4), 1694–1704.
219. W. Septina, R.R. Prabhakar, R. Wick, T. Moehl, S.D. Tilley, *Chem. Mater.* **2017**, *29*, 1735–1743.
220. Y. Xia, L. Yin, *Phys. Chem. Chem. Phys.* **2013**, *15* (42), 18627–18634.

221. Y. Lin, X. Liu, Z. Liu, Y. Xu, *Small* **2021**, *17* (39), 2103348.
222. P.K. Boruah, M.R. Das, *J. Hazard. Mater.* **2020**, *385*, 121516.
223. X. Han, R. Liu, H. Zhang, Q. Zhou, W. Feng, K. Hu, *Chem. Asian J.* **2021**, *16* (12), 1603–1607.
224. C. Cao, H. Zou, N. Yang, H. Li, Y. Cai, X. Song, J. Shao, P. Chen, X. Mou, W. Wang, X. Dong, *Adv. Mater.* **2021**, *33* (52), 2106996.
225. H. Han, X. Xu, H. Kan, Y. Tang, C. Liu, H. Wen, L. Wu, Y. Jiang, Z. Wang, J. Liu, F. Wang, *J. Colloid Interface Sci.* **2022**, *616*, 304–315.
226. J. Shan, K. Yang, W. Xiu, Q. Qiu, S. Dai, L. Yuwen, L. Weng, Z. Teng, L. Wang, *Small* **2020**, *16* (40), 2001099. https://doi.org/10.1002/smll.202001099.
227. R.P. Ambler, *Biochim. Biophys. Acta Bioenerg.* **1991**, *1058* (1), 42–47.
228. D.A. Mavridou, S.J. Ferguson, J.M. Stevens, Cytochrome c assembly, *IUBMB Life* **2013**, *65* (3), 209–216.
229. S. Ma, R. Ludwig, *ChemElectroChem* **2019**, *6* (4), 958–975.
230. H.A. Kalpage, J. Wan, P.T. Morse, M.P. Zurek, A.A. Turner, A. Khobeir, N. Yazdi, L. Hakim, J. Liu, A. Vaishnav, T.H. Sanderson, M.A. Recanati, L.I. Grossman, I. Lee, B.F.P. Edwards, M. Hüttemann, *Int. J. Biochem. Cell Biol.* **2020**, *121*, 105704.
231. B.G. Malström, *BBA Rev. Bioenerg.* **1979**, *549* (3), 281–303.
232. Z. Deng, Y. Gong, Y. Luo, Y. Tian, *Biosens Bioelectron.* **2009**, *24* (8), 2465–2469.
233. M. Chen, Z. Wang, J. Shu, X. Jiang, W. Wang, Z.-H. Shi, Y.-W. Lin, *Chem.* **2017**, *56* (16), 9400–9403.
234. R. Uebe, D. Schüler, *Nat. Rev. Microbiol.* **2016**, *14* (10), 621–637.
235. O. Strbak, P. Hnilicova, J. Gombos, A. Lokajova, P. Kopcansky, *Minerals* **2022**, *12* (11), 1403.
236. B.H. Lower, D.A. Bazylinski, *J. Mol. Microbiol. Biotechnol.* **2013**, *23*, 63–80.
237. W. Lin, G.A. Paterson, Q. Zhu, Y. Wang, E. Kopylova, Y. Li, R. Knight, D.A. Bazylinski, R. Zhu, J.L. Kirschvink, Y. Pan, *Proc. Natl. Acad. Sci. USA* **2017**, *114* (9), 2171–2176.
238. P. Goswami, K. He, J. Li, Y. Pan, A.P. Roberts, W. Lin, *NPJ Biofilms Microb.* **2022**, *8* (1), 43.
239. A. Komeili, Z. Li, D.K. Newman, G.J. Jensen, *Science* **2006**, *311* (5758), 242–245.
240. M.I. Siponen, G. Adryanczyk, N. Ginet, P. Arnoux, D. Pignol, *Biochem. Soc. Trans.* **2012**, *40* (6), 1319–1323.
241. A. Taoka, Y. Eguchi, R. Shimoshige, Y. Fukumori, *Microbiol. Immunol.* **2023**, *67* (5), 228–238.
242. W. Dröge, *Physiol. Rev.* **2002**, *82* (1), 47–95.
243. B. Halliwell, *Plant Physiol.* **2006**, *141* (2), 312–322.
244. H. Sies, *Redox Biol.* **2017**, *11*, 613–619.
245. Y.J. Taverne, D. Merkus, A.J. Bogers, B. Halliwell, D.J. Duncker, T.W. Lyons, *Bioessays* **2018**, *40* (3), 1700158.
246. Y.J. Taverne, A. Caron, C. Diamond, G. Fournier, T.W. Lyons, *Oxidative stress and the early coevolution of life and biospheric oxygen*, in: H. Sies (Ed.), *Oxidative Stress*, Academic Press, 2020, pp. 67–85.
247. I.N. Zelko, T.J. Mariani, R.J. Folz, *Free Radic. Biol. Med.* **2002**, *33* (3), 337–349.
248. M. Zámocký, B. Gasselhuber, P.G. Furtmüller, C. Obinger, *Arch. Biochem. Biophys.* **2012**, *525* (2), 131–144.
249. I. Ślesak, H. Ślesak, P. Zimak-Piekarczyk, P. Rozpądek, *Astrobiology* **2016**, *16* (5), 348–358.
250. A.J. Case, *Antioxidants* **2017**, *6* (4), 82.
251. K.R. Olson, Y. Gao, E.R. DeLeon, M. Arif, F. Arif, N. Arora, K.D. Straub, *Redox. Biol.* **2017**, *12*, 325–339.
252. B. Runnegar, *Glob. Planet Change* **1991**, *5* (1–2), 97–111.

253. J. Castresana, M. Saraste, *Trends Biochem. Sci.* **1995**, *20* (11), 443–448.
254. J. Castresana, M. Lübben, M. Saraste, D.G. Higgins, *EMBO J.* **1994**, *13* (11), 2516–2525.
255. T.M. Lenton, *The coupled evolution of life and atmospheric oxygen*, in: *Evolution on Planet Earth: The Impact of the Physical Environment*, Eds. L. Rothschild, A. Lister, Elsevier Ltd, 2003, pp. 35–53.
256. A. Neubeck, F. Freund, *Astrobiology* **2020**, *20* (5), 670–675.
257. R.B. Frankel, G.C. Papaefthymiou, R.P. Blakemore, W. O'Brien, *BBA Mol. Cell Res.* **1983**, *763* (2), 147–159.
258. D. Faivre, L.H. Böttger, B.F. Matzanke, D. Schüler, *Angew. Chem. Int. Ed.* **2007**, *46* (44), 8495–8499.
259. S. Staniland, B. Ward, A. Harrison, G. Van Der Laan, N. Telling, *Proc. Natl. Acad. Sci. USA* **2007**, *104* (49), 19524–19528.
260. M.L. Fdez-Gubieda, A. Muela, J. Alonso, A. García-Prieto, L. Olivi, R. Fernández-Pacheco, J.M. Barandiarán, *ACS Nano* **2013**, *7* (4), 3297–3305.
261. J. Baumgartner, G. Morin, N. Menguy, T.P. Gonzalez, M. Widdrat, J. Cosmidis, D. Faivre, *Proc. Natl. Acad. Sci. USA* **2013**, *110* (37), 14883–14888.
262. J. Baumgartner, N. Menguy, T.P. Gonzalez, G. Morin, M. Widdrat, D. Faivre, *J. R Soc. Interface* **2016**, *13* (124), 20160665.
263. L. Le Nagard, X. Zhu, H. Yuan, K. Benzerara, D.A. Bazylinski, C. Fradin, A. Besson, S. Swaraj, S. Stanescu, R. Belkhou, A.P. Hitchcock, *Chem. Geol.* **2019**, *530*, 119348.
264. T. Wen, Y. Zhang, Y. Geng, J. Liu, A. Basit, J. Tian, Y. Li, J. Li, J. Ju, W. Jiang, *Biomater. Res.* **2019**, *23* (1), 13.
265. Y.A. Gorby, T.J. Beveridge, R.P. Blakemore, *J. Bacteriol.* **1988**, *170* (2), 834–841.
266. G.J. Doherty, H.T. McMahon, *Annu. Rev. Biochem.* **2009**, *78*, 857–902.
267. C.G. Chilom, B. Zorilă, M. Bacalum, M. Bălăşoiu, R. Yaroslavtsev, S.V. Stolyar, S. Tyutyunnicov, *Chem. Phys. Lipids* **2020**, *226*, 104851.
268. R. Shahzad, S. Bilal, A. Khan, A. Shehzad, E.A. Al-Suhaimi, *Emerging concept on cellular uptake mechanism of nanoparticles*, in: H. Tombuloglu, G. Tombuloglu, E. Al-Suhaimi, A. Baykal, K.R. Hakeem (Eds.), *Molecular Impacts of Nanoparticles on Plants and Algae*, Academic Press, 2024, pp. 31–40.
269. D. Zanella, E. Bossi, R. Gornati, C. Bastos, N. Faria, G. Bernardini, *Sci. Rep.* **2017**, *7* (1), 11413.
270. W. Liu, J. Tian, C. Mao, Z. Wang, J. Liu, R.A. Dahlgren, L. Zhang, X. Wang, *Anal. Chim. Acta* **2020**, *1127*, 246–255.
271. H. Cao, Q. Wang, X. Wang, L. Chen, J. Jiang, L. Gao, *Acc. Mater. Res.* **2023**, *4* (2), 115–132.
272. P. Liu, Y. Zheng, R. Zhang, J. Bai, K. Zhu, K. Benzerara, N. Menguy, X. Zhao, A.P. Roberts, Y. Pan, J. Li, *Natl. Sci. Rev.* **2023**, *10* (1), nwac238.
273. D. Murat, A. Quinlan, H. Vali, A. Komeili, *Proc. Natl. Acad. Sci. USA* **2010**, *107* (12), 5593–5598.
274. J.P. Jolivet, E. Tronc, C. Chanéac, *CR Geosci.* **2006**, *338* (6–7), 488–497.
275. J.P. Jolivet, *Metal Oxide Nanostructures Chemistry: Synthesis from Aqueous Solutions*, Oxford University Press, New York, 2019.
276. F.M. Michel, L. Ehm, S.M. Antao, P.L. Lee, P.J. Chupas, G. Liu, D.R. Strongin, M.A.A. Schoonen, B.L. Phillips, J.B. Parise, *Science* **2007**, *316* (5832), 1726–1729.
277. F.M. Michel, V. Barrón, J. Torrent, M.P. Morales, C.J. Serna, J.F. Boily, Q. Liu, A. Ambrosini, A.C. Cismasu, G.E. Brown Jr, *Proc. Natl. Acad. Sci. USA* **2010**, *107* (7), 2787–2792.
278. A. Manceau, J.M. Combes, *Phys. Chem. Miner.* **1988**, *15* (3), 283–295.
279. T. Girardet, P. Venturini, H. Martinez, J.-C. Dupin, F. Cleymand, S. Fleutot, *Appl. Sci.* **2022**, *12* (16), 8127.

280. R. Grau-Crespo, A.Y. Al-Baitai, I. Saadoune, N.H. De Leeuw, *J. Phys. Condens. Matter* **2010**, *22* (25), 255401.
281. M.P. Morales, S. Veintemillas-Verdaguer, M.I. Montero, C.J. Serna, A. Roig, L.I. Casas, B. Martínez, F. Sandiumenge, *Chem. Mater.* **1999**, *11* (11), 3058–3064.
282. C. Greaves, *J. Solid State Chem.* **1983**, *49* (3), 325–333.
283. A.H. Hill, F. Jiao, P.G. Bruce, A. Harrison, W. Kockelmann, C. Ritter, *Chem. Mater.* **2008**, *20* (15), 4891–4899.
284. N.Y. Dzade, A. Roldan, N.H. De Leeuw, *Minerals* **2014**, *4* (1), 89–115.
285. H. Zhang, W. Sun, X. Xie, J. He, C. Zhang, *ACS Omega* **2023**, *8* (9), 8248–8255.
286. G. Nolze, A. Winkelmann, *Cryst. Res. Technol.* **2014**, *49* (7), 490–501.
287. Z. Li, C. Chanéac, G. Berger, S. Delaunay, A. Graff, G. Lefèvre, *RSC Adv.* **2019**, *9* (58), 33633–33642.
288. M. Usman, J.M. Byrne, A. Chaudhary, S. Orsetti, K. Hanna, C. Ruby, A. Kappler, S.B. Haderlein, *Chem. Rev.* **2018**, *118* (7), 3251–3304.
289. A. Kappler, C. Bryce, M. Mansor, U. Lueder, J.M. Byrne, E.D. Swanner, *Nat. Rev. Microbiol.* **2021**, *19* (6), 360–374.
290. R. Raiswell, D.E. Canfield, *Geochem. Perspect.* **2012**, *1* (1), 1–232.
291. K. Günther, *Biochemistry of Iron*, in: K. Günther (Ed.), *Diet for Iron Deficiency: Metabolism – Bioavailability – Diagnostics*, Springer, Berlin, Heidelberg, 2023, pp. 7–27.
292. J.N. Fitzsimmons, T.M. Conway, *Ann. Rev. Mar. Sci.* **2023**, *15* (1), 383–406.
293. C. Yu, S. Xie, Z. Song, S. Xia, M.E. Åström, *Earth-Sci. Rev.* **2021**, *218*, 103658.
294. M. Usman, M. Abdelmoula, K. Hanna, B. Grégoire, P. Faure, C. Ruby, *J. Solid State Chem.* **2012**, *194*, 328–335.
295. J.-P. Jolivet, E. Tronc, *J. Colloid Interface Sci.* **1988**, *125* (2), 688–701.
296. P. Jungcharoen, M. Pédrot, F. Heberling, K. Hanna, F. Choueikani, C. Catrouillet, A. Dia, R. Marsac, *Environ. Sci. Nano* **2022**, *9* (7), 2363–2371.
297. S.P. Schwaminger, D. Bauer, P. Fraga-García, F.E. Wagner, S. Berensmeier, *Cryst. Eng. Comm.* **2017**, *19* (2), 246–255.
298. M. Creutzburg, K. Sellschopp, R. Gleißner, B. Arndt, G.B. Vonbun-Feldbauer, V. Vonk, H. Noei, A. Stierle, *J. Phys. Condens. Matter* **2022**, *34* (16), 164003.
299. H.-W. Luo, X. Zhang, J.-J. Chen, H.-Q. Yu, G.-P. Sheng, *Environ. Sci. Nano* **2017**, *4* (12), 2395–2404.
300. M. Auffan, W. Achouak, J. Rose, M.A. Roncato, C. Chanéac, D.T. Waite, A. Masion, J.C. Woicik, M.R. Wiesner, J.Y. Bottero, *Environ. Sci. Technol.* **2008**, *42* (17), 6730–6735.
301. S. Liu, F. Lu, R. Xing, J.J. Zhu, *Chem. Eur. J.* **2011**, *17* (2), 620–625.
302. N. Puvvada, P.K. Panigrahi, D. Mandal, A. Pathak, *RSC Adv.* **2012**, *2* (8), 3270–3273.
303. X.-L. Cheng, J.-S. Jiang, D.-M. Jiang, Z.-J. Zhao, *J. Phys. Chem. C* **2014**, *118* (24), 12588–12598.
304. J. Mu, L. Zhang, G. Zhao, Y. Wang, *Phys. Chem. Chem. Phys.* **2014**, *16* (29), 15709–15716.
305. T. Tian, L. Ai, X. Liu, L. Li, J. Li, J. Jiang, *Ind. Eng. Chem. Res.* **2015**, *54* (4), 1171–1178.
306. Y. Peng, Z. Wang, W. Liu, H. Zhang, W. Zuo, H. Tang, F. Chen, B. Wang, *Dalton Trans.* **2015**, *44* (28), 12871–12877.
307. Y. Yang, Z. Mao, W. Huang, L. Liu, J. Li, J. Li, Q. Wu, *Sci. Rep.* **2016**, *6*, 35344.
308. S. Ghosh, P. Roy, N. Karmodak, E.D. Jemmis, G. Mugesh, *Angew. Chem. Int. Ed.* **2018**, *57* (17), 4510–4515.
309. Y. Jiang, T. Xia, L. Shen, J. Ma, H. Ma, T. Sun, F. Lv, N. Zhu, *ACS Catal.* **2021**, *11* (5), 2949–2955.
310. Z. Xu, A. Qu, W. Wang, M. Lu, B. Shi, C. Chen, C. Hao, L. Xu, M. Sun, C. Xu, H. Kuang, *Adv. Healthc. Mater.* **2021**, *10* (23), 2101316.

311. M. Chen, X. Zhou, C. Xiong, T. Yuan, W. Wang, Y. Zhao, Z. Xue, W. Guo, Q. Wang, H. Wang, Y. Li, H. Zhou, Y. Wu, *ACS Appl. Mater. Interfaces* **2022**, *14* (19), 21989–21995.
312. K. Singh, R. Gujju, S. Bandaru, S. Misra, K.S. Babu, N. Puvvada, *ACS Omega* **2022**, *7* (19), 16616–16628.
313. R. Zhang, H. Zhao, K. Fan, *Structure-activity mechanism of iron oxide nanozymes*, in: *Nanozymes: Design, Synthesis, and Applications*, American Chemical Society, 2022, pp. 1–35.
314. H. Dong, W. Du, J. Dong, R. Che, F. Kong, W. Cheng, M. Ma, N. Gu, Y. Zhang, *Nat. Commun.* **2022**, *13* (1), 5365.
315. G.H. Yu, Z.L. Chi, A. Kappler, F.S. Sun, C.Q. Liu, H.H. Teng, G.M. Gadd, *Curr. Biol.* **2020**, *30* (15), 2943–2950.e4.
316. Z.L. Chi, X.Y. Zhao, Y.L. Chen, J.L. Hao, G.H. Yu, B.A. Goodman, G.M. Gadd, *Environ. Microbiol.* **2021**, *23* (2), 893–907.
317. Y. Qi, Y. Sun, D. Song, Y. Wang, F. Xiu, *Talanta* **2023**, *253*, 124039.
318. X. Yu, C.-F. Huo, Y.-W. Li, J. Wang, H. Jiao, *Surf Sci.* **2012**, *606* (9), 872–879.
319. J. Noh, O.I. Osman, S.G. Aziz, P. Winget, J.-L. Brédas, *Chem. Mater.* **2015**, *27* (17), 5856–5867.
320. A. Erlebach, H.-D. Kurland, J. Grabow, F.A. Müller, M. Sierka, *Nanoscale* **2015**, *7* (7), 2960–2969.
321. X. Huang, S.K. Ramadugu, S.E. Mason, *J. Phys. Chem. C* **2016**, *120* (9), 4919–4930.
322. W. Jian, S.-P. Wang, H.-X. Zhang, F.-Q. Bai, *Inorg. Chem. Front.* **2019**, *6* (10), 2660–2666.
323. V.K. Paidi, M. Shepit, J.W. Freeland, D.L. Brewe, C.A. Roberts, J. van Lierop, *J. Phys. Chem. C* **2021**, *125* (48), 26688–26697.
324. S.A. Sanden, R.K. Szilagyi, Y. Li, N. Kitadai, S.M. Webb, T. Yano, R. Nakamura, M. Hara, S.E. McGlynn, *Dalton Trans.* **2021**, *50* (34), 11763–11774.
325. J.M. Domingos, E. Runge, C. Dreher, T.H. Chiu, J. Shuster, S. Fischer, A. Kappler, J.P. Duda, J. Xu, M. Mansor, *Geochem. Perspect. Lett.* **2023**, *26*, 14–19.
326. W. Davison, *Aquat. Sci.* **1991**, *53* (4), 309–329.
327. R.E. Krupp, *Eur. J. Mineral* **1994**, *6*, 265–278.
328. L.G. Benning, R.T. Wilkin, H.L. Barnes, *Chem. Geol.* **2000**, *167* (1), 25–51.
329. B.B. Jørgensen, D.C. Nelson, *Sulfide oxidation in marine sediments: Geochemistry meets microbiology*, in: J.P. Amend, K.J. Edwards, T.W. Lyons (Eds.), *Sulfur Biogeochemistry – Past and Present*, Geological Society of America, 2004.
330. D. Csákberényi-Malasics, J.D. Rodriguez-Blanco, V.K. Kis, A. Rečnik, L.G. Benning, M. Pósfai, *Chem. Geol.* **2012**, *294–295*, 249–258.
331. Y. Lan, A.S. Elwood Madden, E.C. Butler, *Environ. Sci. Process. Impacts* **2016**, *18* (10), 1266–1273.
332. K. Wasmund, M. Mußmann, A. Loy, *Environ. Microbiol. Rep.* **2017**, *9* (4), 323–344.
333. S.H. Li, Y.-H. Chen, J.-J. Lee, H.-S. Sheu, *Phys. Chem. Miner.* **2018**, *45* (1), 27–38.
334. Y. Sano, A. Kyono, Y. Yoneda, N. Isaka, S. Takagi, G.i. Yamamoto, *J. Mineral. Petrol. Sci.* **2020**, *115*, 261–275.
335. R. Bolney, M. Grosch, M. Winkler, J. van Slageren, W. Weigand, C. Robl, *RSC Adv.* **2021**, *11* (51), 32464–32475.
336. R. Dieckmann, *Phil. Mag. A* **1993**, *68* (4), 725–745.
337. G.S. Parkinson, *Surf. Sci. Rep.* **2016**, *71* (1), 272–365.
338. D. Zu, H. Wang, S. Lin, G. Ou, H. Wei, S. Sun, H. Wu, *Nano Res.* **2019**, *12* (9), 2150–2163.
339. S.S. Bhosale, A.P. Tiwari, *Oxygen-deficient iron oxide nanostructures for photocatalytic activities*, in: F.I. Ezema, C.D. Lokhande, R. Jose (Eds.), *Chemically Deposited Nanocrystalline Metal Oxide Thin Films: Synthesis, Characterizations, and Applications*, Springer International Publishing, Cham, 2021, pp. 355–388.

340. S. Datta, P. Maneesha, P.K. Mishra, S. Sen, Role of defects in metal oxide nanostructures, in: V. Kumar, I. Ayoub, V. Sharma, H.C. Swart (Eds.), *Optical Properties of Metal Oxide Nanostructures*, Springer Nature Singapore, Singapore, 2023, pp. 189–221.
341. P.A. Cox, *The Electronic Structure and Chemistry of Solids*, Oxford University Press, Oxford UK, 1987.
342. P.W. Atkins, J.d. Paula, *Physical Chemistry*, 9th ed., Oxford University Presss, Oxford, UK, 2010.
343. S.M. Sze, M.-K. Lee, *Semiconductor Devices: Physics and Technology*, 3rd ed., Wiley, 2012.
344. P. Jamilpanah, H. Pahlavanzadeh, A. Kheradmand, *Heat Mass Transf.* **2017**, *53* (4), 1343–1354.
345. M. Pumera, S. Sánchez, I. Ichinose, J. Tang, *Sens. Actuators B Chem.* **2007**, *123* (2), 1195–1205.
346. L. Shi, H. Dong, G. Reguera, H. Beyenal, A. Lu, J. Liu, H.-Q. Yu, J.K. Fredrickson, *Nat. Rev. Microbiol.* **2016**, *14* (10), 651–662.
347. D.R. Lovley, *Annu. Rev. Microbiol.* **2017**, *71*, 643–664.
348. F.T. Rabouw, C. de Mello Donega, *Top. Curr. Chem.* **2016**, *374* (5), 58.
349. M.C.F. Wander, K.M. Rosso, M.A.A. Schoonen, *J. Phys. Chem. C* **2007**, *111* (30), 11414–11423.
350. F.N. Skomurski, S. Kerisit, K.M. Rosso, *Geochim. Cosmochim. Acta* **2010**, *74* (15), 4234–4248.
351. V. Alexandrov, K.M. Rosso, *J. Chem. Phys.* **2014**, *140* (23), 234701.
352. P. Zarzycki, D.M. Smith, K.M. Rosso, *J. Chem. Theory Comput.* **2015**, *11* (4), 1715–1724.
353. N. Iordanova, M. Dupuis, K.M. Rosso, *J. Chem. Phys.* **2005**, *122* (14), 144305.
354. S. Kerisit, K.M. Rosso, *Geochim. Cosmochim. Acta* **2006**, *70* (8), 1888–1903.
355. J.E. Katz, X. Zhang, K. Attenkofer, K.W. Chapman, C. Frandsen, P. Zarzycki, K.M. Rosso, R.W. Falcone, G.A. Waychunas, B. Gilbert, *Science* **2012**, *337* (6099), 1200–1203.
356. J. Husek, A. Cirri, S. Biswas, L.R. Baker, *Chem. Sci.* **2017**, *8* (12), 8170–8178.
357. L.M. Carneiro, S.K. Cushing, C. Liu, Y. Su, P. Yang, A.P. Alivisatos, S.R. Leone, *Nat. Mater.* **2017**, *16* (8), 819–825.
358. E.M.M. Ibrahim, L.H. Abdel-Rahman, A.M. Abu-Dief, A. Elshafaie, S.K. Hamdan, A.M. Ahmed, *Mater. Res. Bull.* **2018**, *99*, 103–108.
359. J.A. Soltis, A.M. Schwartzberg, P. Zarzycki, R.L. Penn, K.M. Rosso, B. Gilbert, *ACS Earth Space Chem.* **2017**, *1* (4), 216–226.
360. E.J.W. Verwey, *Nature* **1939**, *144*, 327–328.
361. D.S. Tannhauser, *J. Phys. Chem. Solids* **1962**, *23* (1), 25–34.
362. R.A. de Groot, F.M. Mueller, P.G.v. Engen, K.H.J. Buschow, *Phys. Rev. Lett.* **1983**, *50* (25), 2024–2027.
363. J.M.D. Coey, A.E. Berkowitz, L. Balcells, F.F. Putris, F.T. Parker, *Appl. Phys. Lett.* **1998**, *72* (6), 734–736.
364. S. Soeya, J. Hayakawa, H. Takahashi, K. Ito, C. Yamamoto, A. Kida, H. Asano, M. Matsui, *Appl. Phys. Lett.* **2002**, *80* (5), 823–825.
365. Y.S. Dedkov, U. Rüdiger, G. Güntherodt, *Phys. Rev. B Condens. Matter Mater. Phys.* **2002**, *65* (6), 644171–644175.
366. O. Mounkachi, R. Lamouri, M. Hamedoun, H. Ez-Zahraouy, E. Salmani, A. Benyoussef, *J. Supercond. Nov. Magn.* **2017**, *30* (11), 3221–3224.
367. A. Radoń, D. Łukowiec, M. Kremzer, J. Mikuła, P. Włodarczyk, *Materials* **2018**, *11* (5), 735.
368. E.R. Morris, Q. Williams, *J. Geophys. Res. Solid Earth* **1997**, *102* (B8), 18139–18148.

369. M.P. Pasternak, G.K. Rozenberg, G.Y. Machavariani, O. Naaman, R.D. Taylor, R. Jeanloz, *Phys. Rev. Lett.* **1999**, *82* (23), 4663–4666.
370. K. Ohta, K. Hirose, K. Shimizu, Y. Ohishi, *Phys. Rev. B* **2010**, *82* (17), 174120.
371. K. Ohta, R.E. Cohen, K. Hirose, K. Haule, K. Shimizu, Y. Ohishi, *Phys. Rev. Lett.* **2012**, *108* (2), 026403.
372. S. Todo, N. Takeshita, T. Kanehara, T. Mori, N. Môri, *J. Appl. Phys.* **2001**, *89* (11), 7347–7349.
373. M.S. Senn, J.P. Wright, J.P. Attfield, *Nature* **2012**, *481* (7380), 173–176.
374. S.V. Ovsyannikov, M. Bykov, E. Bykova, D.P. Kozlenko, A.A. Tsirlin, A.E. Karkin, V.V. Shchennikov, S.E. Kichanov, H. Gou, A.M. Abakumov, R. Egoavil, J. Verbeeck, C. McCammon, V. Dyadkin, D. Chernyshov, S. van Smaalen, L.S. Dubrovinsky, *Nat. Chem.* **2016**, *8* (5), 501–508.
375. K.H. Hong, A.M. Arevalo-Lopez, J. Cumby, C. Ritter, J.P. Attfield, *Nat. Commun.* **2018**, *9* (1), 2975.
376. S.V. Ovsyannikov, M. Bykov, E. Bykova, K. Glazyrin, R.S. Manna, A.A. Tsirlin, V. Cerantola, I. Kupenko, A.V. Kurnosov, I. Kantor, A.S. Pakhomova, I. Chuvashova, A.I. Chumakov, R. Rüffer, C. McCammon, L.S. Dubrovinsky, *Nat. Commun.* **2018**, *9* (1), 4142.
377. S.J. Cassidy, F. Orlandi, P. Manuel, S.J. Clarke, *Nat. Commun.* **2019**, *10* (1), 5475.
378. S.V. Ovsyannikov, M. Bykov, S.A. Medvedev, P.G. Naumov, A. Jesche, A.A. Tsirlin, E. Bykova, I. Chuvashova, A.E. Karkin, V. Dyadkin, D. Chernyshov, L.S. Dubrovinsky, *Angew. Chem. Int. Ed.* **2020**, *59* (14), 5632–5636.
379. A.A. Qureshi, S. Javed, H.M.A. Javed, M. Jamshaid, U. Ali, M.A. Akram, *Nanomaterials* **2022**, *12* (10), 1635.
380. A.P. Roberts, L. Chang, C.J. Rowan, C.-S. Horng, F. Florindo, *Rev. Geophys.* **2011**, *49* (1), 2010RG000336.
381. M. Wu, J.S. Tse, Y. Pan, *Sci. Rep.* **2016**, *6* (1), 21637.
382. L. Chang, A.P. Roberts, Y. Tang, B.D. Rainford, A.R. Muxworthy, Q. Chen, *J. Geophys. Res. Solid Earth* **2008**, *113* (B6), B06104.
383. A.J. Devey, R. Grau-Crespo, N.H. de Leeuw, *Phys. Rev. B* **2009**, *79* (19), 195126.
384. R.A.D. Pattrick, V.S. Coker, M. Akhtar, M.A. Malik, E. Lewis, S. Haigh, P. O'Brien, P.C. Shafer, G. van der Laan, *Mineral. Mag.* **2017**, *81* (4), 857–872.
385. G. Li, B. Zhang, F. Yu, A.A. Novakova, M.S. Krivenkov, T.Y. Kiseleva, L. Chang, J. Rao, A.O. Polyakov, G.R. Blake, R.A. de Groot, T.T.M. Palstra, *Chem. Mater.* **2014**, *26* (20), 5821–5829.
386. I.S. Lyubutin, S.S. Starchikov, C.-R. Lin, S.-Z. Lu, M.O. Shaikh, K.O. Funtov, T.V. Dmitrieva, S.G. Ovchinnikov, I.S. Edelman, R. Ivantsov, *J. Nanopart. Res.* **2013**, *15* (1), 1397.
387. R. Wu, Y.F. Zheng, X.G. Zhang, Y.F. Sun, J.B. Xu, J.K. Jian, *J. Cryst. Growth* **2004**, *266* (4), 523–527.
388. A. Ennaoui, S. Fiechter, C. Pettenkofer, N. Alonso-Vante, K. Büker, M. Bronold, C. Höpfner, H. Tributsch, *Sol. Energy Mater Sol. Cells* **1993**, *29* (4), 289–370.
389. E. Day-Roberts, T. Birol, R.M. Fernandes, *Phys. Rev. Mater.* **2020**, *4* (5), 054405.
390. M.Z. Rahman, T. Edvinsson, *Joule* **2019**, *3* (10), 2290–2293.
391. D. Rickard, *The sedimentary sulfur system: Biogeochemistry and evolution through geologic time*, in: F.T. Mackenzie (Ed.), *Treatise on Geochemistry*, Elsevier, 2013, pp. 267–326.
392. S. Khalid, E. Ahmed, Y. Khan, S. Nawaz, M. Ramzan, N.R. Khalid, W. Ahmed, *Iron pyrite (FeS2): Sustainable photovoltaic material*, in: M.J. Jackson, W. Ahmed (Eds.), *Micro and Nanomanufacturing*, Volume II, Springer International Publishing, Cham, 2018, pp. 281–318.

393. D.F. Pridmore, R.T. Shuey, *Am. Mineral.* **1976**, *61* (3–4), 248–259.
394. R. Schieck, A. Hartmann, S. Fiechter, R. Könenkamp, H. Wetzel, *J. Mater. Res.* **1990**, *5* (7), 1567–1572.
395. P.K. Abraitis, R.A.D. Pattrick, D.J. Vaughan, *Int. J. Miner. Process.* **2004**, *74* (1), 41–59.
396. S.A. Sorenson, J.G. Patrow, J.M. Dawlaty, *J. Phys. Chem. C* **2016**, *120* (14), 7736–7747.
397. M. Cabán-Acevedo, M.S. Faber, Y. Tan, R.J. Hamers, S. Jin, *Nano Lett.* **2012**, *12* (4), 1977–1982.
398. M. Cabán-Acevedo, D. Liang, K.S. Chew, J.P. DeGrave, N.S. Kaiser, S. Jin, *ACS Nano* **2013**, *7* (2), 1731–1739.
399. X. Zhang, M. Manno, A. Baruth, M. Johnson, E.S. Aydil, C. Leighton, *ACS Nano* **2013**, *7* (3), 2781–2789.
400. S. Takele, G.R. Hearne, *Phys. Rev. B* **1999**, *60* (7), 4401–4403.
401. D.M. Roberts, A.R. Landin, T.G. Ritter, J.D. Eaves, C.R. Stoldt, *Sci. Rep.* **2018**, *8* (1), 6591.
402. S. Sakkopoulos, E. Vitoratos, T. Argyreas, *J. Phys. Chem. Solids* **1984**, *45* (8), 923–928.
403. L. Sagnotti, *Iron sulfides*, in: D. Gubbins, E. Herrero-Bervera (Eds.), *Encyclopedia of Geomagnetism and Paleomagnetism*, Springer Netherlands, Dordrecht, 2007, pp. 454–459.
404. A. Živković, H.E. King, M. Wolthers, N.H. de Leeuw, *J. Phys. Condens. Matter* **2021**, *33* (46), 465801.
405. X. Lai, H. Zhang, Y. Wang, X. Wang, X. Zhang, J. Lin, F. Huang, *J. Am. Chem. Soc.* **2015**, *137* (32), 10148–10151.
406. S.J. Kuhn, M.K. Kidder, D.S. Parker, C. dela Cruz, M.A. McGuire, W.M. Chance, L. Li, L. Debeer-Schmitt, J. Ermentrout, K.C. Littrell, M.R. Eskildsen, A.S. Sefat, *Physica C Supercond.* **2017**, *534*, 29–36.
407. Z. Guo, F. Sun, B. Han, K. Lin, L. Zhou, W. Yuan, *Phys. Chem. Chem. Phys.* **2017**, *19* (13), 9000–9006.
408. S. Chen, Z. Kang, X. Zhang, J. Xie, H. Wang, W. Shao, X. Zheng, W. Yan, B. Pan, Y. Xie, *ACS Cent. Sci.* **2017**, *3* (11), 1221–1227.
409. P.P. Edwards, H.B. Gray, M.T.J. Lodge, R.J.P. Williams, *Angew. Chem. Int. Ed.* **2008**, *47* (36), 6758–6765.
410. D.V. Matyushov, *J. Chem. Phys.* **2013**, *139* (2), 025102.
411. S. Rafiq, G.D. Scholes, *J. Am. Chem. Soc.* **2019**, *141* (2), 708–722.
412. J.J. Warren, H.B. Gray, *Electron transfer proteins*, in: E.C. Constable, G. Parkin, L. Que Jr (Eds.), *Comprehensive Coordination Chemistry III*, Elsevier, Oxford, 2021, pp. 3–18.
413. B. Giese, *Electron hopping in biomolecules*, in: G. Roberts, A. Watts (Eds.), *Encyclopedia of Biophysics*, Springer Berlin Heidelberg, 2018, pp. 1–3.
414. Y. Xu, Z. Zhou, N. Deng, K. Fu, C. Zhu, Q. Hong, Y. Shen, S. Liu, Y. Zhang, *Sci. China Chem. 66* **2023**, 1318–1335.
415. E. Wierzbinski, R. Venkatramani, K.L. Davis, S. Bezer, J. Kong, Y. Xing, E. Borguet, C. Achim, D.N. Beratan, D.H. Waldeck, *ACS Nano* **2013**, *7* (6), 5391–5401.
416. R. Venkatramani, E. Wierzbinski, D.H. Waldeck, D.N. Beratan, *Faraday Discuss.* **2014**, *174* (0), 57–78.
417. D.A. Wink, R.W. Nims, J.E. Saavedra, W.E. Utermahlen, Jr., P.C. Ford, *Proc. Natl. Acad. Sci. USA* **1994**, *91* (14), 6604–6608.
418. C.E. Valdez, Q.A. Smith, M.R. Nechay, A.N. Alexandrova, *Acc. Chem. Res.* **2014**, *47* (10), 3110–3117.
419. B. Yuan, H.-L. Chou, Y.-K. Peng, *ACS Appl. Mater. Interfaces* **2022**, *14* (20), 22728–22736.

420. Y. Qiu, B. Yuan, H. Mi, J.-H. Lee, S.-W. Chou, Y.-K. Peng, *J. Phys. Chem. Lett.* **2022**, *13* (38), 8872–8878.
421. T.L. Poulos, J. Kraut, *J. Biol. Chem.* **1980**, *255* (17), 8199–8205.
422. G. Wächtershäuser, *Prog. Biophys. Mol. Biol.* **1992**, *58* (2), 85–201.
423. M.J. Russell, A.J. Hall, *J. Geol. Soc. London* **1997**, *154* (3), 377–402.
424. W. Nitschke, S.E. McGlynn, E.J. Milner-White, M.J. Russell, *Biochim. Biophys. Acta Bioenerg.* **2013**, *1827* (8–9), 871–881.
425. M.J. Russell, *Life* **2018**, *8* (3), 35.
426. Y. Li, N. Kitadai, R. Nakamura, *Life* **2018**, *8* (4), 46.
427. M.J. Russell, A. Ponce, *Life* **2020**, *10* (11), 291.
428. S. Duval, E. Branscomb, F. Trolard, G. Bourrié, O. Grauby, V. Heresanu, B. Schoepp-Cothenet, K. Zuchan, M.J. Russell, W. Nitschke, *Appl. Clay Sci.* **2020**, *195*, 105737.
429. S. Duval, K. Zuchan, F. Baymann, B. Schoepp-Cothenet, E. Branscomb, M.J. Russell, W. Nitschke, *Minerals and the emergence of life*, in: P. Kroneck, M. Sosa Torres (Eds.), *Metals, Microbes, and Minerals – The Biogeochemical Side of Life*, De Gruyter, Berlin, Boston, 2021, pp. 135–157.
430. M.J. Russell, *Front. Microbiol.* **2023**, *14*, 1145915.
431. C. Sanchez, F. Babonneau, R. Morineau, J. Livage, J. Bullot, *Philos. Mag. B* **1983**, *47* (3), 279–290.
432. C. Sanchez, R. Morineau, J. Livage, *Phys. Status Solidi.* **1983**, *76* (2), 661–666.
433. R. André, F. Natálio, M. Humanes, J. Leppin, K. Heinze, R. Wever, H.C. Schröder, W.E.G. Müller, W. Tremel, *Adv. Funct. Mater.* **2011**, *21* (3), 501–509.
434. A.A. Vernekar, D. Sinha, S. Srivastava, P.U. Paramasivam, P. D'Silva, G. Mugesh, *Nat. Commun.* **2014**, *5*, 5301.
435. Y. Ding, G. Ren, G. Wang, M. Lu, J. Liu, K. Li, Y. Lin, *Anal. Chem.* **2020**, *92* (6), 4583–4591.
436. Z. Chen, *Coord. Chem. Rev.* **2022**, *457*, 214404.
437. L. Li, S. Cao, Z. Wu, R. Guo, L. Xie, L. Wang, Y. Tang, Q. Li, X. Luo, L. Ma, C. Cheng, L. Qiu, *Adv. Mater.* **2022**, *34* (17), 2108646.
438. S.E. Dorris, T.O. Mason, *J. Am. Chem. Soc.* **1988**, *71* (5), 379–385.
439. N. Singh, M. Geethika, S.M. Eswarappa, G. Mugesh, *Chemistry* **2018**, *24* (33), 8393–8403.
440. J. Yao, Y. Cheng, M. Zhou, S. Zhao, S. Lin, X. Wang, J. Wu, S. Li, H. Wei, *Chem. Sci.* **2018**, *9* (11), 2927–2933.
441. S. Devaraj, N. Munichandraiah, *J. Phys. Chem. C* **2008**, *112* (11), 4406–4417.
442. M.U. Farooq, Z. Muhammad, S. Khalid, K. Fatima, B. Zou, *J. Mater. Sci. Mater. Electron.* **2019**, *30* (3), 2802–2808.
443. Y. Huang, Z. Liu, C. Liu, E. Ju, Y. Zhang, J. Ren, X. Qu, *Angew. Chem. Int. Ed.* **2016**, *55* (23), 6646–6650.
444. M. Tang, Z. Zhang, T. Sun, B. Li, Z. Wu, *Adv. Healthc. Mater.* **2022**, *11* (21), 2201733.
445. A. Ramírez, P. Hillebrand, D. Stellmach, M.M. May, P. Bogdanoff, S. Fiechter, *J. Phys. Chem. C* **2014**, *118* (26), 14073–14081.
446. H. Li, N. Yuan, J. Qian, B. Pan, *Environ. Sci. Technol.* **2022**, *56* (7), 4498–4506.
447. Z.J. Chen, Z. Huang, Y.M. Sun, Z.L. Xu, J. Liu, *Chemistry* **2021**, *27* (37), 9597–9604.
448. C.-S. Cheng, M. Serizawa, H. Sakata, T. Hirayama, *Mater. Chem. Phys.* **1998**, *53* (3), 225–230.
449. H.H. Pham, M.-J. Cheng, H. Frei, L.-W. Wang, *ACS Catal.* **2016**, *6* (8), 5610–5617.
450. J. Mu, Y. Wang, M. Zhao, L. Zhang, *Chem. Commun.* **2012**, *48* (19), 2540–2542.
451. Q. Wang, J. Chen, H. Zhang, W. Wu, Z. Zhang, S. Dong, *Nanoscale* **2018**, *10* (40), 19140–19146.
452. W. Li, J. Wang, J. Zhu, Y.-Q. Zheng, *J. Mater. Chem. B* **2018**, *6* (42), 6858–6864.

453. B. Radisavljevic, A. Radenovic, J. Brivio, V. Giacometti, A. Kis, *Nat. Nanotechnol.* **2011**, *6* (3), 147–150.
454. H. Qiu, T. Xu, Z. Wang, W. Ren, H. Nan, Z. Ni, Q. Chen, S. Yuan, F. Miao, F. Song, G. Long, Y. Shi, L. Sun, J. Wang, X. Wang, *Nat. Commun.* **2013**, *4* (1), 2642.
455. T. Chen, H. Zou, X. Wu, C. Liu, B. Situ, L. Zheng, G. Yang, *ACS Appl. Mater. Interfaces* **2018**, *10* (15), 12453–12462.
456. Y. Yu, L. Lu, Q. Yang, A. Zupanic, Q. Xu, L. Jiang, *ACS Appl. Nano Mat.* **2021**, *4* (8), 7523–7537.
457. T. Ito, H. Takagi, S. Ishibashi, T. Ido, S. Uchida, *Nature* **1991**, *350* (6319), 596–598.
458. F.-C. Hsu, J.-Y. Luo, K.-W. Yeh, T.-K. Chen, T.-W. Huang, P.M. Wu, Y.-C. Lee, Y.-L. Huang, Y.-Y. Chu, D.-C. Yan, M.-K. Wu, *Proc. Natl. Acad. Sci. USA* **2008**, *105* (38), 14262–14264.
459. N.S. Sidorov, A.V. Palnichenko, S.S. Khasanov, *Physica. C Supercond.* **2011**, *471* (7–8), 247–249.
460. A.K. Dutta, S.K. Maji, A. Mondal, B. Karmakar, P. Biswas, B. Adhikary, *Sens. Actuators B Chem.* **2012**, *173*, 724–731.
461. T. Liu, B. Xiao, F. Xiang, J. Tan, Z. Chen, X. Zhang, C. Wu, Z. Mao, G. Luo, X. Chen, J. Deng, *Nat. Commun.* **2020**, *11* (1), 2788.
462. Y. Zhu, Z. Zhang, X. Song, Y. Bu, *J. Mater. Chem. B* **2021**, *9* (16), 3533–3543.
463. M. Setvin, C. Franchini, X. Hao, M. Schmid, A. Janotti, M. Kaltak, C.G. Van De Walle, G. Kresse, U. Diebold, *Phys. Rev. Lett.* **2014**, *113* (8), 086402.
464. Z.F. Xu, C.J. Tong, R.T. Si, G. Teobaldi, L.M. Liu, *J. Phys. Chem. Lett.* **2022**, *13* (3), 857–863.
465. K. Liu, X. Ma, S. Xu, Y. Li, M. Zhao, *NPJ Comput. Mater.* **2023**, *9* (1), 16.
466. D. Suri, R.S. Patel, *Appl. Phys. Lett.* **2017**, *110*, 233108.
467. F. Qiao, L. Chen, X. Li, L. Li, S. Ai, *Sens. Actuators B Chem.* **2014**, *193*, 255–262.
468. X. Wu, T. Chen, J. Wang, G. Yang, *J. Mater. Chem. B* **2017**, *6* (1), 105–111.
469. L. Zhang, L. Han, P. Hu, L. Wang, S. Dong, *Chem. Commun.* **2013**, *49* (89), 10480–10482.
470. H.L. Tuller, A.S. Nowick, *J. Phys. Chem. Solids* **1977**, *38* (8), 859–867.
471. S. Kim, J. Maier, *J. Electrochem. Soc.* **2002**, *149* (10), J73.
472. F. Esch, S. Fabris, L. Zhou, T. Montini, C. Africh, P. Fornasiero, G. Comelli, R. Rosei, *Science* **2005**, *309* (5735), 752–755.
473. T. Montini, M. Melchionna, M. Monai, P. Fornasiero, *Chem. Rev.* **2016**, *116* (10), 5987–6041.
474. G. Xiao, H. Li, Y. Zhao, H. Wei, J. Li, H. Su, *ACS Appl. Nano Mat.* **2022**, *5* (10), 14147–14170.
475. Y. Ma, Z. Tian, W. Zhai, Y. Qu, *Nano Res.* **2022**, *15* (12), 10328–10342.
476. K. Korschelt, R. Schwidetzky, F. Pfitzner, J. Strugatchi, C. Schilling, M. von der Au, K. Kirchhoff, M. Panthöfer, I. Lieberwirth, M.N. Tahir, C. Hess, B. Meermann, W. *Nanoscale* **2018**, *10* (27), 13074–13082.
477. H. Liu, J. Liu, *Chem. Nano. Mat.* **2020**, *6* (6), 947–952.
478. T. Yao, Z. Tian, Y. Zhang, Y. Qu, *ACS Appl. Mater. Interfaces* **2019**, *11* (1), 195–201.
479. Z. Wang, R. Zhang, X. Yan, K. Fan, *Mater. Today* **2020**, *41*, 81–119.
480. P. Muhammad, K. Fan, *Adv. Funct. Mater.* **2024**, *34*, 2314686.
481. Z. Zhou, Y. Wang, F. Peng, F. Meng, J. Zha, L. Ma, Y. Du, N. Peng, L. Ma, Q. Zhang, L. Gu, W. Yin, Z. Gu, C. Tan, *Angew. Chem. Int. Ed.* **2022**, *61* (16), e202115939.
482. B. Xu, S. Li, H. Liu, *Photoresponsive nanozymes*, in: X. Wang (Ed.), *ACS Symposium Series*, American Chemical Society, 2022, pp. 163–187.
483. J. Zhang, J. Liu, *Nanoscale* **2020**, *12* (5), 2914–2923.
484. Y. Liu, X. Wang, H. Wei, *Analyst* **2020**, *145* (13), 4388–4397.

485. C. Wang, H. Wang, B. Xu, H. Liu, *VIEW* **2021**, *2* (1), 20200045.
486. J. Xu, R. Cai, Y. Zhang, X. Mu, *Colloids Surf. B Biointerfaces* **2021**, *200*, 111575.
487. Y. Lindqvist, E. Johansson, H. Kaija, P. Vihko, G. Schneider, *J. Mol. Biol.* **1999**, *291* (1), 135–147.
488. G. Schenk, N. Mitić, G.R. Hanson, P. Comba, *Coord. Chem. Rev.* **2013**, *257* (2), 473–482.
489. N. Mitić, S.J. Smith, A. Neves, L.W. Guddat, L.R. Gahan, G. Schenk, *Chem. Rev.* **2006**, *106* (8), 3338–3363.
490. C. Leblanc, H. Vilter, J.B. Fournier, L. Delage, P. Potin, E. Rebuffet, G. Michel, P.L. Solari, M.C. Feiters, M. Czjzek, *Coord. Chem. Rev.* **2015**, *301–302*, 134–146.
491. W. Plass, M. Bangesh, S. Nica, A. Buchholz, *Model studies of vanadium-dependent haloperoxidation: Structural and functional lessons synthetic and computational models of supramolecular interactions and the formation of peroxo species*, in: *Vanadium, the versatile metal*, K. Kustin, J.C. Pessoa, D.C. Crans (Eds.), ACS Symposium Series, 2007, pp. 163–177. https://doi.org/10.1021/bk-2007-0974.
492. W. Hemrika, R. Renirie, H.L. Dekker, P. Barnett, R. Wever, *Proc. Natl. Acad. Sci. USA* **1997**, *94* (6), 2145–2149.
493. R.R. Langeslay, D.M. Kaphan, C.L. Marshall, P.C. Stair, A.P. Sattelberger, M. *Chem. Rev.* **2019**, *119* (4), 2128–2191.
494. D. Xu, L. Wu, H. Yao, L. Zhao, *Small* **2022**, *18* (37), 2203400.
495. A.L. Sessions, D.M. Doughty, P.V. Welander, R.E. Summons, D.K. Newman, *Curr. Biol.* **2009**, *19* (14), R567–R574.
496. J. Farquhar, A.L. Zerkle, A. Bekker, *Photosyn. Res.* **2011**, *107* (1), 11–36.
497. F.E. Jenney Jr, M.F.J.M. Verhagen, X. Cui, M.W.W. Adams, *Science* **1999**, *286* (5438), 306–309.
498. F. Wolfe-Simon, D. Grzebyk, O. Schofield, P.G. Falkowski, *J. Phyco.* **2005**, *41* (3), 453–465.
499. M. Harada, A. Akiyama, R. Furukawa, S.I. Yokobori, E. Tajika, A. Yamagishi, *J. Mol. Evol.* **2021**, *89* (8), 527–543.
500. F.C. Tompkins, *Nature* **1960**, *186* (4718), 3–6.
501. C.J. Kevane, *Phys. Rev.* **1964**, *133* (5A), A1431–A1436.
502. C.M. Wang, D.R. Baer, L.E. Thomas, J.E. Amonette, J. Antony, Y. Qiang, G. Duscher, *J. of Appl. Phys.* **2005**, *98* (9), 094308.
503. M.V. Ganduglia-Pirovano, A. Hofmann, J. Sauer, *Surf. Sci. Rep.* **2007**, *62* (6), 219–270.
504. F. Gunkel, D.V. Christensen, Y.Z. Chen, N. Pryds, *Appl. Phys. Lett.* **2020**, *116* (12), 120505.
505. A. Cabot, V.F. Puntes, E. Shevchenko, Y. Yin, L. Balcells, M.A. Marcus, S.M. Hughes, A.P. *J. Am. Chem. Soc.* **2007**, *129* (34), 10358–10360.
506. S. Mostajabi Sarhangi, D.V. Matyushov, *ACS Omega* **2023**, *8* (30), 27355–27365.
507. B. Giese, M. Graber, M. Cordes, *Curr. Opin. Chem. Biol.* **2008**, *12* (6), 755–759.
508. C. Shipps, H.R. Kelly, P.J. Dahl, S.M. Yi, D. Vu, D. Boyer, C. Glynn, M.R. Sawaya, D. Eisenberg, V.S. Batista, N.S. Malvankar, *Proc. Natl. Acad. Sci. USA* **2021**, *118* (2), e2014139118.
509. S.L. Miller, *Science* **1953**, *117* (3046), 528–529.
510. J. OrÓ, *Nature* **1961**, *191* (4794), 1193–1194.
511. M. Ferus, F. Pietrucci, A.M. Saitta, A. Knížek, P. Kubelík, O. Ivanek, V. Shestivska, S. Civiš, *Proc. Natl. Acad. Sci. USA* **2017**, *114* (17), 4306–4311.
512. M. Frenkel-Pinter, M. Bouza, F.M. Fernández, L.J. Leman, L.D. Williams, N.V. Hud, A. Guzman-Martinez, *Nat. Commun.* **2022**, *13* (1), 2569.
513. J. Halpern, L.E. Orgel, *Discuss. Faraday Soc.* **1960**, *29*, 32–41.
514. J.J. Hopfield, *Proc. Natl. Acad. Sci. USA* **1974**, *71* (9), 3640–3644.

515. J.J. Warren, M.E. Ener, A. Vlček, Jr., J.R. Winkler, H.B. Gray, *Coord. Chem. Rev.* **2012**, *256* (21–22), 2478–2487.
516. H.B. Gray, J.R. Winkler, *Proc. Natl. Acad. Sci. USA* **2015**, *112* (35), 10920–10925.
517. M. Uzun, *Horseradish Peroxidase: Structure, Functions and Applications*, Hauppauge, Nova Science Publishers, 2019.
518. G.I. Berglund, G.H. Carlsson, A.T. Smith, H. Szöke, A. Henriksen, J. Hajdu, *Nature* **2002**, *417* (6887), 463–468.
519. P.C. Loewen, X. Carpena, C. Rovira, A. Ivancich, R. Perez-Luque, R. Haas, S. Odenbreit, P. Nicholls, I. Fita, *Biochemistry* **2004**, *43* (11), 3089–3103.
520. M.S. Lah, M.M. Dixon, K.A. Pattridge, W.C. Stallings, J.A. Fee, M.L. Ludwig, *Biochemistry* **1995**, *34* (5), 1646–1660.
521. S. Shaik, D. Kumar, S.P. de Visser, A. Altun, W. Thiel, *Chem. Rev.* **2005**, *105* (6), 2279–2328.
522. J.T. Groves, *F1000Research* **2015**, *178* (4), F1000 Faculty Rev-178.
523. K. Fan, H. Wang, J. Xi, Q. Liu, X. Meng, D. Duan, L. Gao, X. Yan, *Chem. Commun.* **2017**, *53* (2), 424–427.
524. X. Niu, X. Xu, X. Li, J. Pan, F. Qiu, H. Zhao, M. Lan, *Chem. Commun.* **2018**, *54* (95), 13443–13446.
525. W. Wu, Q. Wang, J. Chen, L. Huang, H. Zhang, K. Rong, S. Dong, *Nanoscale* **2019**, *11* (26), 12603–12609.
526. N.V.S. Vallabani, A. Vinu, S. Singh, A. Karakoti, *J. Colloid Interface Sci.* **2020**, *567*, 154–164.
527. J. Han, Q. Zou, W. Su, X. Yan, *Chem. Eng. J.* **2020**, *394*, 124987.
528. X. Niu, X. Li, Z. Lyu, J. Pan, S. Ding, X. Ruan, W. Zhu, D. Du, Y. Lin, *Chem. Commun.* **2020**, *56* (77), 11338–11353.
529. J. Chen, Q. Ma, M. Li, W. Wu, L. Huang, L. Liu, Y. Fang, S. Dong, *Nanoscale* **2020**, *12* (46), 23578–23585.
530. J. Han, H. Gong, X. Ren, X. Yan, *Nano Today* **2021**, *41*, 101295.
531. R. Geng, R. Chang, Q. Zou, G. Shen, T. Jiao, X. Yan, *Small* **2021**, *17* (19), **2008**114.
532. D. Sun, T. Liu, Y. Dong, Q. Liu, X. Wu, G.-L. Wang, *Sens. Actuators B Chem.* **2021**, *327*, 128859.
533. W. Xu, W. He, Z. Du, L. Zhu, K. Huang, Y. Lu, Y. Luo, *Angew. Chem. Int. Ed.* **2021**, *60* (13), 6890–6918.
534. S. Wang, X. Wang, B. Du, Y. Jin, W. Ai, G. Zhang, T. Zhou, F. Wang, Z. Zhang, *Spectrochim. Acta A Mol. Biomol. Spectrosc.* **2023**, *287* (Pt 2), 122084.
535. H. Liang, F. Lin, Z. Zhang, B. Liu, S. Jiang, Q. Yuan, J. Liu, *ACS Appl. Mater. Interfaces* **2017**, *9* (2), 1352–1360.
536. C. Castillo-Caceres, E. Duran-Meza, E. Nova, R. Araya-Secchi, O. Monasterio, R. Diaz-Espinoza, *Biochim. Biophys. Acta Gen. Subj.* **2021**, *1865* (1), 129729.
537. P. Makam, S.S.R.K.C. Yamijala, V.S. Bhadram, L.J.W. Shimon, B.M. Wong, E. Gazit, *Nat. Commun.* **2022**, *13* (1), 1505.
538. S. Li, Z. Zhou, Z. Tie, B. Wang, M. Ye, L. Du, R. Cui, W. Liu, C. Wan, Q. Liu, S. Zhao, Q. Wang, Y. Zhang, S. Zhang, H. Zhang, Y. Du, H. Wei, *Nat. Commun.* **2022**, *13* (1), 827.
539. D. Wang, D. Jana, Y. Zhao, *Acc. Chem. Res.* **2020**, *53* (7), 1389–1400.
540. G. Li, W. Ma, Y. Yang, C. Zhong, H. Huang, D. Ouyang, Y. He, W. Tian, J. Lin, Z. Lin, *ACS Appl. Mater. Interfaces* **2021**, *13* (41), 49482–49489.
541. Y. Zhang, F. Wang, C. Liu, Z. Wang, L. Kang, Y. Huang, K. Dong, J. Ren, X. Qu, *ACS Nano* **2018**, *12* (1), 651–661.
542. X. Han, Y. Li, Y. Zhou, Z. Song, Y. Deng, J. Qin, Z. Jiang, *Mater. Des.* **2021**, *204*, 109646.

543. L. Huang, D.W. Sun, H. Pu, *Small* **2022**, *18* (20), 2200178.
544. I.A. de Castro, R.S. Datta, J.Z. Ou, A. Castellanos-Gomez, S. Sriram, T. Daeneke, K. Kalantar-Zadeh, *Adv. Mater.* **2017**, *29* (40), 1701619.
545. R. Ragg, F. Natalio, M.N. Tahir, H. Janssen, A. Kashyap, D. Strand, S. Strand, W. Tremel, *ACS Nano* **2014**, *8* (5), 5182–5189.
546. D.A. Dixon, W.N. Lipscomb, *J. Biol. Chem.* 1976, *251* (19), 5992–6000.
547. D.S. Dwyer, *Proteins* **2006**, *63* (4), 939–948.
548. L. Berstis, G.T. Beckham, M.F. Crowley, *J. Chem. Phys.* **2015**, *143* (22), 225102.
549. L. Sepunaru, S. Refaely-Abramson, R. Lovrinčić, Y. Gavrilov, P. Agrawal, Y. Levy, L. Kronik, I. Pecht, M. Sheves, D. Cahen, *J. Am. Chem. Soc.* **2015**, *137* (30), 9617–9626.
550. E.I. Solomon, D.W. Randall, T. Glaser, *Coord. Chem. Rev.* **2000**, *200–202*, 595–632.
551. D.N. Beratan, J.N. Betts, J.N. Onuchic, *Science* **1991**, *252* (5010), 1285–1288.
552. Y. Bromberg, A.A. Aptekmann, Y. Mahlich, L. Cook, S. Senn, M. Miller, V. Nanda, D.U. Ferreiro, P.G. Falkowski, *Sci. Adv.* **2022**, *8* (2), eabj3984.
553. N.V. Grishin, *J. Struct. Biol.* **2001**, *134* (2–3), 167–185.
554. H. Raanan, S. Poudel, D.H. Pike, V. Nanda, P.G. Falkowski, *Proc. Natl. Acad. Sci. USA* **2020**, *117* (13), 7193–7199.
555. R.H. Holm, P. Kennepohl, E.I. Solomon, *Chem. Rev.* **1996**, *96* (7), 2239–2314.
556. C.L. Drennan, J.W. Peters, *Curr. Opin. Struct. Biol.* **2003**, *13* (2), 220–226.
557. H. Beinert, R.H. Holm, E. Münck, *Science* **1997**, *277* (5326), 653–659.
558. K. Brzóska, S. Meczyńska, M. Kruszewski, *Acta Biochim. Pol.* **2006**, *53* (4), 685–691.
559. R. Lill, *Nature* **2009**, *460* (7257), 831–838.
560. C. Andreini, S. Ciofi-Baffoni, *Basic iron-sulfur centers*, in: M. Sosa Torres, P. Kroneck (Eds.), *Transition Metals and Sulfur – A Strong Relationship for Life*, De Gruyter, Berlin, Boston, 2020, pp. 199–256.
561. J.C. Crack, N.E. Le Brun, *Coord. Chem. Rev.* **2021**, *448*, 214171.
562. J.J. Braymer, S.A. Freibert, M. Rakwalska-Bange, R. Lill, *Biochim. Biophys. Acta Mol. Cell Res.* **2021**, *1868* (1), 118863.
563. W. Shen, R.D. Teo, D.N. Beratan, J.J. Warren, *J. Phys. Chem. B* **2022**, *126* (19), 3522–3529.
564. K.J. Waldron, J.C. Rutherford, D. Ford, N.J. Robinson, *Nature* **2009**, *460* (7257), 823–830.
565. R.P. Hong Enriquez, T.N. Do, *Life* **2012**, *2* (4), 274–285.
566. T. Dudev, C. Lim, *Chem. Rev.* **2014**, *114* (1), 538–556.
567. E.K. Moore, B.I. Jelen, D. Giovannelli, H. Raanan, P.G. Falkowski, *Nat. Geosci.* **2017**, *10* (9), 629–636.
568. H. Eom, W.J. Song, *J. Biol. Inorg. Chem.* **2019**, *24* (4), 517–531.
569. J. Wade, D.J. Byrne, C.J. Ballentine, H. Drakesmith, *Proc. Natl. Acad. Sci. USA* **2021**, *118* (51), e2109865118.
570. S.K. Aulakh, S.J. Varma, M. Ralser, *Curr. Opin. Genet. Dev.* **2022**, *77*, 101987.
571. R.V. Eck, M.O. Dayhoff, *Science* **1966**, *152* (3720), 363–366.
572. S. Scintilla, C. Bonfio, L. Belmonte, M. Forlin, D. Rossetto, J. Li, J.A. Cowan, A. Galliani, F. Arnesano, M. Assfalg, S.S. Mansy, *Chem. Commun.* **2016**, *52* (92), 13456–13459.
573. D.O. Hall, R. Cammack, K.K. Rao, *Nature* **1971**, *233* (5315), 136–138.
574. H. Beinert, *J. Biol. Inorg. Chem.* **2000**, *5* (1), 2–15.
575. M.C. Weiss, F.L. Sousa, N. Mrnjavac, S. Neukirchen, M. Roettger, S. Nelson-Sathi, W.F. Martin, *Nat. Microbiol.* **2016**, *1* (9), 16116.
576. J.A. Baross, S.E. Hoffman, *Orig. Life Evol. Biosph.* **1985**, *15* (4), 327–345.
577. D. Zhao, S. Bartlett, Y.L. Yung, *Life* **2020**, *10* (12), 1–9.

578. K.N. McGuinness, G.W. Klau, S.M. Morrison, E.K. Moore, J. Seipp, P.G. Falkowski, V. Nanda, *Orig. Life Evol. Biosph.* **2022**, *52* (4), 263–275.
579. G. Caserta, L. Zuccarello, C. Barbosa, C.M. Silveira, E. Moe, S. Katz, P. Hildebrandt, I. Zebger, S. Todorovic, *Coord. Chem. Rev.* **2022**, *452* (1), 214287.
580. H.M. Holden, B.L. Jacobson, J.K. Hurley, G. Tollin, B.H. Oh, L. Skjeldal, Y.K. Chae, H. Cheng, B. Xia, J.L. Markley, *J. Bioenerg. Biomembr.* **1994**, *26* (1), 67–88.
581. R.J. Gurbiel, P.E. Doan, G.T. Gassner, T.J. Macke, D.A. Case, T. Ohnishi, J.A. Fee, D.P. Ballou, B.M. Hoffman, *Biochemistry* **1996**, *35* (24), 7834–7845.
582. J.A. Imlay, *Mol. Microbiol.* **2006**, *59* (4), 1073–1082.
583. J. Dupuy, A. Volbeda, P. Carpentier, C. Darnault, J.M. Moulis, J.C. Fontecilla-Camps, *Structure* **2006**, *14* (1), 129–139.
584. D.R. Breiter, T.E. Meyer, I. Rayment, H.M. Holden, *J. Biol. Chem.* **1991**, *266* (28), 18660–18667.
585. E.T. Adman, L.C. Sieker, L.H. Jensen, *J. Mol. Biol.* **1991**, *217* (2), 337–352.
586. F. Liu, J. Geng, R.H. Gumpper, A. Barman, I. Davis, A. Ozarowski, D. Hamelberg, A. Liu, *J. Biol. Chem.* **2015**, *290* (25), 15621–15634.
587. M. Bruschi, F. Guerlesquin, *FEMS Microbiol. Lett.* **1988**, *54* (2), 155–175.
588. J.P. Stokkermans, A.J. Pierik, R.B. Wolbert, W.R. Hagen, W.M. Van Dongen, C. Veeger, *Eur. J. Biochem.* **1992**, *208* (2), 435–442.
589. M.R. Reyda, C.J. Fugate, J.T. Jarrett, *Biochemistry* **2009**, *48* (45), 10782–10792.
590. L. Castro, V. Tórtora, S. Mansilla, R. Radi, *Acc. Chem. Res.* **2019**, *52* (9), 2609–2619.
591. I. Span, K. Wang, W. Wang, Y. Zhang, A. Bacher, W. Eisenreich, K. Li, C. Schulz, E. Oldfield, M. Groll, *Nat. Commun.* **2012**, *3* (1), 1042.
592. H. Seino, M. Hidai, *Chem. Sci.* **2011**, *2* (5), 847–857.
593. A.C. Brown, N.B. Thompson, D.L.M. Suess, *J. Am. Chem. Soc.* **2022**, *144* (20), 9066–9073.
594. C.C. Lee, Y. Hu, M.W. Ribbe, *Science* **2010**, *329* (5992), 642–642.
595. J.G. Rebelein, Y. Hu, M.W. Ribbe, *ChemBioChem* **2015**, *16* (14), 1993–1996.
596. V. Waser, M. Mukherjee, R. Tachibana, N.V. Igareta, T.R. Ward, *J. Am. Chem. Soc.* **2023**, *145* (27), 14823–14830.
597. C. De Duve, *Blueprint for a Cell: The Nature and Origin of Life*, Neil Patterson Publishers, Burlington, North Carolina, 1991.
598. I.A. Berg, D. Kockelkorn, W.H. Ramos-Vera, R.F. Say, J. Zarzycki, M. Hügler, B.E. Alber, G. Fuchs, *Nat. Rev. Microbiol.* **2010**, *8* (6), 447–460.
599. W. Wang, B. Yang, Y. Qu, X. Liu, W. Su, *Astrobiology* **2011**, *11* (5), 471–476.
600. A.L. De Aldecoa, F.V. Roldán, C. Menor-Salván, *Life* **2013**, 502–517.
601. C.E. Mitchell, U. Terranova, A.M. Beale, W. Jones, D.J. Morgan, M. Sankar, N.H. de Leeuw, *Catal. Sci. Technol.* **2021**, *11* (3), 779–784.
602. G. Wächtershäuser, *Microbiol. Rev.* **1988**, *52* (4), 452–484.
603. G. Wächtershäuser, *Proc. Natl. Acad. Sci. USA* **1990**, *87* (1), 200–204.
604. T.A. Doane, *ACS Earth Space Chem.* **2017**, *1* (7), 411–421.
605. R.L. Mancinelli, C.P. McKay, *Orig. Evol. Biosphere* **1988**, *18* (4), 311–325.
606. C.F. Garibello, A.N. Simonov, S.L.Y. Chang, B. Johannessen, F. Malherbe, D.S. Eldridge, R.K. Hocking, *Inorg. Chem.* **2023**, *62* (24), 9379–9390.
607. Y. Li, A. Yamaguchi, M. Yamamoto, K. Takai, R. Nakamura, *J. Phys. Chem. C* **2017**, *121* (4), 2154–2164.
608. C.F. Garibello, D.S. Eldridge, F. Malherbe, R.K. Hocking, *Inorg. Chem. Front.* **2023**, *10* (23), 6792–6811.
609. C.F. Garibello, A.N. Simonov, D.S. Eldridge, F. Malherbe, R.K. Hocking, *ChemCatChem.* **2022**, *14* (12), e202200270.

610. J.A. Brandes, N.Z. Boctor, G.D. Cody, B.A. Cooper, R.M. Hazen, H.S. Yoder, *Nature* **1998**, *395* (6700), 365–367.
611. H.C.B. Hansen, C.B. Koch, H. Nancke-Krogh, O.K. Borggaard, J. Sørensen, *Environ. Sci. Technol.* **1996**, *30* (6), 2053–2056.
612. H.C.B. Hansen, S. Guldberg, M. Erbs, C. Bender Koch, *Appl. Clay Sci.* **2001**, *18* (1–2), 81–91.
613. J. Choi, B. Batchelor, *Chemosphere* **2008**, *70* (6), 1108–1116.
614. D. Guerbois, G. Ona-Nguema, G. Morin, M. Abdelmoula, A.M. Laverman, J.M. Mouchel, K. Barthelemy, F. Maillot, J. Brest, *Environ. Sci. Technol.* **2014**, *48* (8), 4505–4514.
615. H.C.B. Hansen, C.B. Koch, H. Nancke-Krogh, O.K. Borggaard, J. Sørensen, *Environ. Sci. Technol.* **1996**, *30* (6), 2053–2056.
616. D. Guerbois, G. Ona-Nguema, G. Morin, M. Abdelmoula, A.M. Laverman, J.M. Mouchel, K. Barthelemy, F. Maillot, J. Brest, *Environ. Sci. Technol.* **2014**, *48* (8), 4505–4514.
617. F. Trolard, J.M.R. Génin, M. Abdelmoula, G. Bourrié, B. Humbert, A. Herbillon, *Geochim. Cosmochim. Acta* **1997**, *61* (5), 1107–1111.
618. J.D. Lipscomb, *Annu. Rev. Microbiol.* **1994**, *48*, 371–399.
619. R. Banerjee, J.C. Jones, J.D. Lipscomb, *Annu. Rev. Biochem.* **2019**, *88*, 409–431.
620. S. Buessecker, H. Imanaka, T. Ely, R. Hu, S.J. Romaniello, H. Cadillo-Quiroz, *Nat. Geosci.* **2022**, *15* (12), 1056–1063.
621. O. Farr, N. Gaudu, G. Danger, M.J. Russell, D. Ferry, W. Nitschke, S. Duval, *J. R. Soc. Interface* **2023**, *20* (206), 20230386.
622. J. Choi, B. Batchelor, C. Won, J. Chung, *Chemosphere* **2012**, *86* (8), 860–865.
623. G. Ona-Nguema, D. Guerbois, C. Pallud, J. Brest, M. Abdelmoula, G. Morin, *Minerals* **2020**, *10* (9), 818.
624. X. Wang, N.S. Wells, W. Xiao, J.L. Hamilton, A.M. Jones, R.N. Collins, *Geochim. Cosmochim. Acta* **2023**, *347*, 28–41.
625. F. Picardal, *Front. Microbiol.* **2012**, *3*, 112.
626. X. Zhu-Barker, A.R. Cavazos, N.E. Ostrom, W.R. Horwath, J.B. Glass, *Biogeochemistry* **2015**, *126* (3), 251–267.
627. M. Yu, Y. Hu, D. Liang, G. Wang, X. Zhu, J. Xie, *Water (Switzerland)* **2023**, *15* (12), 2158.
628. Y. Bai, H. Hu, P.H. Lee, A. Zhussupbekova, I.V. Shvets, B. Du, A. Terada, X. Zhan, *Sci. Total Environ.* **2023**, *901*, 165908.
629. Y. Bai, S. Wang, A. Zhussupbekova, I.V. Shvets, P.H. Lee, X. Zhan, *Water Res.* **2023**, *231*, 119619.
630. F. Feng, Z. Liu, X. Tang, X. Wu, C. Qu, S.W. How, D. Wu, R. Xiao, C.J. Tang, Z. Lin, L. Chai, G.H. Chen, *Water Res.* **2023**, *229*, 119393.
631. Y. Hu, G. Wu, R. Li, L. Xiao, X. Zhan, *Water Res.* **2020**, *179*, 115914.
632. R. Li, Y. Zhang, M. Guan, *Water Res.* **2022**, *221*, 118763.
633. B. Zhu, R. Yuan, S. Wang, H. Chen, B. Zhou, Z. Cui, C. Zhang, *J. Water Process Eng.* **2024**, *59*, 104952.
634. J.D. Bernal, *The Physical Basis of Life*, Routledge and Kegan Paul, London, 1951.
635. M.F. Hochella, D. Aruguete, B. Kim, A.S. Madden, *Naturally occurring inorganic nanoparticles: General assessment and a global budget for one of earth's last unexplored major geochemical components*, in: H. Guo, A. Barnard (Eds.), *Nature's Nanostructures*, Pan Stanford Publishing, Victoria/Australia, 2012, pp. 1–42.
636. V.K. Sharma, J. Filip, R. Zboril, R.S. Varma, *Chem. Soc. Rev.* **2015**, *44* (23), 8410–8423.
637. A. Navrotsky, L. Mazeina, J. Majzlan, *Science* **2008**, *319* (5870), 1635–1638.

638. H. Konishi, H. Xu, H. Guo, *Nanostructures of natural iron oxide nanoparticles*, in: H. Guo, A. Barnard (Eds.), *Nature's Nanostructures*, Pan Stanford Publishing, Victoria/Australia, 2012, pp. 75–113.
639. H. Guo, A.S. Barnard, *J. Mater. Chem. A* **2013**, *1* (1), 27–42.
640. T. Hiemstra, *Geochim. Cosmochim. Acta* **2015**, *158*, 179–198.
641. C. Claudio, E. Di Iorio, Q. Liu, Z. Jiang, V. Barrón, *J. Nanosci. Nanotechnol.* **2017**, *17* (7), 4449–4460.
642. X.L. Huang, *Introduction chapter, incredible spicy iron oxide nanoparticles*, in: X.-L. Huang (Ed.), *Iron Oxide Nanoparticles*, intechopen.com, London, 2022, pp. 1–20.
643. N.R. Posth, D.E. Canfield, A. Kappler, *Earth-Sci. Rev.* **2014**, *135*, 103–121.
644. B.A. Maher, *Quat. Sci. Rev.* **2016**, *154*, 23–84.
645. M. Pósfai, C. Lefèvre, D. Trubitsyn, D. Bazylinski, R. Frankel, *Front. Microbiol.* **2013**, *4*, 344.
646. J.R. Hawkings, J.L. Wadham, M. Tranter, R. Raiswell, L.G. Benning, P.J. Statham, A. Tedstone, P. Nienow, K. Lee, J. Telling, *Nat. Commun.* **2014**, *5* (1), 3929.
647. T. Gonet, B.A. *Maher*, Airborne, *Environ. Sci. Technol.* **2019**, *53* (17), 9970–9991.
648. Y. Yang, M. Vance, F. Tou, A. Tiwari, M. Liu, M.F. Hochella, *Environ. Sci. Nano* **2016**, *3* (3), 534–544.
649. A.J.M. Lough, D.P. Connelly, W.B. Homoky, J.A. Hawkes, V. Chavagnac, A. Castillo, M. Kazemian, K.-i. Nakamura, T. Araki, B. Kaulich, R.A. Mills, *Front. Mar. Sci.* **2019**, *6*, 329. https://doi.org/10.3389/fmars.2019.00329.
650. M. Yücel, S. Sevgen, N. Le Bris, *Front. Microbiol.* **2021**, *12*, 631885.
651. Z. Tao, Q. Zhou, T. Zheng, F. Mo, S. Ouyang, *J. Hazar. Mater.* **2023**, *459*, 132107.
652. W. Xie, Z. Guo, F. Gao, Q. Gao, D. Wang, B.S. Liaw, Q. Cai, X. Sun, X. Wang, L. Zhao, *Theranostics* **2018**, *8* (12), 3284–3307.
653. M. Pósfai, P.R. Buseck, D.A. Bazylinski, R.B. Frankel, *Am. Mineral.* **1998**, *83* (11–12 PART 2), 1469–1481.
654. T. Subramani, K. Lilova, M. Abramchuk, K.D. Leinenweber, A. Navrotsky, *Proc. Natl. Acad. Sci. USA* **2020**, *117* (46), 28645–28648.
655. X. Gu, P.J. Heaney, F.D.A.A. Reis, S.L. Brantley, *Science* **2020**, *370* (6515), eabb8092.
656. N. Karimian, S.G. Johnston, E.D. Burton, *Chemosphere* **2018**, *197*, 803–816.
657. A.J. Findlay, E.R. Estes, A. Gartman, M. Yücel, A. Kamyshny, G.W. Luther, *Nat. Commun.* **2019**, *10* (1), 1597.

6 Application of Electron Paramagnetic Resonance Spectroscopy for Studies of Prebiotic Metals and Complexes

Seneca J. Velling
NASA Jet Propulsion Laboratory,
California Institute of Technology
Division of Engineering & Applied Science,
California Institute of Technology
Kavli Nanoscience Institute,
California Institute of Technology

Laura E. Rodriguez
Lunar and Planetary Institute, Universities
Space Research Association

Paul H. Oyala
Division of Chemistry & Chemical Engineering,
California Institute of Technology

*John-Paul Jones, Laura M. Barge, and Jessica M. Weber**
NASA Jet Propulsion Laboratory,
California Institute of Technology
jessica.weber@jpl.nasa.gov

* Corresponding author

DOI: 10.1201/9781003459910-6

CONTENTS

ABSTRACT

Metals and metal-organic mixtures are hypothesized to have played a crucial role in life's emergence, particularly in metal-rich geological settings such as hydrothermal vents. Electron paramagnetic resonance (EPR) is an important tool in chemical analysis of materials or complexes with unpaired electrons. Owing to its sensitivity to spin-orbit coupling, electron-electron magnetic interaction systems with multiple unpaired electrons, and hyperfine interactions with nuclear magnetic moments, this spectroscopic technique provides detailed information regarding the spatial arrangement of coordinating ligand fields and the hydration environments of paramagnetic species. Herein, we introduce EPR methods relevant to prebiotic chemistry studies, particularly to examine organo-metallic interactions relevant to prebiotic and early biological systems. We report on the analysis of metal-organic mixtures using primarily continuous wave EPR spectroscopy with some supplementary pulsed EPR approaches, identifying what impacts organics have on mineral analysis under early Earth relevant conditions, and examine specific ligand coordination (amino acids and quinones, among others) of three abundant paramagnetic ions of relevance to prebiotic geochemical origins of life (Fe^{2+}, Fe^{3+}, Mn^{2+}). We conclude with a detailed discussion of the impacts and implications of this work on the use of EPR with laboratory-scale origins of life studies, while suggesting future directions of research within the field.

KEYWORDS

electron paramagnetic resonance, organo-metallic, coordination complexes, prebiotic chemistry

1 INTRODUCTION

Electron paramagnetic resonance (EPR) is a technique used in a variety of fields including chemistry, materials, and geology to characterize species with unpaired electrons (e.g., metal/mineral complexes and radicals). The use of EPR in mineralogy and geochemistry has to date been largely focused on the solid-state and dating [1,2]. In the astrobiology context, the use of EPR has primarily been confined to the identification and characterization of insoluble organic matter (IOM) in an inorganic (mineral) matrix such as cherts and carbonaceous chondrites [3–6] or in the study of local coordination environments and use of lineshape changes in native metal oxides to empirically distinguish between biosignatures and abiotic background [7–11]. However, the use of EPR in biochemistry and the life sciences [12–14], organo-metallic chemistry [15,16], and chemical engineering [17] quite commonly employs aqueous species and their complexes—from small molecules to biomacromolecules. Similar methods may be applied to origin of life experiments and even in analysis of space mission returned samples which might have residual organics.

1.1 Metals and EPR in Astrobiology

Metal and metal-organic complexes (either in the form of solubilized ions or minerals) are hypothesized to have played a central role in the emergence of life on early Earth [18–23]. Therefore, studying such species can inform our understanding of prebiotic chemistries that may have led to life on Earth as well as possibly elsewhere in our Solar System. Specifically, metal-rich geological settings, such as hydrothermal vents [24–29] and terrestrial hydrothermal ponds/hot springs [30–32], may have been conducive to the emergence of prebiotic chemistry. Hydrothermal environments believed to be present during the Hadean period [33–36], when the Earth's crust was rich in ultramafic and mafic minerals [37–40], could have served as sources of electron donors, e.g., H_2 [41,42]. Along with other, available reactants such as oceanic metal salts and acids (e.g., CO_2/CO_3^{2-}), vent environments may have been capable of driving abiotic/prebiotic organic reactions [43] via inherent electrochemical redox, and pH, and thermal gradients [44–48]. Lab-scale simulations of seafloor hydrothermal minerals and metals have successfully demonstrated reductive amination via mineral reactions [49,50], Fischer-Tropsch type (FT) synthesis of alkanes and alkenes involving metal catalysts [51], Formose reactions catalyzed by divalent metal salts [52,53], and even polymerization of peptides through surface catalysis via iron-bearing minerals [54]. These examples highlight the key role of metal salts, complexes, and minerals in putative prebiotic pathways.

Here we used EPR spectroscopy to probe three paramagnetic metals—Mn^{2+} (d^5), Fe^{2+} (d^4), and Fe^{3+} (d^5)—in the presence of a variety of prebiotically relevant organic species, namely, *p*-benzoquinone, nicotinamide adenine dinucleotide (NAD^+), cysteine, glycine, as well as inorganic phosphorous species (phosphate/phosphite). In addition, all species were chosen for their capability to undergo redox reactions.

Iron, being one of the most abundant elements on Earth, forms species that are relevant to a variety of proposed prebiotic pathways. Ferrous iron (Fe^{2+}) was enriched in the Hadean ocean [55–61] and is of direct relevance to Fenton reactions as both an oxidant and reductant [62,63]; mixed valence iron oxides can drive the reduction of nitrite and nitrate to ammonia under mild alkaline conditions [64]; as an alternative (though less effective) catalyst to native iron (Fe^0) in the Haber-Bosch process [65]; and iron can be involved in proto-enzymatic activity in FeS metal clusters [66–70]. Ferric iron (Fe^{3+}) is the product of Fenton hydroxide generation and pertinent to the formation of phosphite radicals [71]; it is also a known trivalent catalyst of the Formose reaction [72–74], and as an oxide used to catalyze FT catalyst [75–80]. Iron would have predominantly existed as ferrous iron in Earth's early anoxic ocean [55,60,61], with minor amounts of ferric iron generated via photolysis or reaction with geological oxidants [81–83]. Ferrous iron minerals were also common throughout Hadean geology in the form of olivine (fayalite), Fe^{3+} substituted pyroxenes, and hydrous phyllosilicates [84–86].

Manganese is of particular interest as the kinetically frustrated oxidation of Mn(II) to Mn(IV) can help inform past redox conditions, for instance, the strongly oxidizing conditions within a shallow lake that existed at Gale Crater, Mars 3.6 Ga [87,88]. On Earth, Mn(IV) oxides can be diagnostic of biological activity as few abiotic processes are capable of readily facilitating this oxidation quickly, while bacterial and fungal redox efficiently undergo this oxidation [7,8]. Further, Mn^{2+} salts are useful alternatives in Fenton-like reactions in the ultimate production of phosphates [89]. Manganese is also abundant in environments that may have been conducive to the origin of life, namely, hydrothermal fluids in terrestrial hot springs [90], and is present in submarine hydrothermal vents [91–93].

Regarding our choice of organics for this study, glycine and cysteine are amino acids utilized by life. Glycine is the simplest amino acid and is likely the most prebiotically abundant in abiotic systems given its prevalence in meteorites and prebiotic experiments [49,94–98]. Cysteine, although not yet detected in meteorites, may also be prebiotically relevant [99,100]. Notably, cysteine—with its highly reactive nucleophilic end group, metal-ligating sulfur moiety, and redox activity—is of particular interest in organometallic studies, especially in metalloenzymes. Cysteine is thought to be challenging to synthesize abiotically, and therefore often regarded as an evolutionary product; however, Foden et al. have demonstrated cysteine synthesis using a different prospective biomimetic pathway [100,101].

With regard to metal cation coordination, glycine is predicted to form a mixture of complexes with ferric iron, ferrous iron, and manganese by binding to the metal via the amine nitrogen and carboxylate oxygen [102,103]. Meanwhile, cysteine is known to form complexes with ferrous and ferric iron in weakly acid media, at which pH the amino acid is also zwitterionic [104]. Cysteine is also known to undergo abiotic disulfide formation [105] to produce cystine under oxidizing conditions. Iron has been shown to serve as an excellent mediator of this conversion, whereby two complexed cysteines reduce ferric to ferrous iron while forming the disulfide linkage [106–108].

The additional two organics, 1,4-benzoquinone or *p*-benzoquinone, and NAD^+, were selected because they are both redox active cofactors and analogs to cofactors that perform critical roles in metabolic and biological transfer reactions, particularly pre-protein and non-protein proto-enzymatic reactions [109–111]. NAD^+, being ubiquitous in life, is predicted to be evolutionarily ancient and has been demonstrated to show some abiotic reactivity with FeS minerals [111,112]. Finally, *p*-benzoquinone is the major redox component of the cofactor ubiquinone. Quinones are lipophilic redox compounds necessary for electron transport in simple organisms. They have been shown to participate in both one and two electron concerted redox reactions [113,114]. *p*-Benzoquinone is especially relevant to prebiotic chemistry owing to its resistance to pyrolytic damage up to 800 °C [115,116] and its formation under astrophysical conditions (UV irradiation, 10 K) as evidenced by its presence in icy dust grains and carbonaceous meteorites [117,118].

Phosphorus (P) is a bio-essential element that is thought to have been critical for the origins of life. The dominant oxidation state of phosphorus in Earth's Hadean oceans is predicted to have been phosphate (HPO_4^{2-}/PO_4^{3-}) [119–123]; phosphates are also the only confirmed form of phosphorus that has been detected on Mars and Enceladus [124–126]. The reduced P form, phosphite (HPO_3^{2-}), would have also been stable on the anoxic Earth and has been identified in 3.52-billion-year-old carbonate rocks [121]. Phosphite could have been sourced from corrosion of phosphide minerals that were delivered via iron-stony meteorites or generated during impact events and lightning strikes [62,63]. Dissolved phosphate readily chelates with cations in solutions including Fe^{2+} to form vivianite and Ca^{2+} to form apatite; other phosphorus minerals which are predicted to have been prebiotically feasible include schreibersite ($(Fe,Ni)_3P$) and merrillite ($Ca_9NaMg(PO_4)_7$) [127]. Depending on the pH, Fe^{3+} can also chelate with phosphate anions in solution [62,63]. The chelation of phosphate by dissolved cations is expected to have hindered its availability for prebiotic chemistry, although competing anions like carbonate may have offset precipitation reactions [128]. Alternatively, phosphite (which is more soluble than phosphate) could have reached higher concentrations in the ocean [62,63]. Phosphite subjected to UV irradiation, or a strong oxidant, could be oxidized via Fenton chemistry to phosphate and polyphosphates, the latter of which can go on to phosphorylate organics [62,63,129].

The objectives of this study were to characterize abiotic organic-mineral mixtures via continuous wave and pulsed EPR approaches in order to identify what impacts organics have on mineral analysis under early Earth relevant conditions, and examine specific ligand coordination (amino acids and quinones, among others) of three abundant paramagnetic ions of relevance to prebiotic geochemical origins of life (Fe^{2+}, Fe^{3+}, Mn^{2+}). To do this we analyzed Fe and Mn of various oxidation states using EPR. We then tested how the addition of amino acids, cofactors, and phosphorus species impacted the EPR spectrum.

1.2 Background on EPR

EPR is a nondestructive analytical spectroscopic method comparable to nuclear magnetic resonance (NMR). It is a common technique for analyzing organic materials that is useful for study of oxidative and reductive processes, reactions involving radicals, paramagnetic species in organometallic chemistry, and metallocofactors in biological enzymes. EPR relies on the presence of unpaired electrons, which in the presence of an applied magnetic field (B_0) exhibit Zeeman splitting based on alignment of spin quanta (m_s) with the external B-field, thereby lowering electron spin degeneracy. This technique measures the transitions between the electron spin energy levels induced by the appropriate frequency electromagnetic radiation—typically in the microwave frequency range (1–260 GHz), with the most commonly utilized frequency being X-band (9 GHz). Unlike NMR, the frequency of radiation is typically held constant, while B_0 is swept to generate absorption spectra. Absorption lines are detected when the separation of energy levels between projected spin states $(M_s = \pm\frac{1}{2})$ approaches or equals the frequency of the incident photons such that:

$$\Delta E = g\mu_B B_{\text{eff}} = h\nu \tag{1}$$

for transitions between each split energy level with

$$E = g\mu_B B_{\text{eff}} M_s \tag{2}$$

where g = Landé g-factor, μ_B = Bohr magneton, h = Planck constant, and ν = frequency [130].

The specific field for resonance is non-unique as absorption may occur at a variety of microwave frequencies. However, the isotropic proportionality g-factor—capturing coupled spin and orbit angular momenta without respect to orientation—will deviate from that of an ideal free electron in vacuum ($g_e = 2.00232$). These deviations imply that unpaired electrons have gained or lost angular momentum. Typically, for transition metal complexes large variations are expected due to zero-field splitting and spin-orbit coupling, whereas small deviations are expected for organic molecules. The full g-factor is a diagonalizable tensor which may be used to probe local fields (e.g., atomic arrangements of ligands surrounding a metal center) by decomposition based on B-field components along each principal axis: $g_{11}B_1 + g_{22}B_2 + g_{33}B_3$, exhibiting a characteristic peak for each orientation. Lowering temperature allows these peaks to be more clearly distinguished and raising temperature coalesces anisotropic signal into one isotropic peak by addition in quadrature [130–132].

This sensitivity to the local electromagnetic environment of spins makes EPR sensitive to coupling of angular momentum with the dipole moment of neighboring nuclei in a complex, generating a hyperfine interaction. These hyperfine interactions provide detail into the distance of nuclei from unpaired electrons as well as coordination environment (number and identity). Interactions of unpaired spins with neighboring nuclei modify Eqn 2 to the following form:

$$E = g\mu_B B_{\text{eff}} M_s + AM_s m_I \tag{3}$$

where m_I is the nuclear spin quantum number of neighboring nuclei and a is the hyperfine coupling constant which is the difference in B-field between split spectral lines [11]. The selection rules associated with EPR for hyperfine interactions are identical to those in NMR, based on the atomic number and mass of the interacting nuclei. Similarly, the number of spectral lines arising from hyperfine interactions will be $2N \cdot I + 1$ for N equivalent nuclei and I net nuclear spin. After approximately 2–3 bond lengths distance from the unpaired electron, the signal intensity of coupling interactions will have decayed beyond any effect on the field-swept spectrum [133,134].

Additionally, for high spin systems with $S > 1/2$ and non-spherical symmetry, the spin-spin interaction of electrons in the same molecule will eliminate the degeneracy of the ground state energies. These splittings are present even in the absence of an external applied magnetic field, thus being known as the zero-field splitting interaction:

$$\hat{\text{H}}_{ZFS} = \hat{S}^T D\hat{S} \tag{4}$$

Here $\hat{D}$ represents the zero-field splitting interaction tensor and in EPR conventions is typically traceless (the sum of its diagonal elements is zero). In the principal axes system, $\hat{D}$ can be represented as:

$$\hat{H}_{ZFS} = D_x S_x^2 + D_y S_y^2 + D_z S_z^2 = D[S_z^2 - \frac{1}{3}S(S+1)] + E(S_X^2 - S_y^2) \quad (5)$$

The two primary components that parametrize the zero-field splitting interaction here, D and E, are defined as follows:

$$D = \frac{3}{2}D_z = -\frac{3}{2}(D_x - D_y), E = \frac{1}{2}(D_x - D_y) \quad (6)$$

Though E and D can both vary from system to system, in general their relationship is constrained by the limits $0 \leq \left|\frac{E}{D}\right| \leq 1/3$. These two components arise from the symmetry of the zero-field interaction, varying from $E = 0$ for axial symmetry to $\left|\frac{E}{D}\right| = 1/3$ for rhombic systems.

The two most common techniques for EPR are pulse and continuous wave (CW). Different information can be gleamed from these two techniques. This study focuses primarily on CW-EPR, using some pulse methods to corroborate the hydration shell about metal centers discussed in brief in the supplementary information. EPR is typically conducted with the magnetic field component of the microwave radiation aligned perpendicular to the orientation of the generated magnetic field, which provides selection rules of $\Delta m_s = \pm 1$ for electron spin transitions and $\Delta m_I = 0$ for nuclear spins which are magnetically coupled to the electron spins. Alternatively, using specialized resonant cavities such as the Bruker dual mode resonator utilized in this study, EPR may also be conducted in parallel mode, where the magnetic field component of the microwaves is oriented parallel to the applied magnetic field, changing the selection rules so that higher multiplicity spin state transitions ($\Delta m_s \geq \pm 2$) become allowed [135]. Parallel mode is generally used to probe electronic environments of "non-Kramers" systems with integer spin which can be challenging to interrogate with perpendicular mode EPR.

2 METHODS AND MATERIALS

All solutions were prepared using Milli-Q water (18.2 MΩ •cm) sparged with nitrogen (N_2) or argon (Ar) gas to ensure the resulting solutions were anoxic (30 min per 100 mL of water). To ensure the system remained anoxic, all samples were prepared and stored, if needed, in a nitrogen-filled glovebox. For transfer and analysis, initially samples were transported in nitrogen-filled bags; however, owing to complex/particle clustering and concerns regarding oxidation, all subsequent samples were prepared in nitrogen-filled capped EPR tubes kept at 77 K in a cryogenic Dewar prior to EPR.

2.1 EPR Sample Preparation

Unless otherwise noted, samples of the respective metal and prebiotically relevant organics, with concentrations described in Table 1, were prepared as 10 mL final quantities into which 10% glycerol by volume was thoroughly mixed as a glassing agent. Then, 250 μL of these sample solutions were extracted via micropipette and dispensed into 4 mm quartz EPR tubes. EPR tubes were slowly submerged in liquid nitrogen (77 K) from base upward, retarding the freezing rate to ensure glassy and homogeneous solidification and prevent particle aggregation. Prior to EPR spectroscopy, samples were stored in liquid nitrogen.

The frozen samples were analyzed primarily using a Bruker EMX X-band CW-EPR spectrometer either at 77K (liquid N_2) using a quartz immersion dewar or 5K (liquid He) using an Oxford Instruments ESR900 helium flow cryostat. For 77 K measurements, a Bruker SHQE resonator was used, while for 5 K measurements, a Bruker ER 4116 DM dual mode resonator was used, tuned to either perpendicular ($B_1 \perp B_0$, ~ 9.63 GHz) or parallel mode (B_1 B_0, ~ 9.43 GHz). Iron and manganese solutions were synthesized by dissolving $FeCl_2 \cdot 4H_2O$, $FeCl_3 \cdot 6H_2O$, $MnCl_2 \cdot 6H_2O$ (Fisher Scientific) or their mixtures in ultrapure water. KH_2PO_4 (monobasic) and $NaPHO_3 \cdot 5H_2O$ (Fischer Scientific) were used for the P additives (phosphate and phosphite, respectively). We also explored the effect of oxidizing the metals using hydrogen peroxide (H_2O_2) prepared from hydrogen peroxide solution (30% w/w). For this reaction, we wanted to test whether abiotic oxidation could generate vacant sites in the metal precipitate that would simulate those generated when minerals that are precipitated biologically. The effect of the presence of organics on EPR spectra of each paramagnetic ion was explored using glycine (Fischer Scientific), cysteine (Fischer Scientific), 1,4-benzoquinone (Sigma Aldrich), and nicotinamide dinucleotide (NAD^+) (Sigma Aldrich). Each of the organics was used without any further purification.

All EPR samples were prepared so that the final concentration of organics, metals, and phosphorus species was 1 mM each, unless otherwise outlined in Table 1. Initial controls of all 3 metals were run, as well as combinations of the metal ions, all with a maximum total concentration of 1 mM.

The subsequent EPR spectra were processed using EasySpin software [136] on MATLAB; plots comparing spectra were made using MATLAB using the PlotPub package [137], as modified [138].

2.2 UV-Visible Spectroscopy

Ultraviolet-visible spectroscopy (UV-Vis) was carried out on samples as the ground truth of anticipated oxidation state and to inform organic coordination with metal ions owing to shifts in d-d and ligand-to-metal/metal-to-ligand charge transfer bands. In all cases, UV-Vis was carried out on VWR® UV-6300PC dual beam UV-Visible

TABLE 1
EPR sample combinations for each organic, phosphorus additive (phosphite or phosphate), or co-factor in the presence of their respective paramagnetic metal centers and those additives' respective concentrations compared to the concentrations of listed metal species

	Glycine	Cysteine	*p*-Benzoquinone	NAD^+	PO_3H/PO_4	Water
Ferrous Iron (Fe^{2+})	1 mM	1 mM	1 mM	1 mM	1 mM	-
1 mM	1 M	1 M	100 mM			
Ferric Iron (Fe^{3+})	1 mM	1 mM	1 mM	1 mM	1 mM	-
1 mM	1 M	1 M	100 mM			
Manganese (Mn^{2+})	1 mM	1 mM	1 mM	1 mM	1 mM	-
1 mM	1 M	1 M	100 mM			
Synthetic Regolith	1 mM	1 mM	-	-	-	-
(1 mM Fe^{2+}, 0.29 mM Fe^{3+}, 0.01 mM Mn^{2+})						

spectrophotometer from 190 to 1100 nm at 100 nm/min scan rate. Samples were interrogated in a standard quartz cuvette relative to a sparged ultrapure water background. UV-Vis spectra and Beer-Lambert calibration curves are provided in the Supplementary Information (Section 1).

2.3 Geologically Relevant Synthetic Mineral Generation and Acid Work-Up

For relation to relevant geological conditions that are appropriate in the study of prebiotic chemistry and the origin of life, analysis of mineral phases was required. We sought to test if there was a noticeable difference in the EPR spectrum between a mineral of mixed metals and a solution resulting from the same mineral's dissolution in acid. A synthetic geologic mineral composed of our three paramagnetically active species was chosen based on compositional estimates of the Martian surface [139–146]. Correspondingly, initial solutions of individual metal salts and the synthetic mixture based on regolith estimates composed of 1 mM Fe^{2+}, 0.29 mM Fe^{3+}, and 0.01 mM Mn^{2+} were prepared. Under anoxic conditions, an equivalent of glycine or cysteine was added to each 1 mM solution or their mixture, before being titrated to pH $\geq$ 10 by addition of 1 M NaOH solution. Solutions were mixed by inversion while readily precipitating; samples were then sealed with rubber stoppers and allowed to continue precipitating for seven days. After one week, the iron and manganese (oxy) hydroxide precipitate and the supernatant were lyophilized in a 72-hour run cycle, leaving the residual synthetic regolith. Then, samples of the dried minerals were prepared for EPR spectroscopy. The entirety of the dried mineral was reacted with 50 mL of 1 mM HCl to redissolve metals and organics at the intended concentrations (1 mM Fe^{2+}, 0.29 mM Fe^{3+}, and 0.01 mM Mn^{2+}). Solutions were recapped to avoid evaporative losses and moved to a shaker table to agitate and increase dissolution. This dissolution was carried out for 24 hours before 10%v/v glycerol (5.55 mL) was added to each solution using a syringe (55.55 mL total). Remaining fluids from the lyophilized samples were frozen at – 20 °C for later spectrophotometry. In all other respects, experiments involving synthetic minerals in the presence and absence of organics following acid work-up were identical to aforementioned EPR methods.

2.4 ^{1}H Nuclear Magnetic Resonance (NMR) Controls for Organics

^{1}H NMR controls were run on prepared organic solutions, organic-metal solutions for EPR, and the supernatant from the mineral experiment using a Brucker Verona 400 MHz Spectrometer equipped with an autosampler with water suppression using excitation sculpting. Liquid samples were transferred to clean NMR tubes and analyzed after addition of 10% D_2O and sodium trimethylsilylpropanesulfonate (DSS) chemical shift standard for proton NMR of aqueous solution (0.06 mL of D_2O/DSS added to 0.54 mL of sample solution).

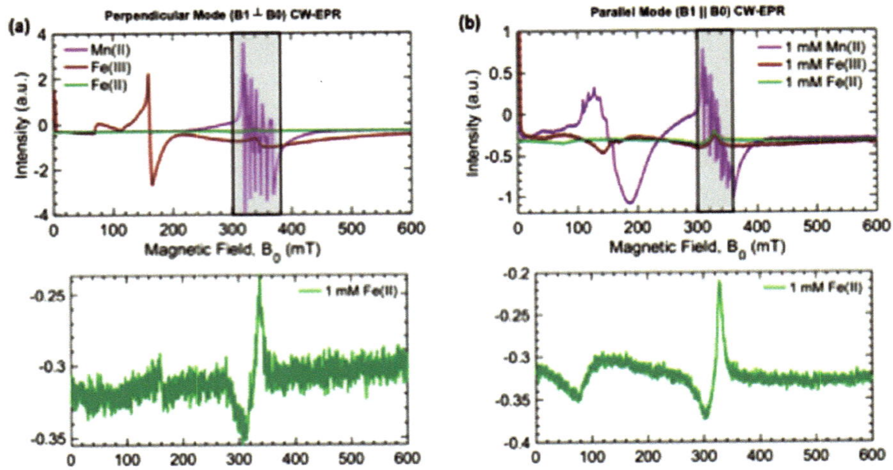

FIGURE 1 Aqua complexes of 1mM concentrations of Fe^{2+}, Fe^{3+}, Mn^{2+} paramagnetic species under (a) perpendicular mode (B1 ⊥ B0) and (b) parallel mode (B1 ⊥ B0) CW-EPR with outsets beneath weak ferrous signal. Distinct metal center spectra evidence the utility of this technique in identification of unknown sample composition, even at low concentrations.

3 EXPERIMENTAL RESULTS

While EPR spectra were successfully obtained in all cases, many factors controlled the quality of the spectra. Samples run at lower temperatures (5 K – liquid He) yielded the best signal-to-noise ratio compared with those at higher temperatures (77 K – liquid N_2 or 300 K – room temperature). The metal ions explored also varied in signal intensity, primarily dependent on the magnitude and anisotropy of the zero-field splitting, which is much larger for Fe^{3+} and Fe^{2+} than $Mn(II)(H_2O)_6$. As shown in Figure 1, the Mn^{2+} signal was more intense than that of Fe^{3+} which in turn is far better resolved and more intense than Fe^{2+}. As a consequence, in samples containing equal proportions of all three metals, the features from Mn^{2+} overwhelmed that of the other peaks; only the Mn^{2+} was distinguishable where spectra overlapped. However, the Fe^{3+}, Fe^{2+}, and Mn^{2+} all produced distinct peaks from each other identifiable up- or down-field from the primary Mn^{2+} band. Interestingly, Mn^{2+} was so easily detected by EPR that even trace amounts from glassware or contaminants from within the ferrous/ferric chloride powders were readily detected if present, such as in the case of NAD^+-metal coordination where trace manganese contamination was visible from the hyperfine spectra centered about 350 mT. These signals while overlapping some rhombic iron signatures in the similar region neither overlapped the main ferric iron resonance or zero-field splitting nor were visible in parallel mode critical for Fe(II) signatures.

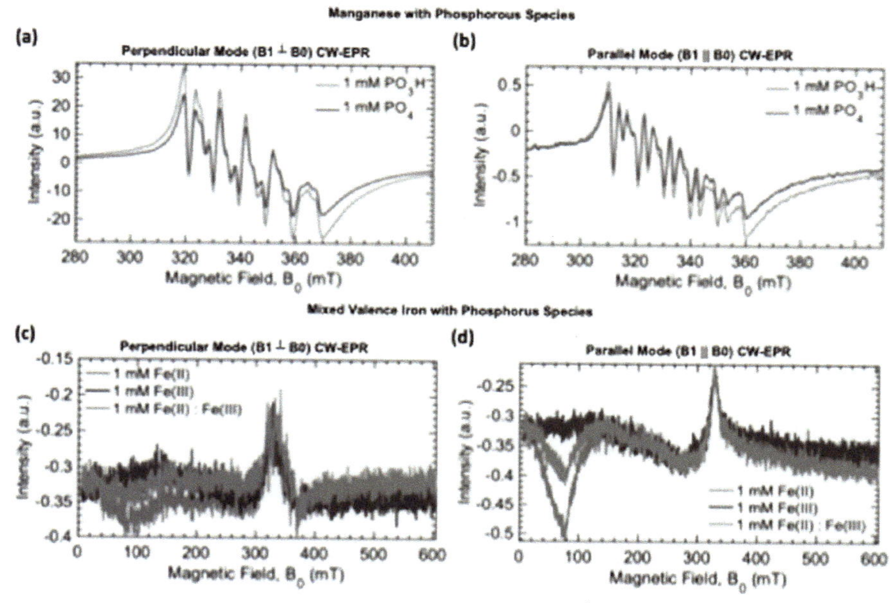

FIGURE 2 EPR spectra of manganese in the presence of phosphate or phosphite observed under (a) perpendicular mode and (b) parallel mode. EPR spectra of ferrous and/or ferric solutions in the presence of phosphate and a 50:50 mixed valence iron solution containing phosphite, under (c) perpendicular mode and (d) parallel mode. Note that the characteristic signal for ferric iron in perpendicular mode at ~150 mT (Figure 1) disappears upon addition of the phosphite or phosphate; neither the signal from Mn^{2+} nor Fe^{2+} was impacted by either phosphorus species.

3.1 Phosphite/Phosphate

We found that the addition of phosphate or phosphite only impacted the EPR spectra of ferric iron (Fe^{3+}) solutions: the signal of ferric iron in perpendicular mode disappears upon the addition of phosphite or phosphate; no new signals or apparent changes in existing spectra were observed in parallel mode, as shown in Figure 2. This suggests that both phosphite and phosphate form clusters with ferric iron, even at low pH (3–4). The resulting clusters are EPR silent, hence the disappearance of the ferric iron signal. Thus, should either phosphite or phosphate be present within samples of dissolved ferric iron minerals, EPR may not be able to detect such ferric iron. The manganese spectra similarly demonstrated only minor changes in signal intensity with no significant difference in spectra in either parallel or perpendicular mode, indicating Mn^{2+} remains a hexa-aqua complex.

3.2 Amino Acids

Adding glycine to the metal solutions decreased the signal intensity of the starting iron species; there was only a slight decrease in the Mn^{2+} signal, and only at high

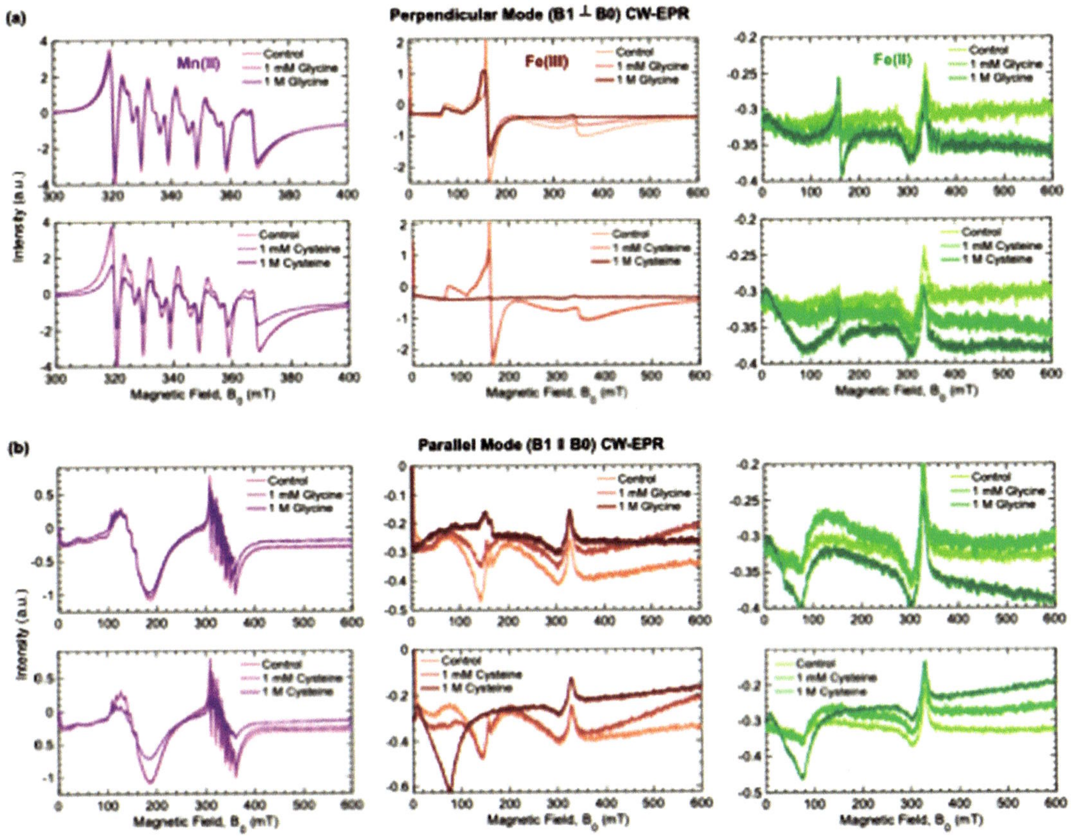

FIGURE 3 Comparison of glycine and cysteine interactions on selected metal species in (a) perpendicular mode (B1 ⊥ B0) CW-EPR, and (b) parallel mode (B1 ∥ B0) CW-EPR. Iron samples collected at 2mW and 8mW, while manganese samples collected at 2uW and 8mW, in each mode, respectively. Control is the EPR spectrum of the metal solution with no added organics.

concentrations of glycine (Figure 3); interestingly, for the Fe^{2+} solutions, a new signal is detected around 1800 G ($g = 4.3$) in perpendicular mode, around where we would expect a signal for high spin Fe^{3+} in a low-symmetry ligand geometry. For Fe^{3+} solutions, glycine addition not only dampened but also broadened the primary band at 1500 G while narrowing the central Fe^{3+} peak. As the solutions are slightly acidic (pH 4–6), glycine is predicted to form a mixture of complexes with ferric iron, ferrous iron, and manganese by binding to the metal via the amine nitrogen and carboxylate oxygen [102,103]. Over longer time periods, glycine appeared to form silent clusters with Fe^{3+} when probed in both CW and pulsed EPR using both D_2O and H_2O solvents (see SI-6 and SI-7), observed in the loss of signal intensity when freezing in liquid N_2 was performed after transport. While exploration of this feature was beyond the scope of this work, we suggest that perhaps clustering or dimer formation by glycine samples is limiting detection via pulse EPR uniquely without affecting detection via CW-EPR. Alternatively, potential oligomerization of iron with glycine could yield similar difficulties. However, the lack of precipitates in all glycine conditions suggests these clusters are limited in size and remain suspended or are not precipitate forming, staying solubilized. CW spectra for Fe^{3+} with 1 M glycine showed significant loss in high spin Fe^{3+} intensity, change in the lineshape, and significant loss of intensity of the "zero field" transition. This corroborates our hypothesis that there is the formation of some EPR-silent species in the presence of 1 M glycine, perhaps antiferromagnetically coupled dimers; however, specific determination is challenging owing to broad linewidths and large variability from baseline contribution affecting peak integration. This limits the analytical usefulness in any quantitative sense.

Upon addition of cysteine to the metal solutions, we found that the peak for ferric iron almost completely disappears in perpendicular mode (with new signals detected in parallel mode). In Mn^{2+} solutions, the signal for the Mn hexa-aqua complex was significantly dampened, but we did not identify any new peaks corresponding to a Mn-cysteine complex, which suggests that this product may be EPR silent (Figure 3). Upon addition of cysteine to the Fe^{3+} solution, we observed the disappearance of the Fe^{3+} signal in perpendicular mode only after 1 M of cysteine was added (there was only a minor decrease in the strength of the Fe^{3+} signal when 1 mM of cysteine was used). The loss of Fe^{3+} is likely due to the reduction of Fe^{3+} by cysteine to Fe^{2+}. Accordingly, EPR in parallel mode of these solutions shows an additional signal consistent with Fe^{2+} only in Fe^{3+} solutions with increasing cysteine concentration. Notably, the parallel mode trough at ~140 mT for ferric iron disappears with increasing cysteine content and a ferrous signal at ~80 mT emerges, consistent with ongoing formation of cys-metal coordination complexes leading to reduction of ferric iron. Careful analysis of the spectra demonstrates that another signal could be present as well, perhaps suggesting continued reduction or coordination by an oxidized cystine complex. This result can be clearly observed in Figure 3, which demonstrates that with increasing concentration of cysteine, the perpendicular mode Fe^{3+} spectral character changes significantly. Regarding the effect of cysteine on dissolved Fe^{2+}, we observed a sharpening of the Fe^{2+} signal in parallel mode and pronouncement of a larger signal at ~90 mT, suggesting the possible formation of Fe(II)-cysteine complexes.

3.3 Organic Cofactor and Analogs

NAD^+ was examined in the presence of the metal ions. A control of NAD^+ without any metal or mineral was run and no EPR peaks were visible, as was expected. Both ferric and ferrous iron peaks appeared unaffected in the sample containing Fe^{2+}:Fe^{3+} in a 1:1 ratio. Significantly distinct from the behavior of iron species, the addition of NAD^+ to Mn^{2+} samples resulted in major reduction in signal intensity with changes to the hyperfine spectrum via the elimination of shouldering and smoothening of features (cf. Mn^{2+} hexa-aqua complex control in Figure 4a and 4c). This drastic reduction in signal intensity (≥ 90%) is observed in both perpendicular and parallel modes with a notable broadening of linewidths, cf. Mn^{2+} aquo complex control.

p-Benzoquinone was examined in the presence of the three metal ions. Unlike NAD^+ which appeared to only weakly impact the Mn^{2+} hyperfine spectrum, but not the ferric or ferrous iron spectra, surprisingly the opposite was observed upon introduction of benzoquinone to each metal species: several new peaks were identified only in solutions containing iron, where in manganese samples only an insignificant difference in the features between 340 and 350 mT was observed. As the concentration of *p*-benzoquinone increased, much of the Fe^{3+} signal was lost, and a significant amount of an organic radical signal at $g = 2$ (presumably the semiquinone radical) was observed. Quantitative CW-EPR of the radical species performed at 77 K with comparison to a TEMPO spin standard in toluene indicates a reproducible organic

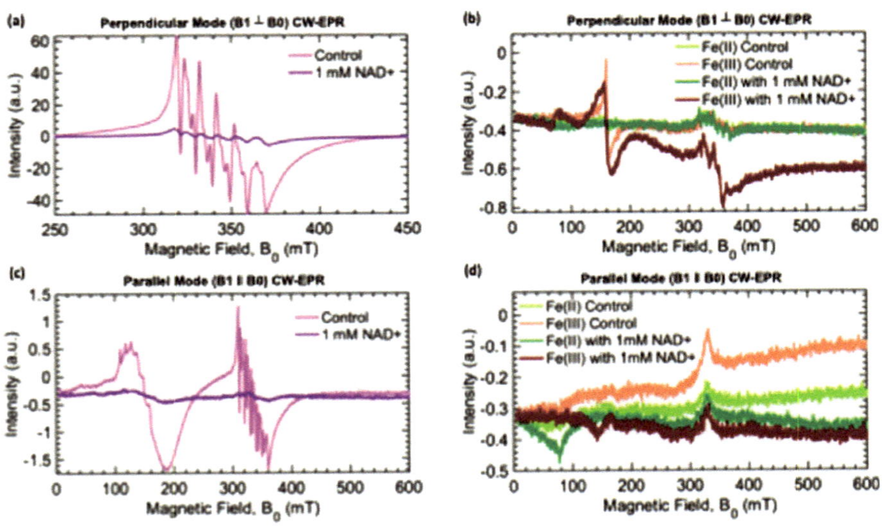

FIGURE 4 NAD^+ effect on metal species is predominantly limited to subtle changes in lineshape at 5K in CW-EPR: (a-b) perpendicular mode of noted metal species; (c-d) parallel mode of noted metal species.

radical concentration of $\approx 10-20$ uM. The disappearance of Fe^{3+} is concomitant with the appearance of Fe^{2+} signals in parallel mode at ~90 mT, though these signals appear somewhat distinct from the Fe^{2+} in water and those in the presence of glycine and cysteine.

3.4 EPR of Organics in Geologically Relevant Synthetic Minerals

Next, we evaluated whether EPR could provide additional information on the redox state of metals and the presence of organics on samples that have been prepared for inductively coupled plasma (ICP) mass spectrometry (MS) or optical emission spectroscopy (OES). ICP-MS/OES both provide information on the abundance of elements, but cannot be used to determine redox information of metals present or the presence/absence of organic matter. In these experiments planetary sample analogs were prepared and then dissolved in hydrochloric acid to liberate metals and trapped organics. The planetary analogs were precipitates generated from (1 mM Fe^{2+}, 0.29 mM Fe^{3+}, and 0.01 mM Mn^{2+} and 1 mM glycine or cysteine). EPR of the dissolved samples have weak manganese hyperfine splitting signals in perpendicular mode, with clear ferric iron features about 160 mT corresponding to mononuclear $S = 5/2$ Fe^{3+} ions with a moderately rhombic symmetry to the zero-field splitting (ZFS) interaction as well as ferrous iron signatures. These samples show primarily characteristics of ferric and manganesian features with some residual (muted) features of ferrous iron, suggesting lyophilization and acid work-up had an oxidative effect on ferrous species, even in presence of a competitive reductive species (cysteine). Samples with cysteine show lower intensity ferric signal than those in the water control or glycine case, while showing emergence of ferrous iron, indicating the iron-mediated cysteine reduction to cystine is in competition with environmental oxidation. Parallel mode more clearly demonstrates the limited presence of Fe^{2+} signals, including the sizable reduction in intensity at 75 mT and shared Fe^{2+} and peak between 320 and 340mT. Importantly, key features of the spectra with glycine and cysteine are relatively well preserved in mineral work-up cases, demonstrating similarity in peak attenuation and lineshape changes. Manganesian signatures are similarly intense as the control case with intensities consistent with their low concentration. Line shapes observed are broadly consistent with those from controlled experiments involving purchased hydrate salt precursors, indicating the utility of the foregoing strictly controlled experiments to unknown paramagnetic-species containing minerals through our acid work-up procedure.

4 DISCUSSION

4.1 Competition with Water Coordination

Owing to the relative molar concentration of water (55.56 M) being far in excess of concentrations of the selected ligands (organics, phosphate/phosphites, etc.), ligand coordination was anticipated to be muted under. Competition in the ligation sphere by water was indeed observed, with little difference between water control samples

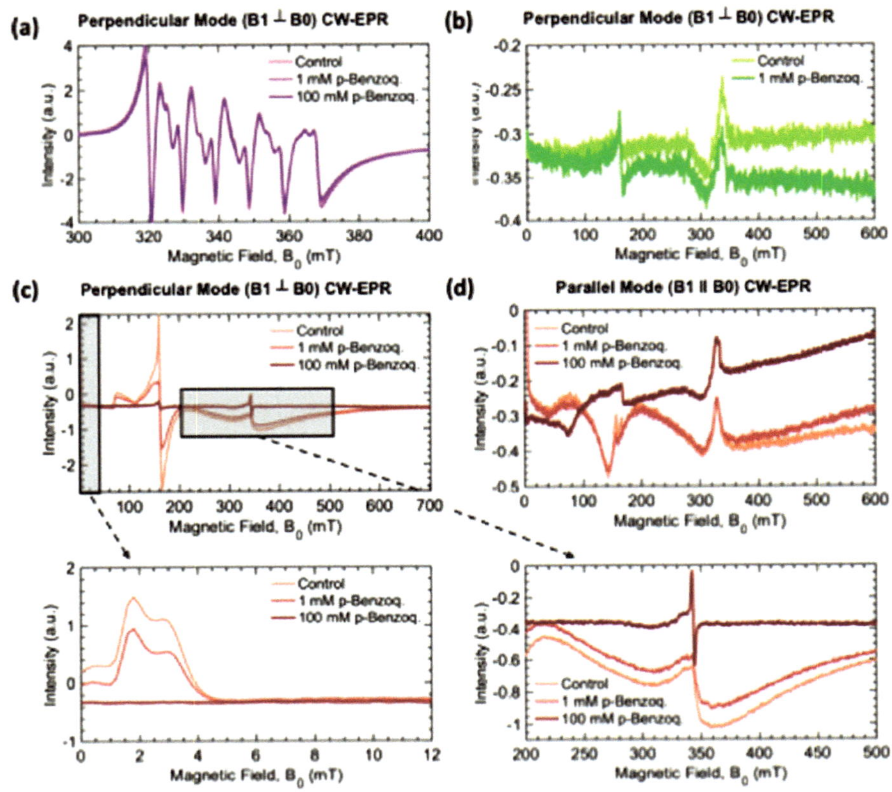

FIGURE 5 Interaction of *p*-benzoquinone on metal species in perpendicular mode (B1 ⊥ B0) CW-EPR or parallel mode (B1 ǀ B0) CW-EPR: (a) Mn^{2+} showing little effect save on line shape other than the emergence of a small feature around 344 mT; (b) Fe^{2+} shows some sharpening of the feature around 158 mT, but otherwise no feature changes beyond sample variation; (c) Fe^{3+} is reduced by *p*-benzoquinone as demonstrated by the collapse of the rhombic signal at 158 mT and the zero-field splitting signal as shown in the outset (left) and emergence of a sharp radical peak in the outset (right); (d) parallel mode on Fe^{3+} shows distinct Fe^{2+} signatures at ~190 mT.

and organic-containing samples except where reactions are noted. Broad lineshape changes made exact determination challenging. To ascertain the effect of increasing concentration, 1 mM and 1 M concentrations of each ligand species relative to 1 mM metal species were run under continuous wave method as shown in Figures 3 and 5. Where possible, D_2O/H_2O ^2H-ESEEM experiments were conducted to determine water coordination of the paramagnetic center, see Supplementary Information (SI-6 and SI-7) for details. As discussed in further detail below, these results largely are consistent with water-coordinated metal complexes.

4.2 Manganese Complexes and Reactions

In the case of Mn^{2+}, the primary features in CW-EPR spectra are all consistent with $Mn(II)\cdot(H_2O)_6$ being the predominant species present. With the exception of NAD^+ and cysteine, we do not see any significant proportion of solution-based Mn^{2+} complexation evident for any of the other species examined, up to their approximate solubility limit in unbuffered water at room temperature (when prepared, prior to freezing). In case of the noted exceptions, we see significant dampening of signal. With cysteine a more gradual loss of signal is seen as a function of increasing concentration by more than approximately half the peak-to-through intensity, which we ascribe to the formation of EPR silent cluster formation. In the case of NAD^+ even its minor addition (≤1 mM) results in major dampening of signal intensity and broadening which we attribute to complexation with Mn^{2+} ions.

Mn^{2+} is a high spin d^5 metal (S = 5/2), but because of the high degree of octahedral symmetry of the hexa-aqua complex and low d-d orbital overlap, the zero-field splitting (ZFS) for $Mn(II)\cdot(H_2O)_6$ is quite small (D ≈ 500 MHz) and at basic fitting and simulation exhibits a highly rhombic character (E/D = 0.27). This estimate of ZFS rhombicity is not strictly rigorous because of the broadness of linewidths observed. This complexes' signal is centered near the free electron g-value at g = 2.004 and split into (2I+1) lines by the hyperfine interaction 100% abundant ^{55}Mn (I=5/2) nucleus, with $a_{iso}(^{55}Mn)$ = 263 MHz. The ZFS primarily contributes to the linewidth and lineshapes of the trailing "wings" of the spectrum. A secondary factor affecting the linewidth and lineshape is the "D-strain" which arises from a distribution of ZFS values due to some degree of heterogeneity in H_2O coordination of the metal center.

There are low-intensity transitions in between these six hyperfine lines representing the spin-forbidden transitions in which both the electron spin and ^{55}Mn nuclear transitions occur (i.e., $\Delta m_s = \pm 1$ and $\Delta m_I = \pm 1$). The intensity of these forbidden transitions is quite sensitive to changes in the magnitude of the ZFS and ^{55}Mn hyperfine. Owing to this fact, we therefore conclude that the lack of measurable change in these features in the presence of most of the species tested save the case of NAD^+ and a minor lineshape changes at 344 mT and signal dampening with cysteine. Barring the case of NAD^+, this indicates that chelation/complexation of the Mn^{2+} cation does not occur as readily with phosphorus species or organics as compared to iron cations. Complexation would be expected to lower the symmetry of the Mn^{2+} and increase the ZFS magnitude (and perhaps introduce asymmetry), and possibly change the ^{55}Mn hyperfine due to different metal-ligand (M-L) covalency. We observe no changes, so there must be no appreciable interaction under the experimental conditions tested. The D_2O/H_2O ^2H-ESEEM modulation experiments shown in the Supplementary Information (SI-7) further confirm this, as there is actually a very slight increase (< 5%) in the ^2H-ESEEM modulation depth in the presence of glycine and cysteine, perhaps indicating a slight shift downward in pH (pD in this case) or a minor change in the hydration environment around the $Mn(II)\cdot(D_2O)_6$ complexes [13].

NAD^+ presents an interesting exception to the general behavior of Mn^{2+} with the other prebiotically relevant ligands explored in this study. NAD^+ derivatives and

NAD^+'s phosphorylated compliment, nicotinamide adenine dinucleotide phosphate ($NADP^+$), are known to complex and dimerize with Mn^{2+} and mono- and di-valent ions on very short timescales (10^{-10} s based on measures of relaxation time ratios) in aqueous solutions, whereby the oxygens of the phosphate backbone, phosphate or alcohol moieties on the ribose, and adenine N(7) sites coordinate with the metal center [148–152]. The coordination of manganese(II) ions with NAD^+ is reasonably expected given the role Mn^{2+} plays as an important activator for a large variety of enzymes, from catalyzing redox reactions to decarboxylation and hydrolysis [152–153]. Given that changes in manganese oxidation state can be diagnostic of biological activity [7,8] and the role Mn^{2+} salts can serve in Fenton-like reactions for the production of phosphates [89], the substantial changes we observe in CW-EPR signal intensity of Mn^{2+} on the addition of NAD^+, shown in Figure 4, compared to its relative non-interaction with other systems is particularly exciting. This loss of intensity would suggest either an oxidation state change, loss of soluble fraction of Mn(II), or the formation of antiferromagnetically coupled EPR-silent Mn^{2+}-NAD^+ dimers or clusters. While we believe it is implausible that there has been an oxidation of Mn^{2+} by NAD^+ on its own given the unfavorable redox potential of this reaction [154], the presence of some ligands such as carbonate has been shown to lower the potential of the Mn(II/III) redox couple by nearly a volt, acting as stimulants for electron donation from exogenous Mn^{2+} [155–158]. Were NAD^+-induced oxidation occurring, we would expect to observe the emergence of a new $S = 2$ Mn^{3+} signal in parallel mode, which we do not [157–159]. With that said, any Mn^{3+} generated would be unstable, likely undergoing disproportionation into Mn^{2+} and Mn^{4+} [160–162] with the later forming an insoluble oxide that would attenuate EPR signal as well. We believe it is, however, unlikely that the presence of alcohol groups on NAD^+ as well as those from our glycerol glassing agent would so significantly modulate the redox properties of manganese. Therefore, the loss of intensity observed in Figure 4 on the addition of NAD^+ is likely owed to either the insolubility and thereby precipitation of Mn-NAD^+ complexes, the aggregation of Mn-NAD^+ complexes to form larger chains which cause significant dipolar broadening of the Mn(II) EPR signal because of the close proximity of the cations, or the binding of Mn^{2+} to NAD^+ occurring in such a manner as to form antiferromagnetically coupled dimers which are effectively diamagnetic. In all of these more likely cases, the smaller amount of Mn^{2+} that remains in solution in aquo or monomeric forms is all that is left to see, hence the substantial reduction in signal intensity. The exact interpretation of this signature is muddled given the 10%v/v (corresponding to 1.4 Molar) glycerol content; its participation in possible coordination cannot be entirely ruled out even with the control experiments performed. Notably, the residual (weak) Mn^{2+} signal is characteristically identical to that of the Mn^{2+} hexa-aqua complex, but with major broadening of linewidths. The ZFS is not significantly larger, which by comparison was the primary change we observe in the Mn^{2+} EDTA complex because of its distorted ligand field enforced by the EDTA ligand (see Supplementary Information SI-6 and SI-7). Similarly, a monomeric NAD^+ ligand could similarly be expected to impart significant lowering of the ligand field symmetry at the Mn^{2+} ion compared to the aquo complex. Together this absence of change to ZFS and similarity of the hyperfine spectrum leads us to believe the Mn^{2+} we weakly observe in CW-EPR is likely not Mn(II)·(NAD^+) complex formation but rather a small Mn^{2+} signal from a precipitate or some residual Mn^{2+} in solution if not all the 1 mM NAD^+ was fully dissolved.

4.3 Ferrous Iron Complexes

Ferrous iron is a d^4 metal and in the presence of the tested weak-field ligands is expected to have high spin with integer value S=2. In situations where the ZFS is large relative to the microwave excitation energy (9 GHz ≈ 0.3 cm^{-1}), these "non-Kramers" systems are often only observed in parallel-mode EPR, where the selection rules of the EPR transitions are changed by orienting the oscillating magnetic field component of the MW radiation (B_1) to be oriented parallel to the static applied magnetic field (B_0) which is swept to collect the spectrum as briefly discussed in the introduction. In this mode, higher spin transitions ($\Delta m_s \geq \pm 2$) become increasingly allowed compared to the $\Delta m_s = \pm 1$ transitions, thus placing the resonance condition at lower fields achievable by conventional electromagnets. For Fe(II)·$(H_2O)_6$, we observe a largely negative derivative signal centered around $g_{eff} = 9.15$ which corresponds to the $\Delta m_s = \pm 2$ transition.

The lineshapes and linewidths in the case of Fe^{2+} show an increase in intensity in the presence of cysteine and to a lesser extent glycine with no other species showing changes within the measurement error of the instrument. The reductive potential of cysteine likely drives the intensity variation in this case. From the data available, we make no comment as to whether there is any coordination of the Fe^{2+} by anything of the prebiotically relevant organic species. Unfortunately, owing to the low intensity and breadth of ZFS heterogeneity in non-Kramers spectra even at liquid helium (5 K) temperatures [14], the analytical conclusiveness of ferrous iron interactions from EPR under our experimental conditions is limited.

4.4 Ferric Iron Complexes and Reactions

Like Mn^{2+}, Fe^{3+} is a high-spin d^5 metal as an aquo complex with S = 5/2. However, in this case of ferric iron, the ZFS is quite a bit larger and asymmetric due to a more distorted ligand field. In the absence of any added organics, there is a relatively high-intensity signal at near zero-field, indicating that the ZFS magnitude is approximately half that of the microwave excitation energy (9 GHz ≈ 2D ≈ 0.3 cm^{-1}). The rest of the control EPR spectrum consists of a rhombic lineshape with primary turning points at $g = 9.25$, 4.26, and 1.99. These effective g-values are consistent with a fairly rhombic ZFS interaction with D ≈ 0.15 cm^{-1} and E/D ≈ 0.15, but in reality, the rather broad linewidths and somewhat odd lineshape of these signals are likely reflective of a distribution of D and E/D values due to heterogeneity in the H_2O coordination microenvironments.

The addition of one molar (1 M) glycine appears to cause some shifts in the lineshapes and primary intensity of the transitions near g = 4.26, as well as a loss of overall intensity in perpendicular mode and a modest increase in Fe^{2+} as observed in the parallel mode EPR of these samples—the extent to which seems to increase with time taken to freeze the samples. The former observation of lineshape changes suggests a coordination by glycine, and the latter loss of observed intensity suggests the formation of EPR-silent polynuclear Fe-Glycine complexes and some reduction to Fe^{2+}.

In the case of 1 M cysteine, we see this process accelerated with a near *total* loss of Fe^{3+} and significant increase in Fe^{2+} signal in parallel mode across numerous repetitions, indicating a significant reduction of the Fe^{3+} by cysteine. With regard to the redox reactivity observed, our results corroborate proposed iron-mediated cysteine self-redox pathways, whereby cysteines form shared disulfide bonds, that in turn reduce nearby (complexed) Fe^{3+} to Fe^{2+} [107,108]. Of relevance to origins of life chemistry, this is critical as it is an example where an organic species present participates in metal/mineral mediated electrochemistry. As the minerals present facilitate potential prebiotic chemistry, knowledge of such redox reactions is important for constraining what reactivity could be present in an environment.

In the presence of high concentrations of *p*-benzoquinone, the intensity of the Fe^{3+} signal is lost, and a significant amount of an organic radical signal at $g = 2$ is observed (presumably the semiquinone radical) at an approximate concentration of 10–20 uM. This concentration is reasonably consistent with those for the formation quinone radical in ferric iron-bearing minerals in icy grains [163] and chondritic meteorites [164,165] with implications for the origins of life as electron shuttles [166]. Fe^{2+} is also observed via parallel mode, but the signals appear somewhat distinct from the Fe^{2+} control in water and those in the presence of amino acids. On the whole, this indicates there is significant redox cycling between Fe(II/III) and the quinone and semiquinone/hydroquinone generated in solution. These results indicate a detection of an organic radical species, based on the addition of *p*-benzoquinone, which is known to form a stable organic radical and act as a radical trap [167–172], we hypothesize that this is a quinone radical species. Of all the results presented in this manuscript, perhaps the most compelling for future instrumentation useful for analysis of prebiotically relevant organic compounds and their interactions with mineral species in extraterrestrial environments is this: the signal intensity of an organic radical, even at the ambient temperatures on Mars (120–300 K) [172–174], the Jovian and Saturnian moons (e.g., surface temperature of 50–100 K on Enceladus, Titan, or Europa) [175–179], or asteroids (e.g. 150–350 K on the surface of Bennu or 130–240 K on Ceres) [180,181] would be observable on-mission with a miniaturized EPR spectrometer. We welcome the use of this information for consideration in potential space mission-related instrumentation for identification of potential prebiotic and biosignatures.

4.5 The Case of Phosphates/Phosphites

Phosphite (HPO_3) and phosphate (PO_4) did not affect manganesian lineshapes but their addition did result in the extinguishing and reduction of ferric and ferrous iron signals, respectively. We hypothesize that this is due to the Fe^{3+} cations chelating with the phosphate and phosphite anions—forming EPR silent oligonuclear or polynuclear clusters, or perhaps *solid* precipitate clusters.

Both ferric and ferrous iron chelate phosphates under our experimental conditions (pH 3–4) [62,63]. However, iron phosphites have substantially higher solubility ($K_{sp} \geq 10^{-4} - 10^{-7}$) [182] than iron phosphates ($K_{sp} \geq 10^{-30}$) [183] . Further, the precipitation of ferric phosphates is more thermodynamically favorable than that of ferrous phosphate, as its solubility is two orders of magnitude lower, resulting in its precipitation on a time-scale of minutes-to-hours compared to hours-to-days characteristic of ferrous phosphates ($K_{sp}(Fe^{3+}) \sim 3 \cdot 10^{-6} - 8 \cdot 10^{-7}$ vs $K_{sp}(Fe^{2+}) \sim 1 \cdot 10^{-4} - 5 \cdot 10^{-6}$) [182,184,185]. The precipitation of these iron minerals is predicted to have been initiated with the formation and subsequent oligomerization of Fe^{3+}-phosphate clusters in solution. Indeed, previous work has shown that polynuclear clusters containing ferric iron bridged by oxygen ions—akin to Fe^{3+}-phosphate complexes—are EPR silent [186,187]; Fe^{2+} and Fe^{3+} solutions containing phosphate in 1:1 ratios at 1 mM concentrations also generated clusters that were EPR silent [188]. Notably, Cheng and Chasteen found that the binding of phosphate to Fe^{2+} lowers the redox potential of the Fe^{2+}, facilitating Fe^{2+} oxidation by dissolved O_2; within 15 mins they found that ~62% of the starting Fe^{2+} was oxidized to Fe^{3+} [188]. Although our samples were prepared under anoxic conditions, they may have been exposed to oxygen during transfer from the anaerobic chamber to the EPR facility (>15 mins travel time) and consequently partially oxidized. We can thus not determine whether the reduction in the Fe^{2+} signal with phosphate was due to the formation of minor amounts of EPR silent Fe^{2+}-phosphate complexes or the oxidation of Fe^{2+} and the subsequent formation of polynuclear EPR silent Fe^{3+}-phosphate complexes.

In the mixed valence iron phosphite case, Fe^{2+}-phosphite similarly possesses much higher solubility than Fe^{3+}-phosphite; phosphite is also significantly more soluble than phosphate. Nevertheless, the addition of phosphite nearly erases the Fe^{3+} EPR signal and slightly reduces the Fe^{2+} signal. This suggests that Fe^{3+}-phosphate complexes are also likely EPR silent. Fe^{2+}-phosphite complexes may also be EPR silent, however, it remains possible that phosphite, like phosphate, facilitates Fe^{2+} oxidation and the subsequent formation of Fe^{3+}-complexes. If so, then the Fe^{2+} oxidation could trigger Fenton chemistry and the oxidation of phosphite to phosphate, resulting in the formation of some polynuclear and EPR silent Fe^{3+}-phosphate complexes. With that said, the ferrous iron signal to noise was weak and lineshapes broad, limiting identification of any ferrous signal. In any case, solubility would of course worsen with freezing, as generally solubility decreases exponentially with temperature, $K_{sp} = {}^{-\Delta G}\!/_{RT}$, due to a reduction in Gibbs free energy of solvation [189].

4.6 Assessment of Organics in Synthetic Mineral Samples

Mineral dissolution proved to be a feasible method of probing elemental and organic composition as well as evidence of redox activity in synthetic mineral samples. As highlighted in Figure 6, the examined mineral samples, relative to their pure controls (see Supplementary Information, Section 3), showed excellent resolution of signal following lyophilization and acid work-up. Even at extremely low concentrations of manganese, the Mn^{2+} signal was visible in the neighborhood of 162 mT and 340 mT. This is beneficial, as it enables identification and quantification of paramagnetic

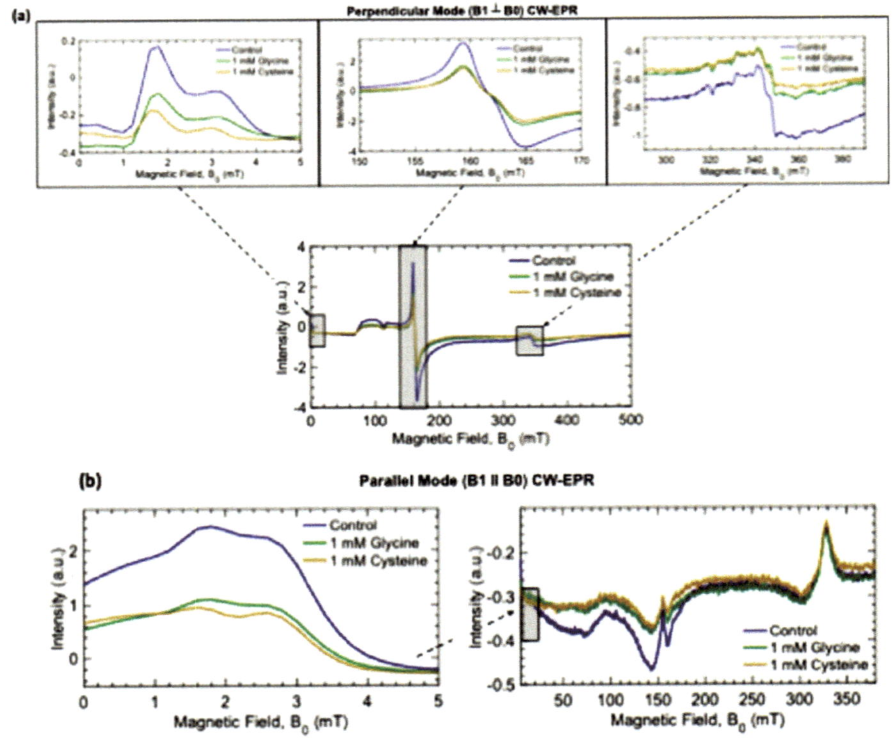

FIGURE 6 Synthetic mineral experiment using mixed paramagnetic species with 1 mM amino acids dissolved in acid: (a) perpendicular mode and (b) parallel mode. Predominant ferric, ferrous, and weak manganesian features are evident. The mineral control is the worked-up mineral species absent any organics.

metal species. Quantitative ratios of various metals are an important geochemical context providing insight into the particular geological conditions that might affect abiotic and prebiotic chemistries. This is of particular relevance when considering applications in astrobiology and studies of the origins of life, where this context plays a crucial role in both defining (with experimental analogs) and identifying (in real systems) the markers of habitability, biotic vs. abiotic/prebiotic signatures, and probable reaction pathways or their intermediates. Ergo, even the detection of trace quantities in our mixed mineral sample is a strong indicator of the utility of EPR as qualitative identification and quantifiable method.

Importantly, the lyophilization process used to generate the synthetic minerals before work-up is not completely anoxic; oxygen from the atmosphere could have diffused through the solution and the precipitated mineral likely changing iron oxidation state. Remarkably, the reductive capacity of cysteine in the formation of cystine via Fe(III)-Fe(II) mediation was still weakly observed in those mixed mineral samples containing cysteine. In the case of cysteine, relative to the control (no organics) and to that with glycine, a large attenuation in Fe^{3+} signal is observed in Figure 6, in both perpendicular (7a) and parallel (7b) modes, consistent with the signal intensity reduction seen in pure salt systems with 1 mM cysteine:1 mM Fe^{3+}, shown in Figure 3. In parallel mode, only a faint signal of Fe^{2+} is visible at ~180 mT, but it is overshadowed by the Mn^{2+} signal in the same neighborhood. Other factors could reasonably explain these results as well, consistent with our discussion of ferric complexes, that Fe^{3+} formed EPR silent clusters with cysteine. Alternatively, cysteine reduction could have been inhibited due to oxidation of cysteine if any atmospheric oxygen was present which would produce cystine without reducing Fe^{3+}. Our results suggest that this work-up method to generate aqueous samples from solid mineral samples preserves potential redox behavior of organic species mediated/catalyzed by metal centers and mineral species, provided hydrogen activity remains within the limits that do not affect the reaction products based on the Pourbaix diagram [190] or relative protonation of the organic species [191].

In this regard, for the use of this method in future experimentation on real or synthetic geological samples one should be conscious of the effect of pH on amino acids. While 1 mM HCl provides 1 mEq of H^+ and therefore is $p\text{H} = -\log(10^{-3}) = 3$, use of more concentrated acid (e.g., 1 M HCl, $p\text{H} = 0$) might result in pH dropping below the pK_a of important functional groups that are either already chelating a metal center or might otherwise in solution if not protonated. It is therefore imperative to buffer solutions or control acid concentration to limit the impact on organic species present (e.g., pK_a of carboxylate is 1.96/2.34 for Cys/Gly, respectively) [191,192] while still permitting dissolution of the mineral phase. In our case, the work-up with 1 mM hydrochloric acid was specifically designed so as not to affect the zwitterionic nature of the amino acids, enabling the relative comparison to solutions prepared from pure ferric, ferrous, and manganesian salts.

Beyond the caveats noted with respect to pH and potential for oxidation during mineral synthesis, we observed no other significant differences between samples prepared from pure salts (ideal) and those worked up from a solid synthetic mineral sample imitating a "geological" origin. Zero field splitting, lineshape and broadening, and relative peak intensities are all quite similar to the idealized samples from Figure 3 (cf. worked-up mineral results shown in Figure 6 and Supplementary Figure 3.1).

5 IMPLICATIONS AND RECOMMENDATIONS

5.1 Complimentary Methods for Orthogonal Characterization

Electron paramagnetic resonance has useful applications in quantification of radical species, identification of reaction products from and interactions of organic species with metal ions/minerals, and offers a useful probe for elemental and organic composition prior to other, more destructive work-up methods (e.g., inductively coupled plasma spectroscopy (ICP) to look at elemental composition or gas chromatography–mass spectrometry (GC-MS) to explore organic inventory). With that said, other orthogonal techniques are important to both qualitatively corroborate species identification, quantify concentrations independently, and serve as an alternative identifier of chelation/coordination of metal centers by organics/inorganic additives. The latter is particularly relevant given that broad lineshape changes in EPR are difficult to interpret and pulsed EPR methods like ESEEM and ENDOR are limited, performing better with some species relative to others. Direct alternatives to interrogate functional groups, electronic transitions and crystal field splitting environment, and equilibrium thermodynamics of complexation should serve as orthogonal methods for interpretation of datasets.

Local functional group modes may be probed with Fourier transform infrared spectroscopy (FTIR) and Raman spectroscopy, which can provide identification of mineral species, organics present, and in some cases bonding environment if within range of the instrumentation. These direct measures of bonding (vibrational and torsional modes) offer insight into local coordination. They are also quantifiable, in some cases directly or in others with standards, and have been used to characterize metal coordination to organics (e.g., metal-ion and mineral interactions with carboxylates, relevant to organic electrolytes and amino acids) [138,193,194].

UV-visible spectroscopy offers direct insight into electronic transitions in many organic [195] and mineral systems [196] providing useful information as to degrees of conjugation, radical generation and stability of UV-sensitive compounds, associative kinetics, and participation in charge transfer complexes and other ligand field effects that give rise to color of transition metal cations. This information, via calibration standards using the Beer-Lambert law, may also be used for validation of unknown concentrations of species as was performed in this study (see Supplementary Information, Section 1).

For organics and metal species that are UV-Visible transparent for which other local mode characterization methods break down (e.g., bridged metal dimerization or heavy metal chelation being outside the typical range of standard infra-red techniques), mass spectrometry coupled with soft ionization techniques (e.g., electrospray) can support identification of species on a mass basis. Similarly, for species with non-zero nuclear spin, liquid nuclear magnetic resonance (NMR) spectroscopy can be used to explore organic structure and composition as it was in our NMR controls (see Supplementary Sections 3.2 and 3.5). NMR methods are more challenging with paramagnetic species, particularly iron; however, because paramagnetism generates locally inhomogeneous fields which increase the distribution of the Larmor frequencies while additionally shortening τ_1 and τ_2 relaxation times, which in turn leads to peak broadening. Where NMR is possible, however, it can provide meaningful molecular structural information for metal-organic interactions [197,198] and minerals [199,200].

Perhaps the least well known of these orthogonal techniques, but one that provides conclusive associative/bonding kinetics and coordination number, is isothermal titration calorimetry (ITC). By injection of a small volume of low concentration titrant (typically the binding agent) into a known aliquot of low concentration analyte (typically the binding site relevant species), a calorimetric titration curve may be constructed. The heat evolved or absorbed as a result of the injection is measured by cooling or heating the jacketed cell to maintain isothermal conditions. Owing to the isobaric nature of the calorimeter, this heat is precisely the enthalpy, allowing direct measurement of the enthalpy of (de)solvation and reaction. Integration of heats of injection with respect to mole ratio provides a titration curve, the slope at the inflection point of which provides $\Delta G = -RT \cdot \ln(K_a)$ of the reaction and its altitude at the abscissa provides coordination number via molar binding ratio [201–203]. This technique is a particularly powerful method for entirely orthogonal measurements, with no reliance on electronic state information or paramagnetic character. While it cannot discern structural arrangement of coordinating/solvating species, it can identify their number, their Gibbs free energy of binding, and other equilibrium thermodynamics of their association. This specifically makes ITC a potent approach for assessing binding kinetics and reactivity from a purely thermodynamic basis. It provides a complimentary toolset for EPR, which provides electronic information, to conclusively identify the presence (or absence) or chelation/coordination that might be driving line shape differences or broadening without relying on ESEEM or ENDOR methods.

5.2 Uses for EPR in Prebiotic Chemistry and Astrobiology

EPR poses an excellent use case in the study of radical stabilization and electron shuttling molecules, viz. *p*-benzoquinone, and their redox coupling with metal centers. These have direct relevance to cosmochemistry (transformation of organics on and organic inclusions in meteorites and icy grains) and astrobiology (prebiotic chemistry and its transition to biochemistry). Radical pathways are relevant in prebiotic chemistry in relation to the origin of metabolism and understanding the reactivity of quinone-containing cofactors [116,164,170,204,205]. In the solid phase syntheses of organic compounds and aqueous-phase reaction space, be it from electric discharge, irradiation-induced, or mineral-catalyzed radical formation, radicals enable coupled energy transport, facile polymerization kinetics that are not condensation dependent, and increase molecular complexity [206].

Deep within icy interplanetary dust particles and asteroids, there remain traces of ongoing organic reactions [207–209] resulting from persistent radiation exposure (photolysis) and freeze-thaw cycles from inhomogeneous heating profiles [210–212]. These activities generate organic and some inorganic radicals from the generation of N_2O and NH_3 from nitrogen to the breakdown of polycyclic aromatic hydrocarbons (PAHs) like naphthalene and coronene to form aromatic ketones (quinones and derivatives) and alcohols [118,204,213–215]. Of these organic precursors and their products, many are often long-lived radical stabilizers with excellent thermal stability over cycling (e.g., menaquinone) [115,116,204,213], making them of particular interest in the examination of abiotic/prebiotic mechanisms for their formation and cycling as well as in their applicability as biosignatures. Meteor infall containing organics was a likely source of quinones and other PAH derivatives on Earth. Evidence of these asteroid-generated electron carriers is also found in meteorite organic extracts (e.g., Murchison meteorite) where they have been shown to be of prebiotic relevance owing to their capacity to serve as redox couples for proton gradients generated across lipid membranes similar to what exists in cells [164].

Similar to our study and quantification of quinone radicals in this work, EPR spectroscopy could serve to quantify reactive organics in recovered samples and meteorites similar to the meteorite infall which is thought to have brought PAH-derivatives to prebiotic Earth. The coupling with photolysis methods using UV irradiation in situ with EPR would enable direct study of radical species, their lifetimes, concentrations, and transformations. Possible in-situ EPR methods for electrochemical study [216–218] and FTIR or UV-Vis spectroscopy [219–221] offer compelling means to study organic radicals and their interaction with complex redox systems, solvent environments, and even surface reactivity with mineral/metal species. In particular, the study of ubiquinone and other membrane-bound electron carriers that are coupled to proton-motive systems could be approached with these methods. The conversion of ubiquinone (oxidized) to semiquinone radical to ubiquinol through sequential uptake of a proton and electron is an essential component to transport of electrons in the mitochondrial membrane of most life on Earth and couples proton gradients to electron flow [205]. Investigating radical reactions of metal centers or mineral species with organics could help inform which systems may have served as prebiotic analogs to early cell membranes and the coupling of electromotive and proton-motive forces. Here, EPR offers a unique window into the study of these interactions—generation of organic radicals, identification of their lifetimes, and their transformation to other species together with their potential redox activity with paramagnetic metal centers in prebiotically relevant metal cations.

Alternatively, the Fenton reaction [222], involving the catalytic decomposition of hydrogen peroxide by ferrous iron (Fe^{2+}) to generate highly reactive hydroxyl radicals ($^{\bullet}OH$), a proposed mechanism in the formation and degradation of organics on prebiotic Early Earth as well as Mars [223], is an excellent candidate of study for EPR. This reaction, which can transition Fe^{2+} to Fe^{3+}, facilitates the oxidative breakdown and transformation of simple organic molecules into more complex organics. With the generation and consumption of both organic radicals and redox state change of Fe(II/III) species, EPR spectroscopy emerges as a crucial technique for elucidating the mechanistic pathways of Fenton chemistry in such environments, particularly in consideration of the pH dependence of the reaction steps. The Fenton reaction is most efficient under acidic conditions (typically pH 3–4), which optimizes the generation of hydroxyl radicals and impacts the stability and solubility of Fe^{2+} and Fe^{3+} species [224]. EPR's sensitivity to unpaired electrons allows for the direct detection and characterization of transient radical species, such as hydroxyl, hydroperoxyl, and intermediate organic radicals formed during the reaction [225–228]. By providing detailed insights into the electronic states and local environments of F^{e2+} and Fe^{3+} species, EPR provides an operando examination of these dynamic redox processes. Employed under simulated prebiotic conditions, a deeper understanding of the role and efficiency of Fenton reactions in generating complex organics could be gained.

EPR spectroscopy stands out as an ideal "first pass" non-destructive technique for gaining critical information about organic radical species and other paramagnetic species in unknown samples, particularly in prebiotic chemistry and astrobiology experiments where quantification of relatively dilute (<1 mM) organics is necessary. The unparalleled sensitivity of EPR to trace concentrations of paramagnetic species allows for the detection and characterization of these entities at low abundances (as seen in Figures 5 with detection of organic radical derived from 1 mM *p*-benzoquinone and in Figure 6 with detection of 10 μM Mn^{2+}). Since EPR does not alter the sample, it preserves the integrity of the material for subsequent analyses using more destructive methods like mass spectrometry, gas chromatography, and pyrolysis. This sequential analytical approach ensures that preliminary insights into the electronic environments, spin states, and local interactions of paramagnetic species are obtained without compromising the sample. Thus, integrating EPR as a preliminary analysis step may enhance the overall analytical strategy for characterization of an unknown geological or lab-simulated experimental sample.

Perhaps most impactfully, EPR may also be used to track reaction kinetics in prebiotic experiments, elucidating reaction pathways, and examining abiotic metal-organic coordination and reactivity. Using time-resolved EPR, researchers can directly observe the formation and disappearance of paramagnetic species, providing real-time insights into reaction kinetics. Time-resolved EPR measurements allow for the precise determination of rate constants and the identification of intermediate species, shedding light on the underlying reaction mechanisms, such as the possible redox cycling observed with Fe(II/III) in the presence of quinone/semiquinone/hydroquinone or the ferric iron-mediated oxidation of cysteine to cystine.

6 CONCLUSIONS

Metals and metal-organic mixtures are posited as pivotal aspects of origin of life type events, particularly in metal-rich geological settings such as hydrothermal vents and shallow, terrestrial hot springs. Fe^{2+}, Fe^{3+}, and Mn^{2+} are prebiotically and abiotically relevant metal cations for prebiotic chemistry due to their abundance in early Earth environments and their involvement in redox/catalytic activity in numerous reaction pathways. Prebiotic conditions likely required simple mineral-catalyzed/mediated or metal-organic redox coupled mechanisms for formation of early compounds of interest such as amino acids, electron transport molecules, and phosphorylating agents. These crucial biologically relevant molecules are thought to have emerged abiotically and served potential roles as building blocks for early biomolecules. To study the interactions of these organic and inorganic species with metal species that were likely present on Hadean Earth and may be astrobiologically relevant in geochemical contexts in our solar system, we recommend the incorporation of electron paramagnetic resonance (EPR) as an analytical tool for probing metal-organic complexes, redox couples, and organic radical generation. In this study, we present EPR methodologies tailored for investigations pertinent to prebiotic chemistry, focusing on the interrogation of organo-metallic interactions. Employing both continuous wave and pulsed EPR techniques, we examined the influence of organic and phosphite/phosphate species with metal centers under conditions reminiscent of early Earth.

We find predominantly water coordination of metal species, which is largely consistent with expectations based on relative molar concentrations under the conditions utilized. However, we also observed interaction of organics with ferric and ferrous iron. In this work we find little interactivity with manganese (II), which exhibits nearly exclusive hexa-aqua coordinating environment irrespective of chosen additives. However, iron demonstrated redox coupling with organic species to prebiotic chemistry, including facilitating the generation of a quinone-derived radical as an electron shuttle between Fe^{3+} and Fe^{2+}, as well as serving as an oxidant for cysteine in the formation of disulfide-linked cystine. In the case of phosphates/phosphites, the ferrous compounds remain soluble while, on the timescale of experimentation, ferric compounds reacted readily, forming EPR-silent Fe-phosphorus clusters. This appears consistent with geochemical expectations from a solubility thermodynamics lens. Furthermore, we provide a comprehensive discussion on the implications and ramifications of our findings concerning the application of EPR in laboratory-scale studies of the origins of life, applications for geological samples containing organics, and implications for identification of organic radicals initiated by metal species. We propose promising avenues for future research endeavors in this domain ranging from operando experimentation possibilities to abiotic emergence of proton-coupled electron transport mechanisms and abiotically produced astrobiological signatures of interest.

ACKNOWLEDGMENTS

We thank Dr. Sam Kim (JPL), Matthew E. Luna (JPL/Citrus College), and Charlotte Bays (Natural History Museum, London) for helpful discussions. We acknowledge the support of the Beckman Electron Paramagnetic Resonance (EPR) Facility and the Liquid Nuclear Magnetic Resonance (NMR) Facility, Division of Chemistry and Chemical Engineering at the California Institute of Technology. We thank Dr. David G. VanderVelde for running NMR samples on our behalf. This work was funded by a JPL Spontaneous Research and Technology Development (R&TD), A New Magnetic Electron Detection Technique for Astrobiology. This research was carried out at the Jet Propulsion Laboratory, California Institute of Technology, under a contract with NASA (80NM0018D004).

ABBREVIATIONS AND DEFINITIONS

A	hyperfine coupling constant, the difference in B-field between split spectral lines
$\mathbf{B_0}$	applied magnetic field
$\mathbf{B_{eff}}$	effective magnetic field
C	concentration
C.I.	confidence interval
CW	continuous wave EPR
Cys	cysteine
D	zero-field splitting interaction tensor, which may be decomposed along each principal axis.
$\mathbf{D_2O}$	deuterium dioxide, heavy water
d-d	d-orbital transitions
DSS	trimethylsilylpropanesulfonate
DTPA	diethylenetriamine pentaacetate, a strongly binding ligand to metal centers used in this work
E	energy levels associated with Zeeman splitting

EDTA	ethylenediaminetetraacetic acid, a strongly binding ligand to metal centers used in this work
ENDOR	electron nuclear double resonance, a resonance technique for probing the structure and arrangement of paramagnetic centers, using the electron-nuclear hyperfine couplings of unpaired electrons with neighboring EPR-active nuclei
EPR	electron paramagnetic resonance
Eq	molar equivalents of, the ratio of the moles of one compound to the moles of another
ESEEM	electron spin echo envelope modulation, a magnetic resonance effect that arises from state mixing of hyperfine energy levels common to systems with large nuclear quadrupoles relative to their hyperfine coupling constant (A). ESEEM is observable with pulse EPR and can be used to probe the radial distributions of magnetic nuclear spins about the electron spin source of the echo
FT	Fischer-Tropsch
FTIR	Fourier transform infrared spectroscopy
G	Gauss
g	Landé g-factor
g_e	g-factor for ideal free electron in vacuum, $g_e = 2.00232$
$g_{i,j}$	element of $[g]$ tensor decomposed when projected onto principal axes
Gly	glycine
h	Planck constant
$\hat{H}_{ZFS}$	zero-field splitting Hamiltonian
I	intensity
I	net nuclear moment
ICP	inductively coupled plasma
IOM	insoluble organic matter
ITC	isothermal titration calorimetry
K_a / K_d	association/dissociation constant
K_{sp}	solubility product
l	optical path length
M	molar/molarity (moles/liter)
m_I	nuclear spin quantum number
M-L	metal-ligand
MS	mass spectrometry
M_s	projected spin quanta
m_s	spin quanta
mT	millitesla
N	number of equivalent nuclei
NAD	nicotinamide adenine dinucleotide
NMR	nuclear magnetic resonance

Non-Kramers	systems that do not follow Kramers' degeneracy theorem, i.e., that do not require every energy level to be an even number for a half-integer spin
OES	optical emission spectroscopy
***p*-**	*para-*
PAH	polycyclic aromatic hydrocarbons
***p*D**	negative base-10 logarithm of deuterium concentration as a proxy to activity
***p*H**	negative base-10 logarithm or hydrogen concentration as a proxy to activity
$p\mathrm{K_a}$	acid dissociation/ionization constant, describing the strength of acid in solution based on the dissociation product of the acid-base equilibrium reaction
R^2	coefficient of determination
S	total spin
TEMPO	(2,2,6,6-tetramethylpiperidin-1-yl)oxyl or (2,2,6,6-tetramethyl piperidin-1-yl)oxidanyl, a heterocyclic compound with a stable aminoxyl radical used as a spin standard
UV-Vis	UV-visible spectroscopy
ν	frequency
ZFS	zero-field splitting
ΔG	change in Gibbs free energy of solvation
Δt	time interval
ε	molar attenuation coefficient (absorptivity)
$\lambda = E/D$	rhombicity, min λ=0 (axial), max λ=1/3 (rhombic), on an appropriate choice of axes
μ_B	Bohr magneton
Ω	Ohm, a measure of resistance

APPENDIX: SUPPLEMENTARY INFORMATION

SUPPLEMENTARY 1. UV-VIS CALIBRATION CURVE FOR CONFIRMATION OF SALT CONCENTRATIONS

UV-visible (UV-Vis) photospectrometry was used to confirm the concentration of each metal species in solution, typically prepared by serial dilution from appropriate stocks. Stocks were prepared using metal salt hydrates indicated in the methods at 1 M and diluted in series down to 1 mM for electron paramagnetic resonance (EPR) spectroscopy. Calibrations provide a clear reference for concentration at the indicated wavelengths from SI-1.1 using calibrations constructed in Sup. Fig. 2. Molar attenuation coefficients were determined using the slope of the calibration curves from the Beer-Lambert law,

$$A = \log_{10}\left(\frac{I_0}{I}\right) = \varepsilon \cdot l \cdot C \quad \text{(SI-1)}$$

with A = absorbance, ε = molar attenuation coefficient (absorptivity), l = optical pathlength, and C = species concentration. These values, and the wavelengths of maximum intensity they are based off of, agree well with literature tabulations for these compounds [229–236].

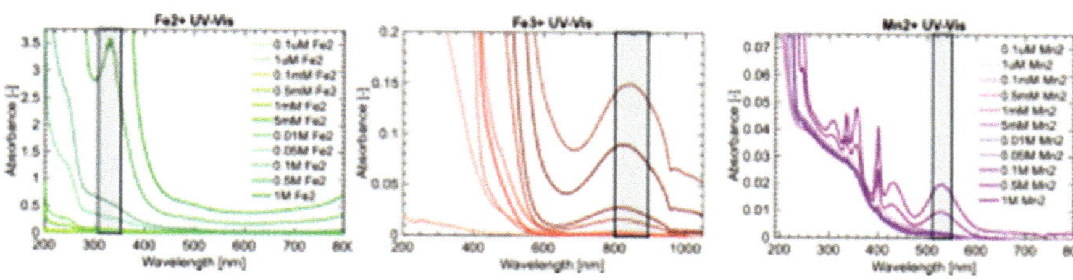

SI-1.1 UV-visible photospectrometry of Fe(II), Fe(III), and Mn(II) hexa-aqua complexes at concentrations as indicated, from 0.1 μM to 1 M corresponding to darkening in color. Center of shaded regions indicates intensity for construction of calibration curves.

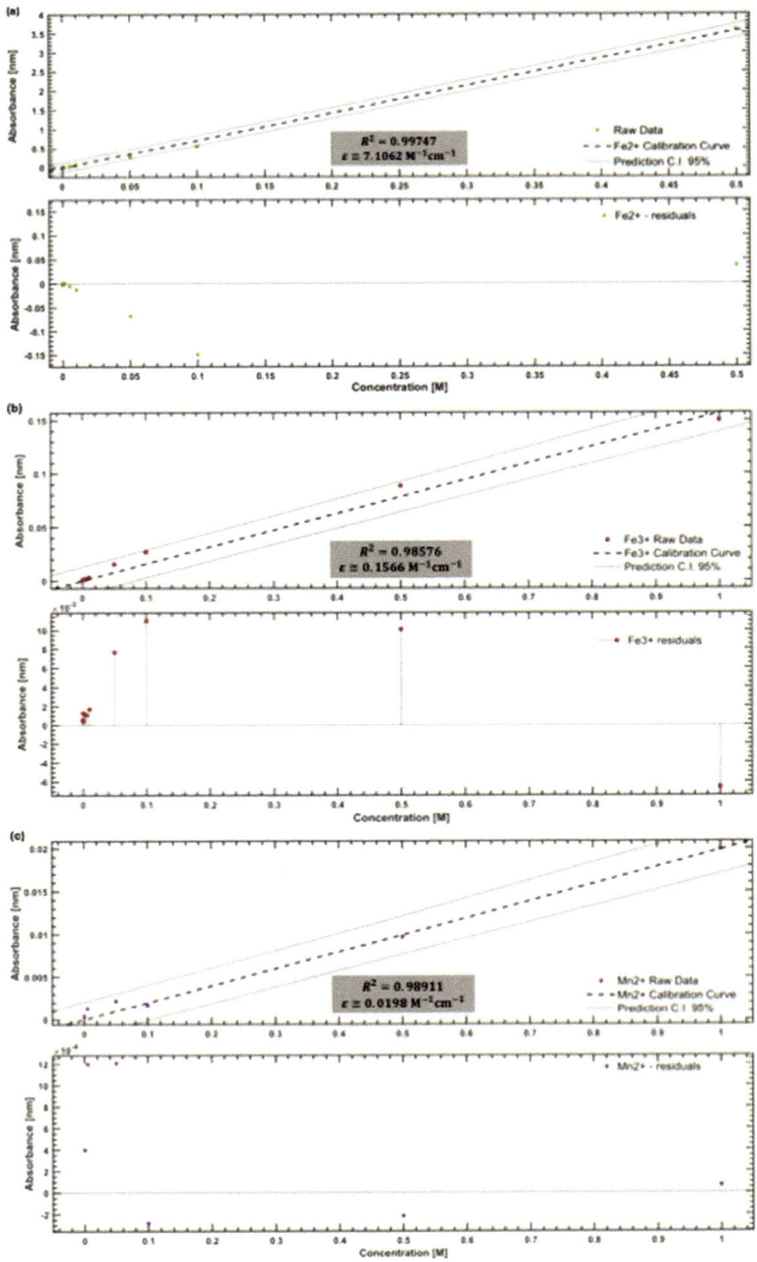

SI-1.2 Beer-Lambert Law calibration curves (a) Fe(II), (b) Fe(III), and (c) Mn(II) hexa-aqua complexes constructed at wavelengths indicated in Sup. Fig. 1. Top plot in each subfigure corresponds to calibration curve with prediction confidence interval (C.I.) 95%; bottom plot corresponds to root-mean square residuals. Coefficient of determination (R^2) shown in each case together with molar absorptivity (slope) for standard 10 mm pathlength.

Supplementary 2. FeS-Peroxide Experiments

Sample Preparation: For solid FeS reactions, approximately 5 mg of mineral was placed directly inside the EPR tube. In the cases where FeS was reacted with organics, a sample from the liquid portion was taken and diluted such that the organic concentration was ~1 mM.

Peroxide: The addition of peroxide to solutions of FeS did not produce any detectable changes in EPR; however, FeS was also not detected. FeS is insoluble in water and it was likely an insufficient quantity was captured when diluting solutions to 1 mM. We also measured solutions of peroxide by itself but did not detect any signals. Even though peroxide generates radicals, these radicals are too short-lived to be detected by EPR in the absence of a spin trapping reagent.

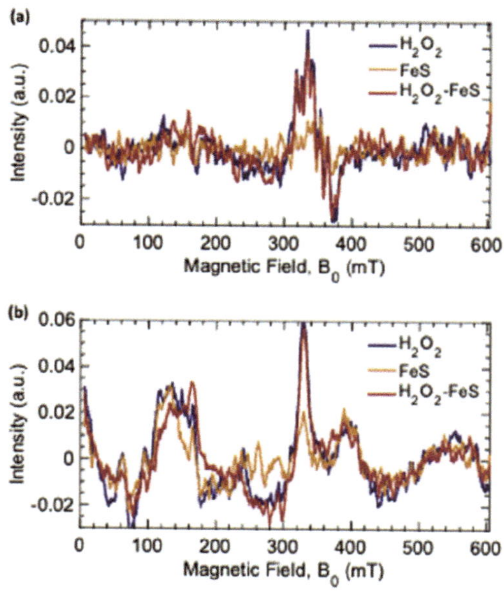

SI- 2.1 1 mM iron(II) sulfide (FeS) EPR spectra in the absence and after treatment by 1mM hydrogen peroxide (H_2O_2), relative to peroxide control. Insufficient FeS may be cause for lack of significant spectral signature, particularly given the baseline intensity of H_2O_2 control has thrice the intensity of untreated FeS mineral.

Supplementary 3. Mineral Work-Ups vs. Salt Systems of Equivalent Concentration

3.1 Synthetic Mineral Controls

For the geologically relevant synthetic minerals with organic inclusions that underwent acid work-up for EPR, 9 controls were run: ferrous iron, ferric iron, and manganese(II) with each of water, 1 mM glycine, 1 mM cysteine, separately. The processing to produce mineral analogs with lyophilization could not be performed fully anoxically, leading to much of the ferrous iron oxidizing to ferric iron (Sup. Fig. 3.1.1). However, as shown in Sup. Fig. 3.1.2, some residual ferrous iron signal is present in parallel mode. Ferric iron and manganese stayed largely unchanged; however, Mn(II) at 10 μM concentrations was too dilute to make out much hyperfine lineshape with a good signal-to-noise ratio.

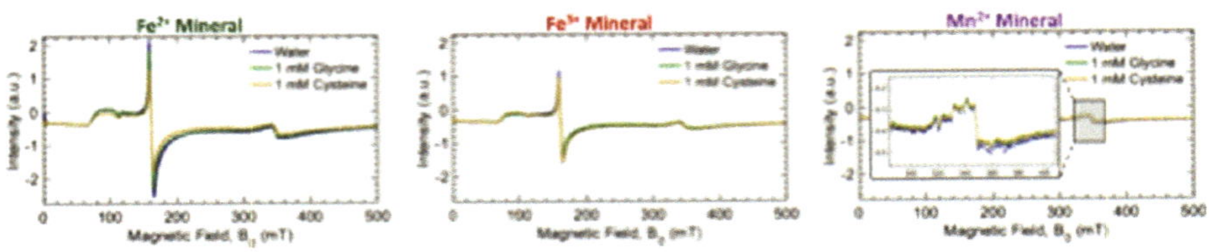

SI-3.1.1 EPR spectra for analog prepared iron and manganesian mineral in the presence of water, glycine, and cysteine. Concentrations of Fe(II), Fe(III), Mn(II) equivalent to the mixed mineral case: 1 mM, 0.29 mM, and 10 μM, respectively.

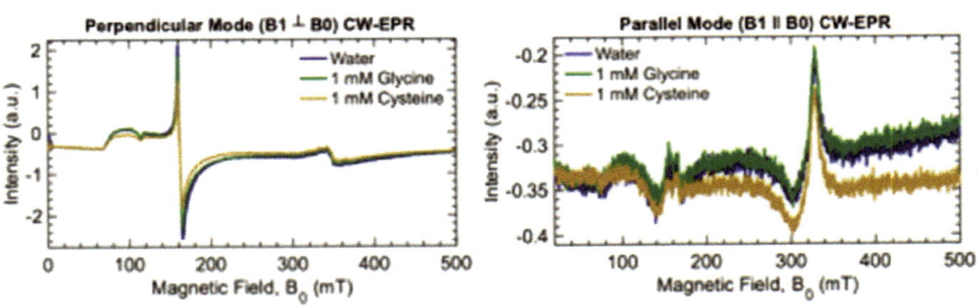

SI-3.1.2 EPR spectra of 1 mM Fe(II) specifically, examined in both perpendicular mode (reprinted from Sup. Fig. 3.1.1.) and parallel mode, in the presence of water, glycine, and cysteine. Post-lyophilization, Fe(II) is still visible in parallel mode, but as seen in perpendicular much of the iron has oxidized to Fe(III) state.

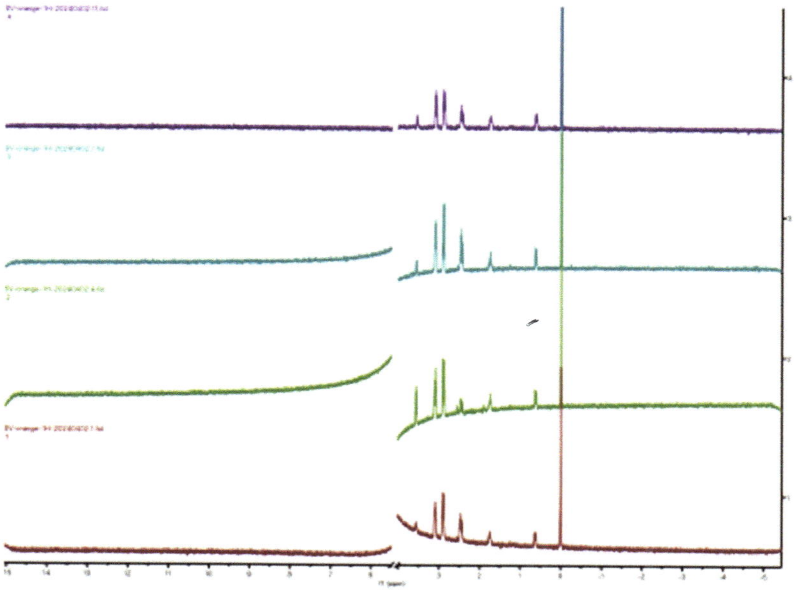

SI-3.2.1 NMR controls of metals in the presence of cysteine. Spectra top to bottom sampled from supernatant of: mixed mineral Fe(II), Fe(III), Mn(II), Mn(II), Fe(III), and Fe(II). In all cases, mineral had fully precipitated out after seven days as described in the methods.

1 mM cysteine behavior, in particular, shows good agreement with pure salt-prepared EPR solutions, cf. Figure 3 in the main text. A significant dampening of the intensity of the main ferric band at 150 mT together indicates the reduction to ferrous iron by cysteine still occurs notwithstanding potentially oxic conditions.

3.2 ^{1}H Nuclear Magnetic Resonance (NMR) Controls for Organics in Mineral Analogues

The supernatant of mineral samples precipitated out with NaOH was sampled anoxically after seven days, as described in the methods. ^{1}H NMR was run on these fluids to confirm presence of organics and that they were unchanged after this waiting period. The spectra for each of Fe(II), Fe(III), and Mn(II) for each of the two amino acids and a blank water reference are shown below. Comments for each are described under the figures. In all cases, features centered about 4.5–5.5 ppm chemical shift correspond to water/deuterated water which has been cut from the spectra for readability.

Spectra for metals with cysteine appear consistent with cysteine reference data, with cluster midpoints near 3.97 ppm and 3.06 ppm under 1D, 500 MHz, H_2O conditions with a DSS (peak at 0.00 ppm) in D_2O reference (HMDB0000574) [237]. In the case of ferric iron, there appear to be additional smaller peaks and shouldering in the vicinity of cluster midpoint 3.85, 3.24, and 3.05 ppm which *prima facie* appear consistent with L-cystine (BMSE000035 / HMDB0000192) [237,238].

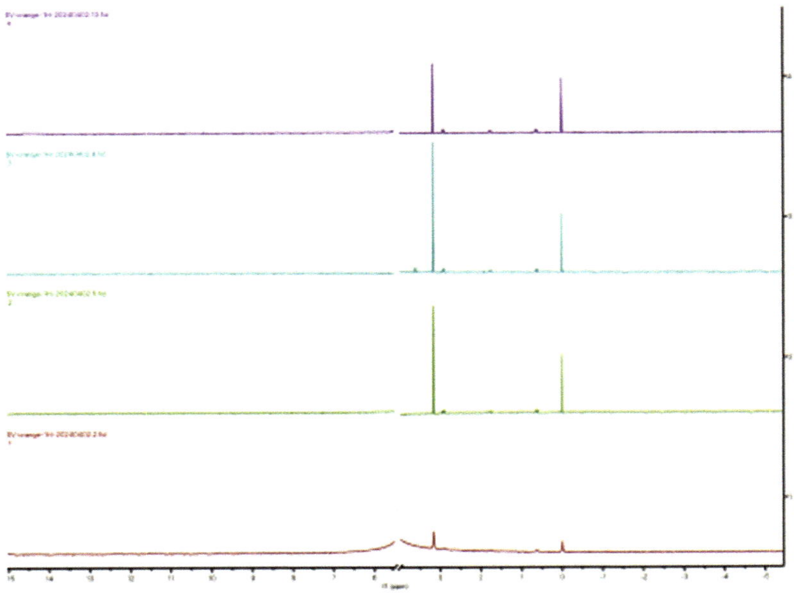

SI-3.2.2 NMR controls of metals in the presence of glycine. Spectra top to bottom sampled from supernatant of: mixed mineral Fe(II), Fe(III), Mn(II), Mn(II), Fe(III), and Fe(II). In all cases, mineral had fully precipitated out after seven days as described in the methods.

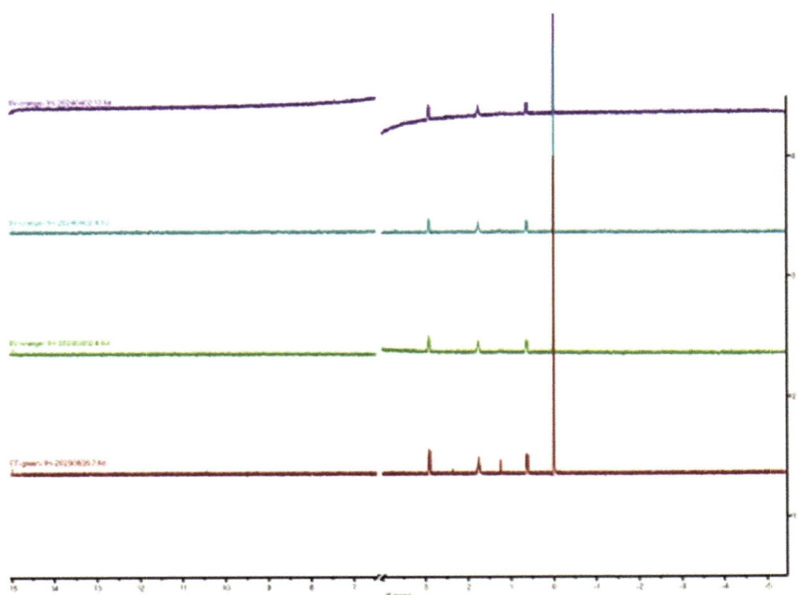

SI-3.2.3 NMR controls of metals in the absence of any organics. Spectra top to bottom sampled from supernatant of: mixed mineral Fe(II), Fe(III), Mn(II), Mn(II), Fe(III), and Fe(II). In all cases, mineral had fully precipitated out after seven days as described in the methods.

Spectra for metals with glycine appear consistent with glycine reference data, with cluster midpoint at approximately 3.54 ppm for the alkyl group ^1H moieties under 1D, 500 MHz, H_2O conditions with a DSS in D_2O reference (HMDB0000123) [237]. Note that amine group ^1H moieties while technically encapsulated by the water peak from the D_2O and therefore invisible in this spectrum.

Spectra for metals in the absence of any amino acids appear consistent with expectations for sodium 3-trimethylsilyl-1-propanesulfonate (DSS) reference data, with cluster midpoints near 2.46, 1.59, 0.50, and 0 under 1D, 500 MHz, H_2O conditions with D_2O (Spectrabase 8e6IL8WPuth) [239]. No additional unexpected features are present.

SUPPLEMENTARY 4. pH VALUES OF SOLUTIONS

At low concentrations amino acids tended toward neutrality (pH = 7). Values largely agreed with values from literature and theory. In the case of amphiprotic molecules (e.g., amino acids), values of pH agree well with their physical chemistry pH estimates:

$$H_3N^+ - R - COOH \xleftarrow{+H^+} H_2N - R - COOH \xrightarrow{-H^+} H_2 - R - COO^- \tag{SI-2}$$

$AA \rightleftharpoons AA^- + H^+$ and $AA + H^+ \rightleftharpoons AAH^+$ at total chemical equilibrium $\Rightarrow [AAH^+] = [AA^-]$

Recalling the defined $K_{a,i}$ values for each of these two equilibria in either of the amphoteric extremes:

$$K_{a,1} \equiv \frac{[AA][H^+]}{[AAH^+]} \text{and}\, K_{a,2} \equiv \frac{[AA^-][H^+]}{[AA]} \tag{SI-3}$$

In total, ($AA \rightleftharpoons AA^- + H^+$) + ($AA + H^+ \rightleftharpoons AAH$) one finds,

$$\frac{[AA^-][H^+]^2}{[AAH^+]} = K_{a,1}K_{a,2} \tag{SI-4}$$

At total chemical equilibrium, substituting for $[AAH^+] = [AA^-]$

$$[H^+]^2 = K_{a,1}K_{a,2} \tag{SI-5}$$

By definition of pH as the reciprocal logarithm of hydronium ions one finds,

$$pH = -\log_{10}([H^+]) = \frac{1}{2}\log_{10}\left(K_{a,1}K_{a,2}\right) = \frac{1}{2}\left(pK_{a,1} + pK_{a,2}\right) \tag{SI-6}$$

We can generalize this for n such dissociation interactions with ease,

TABLE S-4.1
Relevant pH values of metal, phosphate/phosphite, and organic species

Metal Species	Concentration	pH Value
Fe^{3+}	1 mM	2.95
Fe^{2+}	1 mM	4.85
50:50 Fe^{2+}:Fe^{3+}	1 mM:1mM	2.14
Mn^{2+}	1 mM	5.41
Fe^{2+}:Fe^{3+}:Mn^{2+}	1 mM:1mM:1 mM	3.44
Mn^{2+} with HPO_3	1 mM:1 mM	6.38
Mn^{2+} with PO_4	1 mM:1 mM	5.24
Fe^{2+} with PO_4	1 mM:1 mM	4.90
Fe^{3+} with PO_4	1 mM:1 mM	3.17
Fe^{2+}:Fe^{3+} with HPO_3	1 mM:1 mM:1 mM	4.27
Glycine	1M (before serial dilution)	6.00
Cysteine	~1M (before serial dilution)	6.86

TABLE S-4.2
Amino acid reference data including {pK_a}s for a pure (MilliQ deionized) water environment. p$K_{a,1}$ is the dissociation term for -COOH group; p$K_{a,2}$ is the dissociation term for the -NH3; p$K_{a,3}$ is the dissociation term for any other group on the molecule (in the case of Cys, -SH group); and pI is the isoelectric point [191, 192]

Name	MW [g/mol]	Formula	p$K_{a,1}$	p$K_{a,2}$	p$K_{a,3}$	pI
Glycine	75.07	$C_2H_5NO_2$	2.34	9.60	-	5.97
Cysteine	121.16	$C_3H_7NO_2S$	1.96	10.28	8.18	5.07

$$\mathrm{pH} = -\log_{10}([\mathrm{H}^+]) = \frac{1}{n}\log_{10}(K_{a,1}K_{a,2}\cdots K_{a,n})$$
$$= \frac{1}{n}(\mathrm{p}K_{a,1} + \mathrm{p}K_{a,2} + \ldots + \mathrm{p}K_{a,n}) \qquad \text{(SI-7)}$$

Computing these values for Gly and Cys, one finds the expected {pH}s are 5.97 and 6.81, respectively. These values agree well with measurements made in 1 M solutions noted above. Deviations in pH occur as concentration decreases, with values tending to that of neutral water, as expected. The increased difference in the case of cysteine from expected value may reflect drift in the pH probe after numerous measurements or perhaps the lower solubility of cysteine meaning some amino acid was crashing out of solution.

SUPPLEMENTARY 5. OTHER ^{1}H NUCLEAR MAGNETIC RESONANCE (NMR) CONTROLS

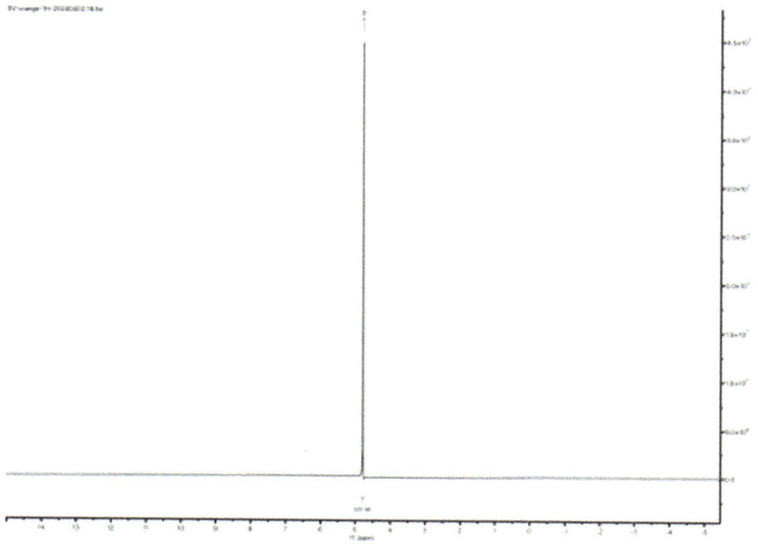

SI-5.1 NMR control of blank water under 1D, 500 MHz, H_2O conditions with a DSS in D_2O reference. The primary peak position and integration are with respect to normalized DSS reference at 0.00 ppm.

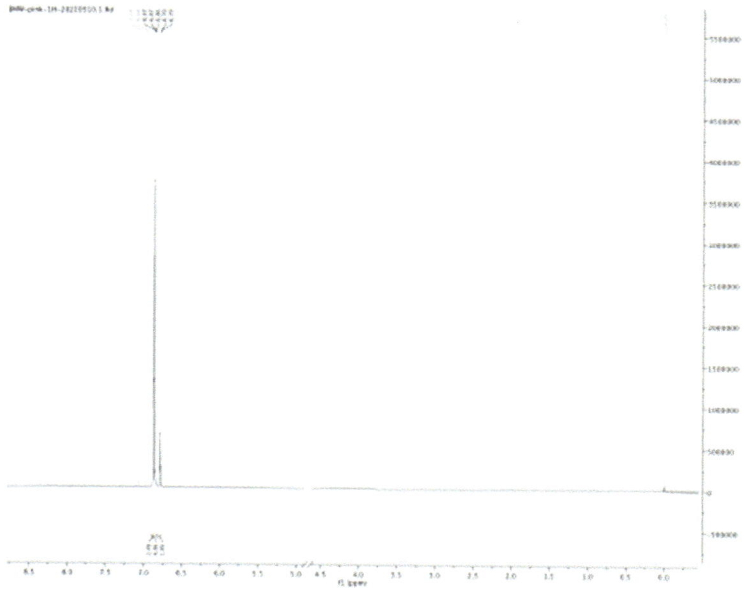

SI-5.2 NMR controls of *p*-benzoquinone under 1D, 500 MHz, H_2O conditions with a DSS in D_2O reference. Spectrum is consistent with 1,4-benzoquinone (Spectrabase 7TpGh1Nd4td) and portions with hydroquinone (HMDB0002434) [237, 240]. The primary peak positions and integrations are with respect to normalized DSS reference at 0.00 ppm.

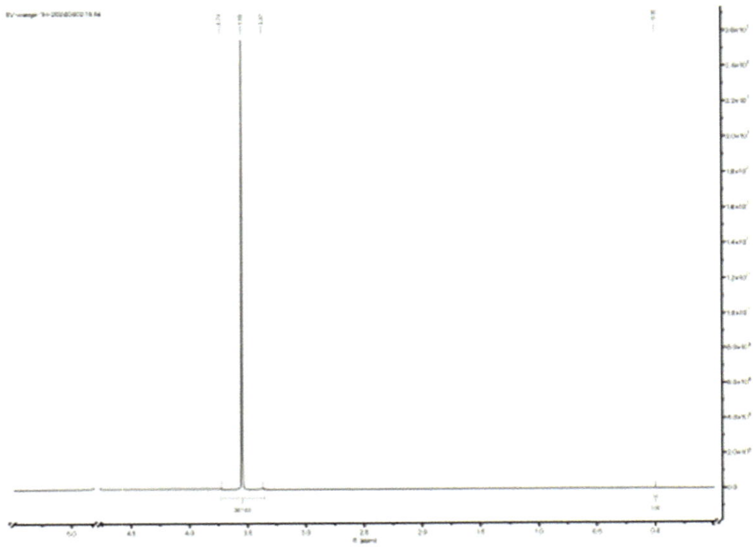

SI-5.3 NMR control of glycine under 1D, 500 MHz, H_2O conditions with a DSS in D_2O reference (HMDB0000123) [237]. The primary peak positions and integrations are with respect to normalized DSS reference at 0.00 ppm.

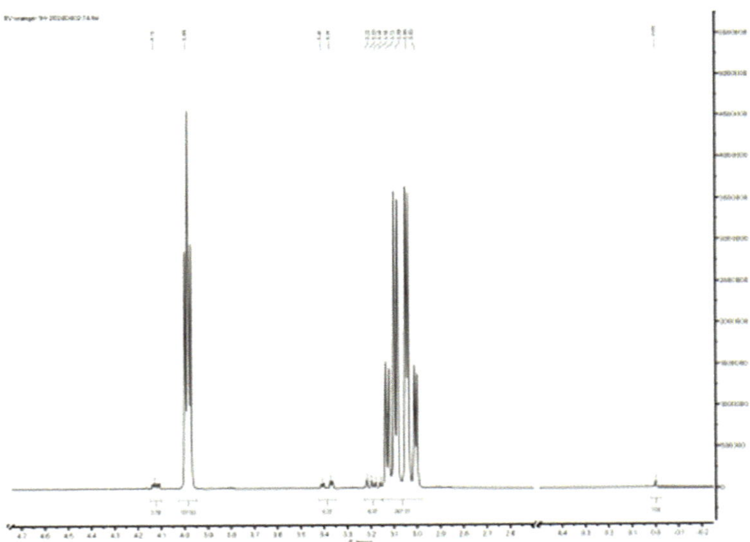

SI-5.4 NMR control of cysteine under 1D, 500 MHz, H_2O conditions with a DSS in D_2O reference (HMDB0000574) [237]. The primary peak positions and integrations are with respect to normalized DSS reference at 0.00 ppm.

Supplementary 6. Deuterium Oxide (D_2O) CW-EPR Controls for Pulsed EPR

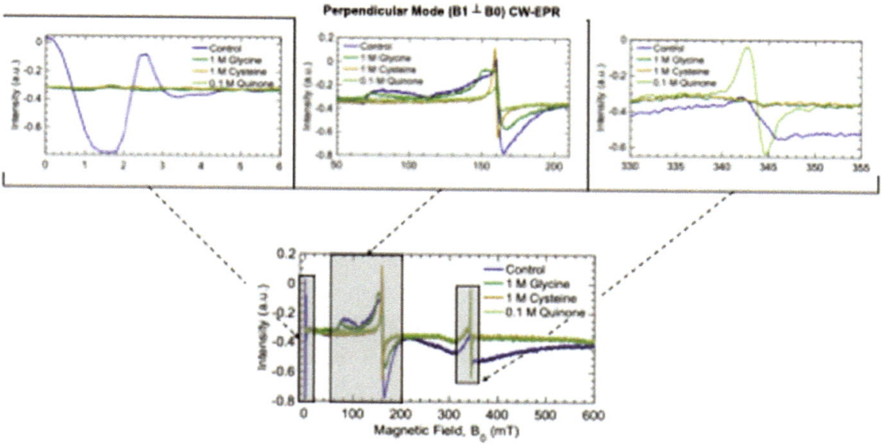

SI-6.1 Perpendicular mode (B1 ⊥ B0) CW-EPR for Fe(III) at the solubility limits of relevant organics of interest. Control case is 99% pure deuterium oxide (D_2O) all other cases have organics at concentrations listed in D_2O.

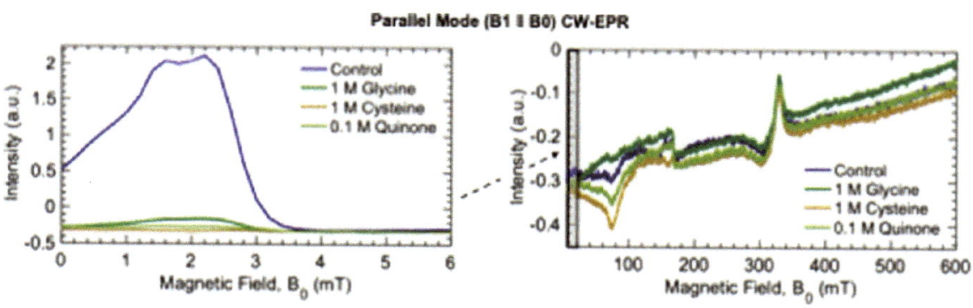

SI-6.2 Parallel mode (B1 ∥ B0) CW-EPR for Fe(III) species at the solubility limits of relevant organics of interest. Feature loss of ferric iron peak from Sup. Fig. 6.1 corresponds to increasing ferrous iron character in parallel mode. Control case is 99% pure deuterium oxide (D_2O) all other cases have organics at concentrations listed in D_2O.

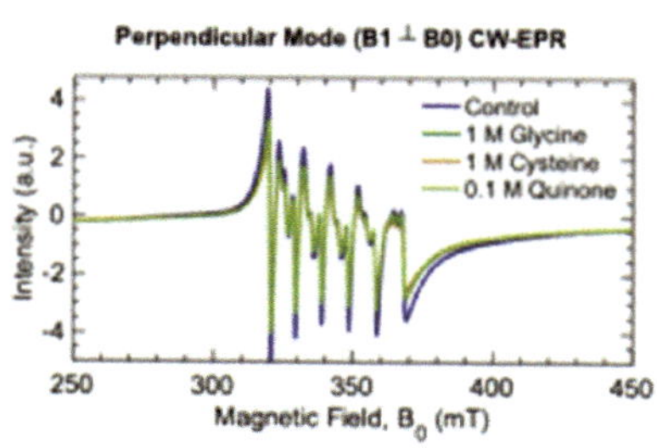

SI-6.3 Perpendicular mode (B1 ⊥ B0) CW-EPR for Mn(II) species at the solubility limits of organics of interest. Control case is 99% pure deuterium oxide (D_2O) all other cases have organics at concentrations listed in D_2O.

Supplementary 7. Pulsed EPR Methods

The two most common techniques for EPR are pulse and continuous wave (CW). Different information can be gleamed from these two techniques. CW spectra are as described above, where the resonance condition is achieved by sweeping external B-field at a constant microwave frequency ν. In pulse EPR, traditionally a high-power square microwave pulse over a wide range of frequencies $\nu \in (-f, f)$ is irradiated onto the sample under constant B-field for a time Δt. The Fourier Transform of the square pulse generates a sinc function which is apodized in frequency domain after $|\pm 1/\Delta t|$, leaving only the central band of approximately uniform excitation. The pulse length is inversely proportional to the excitation bandwidth in this method, which allows increased frequency range of excitation with shorter pulses. Therefore, excitation selectivity is determined by pulse duration. Instrumentation design differences between CW and pulsed EPR means however that there is a deadtime of the detector which is also proportional to excitation frequency bandwidth (shorter pulses allows for less deadtime). While larger frequency bandwidth results in lower sensitivity compared to the CW method, pulsed EPR allows excitation of much of the spectrum in a single pulse or multiple short pulses. In addition, pulsed EPR must be performed at lower temperatures owing to the short magnetization relaxation times, *c.f.* CW measurement which may occur at room temperature for many systems [241,242].

Ligand competition for metal coordination: Samples were prepared identically to the above except using deuterated water (D_2O) and glycerol (D8). A Bruker ELEXSYS E-580 Q-Band Pulse EPR spectrometer for ^2H electron spin-echo envelope modulation (ESEEM) was used in an attempt to probe the environment coordinating Fe^{3+} and Mn^{2+} paramagnetic centers and determine the hydration level the ions bound using the methods described in Hoogstraten and Britt [13].

Owing to the small zero field splitting of Mn^{2+} the ^2H-ESEEM signal was well resolved. However, it shows only marginal differences from the water control case,

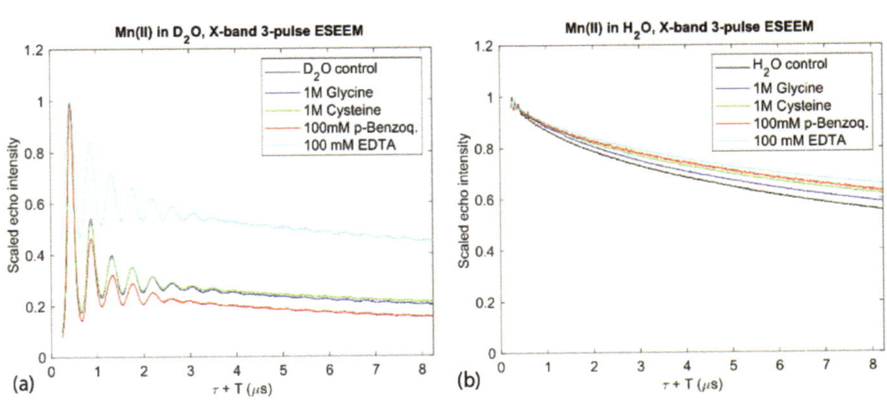

SI-7.1 X-band 3-pulse ESEEM spectra of 1 mM Mn(II) in D_2O (left) and H_2O (right) – deep modulations in the D_2O samples are from weakly coupled ^2H nuclei in the primary coordination sphere – the depth of these modulations are proportional to the number of equivalent coupled nuclei, thus serving as a proxy for the number of coordinated D_2O molecules.

agreeing well with existing continuous wave data reported in the main text. This suggests that there is no significant displacement of H_2O ligands to Mn^{2+} for any of the organics. While run, we have not reported the diethylenetriamine pentaacetate (DTPA) spectrum because its low solubility—worsened by freezing—meant there was insufficient DTPA to be observed in the field swept spectra and the ESEEM modulation depth was nearly unchanged. Instead, we report ethylenediaminetetraacetic acid (EDTA) ESEEM signal. EDTA is another tight binding ligand that, as expected, dropped the modulation depth considerably relative to water.

Meanwhile, we do not report Fe^{3+} results given the poor signal-to-noise observed. We attribute the tremendous loss in high spin Fe^{3+} signal intensity in the presence of most of the organics to the interactions/reactions described in the main text between Fe^{3+} and organics and the linewidth changes we have already discussed regarding the CW-EPR. The significant loss of intensity of the zero-field transition is further indicative of these reactions occurrence, while simultaneously frustrating the utility of ESEEM.

In the case of amino acids, the samples on multiple repetitions showed very little observable spin echo signal, while the non-glycine sample looked identical to the previous samples run in CW. While exploration of this feature was beyond the scope of this work, we suggest that perhaps clustering or antiferromagnetic dimer formation by glycine samples is limiting detection via pulse EPR uniquely without affecting detection via CW-EPR.

Supplementary 8. CW-EPR Cavity Background

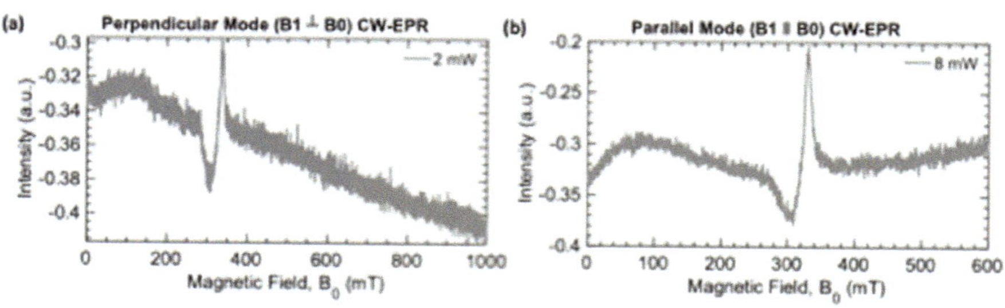

SI-8.1 Perpendicular and parallel mode at 2 mW and 8 mW microwave power, respectively, of cavity background for CW-EPR at 5K, 8GMA.

REFERENCES

MAIN TEXT REFERENCES

1. M. Y. Bakr, M. Akiyama, Y. Sanada. *Org. Geochem.* **1990**, *15* (6), 595–9.
2. L. Le Pape. *Application of EPR in Studies of Archaeological Samples*, in *Modern Magnetic Resonance*, Ed. G. A. Webb. Springer, Cham. 2018, pp. 305–29.
3. M. Bourbin, D. Gourier, S. Derenne, L. Binet, Y. Le Du, F. Westall, B. Kremer, P. Gautret. *Astrobiology* **2013**, *13* (2), 151–62.
4. D. Gourier, L. Binet, H. Vezin. *EPR of Primitive Organic Matter: A Tool for Astrobiology*, in *Applications of EPR in Radiation Research*, Eds A. Lund, M. Shiotani. Springer, Cham. 2014, pp. 541–77.
5. L. Binet, D. Gourier, A. Skrzypczak, S. Derenne, F. Robert. *Insoluble Organic Matter in Carbonaceous Chondrites and Archean Cherts*, in *Astrobiology: Future Perspectives*, Eds P. Ehrenfreund, W. Irvine, T. Owen, L. Becker, J. Blank, J. Brucato, L. Colangeli, S. Derenne, A. Dutrey, D. Despois, A. Lazcano, F. Robert. Springer, Dordrecht. 2004, pp. 333–57.
6. M. Bourbin, S. Derenne, D. Gourier, J.-N. Rouzaud, P. Gautret, F. Westall. *Orig. Life Evol. Biosph.* **2012**, *42* (6), 569–85.
7. S. Huld, S. McMahon, S. Sjöberg, P. Huang, A. Neubeck. *Astrobiology* **2023**, *23* (1), 24–32.
8. S. S. Kim, J. R. Bargar, K. H. Nealson, B. E. Flood, J. L. Kirschvink, T. D. Raub, B. M. Tebo, M. Villalobos. *Astrobiology* **2011**, *11* (8), 775–86.
9. I. A. Shkrob, T. M. Marin, A. Adhikary, M. D. Sevilla. *J. Phys. Chem. C* **2011**, *115* (8), 3393–3403.
10. T. Milojevic, D. Kölbl, L. Ferrière, M. Albu, A. Kish, R. L. Flemming, C. Koeberl, A. Blazevic, Z. Zebec, S. K.-M. R. Rittmann, C. Schleper, M. Pignitter, V. Somoza, M. P. Schimak, A. N. Rupert. *Sci. Rep.* **2019**, *9* (1), 18028.
11. Y. Pan, M. J. Nilges. *Rev. Mineral. Geochem.* **2014**, *78* (1), 655–90.
12. I. D. Sahu, R. M. McCarrick, G. A. Lorigan. *Biochem.* **2013**, *52* (35), 5967–84.
13. C. G. Hoogstraten, R. D. Britt. *RNA* **2002**, *8* (2), 252–60.
14. M. P. Hendrich, P. G. Debrunner. *J. Biophys.* **1989**, *56* (3), 489–506.
15. S. Van Doorslaer. *Continuous Wave and Pulsed EPR Analyses of Metalloproteins*, in *Electron Paramagnetic Resonance*; Ed. B. C. Gilbert; Royal Society of Chemistry: Cambridge, 2008; Vol. 21, pp. 162–183.
16. P. J. Chmielewski, L. Latos-Grażyński. *Inorg. Chem.* **1997**, *36* (5), 840–5.
17. M. G. Bakker, B. Fowler, M. K. Bowman, G. S. Patience. *Can. J. Chem. Eng.* **2020**, *98* (8), 1668–81.
18. W. Nitschke, M. J. Russell. *J. Mol. Evol.* **2009**, *69* (5) 481–96.
19. C. Huber, G Wächtershäuser. *Science* **1998**, *281* (5377), 670–2.
20. G. D. Cody, N. Z. Boctor, T. R. Filley, R. M. Hazen, J. H. Scott, A. Sharma, H. S. Yoder. *Science* **2000**, *289* (5483), 1337–40.
21. M. T. Beck, J. Ling. *Naturwissenschaften* **1977**, *64* (2), 91.
22. T. P. Begley. *Nat. Prod. Rep.* **2006**, *23* (1), 15–25.
23. J. Hao, M. Mokhtari, U. Pedreira-Segade, L. J. Michot, I. Daniel. *ACS Earth Space Chemn.* **2019**, *3* (1), 109–19.
24. N. Lane. *BioEssays* **2017**, *39* (6), 1600217.
25. M. J. Russell, A. J. Hall. *The onset and early evolution of life,* in *Evolution of Early Earth's Atmosphere, Hydrosphere, Biosphere—Constraints from Ore Deposits: Geological Society of America Memoir*, Eds S. E. Kesler, H. Ohmoto, 2006, 198, pp. 1–32.

26. W. Martin, J. Baross, D. Kelley, M. J. Russell. *Nat. Rev. Microbio.* **2008**, *6* (11), 805–14.
27. W. Nitschke, M. J. Russell. *J. Mol. Evol.* **2009**, *69* (5), 481–96.
28. B. T. Burcar, L. M. Barge, D. Trail, E. B. Watson, M. J. Russell, L. B. McGown. *Astrobiology* **2015**, *15* (7), 509–22.
29. E. Branscomb, T. Biancalani, N. Goldenfeld, M. Russell. *Phys. Rep.* **2017**, *677*, 1–60.
30. M. J. Van Kranendonk, R. Baumgartner, T. Djokic, T. Ota, L. Steller, U. Garbe, E. Nakamura. *Astrobiology,* **2021**, *21* (1), 39–59.
31. B. Damer, D. Deamer. *Astrobiology* **2020**, *20* (4), 429–52.
32. S. L. Miller, J. L. Bada. *Nature* **1988**, *334* (6183), 609–11.
33. M. J. Van Kranendonk, D. W. Deamer, T. Djokic. *Sci. Am.* **2017**, *317* (2), 28–35.
34. A. C. Harris, N. C. White, J. McPhie, S. W. Bull, M. A. Line, R. Skrzeczynski, T. P. Mernagh, R. M. Tosdal. *Econ. Geo.* **2009**, *104* (6), 793–814.
35. V. Sojo, B. Herschy, A. Whicher, E. Camprubí, N. Lane. *Astrobiology* **2016**, *16* (2), 181–97.
36. T. Shibuya, M. Yoshizaki, M. Sato, K. Shimizu, K. Nakamura, S. Omori, K. Suzuki, K. Takai, H. Tsunakawa, S. Maruyama. *Prog. Earth Planet. Sci.* **2015**, *2* (1), 46.
37. A. Y. Borisova, A. Nédélec, N. R. Zagrtdenov, M. J. Toplis, W. A. Bohrson, O. G. Safonov, I. N. Bindeman, O. E. Melnik, G. S. Pokrovski, G. Ceuleneer, K. P. Jochum, B. Stoll, U. Weis, A. Y. Bychkov, A. A. Gurenko. *Geology* **2022**, *50* (3), 300–4.
38. T. M. Harrison. *Radionuclide Produced Isotopic Variations in Mantle Rocks*, in *Hadean Earth*, Ed. T. M. Harrison. Springer, Cham, 2020, pp. 39–58.
39. N. Drabon, B. L. Byerly, G. R. Byerly, J. L. Wooden, C. B. Keller, D. R. Lowe. *Proc. Nat. Acad. Sci.* **2021**, *118* (8), e2004370118.
40. J. Korenaga. *Precambrian Res.* **2021**, *359*, 106178.
41. C. Konn, J. L. Charlou, N. G. Holm, O. Mousis. *Astrobiology* **2015**, *15* (5), 381–99.
42. J. M. Petersen, F. U. Zielinski, T. Pape, R. Seifert, C. Moraru, R. Amann, S. Hourdez, P. R. Girguis, S. D. Wankel, V. Barbe, E. Pelletier, D. Fink, C. Borowski, W. Bach, N. Dubilier. *Nature* **2011**, *476* (7359), 176–80.
43. L E. Rodriguez, T. Altair, N. Y. Hermis, T. Z. Jia, T. P. Roche, L. H. Steller, J. M. Weber, *A Geological and Chemical Context for the Origins of Life on Early Earth*, in *The Astrobiology Primer 3.0,* Ed S. L. Cady, Mary Ann Liebert, Inc. Publishers, *Astrobiology* 2024, *24* (S1), S-76-S-106.
44. K. Chin, J. Pasalic, N. Hermis, L. M. Barge. *ChemPlusChem* **2020**, *85* (12), 2619–28.
45. M. Yamamoto, Y. Takaki, H. Kashima, M. Tsuda, A. Tanizaki, R. Nakamura, K. Takai. *ISME J* **2023**, *17* (1), 12–20.
46. A. Yamaguchi, M. Yamamoto, K. Takai, T. Ishii, K. Hashimoto, R. Nakamura. *Electrochimica Acta* **2014**, *141*, 311–18.
47. B. R. T. Simoneit. *Org. Geochem.* **1984**, *6*, 857–64.
48. L. M. Barge, E. Flores, M. M. Baum, D. G. VanderVelde, M. J. Russell. *Proc. Nat. Acad. Sci.* **2019**, *116* (11), 4828–33.
49. L. M. Barge, E. Flores, D. G. VanderVelde, J. M. Weber, M. M. Baum, A. Castonguay. *J. Geophys. Res: Planets* **2020**, *125* (11), e2020JE006423.
50. K. B. Muchowska, S. J. Varma, E. Chevallot-Beroux, L. Lethuillier-Karl, G. Li, J. Moran. *Nat. Ecol. Evol.* **2017**, *1* (11), 1716–21.
51. Z. Zhao, Q. Feng, H. Lu, P. Peng, T. Zhang, C. S. Hsu. *ACS Energ. Fuels* **2021**, *35* (15), 11909–19.
52. A. L. Weber. *J. Mol. Evol.* **1992**, *35* (1), 1–6.
53. D. Kopetzki, M. Antonietti. *N. J. Chem.* **2011**, *35* (9), 1787–94.
54. W. Takahagi, K. Seo, T. Shibuya, Y. Takano, K. Fujishima, M. Saitoh, S. Shimamura, Y. Matsui, M. Tomita, K. Takai. *ACS Earth Space Chem.* **2019**, *3* (11), 2559–2568.

55. G. Macleod, C. McKeown, A. J. Hall, M. J. Russell. *Orig. Life Evol. Biosphere* **1994**, *24* (1), 19–41.
56. C. Z. Jiang, N. J. Tosca. *Earth Planet. Sci. Let.* **2019**, *506*, 231–42.
57. N. J. Tosca, C. Z. Jiang, B. Rasmussen, J. Muhling. *Free Rad. Bio. Med.* **2019**, *140*, 138–53.
58. N. J. Tosca, S. Guggenheim, P. K. Pufahl. *Geo. Soc. Amer. Bul.* **2016**, *128* (3–4), 511–30.
59. I. Halevy, M. Alesker, E. M. Schuster, R. Popovitz-Biro, Y. Feldman. *Nat. Geosci.* **2017**, *10* (2), 135–9.
60. H. D. Holland. *The Geologic History of Seawater*, in *Treatise on Geochemistry*, Elsevier, 2007, pp. 1–46.
61. H. D. Holland. *The Chemical Evolution of the Atmosphere and Oceans*, in *Princeton Series in Geochemistry*, Princeton University Press, 1984.
62. M. A. Pasek. *Geosci. Front.* **2017**, *8* (2), 329–35.
63. M. A. Pasek, M. Gull, B. Herschy. *Chem. Geo.* **2017**, *475*, 149–70.
64. D.-W. Cho, H. Song, F. W. Schwartz, B. Kim, B.-H. Jeon. *Chemosphere* **2015**, *125*, 41–9.
65. L. Jia, H. Xue, F. Xian, Y. Sugahara, N. Sakai, J. Nan, Y. Yamauchi, T. Sasaki, R. Ma. *Small* **2023**, *19* (42), 2303221.
66. W. Wang, B. Yang, Y. Qu, X. Liu, W. Su. *Astrobiology* **2011**, *11* (5), 471–6.
67. G. Caserta, L. Zuccarello, C. Barbosa, C. M. Silveira, E. Moe, S. Katz, P. Hildebrandt, I. Zebger, S. Todorovic. *Coord. Chem. Rev.* **2022**, *452*, 214287.
68. B. Zhang, S. Bandyopadhyay, P. Shakamuri, S. G. Naik, B. H. Huynh, J. Couturier, N. Rouhier, M. K. Johnson. *J. Amer. Chem. Soc.* **2013**, *135* (40), 15153–64.
69. K. H. Ebrahimi, S. Ciofi-Baffoni, P.-L. Hagedoorn, Y. Nicolet, N. E. Le Brun, W. R. Hagen, F. A. Armstrong. *Nat. Chem.* **2022**, *14* (3), 253–66.
70. H. Sawai. *J. Amer. Chem. Soc.* **1976**, *98* (22), 7037–39.
71. B. Herschy, S. J. Chang, R. Blake, A. Lepland, H. Abbott-Lyon, J. Sampson, Z. Atlas, T. P. Kee, M. A. Pasek. *Nat. Comms.* **2018**, *9* (1), 1346.
72. Y. Shigemasa, C. Sakazawa, R. Nakashima, T. Matsuura, in *Origin of Life*, Ed H. Noda, 1978, Center for Academic Publications Japan, Tokyo, pp. 211–16.
73. Y. Shigemasa, O. Nagae, C. Sakazawa, R. Nakashima, T. Matsuura. *J. Am. Chem. Soc.* **1978**, *100* (4), 1309–10.
74. R. Breslow. *Tetrahedron Let.* **1959**, *1* (21), 22–6.
75. E. S. Lox, G. B. Marin, E. De Grave, P. Bussière. *App. Cat.* **1988**, *40*, 197–218.
76. D. R. Milburn, K. V. R. Chary, B. H. Davis. *App. Cat. A: Gen.* **1996**, *144* (1), 121–32.
77. D. R. Milburn, K. V. R. Chary, R. J. O'Brien, B. H. Davis. *App. Cat. A: Gen.* **1996**, *144* (1), 133–46.
78. S. Li, W. Ding, G. D. Meitzner, E. Iglesia. *J. Phys. Chem. B* **2002**, *106* (1), 85–91.
79. R. J. O'Brien, L. Xu, R. L. Spicer, B. H. Davis. *ACS Energy Fuels* **1996**, *10* (4), 921–6.
80. T. M. McCollom, J. S. Seewald. *Chem. Rev.* **2007**, *107* (2), 382–401.
81. N. X. Nie, N. Dauphas, R. C. Greenwood. *Earth Planet. Sci. Let.* **2017**, *458*, 179–91.
82. T. Shibuya, M. J. Russell, K. Takai. *Geochim. Cosmochim. Acta* **2016**, *175*, 1–19.
83. S. Ranjan, Z. R. Todd, P. B. Rimmer, D. D. Sasselov, A. R. Babbin. *Geochem. Geophys. Geosys.* **2019**, *20* (4), 2021–39.
84. L. T. Elkins-Tanton. *Earth Planet. Sci. Let.* **2008**, *271* (1), 181–91.
85. S. Morrison, S. Runyon, R. Hazen. *Life* **2018**, *8* (4), 64.
86. R. M. Hazen. *Am. J. Sci.* **2013**, *313* (9), 807–43.
87. N. L. Lanza, W. W. Fischer, R. C. Wiens, J. Grotzinger, A. M. Ollila, A. Cousin, R. B. Anderson, B. C. Clark, R. Gellert, N. Mangold, S. Maurice, S. Le Mouélic, M. Nachon, M. Schmidt, J. Berger, S. M. Clegg, O. Forni, C. Hardgrove, N. Melikechi, H. E. Newsom, V. Sautter. *Geophys. Res. Let.* **2014**, *41* (16), 5755–63.

88. P. J. Gasda, N. L. Lanza, P.-Y. Meslin, S. N. Lamm, A. Cousin, R. Anderson, O. Forni, E. Swanner, J. L'Haridon, J. Frydenvang, N. Thomas, S. Gwizd, N. Stein, W. W. Fischer, J. Hurowitz, D. Sumner, F. Rivera-Hernández, L. Crossey, A. Ollila, A. Essunfeld, H. E. Newsom, B. Clark, R. C. Wiens, O. Gasnault, S. M. Clegg, S. Maurice, D. Delapp, A. Reyes-Newell. *J. Geophys. Res.: Planets* **2024**, *129* (5), e2023JE007923.
89. L. E. Rodriguez, M. A. Pasek, D. VanderVelde, L. M. Barge, "*Are Mixed Fe/Mn (oxy) hydroxides Key to Enabling Aqueous Phosphorylation via Fenton-Phosphite Chemistry?*", *The Astrobiology Science Conference (AbSciCon),* 2022, pp. 432–404.
90. D. Trail, T. M. McCollom. *Science* **2023**, *379* (6632), 582–586.
91. D. S. Kelley, J. A. Baross, J. R. Delaney. *An. Rev. Earth Planet. Sci.* **2002**, *30* (1), 385–491.
92. M. K. Tivey. *Oceanography* **2007**, *20* (1), 50–65.
93. R. E. Price, D. E. LaRowe, F. Italiano, I. Savov, T. Pichler, J. P. Amend. *Chem. Geo.* **2015**, 407–408, 21–45.
94. D. A. M. Zaia, C. T. B. V. Zaia, H. De Santana. *Origins of Life and Evolution of Biospheres* **2008**, *38* (6), 469–88.
95. S. L. Miller. *Origins of Life* **1974**, *5* (1–2), 139–51.
96. D. P. Glavin, J. P. Dworkin, A. Aubrey, O. Botta, J. H. Doty, Z. Martins, J. L. Bada. *Met. Planet. Sci.* **2006**, *41* (6), 889–902.
97. J. H. Cleaves, J. H. Chalmers, A. Lazcano, S. L. Miller, J. L. Bada. *Origins Life Evol. Biosph.* **2008**, *38* (2), 105–15.
98. M. Ferus, P. Rimmer, G. Cassone, A. Knížek, S. Civiš, J. E. Šponer, O. Ivanek, J. Šponer, H. Saeidfirozeh, P. Kubelík, R. Dudžák, L. Petera, L. Juha, A. Pastorek, A. Křivková, M. Krůs. *Astrobiology* **2020**, *20* (12), 1476–88.
99. E. T. Parker, H. J. Cleaves, J. P. Dworkin, D. P. Glavin, M. Callahan, A. Aubrey, A. Lazcano, J. L. Bada. *Proc. Nat. Acad. Sci.* USA **2011**, *108* (14), 5526–31.
100. C. S. Foden, S. Islam, C. Fernández-García, L. Maugeri, T. D. Sheppard, M. W. Powner. *Science* **2020**, *370* (6518), 865–9.
101. Muchowska, Kamila B, Joseph Moran. *Science* **2020**, *370* (6518), 767–8.
102. V. Cuculić, I. Pižeta, M. Branica. *J. Electroanal. Chem.* **2005**, *583* (1), 140–7.
103. R. F. de Farias. *Química Nova* **2002**, *25* (5), 729–30.
104. N. Tanaka, I. M. Kolthoff, W. Stricks. *J. Am. Chem. Soc.* **1955**, *77* (7), 1996–2004.
105. G. Steinman, A. E. Smith, J. J. Silver. *Science* **1968**, *159* (3819), 1108–9.
106. A. Amirbahman, L. Sigg, U. von Gunten. *J. Coll. Interf. Sci.* **1997**, *194* (1), 194–206.
107. A. Bhattacharyya, E. Stavitski, J. Dvorak, C. E. Martínez. *Geochim. Cosmochim. Acta* **2013**, *122*, 89–100.
108. E. Flores, E. Martinez, L. E. Rodriguez, J. M. Weber, A. Khodayari, D. G. VanderVelde, L. M. Barge. *ACS Earth Space Chem* **2021**, *5* (5), 1048–57.
109. C. Bonfio, E. Godino, M. Corsini, F. F. de Biani, G. Guella, S. S. Mansy. *Nat. Catal.* **2018**, *1* (8), 616–23.
110. S. A. Sanden, R. Yi, M. Hara, S. E. McGlynn. *Chem. Comms.* **2020**, *56* (80), 11989–92.
111. J. M. Weber, B. L. Henderson, D. E. LaRowe, A. D. Goldman, S. M. Perl, K. Billings, L. M. Barge. *Astrobiology* **2022**, *22* (1), 25–34.
112. A. de Aldecoa, F. Roldán, C. Menor-Salván. *Life* **2013**, *3* (3), 502–17.
113. A. Arya, J. Ray, S. Sharma, R. C. Simbron, A. Lozano, H. B. Smith, J. L. Andersen, H. Chen, M. Meringer, H. J. Cleaves. *Chem. Sci.* **2022**, *13* (17), 4838–53.
114. T. Franza, P. Gaudu. *Res. in Microbio.* **2022**, *173* (6), 103953.
115. H. J. Hageman, U. E. Wiersum. *Tetrahedron Let.* **1971**, *12* (45), 4329–32.
116. C. D. Stevenson, L. J. Heinle, R. C. Reiter. *J. Org. Chem.* **2002**, *67* (1), 119–24.
117. L. J. Allamandola, G. G. M. Tielens, J. R. Barker. *Astrophys. J. Sup. Series* **1989**, *71*, 733.

118. M. P. Bernstein, S. A. Sandford, L. J. Allamandola, J. S. Gillette, S. J. Clemett, R. N. Zare. *Science* **1999**, *283* (5405), 1135–8.
119. M. A. Pasek, D. S. Lauretta. *Astrobiology* **2005**, *5* (4), 515–35.
120. M. A. Pasek. *Proc. Nat. Acad. Sci.* USA **2008**, *105* (3), 853–8.
121. M. A. Pasek, J. P. Harnmeijer, R. Buick, M. Gull, Z. Atlas. *Proc. Nat. Acad. Sci.* USA **2013**, *110* (25), 10089–94.
122. M. A. Pasek, A. Omran, T. Feng, M. Gull, C. Lang, J. Abbatiello, L. Garong, R. Johnston, J. Ryan, H. Abbott-Lyon. *Geochim. Cosmochim. Acta* **2022**, *336*, 332–40.
123. N. J. Planavsky, O. J. Rouxel, A. Bekker, S. V. Lalonde, K. O. Konhauser, C. T. Reinhard, T. W. Lyons. *Nature,* **2010**, *467* (7319), 1088–90.
124. C. T. Adcock, E. M. Hausrath. *Astrobiology* **2015**, *15* (12), 1060–75.
125. A. H. Treiman, N. L. Lanza, S. VanBommel, J. Berger, R. Wiens, T. Bristow, J. Johnson, M. Rice, R. Hart, A. McAdam, P. Gasda, P.-Y. Meslin, A. Yen, A. J. Williams, A. Vasavada, D. Vaniman, V. Tu, M. Thorpe, E. D. Swanner, C. Seeger, S. P. Schwenzer, S. Schröder, E. Rampe, W. Rapin, S. J. Ralston, T. Peretyazhko, H. Newsom, R. V. Morris, D. Ming, M. Loche, S. Le Mouélic, C. House, R. Hazen, J. P. Grotzinger, R. Gellert, O. Gasnault, W. W. Fischer, A. Essunfeld, R. T. Downs, G. W. Downs, E. Dehouck, L. J. Crossey, A. Cousin, J. M. Comellas, J. V. Clark, B. Clark III, S. Chipera, G. Caravaca, J. Bridges, D. F. Blake, R. Anderson. *Minerals* **2023**, *13* (9), 1122.
126. F. Postberg, Y. Sekine, F. Klenner, C. R. Glein, Z. Zou, B. Abel, K. Furuya, J. K. Hillier, N. Khawaja, S. Kempf, L. Noelle, T. Saito, J. Schmidt, T. Shibuya, R. Srama, S. Tan. *Nature* **2023**, *618* (7965), 489–93.
127. C. R. Walton, O. Shorttle, F. E. Jenner, H. M. Williams, J. Golden, S. M. Morrison, R. T. Downs, A. Zerkle, R. M. Hazen, M. Pasek. *Earth-Sci. Rev.* **2021**, *221*, 103806.
128. J. D. Toner, D. C. Catling. *Proc. Nat. Acad. Sci.* USA **2020**, *117* (2), 883–8.
129. M. A. Pasek, T. P. Kee, D. E. Bryant, A. A. Pavlov, J. I. Lunine. *Ang. Chem. Int. Ed.* **2008**, *47* (41), 7918–20.
130. J. A. Weil, J. R. Bolton. *Electron Paramagnetic Resonance: Elementary Theory and Practical Applications.* John Wiley & Sons, 2007.
131. F. E. Mabbs. *Chem. Soc. Rev.* **1993**, *22* (5), 313–24.
132. F. E. Mabbs, D. Collison. *Electron Paramagnetic Resonance of d Transition Metal Compounds.* Elsevier, 2013.
133. H. Hosoya. *J. Math. Chem.* **1998**, *23* (1), 169–78.
134. T. A. Shaler. *J. Chem. Edu.* **1991**, *68* (10), 853.
135. A. Abragam, B. Bleaney. *Electron Paramagnetic Resonance of Transition Ions.* Oxford University Press, Oxford, 2012.
136. S. Stoll, A. Schweiger. EasySpin. *J. Mag. Res.* **2006**, *178* (1), 42–55.
137. K. M. Masum Habib. PlotPub - Publication Quality Graphs in MATLAB (https://github.com/masumhabib/PlotPub), GitHub, 2024.
138. S. Lee, P. Walker, S. Velling, A. Chen, Z. Taylor, C. Fiori, V. Gandhi, Z.-G. Wang, J. Greer. *Nat. Comms.* **2024**, *15*, 6850.
139. A. Banin. "The Soils of Mars", in *Lunar and Planetary Inst., Workshop on Mars Sample Return Science* (SEE N89-18288 10-91), 1988, pp. 35–36.
140. A. Banin, T. Ben-Shlomo, L. Margulies, D. F. Blake, R. L. Mancinelli, A. U. Gehring. *J. Geophys. Res. Planets* **1993**, *98* (E11), 20831–53.
141. A. Banin, B. C. Clark, H. Wänke. *Mars*, Eds H. H. Kieffer, B. M. Jakosky, C. W. Snyder, M. S. Matthews. *Chapter 18: Surface Chemistry and Mineralogy.* Space Science Series. University of Arizona Press, 1992.
142. N. C. Foley, T. E. Economou, R. N. Clayton, W. Dietrich. *J. Geophys. Res. Planets* **2003**, *108* (E12), 2002JE002018.

143. R. Gellert, R. Rieder, R. C. Anderson, J. Brückner, B. C. Clark, G. Dreibus, T. Economou, G. Klingelhöfer, G. W. Lugmair, D. W. Ming, S. W. Squyres, C. d'Uston, H. Wänke, A. Yen, J. Zipfel. *Science* **2004**, *305* (5685), 829–32.
144. R. Rieder, R. Gellert, R. C. Anderson, J. Brückner, B. C. Clark, G. Dreibus, T. Economou, G. Klingelhöfer, G. W. Lugmair, D. W. Ming, S. W. Squyres, C. d'Uston, H. Wänke, A. Yen, J. Zipfel. *Science* **2004**, *306* (5702), 1746–9.
145. D. F. Blake , R. V. Morris, G. Kocurek, S. M. Morrison, R. T. Downs, D. Bish, D. W. Ming, K. S. Edgett, D. Rubin, W. Goetz, M. B. Madsen, R. Sullivan, R. Gellert, I. Campbell, A. H. Treiman, S. M. McLennan, A. S. Yen, J. Grotzinger, D. T. Vaniman, S. J. Chipera, C. N. Achilles, E. B. Rampe, D. Sumner, P.-Y. Meslin, S. Maurice, O. Forni, O. Gasnault, M. Fisk, M. Schmidt, P. Mahaffy, L. A. Leshin, D. Glavin, A. Steele, C. Freissinet, R. Navarro-González, R. A. Yingst, L. C. Kah, N. Bridges, K. W. Lewis, T. F. Bristow, J. D. Farmer, J. A. Crisp, E. M. Stolper, D. J. Des Marais, P. Sarrazin, MSL Science Team, C. Agard, J. A. A. Verdasca, R. Anderson, R. Anderson, D. Archer, C. Armiens-Aparicio, R. Arvidson, E. Atlaskin, S. Atreya, A. Aubrey, B. Baker, M. Baker, T. Balic-Zunic, D. Baratoux, J. Baroukh, B. Barraclough, K. Bean, L. Beegle, A. Behar, J. Bell, S. Bender, M. Benna, J. Bentz, G. Berger, J. Berger, D. Berman, J. J. Blanco Avalos, D. Blaney, J. Blank, H. Blau, L. Bleacher, E. Boehm, O. Botta, S. Böttcher, T. Boucher, H. Bower, N. Boyd, B. Boynton, E. Breves, J. Bridges, W. Brinckerhoff, D. Brinza, C. Brunet, A. Brunner, W. Brunner, A. Buch, M. Bullock, S. Burmeister, M. Cabane, F. Calef, J. Cameron, B. Cantor, M. Caplinger, J. Caride Rodríguez, M. Carmosino, I. Carrasco Blázquez, A. Charpentier, D. Choi, B. Clark, S. Clegg, T. Cleghorn, E. Cloutis, G. Cody, P. Coll, P. Conrad, D. Coscia, A. Cousin, D. Cremers, A. Cros, F. Cucinotta, C. d'Uston, S. Davis, M. Day, M. de la Torre Juarez, L. DeFlores, D. DeLapp, J. DeMarines, W. Dietrich, R. Dingler, C. Donny, D. Drake, G. Dromart, A. Dupont, B. Duston, J. Dworkin, M. D. Dyar, L. Edgar, C. Edwards, L. Edwards, B. Ehlmann, B. Ehresmann, J. Eigenbrode, B. Elliott, H. Elliott, R. Ewing, C. Fabre, A. Fairén, K. Farley, C. Fassett, L. Favot, D. Fay, F. Fedosov, J. Feldman, S. Feldman, M. Fitzgibbon, G. Flesch, M. Floyd, L. Flückiger, A. Fraeman, R. Francis, P. François, H. Franz, K. L. French, J. Frydenvang, A. Gaboriaud, M. Gailhanou, J. Garvin, C. Geffroy, M. Genzer, A. Godber, F. Goesmann, D. Golovin, F. Gómez Gómez, J. Gómez-Elvira, B. Gondet, S. Gordon, S. Gorevan, J. Grant, J. Griffes, D. Grinspoon, P. Guillemot, J. Guo, S. Gupta, S. Guzewich, R. Haberle, D. Halleaux, B. Hallet, V. Hamilton, C. Hardgrove, D. Harker, D. Harpold, A.-M. Harri, K. Harshman, D. Hassler, H. Haukka, A. Hayes, K. Herkenhoff, P. Herrera, S. Hettrich, E. Heydari, V. Hipkin, T. Hoehler, J. Hollingsworth, J. Hudgins, W. Huntress, J. Hurowitz, S. Hviid, K. Iagnemma, S. Indyk, G. Israël, R. Jackson, S. Jacob, B. Jakosky, E. Jensen, J. K. Jensen, J. Johnson, M. Johnson, S. Johnstone, A. Jones, J. Jones, J. Joseph, I. Jun, H. Kahanpää, M. Kahre, N. Karpushkina, W. Kasprzak, J. Kauhanen, L. Keely, O. Kemppinen, D. Keymeulen, M.-H. Kim, K. Kinch, P. King, L. Kirkland, A. Koefoed, J. Köhler, O. Kortmann, A. Kozyrev, J. Krezoski, D. Krysak, R. Kuzmin, J. L. Lacour, V. Lafaille, Y. Langevin, N. Lanza, J. Lasue, S. Le Mouélic, E. M. Lee, Q.-M. Lee, D. Lees, M. Lefavor, M. Lemmon, A. Lepinette Malvitte, R. Léveillé, É. Lewin-Carpintier, S. Li, L. Lipkaman, C. Little, M. Litvak, E. Lorigny, G. Lugmair, A. Lundberg, E. Lyness, J. Maki, A. Malakhov, C. Malespin, M. Malin, N. Mangold, H. Manning, G. Marchand, M. Marín Jiménez, C. Martín García, D. Martin, M. Martin, J. Martínez-Frí. *Science* **2013**, *341* (6153), 1239505.
146. S. R. Taylor, S. M. McLennan, *Planetary Crusts: Their Composition, Origin and Evolution*, Cambridge Planetary Science, Cambridge, UK; New York: Cambridge University Press, 2009.
147. M. K. Green, G. Kotowycz. *Can. J. Chem.* **1979**, *57* (18), 2434–43.

148. Y. Furuike, Y. Akita, I. Miyahara, N. Kamiya. *Biochem.* **2016**, *55* (12), 1801–12.
149. S. Yoshiba, T. Ooga, N. Nakagawa, T. Shibata, Y. Inoue, S. Yokoyama, S. Kuramitsu, R. Masui. *J. Bio. Chem.* **2004**, *279* (35), 37163–74.
150. W. Saenger, B. S. Reddy, K. Mühlegger, G. Weimann. *Nature* **1977**, *267* (5608), 225–9.
151. J. L. Czeisler, D. P. Hollis. *Biochem.* **1975**, *14* (12), 2781–5.
152. J. N. Burnell. *The Biochemistry of Manganese in Plants*, in *Manganese in Soils and Plants*, Eds by R. D. Graham, R. J. Hannam, N. C. Uren, 125–37. Springer, Dordrecht, The Netherlands, 1988.
153. D. Shevela, J. J. Eaton-Rye, J.-R. Shen, Govindjee. *Biochim. Biophys. Acta – Bioenerg.* **2012**, *1817* (8), 1134–51.
154. M. Medina, A. Rizo, D. Dinh, B. Chau, M. Omidvar, A. Juarez, J. Ngo, H. A. Johnson. *Front. Microbio.* **2018**, *9*, 2671.
155. Y. N. Kozlov, S. K. Zharmukhamedov, K. G. Tikhonov, J. Dasgupta, A. A. Kazakova, G. C. Dismukes, V. V. Klimov. *Phys. Chem. Chem. Phys.* **2004**, *6* (20), 4905.
156. V. V. Klimov, S. V. Baranov. *Biochim. Biophys. Acta – Bioenerg.* **2001**, *1503* (1–2), 187–96.
157. J. Dasgupta, A. M. Tyryshkin, S. V. Baranov, G. C. Dismukes. *App. Mag. Reson.* **2010**, *37* (1–4), 137–50.
158. J. Dasgupta, A. M. Tyryshkin, G. C. Dismukes. *Ang. Chem.* **2007**, *119* (42), 8174–7.
159. R. Stoyanova, M. Gorova, E. Zhecheva. *J. Phys. Chem. Solids* **2000**, *61* (4), 609–14.
160. T. Zhang, L. Liu, W. Tan, S. L. Suib, G. Qiu. *Chemosphere* **2021**, *262*, 128082.
161. A. Qian, W. Zhang, C. Shi, C. Pan, D. E. Giammar, S. Yuan, H. Zhang, Z. Wang. *Env. Sci. & Tech.* **2019**, *53* (10), 5768–77.
162. J. K. Klewicki, J. J. Morgan. *Env. Sci. & Tech.* **1998**, *32* (19), 2916–22.
163. S. F. M. Ashbourn, J. E. Elsila, J. P. Dworkin, M. P. Bernstein, S. A. Sandford, L. J. Allamandola. *Met. & Planet. Sci.* **2007**, *42* (12), 2035–41.
164. D. Milshteyn, G. Cooper, D. Deamer. *Sci. Rep.* **2019**, *9* (1), 12447.
165. I. R. Kaplan, E. T. Degens, J. H. Reuter. *Geochim. Cosmochim. Acta* **1963**, *27* (7), 805–34.
166. S. G. Henao, V. Karanauskas, S. M. Drummond, L. R. Dewitt, C. M. Maloney, C. Mulu, J. M. Weber, L. M. Barge, P. Videau, M. O. Gaylor. *Astrobiology* **2022**, *22* (2), 197–209.
167. D. Gan, M. Jia, P. P. Vaughan, D. E. Falvey, N. V. Blough. *J. Phys. Chem. A.* **2008**, *112* (13), 2803–12.
168. K.-D. Asmus, *Heteroatom-Centered Free Radicals Some Selected Contributions by Radiation Chemistry*, in *Studies in Physical and Theoretical Chemistry*, Eds by C. D. Jonah, B. S. M. Rao, Elsevier, Radiation Chemistry, 2001, 87, 341–93.
169. A. R. Cook, L. A. Curtiss, J. R. Miller. *J. Am. Chem. Soc.* **1997**, *119* (24), 5729–34.
170. J. Yuasa, T. Suenobu, S. Fukuzumi. *ChemPhysChem.* **2006**, *7* (4), 942–54.
171. B.-Z. Zhu, G.-Q. Shan, C.-H. Huang, B. Kalyanaraman, L. Mao, Y.-G. Du. *Proc. Nat. Acad. Sci.* **2009**, *106* (28), 11466–71.
172. F. Gifford Jr. *Astrophys. J.* **1956**, *123* (123), 154.
173. G. Ohring, J. Mariano. *J. Atm. Sci.* **1968**, *25* (5), 673–81.
174. N. Spanovich, M. D. Smith, P. H. Smith, M. J. Wolff, P. R. Christensen, S. W. Squyres. *Icarus* **2006**, *180* (2), 314–20.
175. M. T. Bland, K. N. Singer, W. B. McKinnon, P. M. Schenk. *Geophys. Res. Let.* **2012**, *39* (17), 2012GL052736.
176. R. H. Brown, R. N. Clark, B. J. Buratti, D. P. Cruikshank, J. W. Barnes, R. M. E. Mastrapa, J. Bauer, S. Newman, T. Momary, K. H. Baines, G. Bellucci, F. Capaccioni, P. Cerroni, M. Combes, A. Cordadini, P. Drossart, V. Formisano, R. Jaumann, Y. Langevin, D. L. Matson, T. B. McCord, R. M. Nelson, P. D. Nicholson, B. Sicardy, A. C. Sotin. *Science* **2006**, *311* (5766), 1425–8.

177. F. M. Flasar, R. K. Achterberg, B. J. Conrath, J. C. Pearl, G. L. Bjoraker, D. E. Jennings, P. N. Romani, A. A. Simon-Miller, V. G. Kunde, C. A. Nixon, B. Bézard, G. S. Orton, L. J. Spilker, J. R. Spencer, P. G. J. Irwin, N. A. Teanby, T. C. Owen, J. Brasunas, M. E. Segura, R. C. Carlson, A. Mamoutkine, P. J. Gierasch, P. J. Schinder, M. R. Showalter, C. Ferrari, A. Barucci, R. Courtin, A. Coustenis, T. Fouchet, D. Gautier, E. Lellouch, A. Marten, R. Prangé, D. F. Strobel, S. B. Calcutt, P. L. Read, F. W. Taylor, N. Bowles, R. E. Samuelson, M. M. Abbas, F. Raulin, P. Ade, S. Edgington, S. Pilorz, B. Wallis, E. H. Wishnow. *Science* **2005**, *307* (5713), 1247–51.
178. E. Heggy, G. Scabbia, L. Bruzzone, R. T. Pappalardo. *Icarus* **2017**, *285*, 237–51.
179. Y. Ashkenazy. *Heliyon* **2019**, *5* (6), e01908.
180. D. N. DellaGiustina, J. P. Emery, D. R. Golish, B. Rozitis, C. A. Bennett, K. N. Burke, R.-L. Ballouz, K. J. Becker, P. R. Christensen, C. Y. Drouet d'Aubigny, V. E. Hamilton, D. C. Reuter, B. Rizk, A. A. Simon, E. Asphaug, J. L. Bandfield, O. S. Barnouin, M. A. Barucci, E. B. Bierhaus, R. P. Binzel, W. F. Bottke, N. E. Bowles, H. Campins, B. C. Clark, B. E. Clark, H. C. Connolly Jr., M. G. Daly, J. de Leon, M. Delbo', J. D. P. Deshapriya, C. M. Elder, S. Fornasier, C. W. Hergenrother, E. S. Howell, E. R. Jawin, H. H. Kaplan, T. R. Kareta, L. Le Corre, J.-Y. Li, J. Licandro, L. F. Lim, P. Michel, J. Molaro, M. C. Nolan, M. Pajola, M. Popescu, J. L. Rizos Garcia, A. Ryan, S. R. Schwartz, N. Shultz, M. A. Siegler, P. H. Smith, E. Tatsumi, C. A. Thomas, K. J. Walsh, C. W. V. Wolner, X.-D. Zou, D. S. Lauretta & The OSIRIS-REx Team. *Nat. Astronomy* **2019**, *3* (4), 341–51.
181. L. F. Lim, T. H. McConnochie, J. F. Bell, T. L. Hayward. *Icarus* **2005**, *173* (2), 385–408.
182. K. Meyberg. "*Synthesis, Characterization, and Thermal Investigation of Metal Phosphites and Potential Implications for Astrobiology.*" Master of Science in Chemical Sciences Theses, 50, 2022. Kennesaw State University, Kennesaw, GA, USA.
183. S. C. Chang, M. L. Jackson. *Soil Sci. Soc. Am. J.* **1957**, *21* (3), 265–9.
184. H. M. Azam, K. T. Finneran. *Chemosphere* **2014**, *97*, 1–9.
185. E. Pierri, D. Tsamouras, E. Dalas. *J. Cryst. Growth* **2000**, *213* (1), 93–98.
186. R. W. Evans, R. Rafique, A. Zarea, C. Rapisarda, R. Cammack, P. J. Evans, J. B. Porter, R. C. Hider. *J. Bio. Inorg. Chem.* **2007**, *13* (1), 57–74.
187. A. M. N. Silva, X.-L. Kong, M. C. Parkin, R. Cammack, R. C. Hider. *Dalton Trans.* **2009**, *40*, 8616.
188. Y. G. Cheng, N. D. Chasteen. *Biochem.* **1991**, *30* (11), 2947–53.
189. E. A. Guggenheim. "*On the Statistical Mechanics of Dilute and of Perfect Solutions.*" *Proceedings of the Royal Society of London. Series A, Containing Papers of a Mathematical and Physical Character*, **1932**, *135* (826), 181–92.
190. M. J. N. Pourbaix, *Atlas d'equilibres Electrochimiques* (Atlas of Electrochemical Equilibria in Aqueous Solutions), Ed Gauthier-Villars, Trans. J. A. Franklin, 2nd Ed., National Association of Corrosion Engineers (NACE), International Centre Belge d'Etude de la Corrosion (Cebelcor), Brussels, 1974.
191. I. G. Darvey. *Biochem. Edu.* **1995**, *23* (2), 80–2.
192. D. R. Lide. *CRC Handbook of Chemistry and Physics: A Ready-Reference Book of Chemical and Phyical Data.* 72nd Ed., Boca Raton, FL, CRC Press, 1991.
193. G. Deacon. *Coord. Chem. Rev.* **1980**, *33* (3), 227–50.
194. L. J. Kirwan, P. D. Fawell, W. Van Bronswijk. *Langmuir* **2003**, *19* (14), 5802–7.
195. P. S. Kalsi. *Spectroscopy of Organic Compounds.* New Age International, 2007.
196. G. R. Rossman. *Rev. Mineral. and Geochem.* **2014**, *78* (1), 371–98.
197. N. C. Li, R. L. Scruggs, E. D. Becker. *J. Am. Chem. Soc.* **1962**, *84* (24), 4650–4.
198. A. P. Arnold, K. S. Tan, D. L. Rabenstein. *Inorg. Chem.* **1986**, *25* (14), 2433–7.
199. K. Begaudeau, Y. Morizet, P. Florian, M. Paris, J.-C. Mercier. *Euro. J. Mineral.* **2012**, *24* (3), 535–50.

200. R. J. Kirkpatrick, K. A. Smith, S. Schramm, G. Turner, W.-H. Yang. *Ann. Rev. Earth Planet. Sci.* **1985**, *13*, 29–47.
201. E. Freire, O. L. Mayorga, M. Straume. *Anal. Chem.* **1990**, *62* (18), 950A–9A.
202. R. J. Falconer. *J. Mol.Recog.* **2016**, *29* (10), 504–15.
203. J.-P. E. Grolier, J. M. del Río. *J. Soln. Chem.* **2015**, *44* (5), 987–1003.
204. M. P. Bernstein, J. P. Dworkin, S. A. Sandford, L. J. Allamandola. *Adv. Space Res.* **2002**, *30* (6), 1501–8.
205. A. Kröger, M. Klingenberg. *Eur. J. Biochem.* **1973**, *34* (2), 358–68.
206. R. W. J. Lim, A. C. Fahrenbach. *Pure App. Chem.* **2020**, *92* (12), 1971–86.
207. C. Potiszil, M. Yamanaka, C. Sakaguchi, T. Ota, H. Kitagawa, T. Kunihiro, R. Tanaka, K. Kobayashi, E. Nakamura. *Life* **2023**, 13 (7), 1448.
208. Q. H. S. Chan, I. A. Franchi, X. Zhao, A. Stephant, I. P. Wright, C. M. O'D. Alexander. *Met. & Planet. Sci.* **2020**, *55* (6), 1320–48.
209. G. Matrajt, G. M. Muñoz Caro, E. Dartois, L. d'Hendecourt, D. Deboffle, J. Borg. *Astro. & Astrophys.* **2005**, *433* (3), 979–95.
210. B. E. Clark, A. Sen, X.-D. Zou, D. N. DellaGiustina, S. Sugita, N. Sakatani, M. Thompson, D. Trang, E. Tatsumi, M. A. Barucci, M. Barker, H. Campins, T. Morota, C. Lantz, A. R. Hendrix, F. Vilas, L. Keller, V. E. Hamilton, K. Kitazato, S. Sasaki, M. Matsuoka, T. Nakamura, A. Praet, S. M. Ferrone, T. Hiroi, H. H. Kaplan, W. F. Bottke, J.-Y. Li, L. Le Corre, J. L. Molaro, R.-L. Ballouz, C. W. Hergenrother, B. Rizk, K. N. Burke, C. A. Bennett, D. R. Golish, E. S. Howell, K. Becker, A. J. Ryan, J. P. Emery, S. Fornasier, A. A. Simon, D. C. Reuter, L. F. Lim, G. Poggiali, P. Michel, M. Delbo, O. S. Barnouin, E. R. Jawin, M. Pajola, L. Riu, T. Okada, J. D. P. Deshapriya, J. R. Brucato, R. P. Binzel, D. S. Lauretta. *Icarus* **2023**, *400*, 115563.
211. M. Yasui, T. Tazawa, R. Hashimoto, M. Arakawa, K. Ogawa. *Comms. Earth Env.* **2021**, *2* (1), 95.
212. M. S. Thompson, R. V. Morris, S. J. Clemett, M. J. Loeffler, D. Trang, L. P. Keller, R. Christoffersen, D. G. Agresti. *Icarus* **2020**, *346*, 113775.
213. M. P. Bernstein, J. P. Dworkin, S. A. Sandford, L. J. Allamandola. *Met. Planet. Sci.* **2001**, *36* (3), 351–8.
214. V. S. Airapetian, A. Glocer, G. Gronoff, E. Hébrard, W. Danchi. *Nat. Geosci.* **2016**, *9* (6), 452–5.
215. M. G. Gonzalez, E. Oliveros, M. Wörner, A. M. Braun. *J. Photochem. Photobio. C: Photochem. Rev.* **2004**, *5* (3), 225–46.
216. M. Toybenshlak, R. Carmieli. *Israel J. Chem.* **2019**, *59* (11–12), 1020–6.
217. E. W. Zhao, E. Jónsson, R. B. Jethwa, D. Hey, D. Lyu, A. Brookfield, P. A. A. Klusener, D. Collison, C. P. Grey. *J. Am. Chem. Soc.* **2021**, *143* (4), 1885–95.
218. E. M. Genies, M. Lapkowski. *J. Electroanal. Chem. Interfac. Electrochem.* **1987**, *236* (1), 199–208.
219. M. Hunger, J. Weitkamp. *Ang. Chem. Int. Ed.* **2001**, *40* (16), 2954–71.
220. A. Petr, L. Dunsch, A. Neudeck. *J. Electroanal. Chem.* **1996**, *412* (1), 153–8.
221. P. Rapta, A. Neudeck, A. Petr, L. Dunsch. *J. Chem. Soc. Faraday Trans.* **1998**, *94* (24), 3625–30.
222. W. H. Koppenol. *Free Rad. Bio. Med.* **1993**, *15* (6), 645–51.
223. S.A Benner, K. G. Devine, L. N. Matveeva, D. H. Powell. *Proc. Natl. Acad. Sci.* USA **2000**, *97* (6), 2425–30.
224. M. L. Kremer. *J. Phys. Chem. A* **2003**, *107* (11), 1734–41.
225. S. Radhakrishna, B. V. R. Chowdari, A. Kasi Viswanath. *Chem. Phys. Let.* **1976**, *40* (1), 134–8.
226. E. Schubert, T. Hett, O. Schiemann, Y. NejatyJahromy. *J. Mag. Res.* **2016**, *265*, 10–15.
227. I. Yamazaki, L. H. Piette. *J. Am. Chem. Soc.* **1991**, *113* (20), 7588–93.

228. B. C. Gilbert, G. Harrington, G. Scrivens, S. Silvester, *EPR Studies of Fenton-Type Reactions in Copper- Peroxide Systems*, in *Free Radicals in Biology and Environment*, Ed. Francesco Minisci, 1997, 49–62. Dordrecht: Springer Netherlands.

APPENDIX/SUPPLEMENTARY INFORMATION REFERENCES

229. C. C. Addison, D. Sutton. *J. Che. Soc. A Inorg. Phys. Theor.* **1966**, 1524–8.
230. D. S. Rustad, N. W. Gregory. *Inorg. Chem.* **1977**, *16* (12), 3036–40.
231. R. C. Turner, K. E. Miles. *Can. J. Chem.* **1957**, *35* (9), 1002–9.
232. M. Windholz, "The Merck Index Online - Chemicals, Drugs and Biologicals", in *The Merck Index*, Online Edition. Royal Society of Chemistry.
233. M.-B. Kime, D. Makgoale. *Chem. Eng. Comms.* *203* (12), 1648–55.
234. R. Zhao, P. Pan. *Can. J. Chem.* **2001**, *79* (2), 131–44.
235. N. J. Susak, D. A. Crerar. *Geochim. Cosmochim. Acta* **1985**, *49* (2), 555–64.
236. W. Liu, B. Etschmann, J. Brugger, L. Spiccia, G. Foran, B. McInnes. *Chem. Geo.* **2006**, *231* (4), 326–49. https://doi.org/10.1016/j.chemgeo.2006.02.005.
237. D. S. Wishart, C. Knox, A. C. Guo, R. Eisner, N. Young, B. Gautam, D. D. Hau, N. Psychogios, E. Dong, S. Bouatra, R. Mandal, I. Sinelnikov, J. Xia, L. Jia, J. A. Cruz, E. Lim, C. A. Sobsey, S. Shrivastava, P. Huang, P. Liu, L. Fang, J. Peng, R. Fradette, D. Cheng, D. Tzur, M. Clements, A. Lewis, A. De Souza, A. Zuniga, M. Dawe, Y. Xiong, D. Clive, R. Greiner, A. Nazyrova, R. Shaykhutdinov, L. Li, H. J. Vogel, I. Forsythe. "HMDB: A Knowledgebase for the Human Metabolome." *Nucleic Acids Research, 37,* no. Database issue (January 2009): D603–610.
238. F. Jofre, M. E. Anderson, J. L. Markley, M. E. Ulrich. "L_cystine." BMRB entry BMSE000035, 2006.
239. Wiley. "3-(Trimethylsilyl)-1-Propanesulfonic Acid Sodium Salt - [1H NMR] - Chemical Shifts - SpectraBase." SpectraBase.
240. Wiley. "P-Benzoquinone - [1H NMR] - Spectrum - SpectraBase." SpectraBase.
241. A. Schweiger. *Ang. Chem. Int. Ed. English* **1991**, *30* (3), 265–92.
242. A. Schweiger, G. Jeschke, *Principles of Pulse Electron Paramagnetic Resonance.* Oxford University Press, 2001.

7 Unique Semiconducting Properties of Layered Mn and Fe Oxide Minerals and Their Influence on Biogeochemical Processes

Vidhya Chakrapani
Howard P. Isermann Department of Chemical and Biological Engineering
Rensselaer Polytechnic Institute,
Troy, NY 12180, USA
chakrv@rpi.edu

CONTENTS

DOI: 10.1201/9781003459910-7

ABSTRACT

Minerals play an important role in the diversity of life on Earth, influencing many biogeochemical processes. This chapter highlights some of the unique semiconductor and catalytic properties of $Fe^{II}Fe^{III}$ hydroxide (green rust) and $MnO_2.xH_2O$ (birnessite family) layered minerals relative to other Fe and Mn minerals and how microorganisms actively exploit these properties for their survival and growth. The usefulness of electrochemical techniques is demonstrated for probing the interrelationships between electronic structure and chemical reactivity and how these factors affect the synergistic existence of minerals and microbes.

KEYWORDS

mineral-microbe interaction, birnessite, green rust, semiconductor, band lineup, photosystem II, nitrogenases

1 INTRODUCTION

The interaction between microorganisms and minerals is an important link between the biosphere and the lithosphere, affecting global cycles of natural elements and profoundly impacting water quality, soil formation, agriculture, and material diversity. The dynamic synergy of the coexistence of minerals and microorganisms is apparent in the fact that minerals provide the energy and nutrients necessary to support the growth and function of microorganisms [1–2]. In turn, many microbes actively regulate their metabolic activities to modulate the surrounding environmental conditions conducive to mineral precipitation (biomineralization), dissolution (bioweathering), and structural or redox transformations for survival and growth. These key processes aid in the cycling of metals and nutrients (C, P, N, S, and O) in the environment and make the Earth habitable for all forms of life, even under extreme conditions. They are also widely exploited industrially for biomining, biorestoration, and bioremediation. Mineral interactions can also have detrimental effects on the microbe, such as when they lead to high metal toxicity or when exposed to highly reactive oxygen species (ROS) formed on mineral surfaces that can disrupt cell functions. This range of dynamic relationships is further expanded when the interacting minerals are electrically conductive (semiconducting), as they can now not only be a source and a sink for electrons but also serve as conduits for transporting electrons [3]. Consequently, such minerals can facilitate redox reactions with and between microbes. An elegant example is the methanogenesis that occurs through electron exchange between syntropic Fe^{II}-oxidizing methanogens and Fe^{III}-reducing bacteria through semiconducting Fe-bearing minerals such as magnetite [4]. Thus, insights into these symbiotic mineral-microbe interactions are of great significance in understanding the evolution of life.

Iron and manganese are two of the most abundant metals found in nature and are widely distributed as ocean nodules, microconcretions, desert coatings, and crusts. Mn is known to exist in more than 30 different crystalline and amorphous oxide minerals [5]. Similarly, iron is known to form at least 16 different crystallographic structures of oxides, hydroxides, and oxyhydroxides of Fe^{II} and Fe^{III} depending on the redox and pH conditions of the aqueous environment [6]. Of particular interest are the layered oxides/hydroxides of Mn (MnO_2, birnessite family) and Fe (double-layered hydroxides (DLH) such as $Fe^{II}Fe^{III}(OH)_2$ known as "green rust"). Because of their high surface area and high surface charge density, they are potent environmental catalysts for heavy metal sorption, nutrient cycling, and the degradation of organic compounds, and consequently exert strong control over the biogeochemical cycles of several elements. Interestingly, birnessite and green rust (GR) also have a contrasting geochemical function [7]; the former is a potent oxidizer and/or absorber, while the latter is a powerful reductant. Birnessite exists in two crystallographic forms, namely in a triclinic and hexagonal structure [8–9]. The basic unit consists of an MnO_6 octahedron that is arranged in layered sheets with various cations and water in the interlayer region. Birnessites are distinguished by their ~7 Å d_{100}-spacing due to the presence of a single sheet of water molecules, while buserite with two sheets of interlayer water has a distinct ~10 Å spacing. The hexagonal form contains many octahedral vacancies in the layers that are capped by Mn^{II} in the interlayer spacing under acidic conditions [9–12]. However, in triclinic form, the octahedral sheets are fully occupied with Mn^{IV} and Mn^{III} cations. The negative charge of Mn^{III} in the octahedral layer of $Mn^{IV}O_2$ is balanced by the interlayer cation, such as Na^+, or K^+, and coordinated by the interlayer O atoms of water and the layer O atoms. Collective Jahn-Teller distortion due to the presence of rows of edge-sharing Mn^{III} octahedra causes the elongation of the a-axis and distorts the perfect hexagonal symmetry to the triclinic symmetry [9]. The hexagonal form has received significant attention because it has long been considered the dominant phase that forms biotically [13–15] and abiotically [16–19]. However, a recent study [20] points to extreme reactivity of the triclinic form against phase transformation to the hexagonal form under laboratory test conditions used for phase analysis. Consistent with this, Ling et al. [8], through multiple characterization techniques, showed that natural birnessites consist of both triclinic and hexagonal intergrowths with increased proportions of triclinic birnessite correlating with higher Mn^{III} concentrations relative to those of Mn^{II}. Several factors, such as electrolyte pH and many Mn-complexing ligands, strongly influence the interconversion between these two crystallographic phases as well as their nature of reactivity. Although the triclinic form is known for its high oxidizing power for the conversion of many heavy metal ions such as Cr^{III} to Cr^{VI} and As^{III} to As^{V}, as well as many organic compounds [21–22], the hexagonal H-birnessite has low oxidative power [23] but is a powerful adsorber for a wide range of aqueous ions due to the presence of a high density of Mn vacancy sites. This behavior contrasts with that of Fe^{II}-containing GR, which is a powerful reductant that can reduce nitrate [24], nitrite [25], chlorinated hydrocarbons [26] (ethane, ethylene), and many heavy metals, including Cr^{VI} to Cr^{III} [27], Se^{VI} to Se^{0} [28], U^{VI} to U^{IV} [29], and Au^{III} [30]. Another striking difference pertains to the crystallographic arrangement. Although both birnessite and GR have a flexible layered structure with a typical 7-11 Å d_{100}-spacing, in birnessite, the interlayer species consists of cations (Na^{I}, K^{I}, and others) with a single sheet of water molecules, while in GR, the interlayer species are anions, such as SO_4^{2-}, CO_3^{2-}, and Cl^-, that compensate the excess positive charges of structural Fe^{III} [31]. Synthetic GR and its natural mineral form, as fougerite, have a pyroaurite-sjögrenite structure with a general formula $[(Fe^{II},Mg^{II})_{1-x}Fe_x^{III}(OH)_2]^{x+}[x/nA^{n-}\cdot mH_2O]^{x-}$, where A^{n-} (e.g., SO_4^{2-}, CO_3^{2-}, Cl^-) present between the layers compensate for the excess positive charges of the layer [31]. They are prevalent in many natural anoxic and suboxic environments, for example, in hydromorphic soils [32–33], groundwater [34], ferruginous lakes [35–36], mofette sites [37], and mine drainage sites [38–39].

The biotic formation of birnessite and GR is known to occur in a wide range of environments [40], ranging from deep-sea hydrothermal vent regions to the surface of desert rocks, where Mn- and Fe-depositing microbes, including bacteria, fungi, algae, and protozoa, have been detected in these oxide deposits. The primary Mn^{IV} oxide produced by microbial oxidation of aqueous Mn^{II}/Mn^{III} ions is the disordered form of birnessite (δ-MnO_2). Since direct and indirect abiotic oxidation [41] of Mn^{II} by O_2 or by mineral-produced ROS such as $O_2^{-\bullet}$ is slow, most of the naturally occurring birnessites are believed to be biogenic in origin [13–15,42]. Similarly, GR is an important metastable intermediate in many phase transformations of Fe in nature. It is the first solid phase to precipitate from dissolved Fe species produced by abiotic and microbially induced corrosion of iron and steel under circumneutral to alkaline conditions in suboxic environments [43–46]. Several dissimilatory Fe^{III}-reducing bacteria precipitate GR through bioreduction of hydrous ferric oxide [47] and lepidocrocite (γ-FeOOH) [48]. GRs are also observed as products of direct anaerobic biooxidation of Fe^{II} by Fe^{II}-oxidizing bacteria [49] and indirectly through oxidation by nitrate-reducing bacteria [50]. GR is also believed to be the precursor during the precipitation of Fe^{III} oxyhydroxides and magnetite (Fe_3O_4) in the deposition of banded iron formation [51–53]. Although several Fe^{II}- and Fe^{III} minerals, such as magnetite, goethite (α-FeOOH), and lepidocrocite are known for their high reductive activity, recent studies [54,55] show that their high reactivity is due to the $Fe(OH)_2$/GR intermediate that forms on these mineral surfaces upon sorption of aqueous Fe^{II}.

One can ask why many biogenic MnO_x is predominantly a birnessite-type phyllomanganate despite the abundance of MnO_x polymorphs in nature or why GR is the first intermediate of Fe^{II} oxidation or Fe corrosion. Why do certain microbial consortiums choose Fe minerals such as magnetite for interspecies electron transfer? Perhaps understanding the unique and selective features of minerals can help identify the nature of microbial community that it can support and hence enable a better understanding of their synergistic co-existence. Some of the answers to these questions may lie in their unique semiconducting properties and their structural redox centers, which render them highly active. Several in-depth investigations have been carried out to elucidate the crystal structure of birnessite [11,20,56–62] and GR [31,45,63–65] and relate these to their catalytic activities. However, except for a few studies [66–70], the electronic properties of these semiconducting minerals and how they could influence geochemical reactions are largely unknown. For example, recent studies [71,72] suggest that birnessite minerals are involved in widespread photoelectric conversion processes on the sunlit surface of the Earth. The present chapter gives a review of our recent results exploring correlations between catalytic activities and the semiconducting nature of various forms of birnessite and GR minerals and provides some insight into how these properties are modulated by environmental redox changes that, in turn, affect biogeochemical functions.

2 ELECTROCHEMICAL TECHNIQUES FOR THE DETERMINATION OF SEMICONDUCTING PROPERTIES

To begin, we discuss the usefulness of electrochemical techniques in determining the semiconducting properties, namely the Fermi energy (E_F) or the work function ($\Phi s = E_{VAC} - E_F$, where $E_{VAC}=0$ is the vacuum energy), in an in situ aqueous environment. The semiconducting properties of minerals are generally understood through the conventional redox Eh-pH diagram, which predicts equilibria among the species involved using the thermodynamic values of the standard Gibbs free energy of formation and is widely used in mineralogical consideration. However, free energy values can vary widely depending on the ligand environment. Deviations from the stoichiometric composition also greatly influence the nature and concentration of charge carriers in a mineral, and hence its reactivity. This information is not commonly represented in the Eh diagrams. Experimental Eh measurement involves measuring changes in the open-circuit potential by employing either a mineral as the working electrode or, alternatively, with the dispersed aqueous mineral particles in contact with platinum as the working electrode against the potential of a known reference electrode (RE). Sharp changes in the Eh values indicate phase transition, and steady-state values are taken to represent the redox potential of the corresponding phase. However, this Eh value does not represent the true electrochemical potential of the electrons of the solid ($\bar{\mu}_{e-,solid}$), which is indicative of its reactivity. When a semiconducting mineral is immersed in an aqueous solution, the difference in the electrochemical potential of electrons ($\bar{\mu}_{e-}$) between the solid and aqueous phases leads to electron transfer that gives rise to a space charge layer (usually a depletion layer) in the semiconductor and a corresponding double layer of adsorbed counter ions in the aqueous phase, which constitutes the Helmholtz layer. Here, $\bar{\mu}_{e-}$ in the semiconductor is given by its E_F and that in the solution is set by the dominant redox couple or by the mixed potential when several different redox couples are operative. The measurement of Eh probes the potential difference between the surface E_F position of the semiconducting mineral and that of the reference electrode under open circuit conditions (Figure 1a). Therefore, this measured Eh value does not correct for the potential drop across the semiconductor space charge region due to the band bending. In contrast, techniques such as photocurrent or Mott-Schottky (M-S) measurements give information about the semiconductor under flat band conditions (Figure 1b), which is obtained with high charge carrier injection either through the excitation by supraband gap light or with the application of an electrical potential under forward biasing conditions, respectively. Therefore, the bulk E_F position, which is derived from the value of the flat band potential (U_{FB}) by converting it to absolute electron energies, represents $\bar{\mu}_{e-,solid}$ and its value relative to that of the

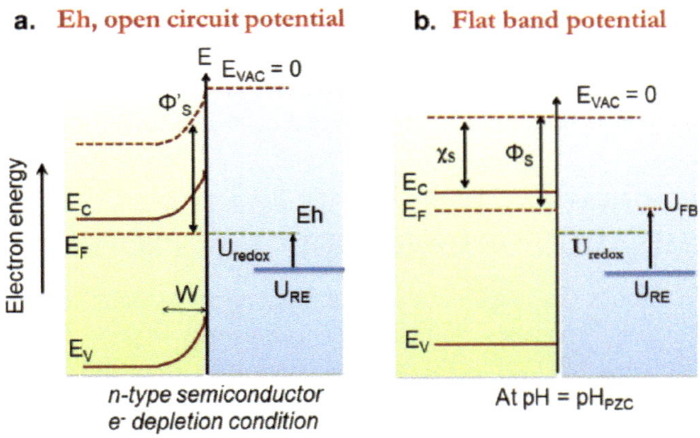

FIGURE 1 Schematics showing the nature of band bending in an n-type semiconductor under (a) open-circuit Eh measurement condition and (b) flat band conditions achieved through charge injection via externally applied voltage or light bias. The example shown here is for $E_F > E_{redox}$ before equilibrium.

electrochemical potential of electrons of the redox couples ($\bar{\mu}_{e-,redox}$) represents the driving force for the charge transfer. In addition to U_{FB}, M-S measurements also provide information on the nature and density of uncompensated charge carriers (type of doping). The experiment is usually performed using a three-electrode setup consisting of a semiconducting mineral as the working electrode, a reference electrode, and a platinum mesh as the counter electrode. The electrochemical impedance of the electrode is determined by using a small AC perturbation voltage at a known frequency as a function of the applied DC bias (U). Because of the high density of charge carriers and the small width of the Helmholtz plane relative to the space charge layer in the semiconductor, the applied potential drop mostly occurs entirely across the space charge region of the semiconductor. The potential drop in the Helmholtz double layer generally remains constant and independent of the applied voltage if there is no significant potential-induced change to the structure of the Helmholtz double layer [73]. Then the measured capacitance (C), which is taken to be the semiconductor space-charge capacitance, is related to the applied potential bias through the M-S relation given by Equation (1):

$$\frac{1}{C^2} = \frac{2\left[\left(U_{FB} - \frac{kT}{q}\right) - U\right]}{q\varepsilon_s N_D} \tag{1}$$

Here, N_D is the concentration of the uncompensated charge carrier, ε_s is the permittivity of the semiconductor, k is the Boltzmann constant, T is the temperature, and q is the elementary charge.

The distinct advantage of the electrochemical technique over other related techniques, such as photoemission spectroscopy that operates under ultrahigh vacuum, is that it allows direct measurement of changes in U_{FB} or Φs of the mineral resulting from surface adsorption/complexation with electrolyte ions under in situ working conditions, which can then be used to correlate chemical activity with band structural changes. This is especially important for hydrous oxides, such as birnessite, which is known to dehydrate in an ultrahigh vacuum environment, leading to the collapse of the interlayer spacing [9,74]. Finally, the values of U_{FB} allow the determination of the absolute position of the conduction (E_C) and valence band (E_V) energies relative to the vacuum level if the values of the band gap (E_G) and the position of E_F relative to the band edge of the majority charge carriers (ΔE_F) are independently known through the relation: $E_C = -U_{FB} + E_{SHE} + \Delta E_F$. Here, E_{SHE} is the energy of the standard hydrogen electrode with respect to E_{VAC} and its value is −4.44 eV [75,76]. The value $\chi = E_{VAC} - E_C$ is the electron affinity of the semiconductor.

3 UNIQUE BAND LINEUPS OF BIRNESSITE AND GREEN RUST PHASES

In our recent studies [77–79], we evaluated the semiconducting properties of various types of natural and synthetic birnessites and GR of varying crystallinity relative to other phases of Mn and Fe oxides and studied how they influence the reactivity of these minerals. The results show that both GR and birnessite have unique band lineups, which can explain their potent chemical activity. Figure 2a shows the representative M-S plots of several forms of birnessite and GR samples. Highly crystalline synthetic alkali ion-rich triclinic, hexagonal, and poorly crystalline Na-disordered birnessites are all seen to be n-type semiconductors (positive slope of the M-S curve) with high positive U_{FB} values of 1.0–1.24 V versus a standard calomel electrode (SCE) when measured in unbuffered $NaNO_3$ or NaCl electrolytes. Natural birnessite, similar to synthetic triclinic birnessite, also shows n-type behavior but with a slightly more negative U_{FB} of ~0.78 V. In contrast, hexagonal H-birnessite shows a p-type semiconducting behavior (negative slope) with a U_{FB} of +1.16 V vs. SCE. These differences in the semiconducting behavior may be the result of varying lattice redox states. The p-type nature of the acidic hexagonal phase is likely due to the presence of native acceptor-type Mn vacancies, which is a result of labile Mn^{II} dissolution from the lattice. On the contrary, the n-type conductivity of alkali ion-rich triclinic and hexagonal (acid-free) phases may be due to high Mn^{III} or associated oxygen-related defects. Like triclinic birnessite, GR also shows n-type behavior but exhibits a highly negative U_{FB} of −1.2 V. The nature of the dopant responsible for the n-type behavior of GR is unknown but could arise from the presence of metallic Fe^0 clusters, which has been suggested to form upon disproportionation of the initial $Fe^{II}(OH)_2$ into GR [80]. Magnetite (Fe_3O_4), which is one of the stable oxidation products of GR, is a p-type semiconductor with a positive U_{FB} of ~1 V (not shown).

Using the U_{FB} value derived from M-S measurements and the band gap (E_G) from absorption spectroscopy, the band lineups of various Fe and Mn oxides can be derived on the absolute energy (vacuum) scale. Here, ΔE_F for both n-type and p-type semiconductors is taken to be ~0.1 eV, which is probably a reasonable assumption

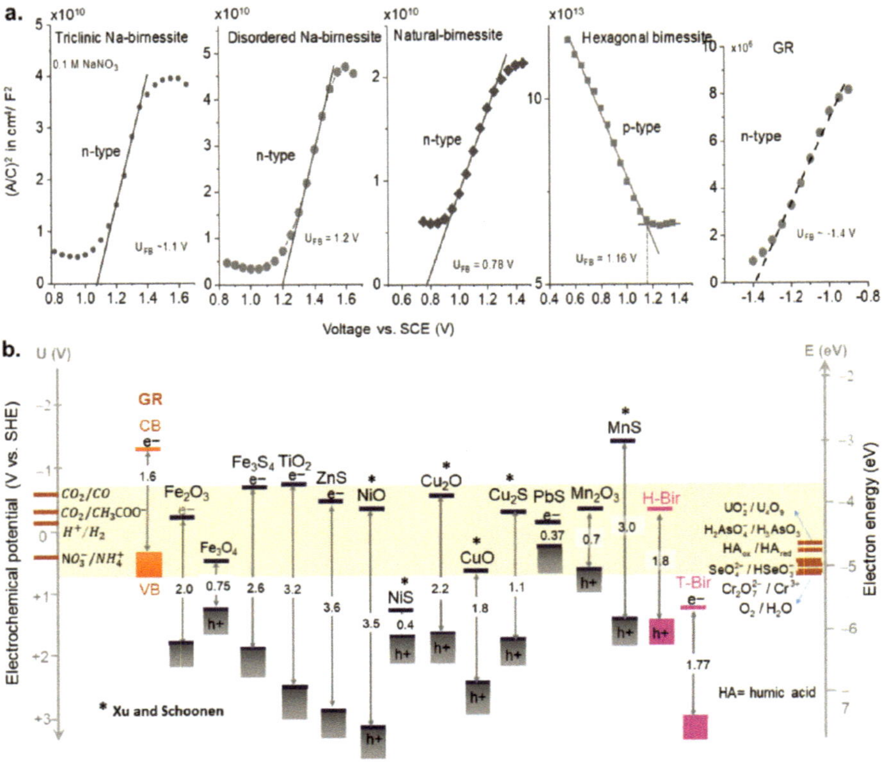

FIGURE 2 (a) Mott-Schottky plots of GR and various types of birnessite samples showing the U_{FB} values and the nature of doping and (b) band lineup of different semiconducting minerals relative to the redox potentials of common aqueous ions. Reprinted and adapted with permission from [77]. Copyright 2023 American Chemical Society. Reprinted/adapted from [131]. Copyright 2024 American Chemical Society.

considering that all minerals show relatively high electrical conductivity. Figure 2b compares the band lineups of birnessite and GR together with the band edge positions of several common semiconductors against the $\bar{\mu}_{e-}$ of several redox couples that are of relevance in the soil and aqueous environments, including that of Cr, As, and humic acid (HA), and O_2 at pH=7. A more extensive list of band lineups of semiconductors can be found in references [78,81,82]. Alkali ion-exchanged triclinic and disordered birnessite have the highest electron affinity values (χ~6.0 eV) among all the common functional binary and ternary oxides and sulfides in the aqueous and soil environment. The high χ puts the CB edge of these birnessites well below the redox potential of geochemical reactions, such as Cr^{III}/Cr^{VI}, As^{III}/As^{V}, O_2/H_2O, humic acid, and NO_3^- / NO_2^- reactions, and thereby promoting their oxidation [83] through injection of e^- into its CB states. This unusually low placement of the band edges and E_F of the n-type birnessites compared to hexagonal H-birnessite and all other minerals is also the reason why only these birnessite minerals are great oxidizers in the natural environment, while H-birnessite is a great absorber but is less oxidatively active due to its more negative U_{FB} and χ. High electron affinity also enables birnessite to serve as terminal electron acceptors for bacterial respiration [84]. The results show that the biomineralization of manganese to birnessite provides protection against nitrite toxicity, which is due to the ability of birnessite to oxidize nitrite to nitrate under physiological conditions [85]. Birnessite minerals have the ability to oxidatively lyse otherwise unusable high-molecular weight humic and fulvic substances to low-molecular weight carbon compounds that microbes can use as a ready source of both energy and carbon substrates for growth [83].

Like birnessite, our results [79] show that GR is also a semiconductor with a unique band lineup, but it has the lowest U_{FB} and electron affinity (χ~3.2 eV) among the FeO_x phases and the most common semiconducting minerals. The high CB energy position, which is well above the reduction potential of gases such as O_2 and CO, as well as heavy metal ions and many organic compounds, in combination with the n-type conductivity, is the reason why it is one of the most potent reductants among all minerals for a wide range of organic and inorganic compounds. However, GR is only stable under anaerobic conditions because exposure of GR to ambient O_2 readily converts it to other stable FeO_x phases like magnetite or goethite depending on the prevailing redox conditions. The ability of GR to reduce nitrate to gaseous N_2 is exploited by heterotrophic nitrate-reducing bacteria to decrease nitrite accumulation and decrease the selectivity toward NH_4^+ formation [86]. The lowest U_{FB} (higher $\bar{\mu}_{e-,solid}$) of GR among all other Fe minerals explains why it is the first intermediate phase to form during Fe corrosion under highly reduced anaerobic conditions and transforms to solids with a higher U_{FB} (lower $\bar{\mu}_{e-,solid}$) when more oxidizing conditions prevail.

Band lineups of the minerals can also be used to explain the bacterial syntropy that occurs via the exchange of electrons between microbial species with a semiconducting mineral as the electron-conducting conduit. Kato et al. [3] showed that electrically conductive magnetite and hematite nanoparticles can facilitate the transfer of electrons between *Geobacter sulfurreducens*, which oxidizes acetate to bicarbonate, and *Thiobacillus denitrificans* that use donated electrons to reduce NO_3^- to NH_4^+. According to the band lineup shown in Figure 2b, magnetite is a conductive p-type semiconductor with a small band gap (0.45 – 0.75 eV) and a large work function of ~5.2 eV [87] to 5.6 eV [79], which puts the E_F between the redox potential of acetate / HCO_3^- (–0.28 V) and that of NO_3^-/NH_4^+ (0.45 V) at pH = 7. Being a small band gap semiconductor, magnetite can easily switch conductivity from p-type (electron acceptor) to n-type (electron donating) upon band filling due to electron injection. Consequently, magnetite (also hematite) can reversibly act as e^- acceptors (magnetite reduction) for *G. sulfurreducens* and an e^- donor (magnetite oxidation) for *T. denitrificans*.

4 FACTORS CONTROLLING THE SEMICONDUCTING BAND LINEUP AND MINERAL REACTIVITY

This raises the question of what factors are responsible for the high/low U_{FB} and χ of the birnessite and GR minerals. Several studies [57,58] have clearly shown that the nature of the interlayer cation affects both the interlayer spacing of the basal sheets and the Mn cation composition of the birnessite minerals. Theoretical calculations [69] show that the nature of the interlayer cation strongly affects the band gap of birnessites and predicted that both the bulk and surface band gaps could be tuned to enhance light capture. Our experimental results [78] with various cation-exchanged birnessite show that both the oxidative activity (Figure 3a) and the electron affinity (Figure 3b) increase with increasing concentrations of Mn^{III} relative to Mn^{II} in the birnessite structure. Here, the ratio represents the concentration of Mn^{III} present within the basal planes and on the edge sites relative to the Mn^{II} concentration that exists in the interlayer region, as steric hindrance within the basal sheets pushes them to the interlayer region. Furthermore, the ionic radii and hydration strengths of the interlayer alkali cations control the interlayer spacing between the basal sheets, which strongly influences both the band gap and the extent of dissolution of labile Mn^{II} out of this region. The band gap of birnessite is seen to increase with an increase in the interlayer spacing (Figure 3c). Alkali cations of smaller ionic radii (e.g., Li^{I}) or high hydration strengths (e.g., Ca^{II}) tend to restrict the Mn^{II} dissolution (Figure 3d) and therefore lead to its accumulation within the birnessite structure. An increase in Mn^{II} relative to Mn^{III} reduces the oxidative activity of birnessite and also causes a decrease in the χ. As a result of this dependence of χ on the Mn^{III}/Mn^{II} ratio, inorganic and organic ligands that can selectively complex with Mn^{II} and Mn^{III} and affect the lability of these Mn cations exert a strong influence on catalytic activity. Common inorganic ligands such as Cl^-, NO_3^-, HCO_3^-, SO_4^{2-}, and $H_2PO_4^-$ selectively complex with Mn^{II} over Mn^{III} to varying degrees. Stronger Mn^{II}-complexing ions (SO_4^{2-}, HCO_3^-, and $H_2PO_4^-$) restrict Mn^{II} lability (dissolution) from birnessite, as seen from the dissolved Mn^{II} concentration shown in Figure 3d, while the weaker Mn^{II}-complexing Cl^- and NO_3^- anions facilitate efficient Mn^{II} dissolution and prevent its accumulation within the birnessite structure. Consequently, similar to the effect of cations, birnessite exposed to weak Mn^{II} complexing anions has a high Mn^{III}/Mn^{II} ratio and consequently exhibits the highest U_{FB} of ~1.2 V vs. SCE (Figure 3e). On the contrary, stronger Mn^{II} complexing anions (SO_4^{2-} and HCO_3^-) shift the U_{FB} to more negative values by more than 400 mV and decrease the oxidative activity of birnessite for both the sorption [77] and water oxidation [88] processes.

In addition to the major anions, natural waters also contain a wide range of organic ligands that include both low molecular weight anions such as carboxylic acid, siderophores, amino acids, and sugars, as well as high molecular weight complex heterogeneous compounds that originate from living organisms and polluting sources. Despite their relatively low concentrations, they form stronger and stable complexes with many metal ions and therefore strongly influence mineral precipitation and reactivity. Redox-inactive pyrophosphate (PP), redox-active citrate, and ethylenediaminetetraacetic acid (EDTA) are all known to adsorb and extract Mn^{III}

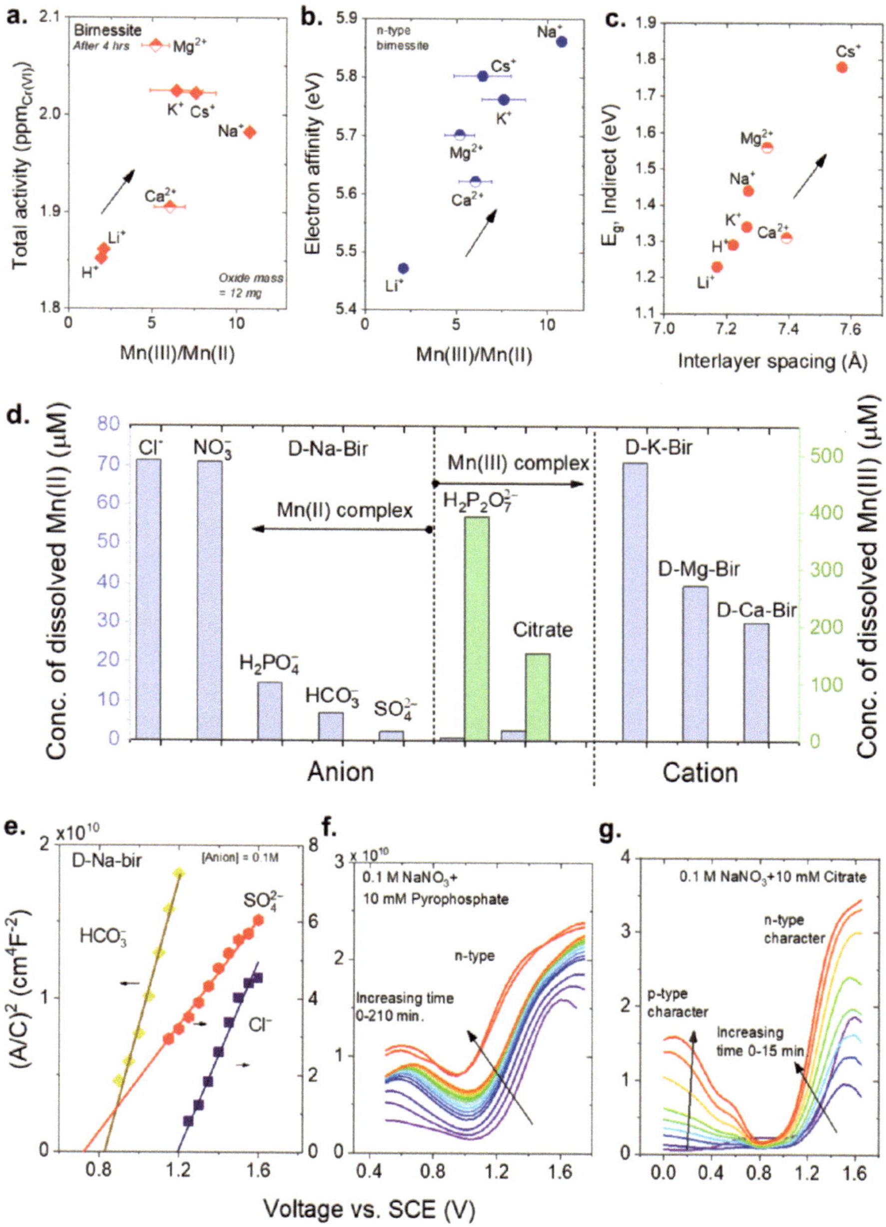

FIGURE 3 (a-c) Effect of electrolyte cations on the composition of structural Mn, oxidative activity (a), electron affinity (b), and band gap (c) of semiconducting birnessite; (d) influence of different electrolyte ions on the dissolution of Mn as Mn^{II} and Mn^{III} from the birnessite structure; (e-g) Mott-Schottky plots of disordered Na-birnessite films showing the shifts in the U_{FB} with different inorganic and organic ligands in the electrolyte of pH = 7. Reprinted with permission from [78]. Copyright 2021 American Chemical Society. Reprinted with permission from [77]. Copyright 2023 American Chemical Society.

from MnO_x mineral surfaces and affect birnessite's oxidative reactivity through coupled effects [17,89–97]. For example, exposure of disordered Na-birnessite (δ-MnO_2) to PP dramatically quenches the oxidative activity for Cr^{III} to Cr^{VI} conversion [98], while the same ligand has been shown to enhance the rate of oxidation of As^{III} to As^{V} in hexagonal birnessite [22]. In contrast, the adsorption of other Mn^{III}-complexing ligands such as citrate [99] or phosphate [100,101] was reported to suppress As^{III} oxidation in hexagonal birnessite. The adsorption of PP also induces a transformation from hexagonal to triclinic phase under alkaline conditions [17]. Studies of these ligands on the semiconducting properties show that strong Mn^{III} complexing ligands (PP, citrate, EDTA) promote Mn^{III} dissolution (Figure 3d), i.e., decrease the Mn^{III}/Mn^{II} ratio and increase the lattice Mn vacancy concentration, and cause a negative shift of the U_{FB} of birnessite (Figure 3f and 3g), and decrease oxidative activity [77]. These results show that birnessite has the highest oxidative activity in electrolytes whose anions enable facile structural dissolution of Mn^{II} into the electrolyte while preventing complexation and dissolution of Mn^{III}.

Although all of the above results point to the importance of Mn^{III} for the high oxidative activity of birnessite, paradoxically, however, the presence of Mn^{III} in the lattice is not the only requirement because other MnO_x phases, such as Mn_2O_3 or MnOOH, with a higher Mn^{III} content than birnessite lattices, show lower activity and lower electron affinity than birnessite. One explanation could be that some other species, likely intercalated H_2O or oxygen defects associated with Mn^{III}, are responsible for the high electron affinity. Adsorbed water on the surface is known to cause an increase in the electron affinity of semiconductors [102]. Unfortunately, the role of adsorbed water in the oxidative properties of birnessite has not been thoroughly investigated. Temperature-programmed dehydration studies [58] show that birnessite dehydration causes phase transformation to more crystalline (lower MnO_x phases such as Mn_3O_4 or Mn_2O_3 that lack structural water). Conversely, both experimental [103] and theoretical [104] studies show that insertion of water molecules into a spinel Mn_3O_4 lattice can transform to the birnessite structure. Furthermore, several researchers have long noted the presence of certain oxygen defects, such as $O_2^{\bullet -}$ possibly associated with Mn^{III} in birnessite. Through dehydration and dehydrogenation studies, Hasegawa et al. [105] attributed that the high oxidative activity of hydrated MnO_2 is directly correlated with the presence of these surface active anionic radicals, $O_2^{\bullet -}$ or $HOO^{\bullet -}$, present even under dark conditions. The results of more recent studies [106,107] confirm ROS formation under dark conditions in α-, β-, γ-, and δ-MnO_2, with birnessite forms, δ-MnO_2, hexagonal and triclinic forms, exhibiting the highest ROS concentration. Thus, interlayer water and/or oxygen defects may play a key role in rendering high oxidative power to birnessite and need to be investigated.

5 IMPLICATION OF BAND STRUCTURE: HIGH OXIDATIVE ACTIVITY, NATURE OF ROS

Highly reactive oxygen species, such as the hydroxyl radical (•OH), the superoxide anion radical ($O_2^{-\bullet}$), and hydrogen peroxide (H_2O_2), which are produced by both biotic and abiotic processes, profoundly influence numerous elemental cycles by affecting various chemical and phase transformations in nature. ROS formation on mineral surfaces under dark and light-induced reactions is recognized as one of the important mechanisms for the abiotic production of ROS, but many heterotrophic bacteria are known to produce extracellular ROS such as $O_2^{-\bullet}$ in the absence of light and understanding their ROS-mediated synergistic interactions is of interest to understand the basis for mineral-microbe co-evolution. For example, direct oxidation of Mn^{II} by molecular O_2 is known to be extremely sluggish, while the oxidization of Mn^{II} to MnO_x can be facile when reacted with microbially produced $O_2^{-\bullet}$ radicals [42]. The band alignments of a semiconducting mineral with respect to the ROS redox reactions determine the nature and mechanism of ROS production/dismutation (Figure 4). Birnessite has been known to demonstrate enzyme catalase-like behavior in the dismutation of H_2O_2 to O_2 [108,109]. This process may involve e^- injection from H_2O_2 $(H_2O_2 \rightarrow O_2 + 2H^+ + 2e^-)$ into the low-lying CB of birnessite. This dismutating ability of birnessite assists in bacterial survival under high oxidative stress conditions. The presence of H_2O_2 in the environment can be fatal to cell survival because H_2O_2 can produce •OH, a powerful indiscriminate oxidizer, when reacted with Fe^{II} through the Fenton reaction. As a survival strategy, bacteria can indirectly oxidize Mn^{II} by enzymatic generation of extracellular $O_2^{-\bullet}$ to produce biogenic birnessite that can then lyse H_2O_2 [42].

Exposure to excessive ROS can lead to cellular toxicity. The extreme positions of the band edges of the GR and birnessite minerals also provide them with a high potency for ROS generation. The photocatalytic ROS activities (Figure 4a and 4b) of several Fe and Mn minerals [79] show that GR has the highest activity for H_2O_2 formation among all investigated minerals, but interestingly only a low to negligible activity for •OH production. On the other hand, triclinic and disordered birnessite minerals have the highest activity for •OH production, but show only low to moderate activity for H_2O_2. Importantly, the highest ROS concentrations measured with these minerals are nearly 2 to 10 times higher than those seen with photoactive TiO_2 and Fe_2O_3. The extremely high placement of the CB (very low χ) of GR (Figure 4c) provides a large driving force for the injection of e^- into O_2 that leads to H_2O_2, as ROS production is nearly quenched under anaerobic conditions (inset in Figure 4a). In birnessite, ROS generation only decreases slightly under anaerobic conditions (inset in Figure 4a), indicating that the reaction with O_2 is not the dominant path for •OH generation but occurs instead through the hole oxidation of OH^-/H_2O.

5.1 Birnessite's Structural and Functional Similarity to the Catalase and Oxygen Evolving Complex of Photosystem II

Metalloenzymes known to produce molecular oxygen in nature are superoxide dismutase, catalase, and the Mn_4O_xCa oxygen-evolving complex (OEC) of the photosystem II (PS-II), which produce molecular oxygen from superoxide, H_2O_2, and H_2O, respectively, as substrates. Interestingly, the active site for all three enzymes consists of Mn clusters that undergo a reversible valence change between Mn^{II}, Mn^{III}, and Mn^{IV} upon binding and sequential deprotonation of substrate molecules. The

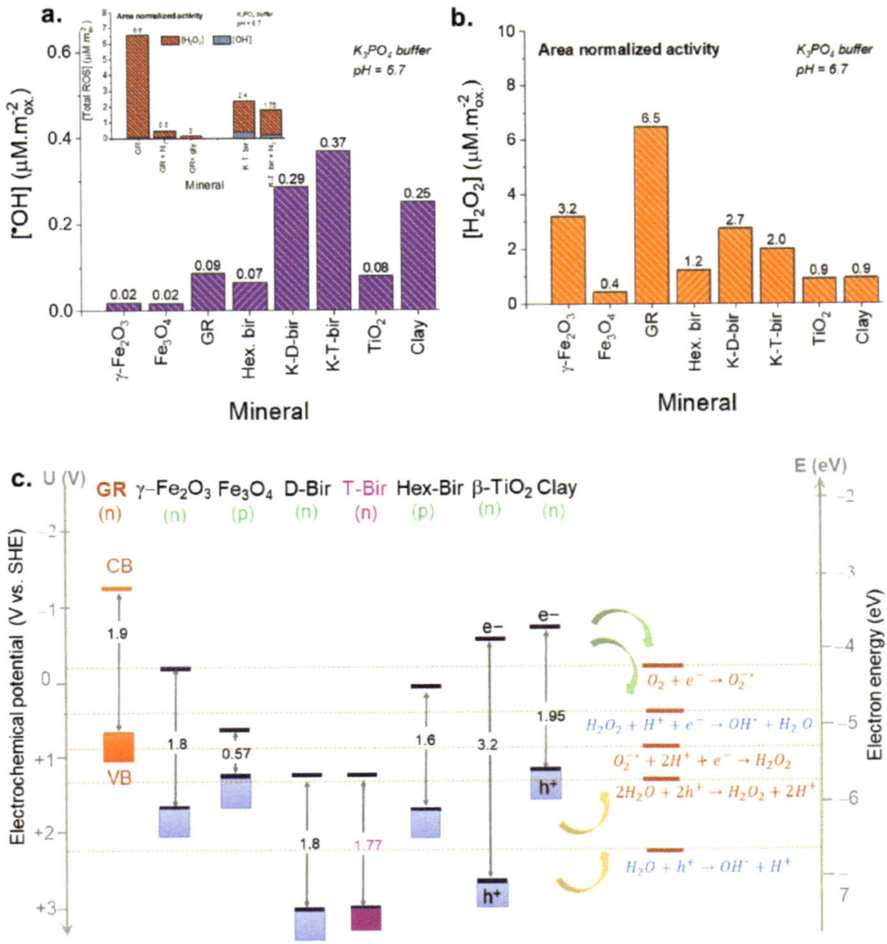

FIGURE 4 (a-b) Comparison of the surface area-normalized concentrations of photocatalytically produced *OH (a) and H_2O_2 (b) on various oxide minerals suspended in K_3PO_4 buffer of pH = 6.7 under aerobic conditions. Inset in (a) comparison of the total ROS concentration (•OH and H_2O_2) produced on GR and triclinic birnessite under aerobic and anaerobic conditions. For GR, the data with glycerol was taken under aerobic conditions; (c) various ROS generation pathways on oxide minerals through comparison of their band lineups with the $\bar{\mu}_{e-,redox}$ of reactions involving ROS in the absolute vacuum energy scale as well as on the electrochemical scale. Reprinted with permission from [79] Copyright 2023 American Chemical Society.

structure of catalase and OEC consists of Mn atoms bridged using µ-oxido ligands. Several synthetic phases of MnO_x, including birnessite, show catalase-like behavior in effectively disproportioning H_2O_2 to O_2 [108]. However, only a few are active for water oxidation. Figure 5 shows the multielectron cascade steps of oxidation of water to the O_2 mechanism that occurs in PS-II after light excitation. This reaction

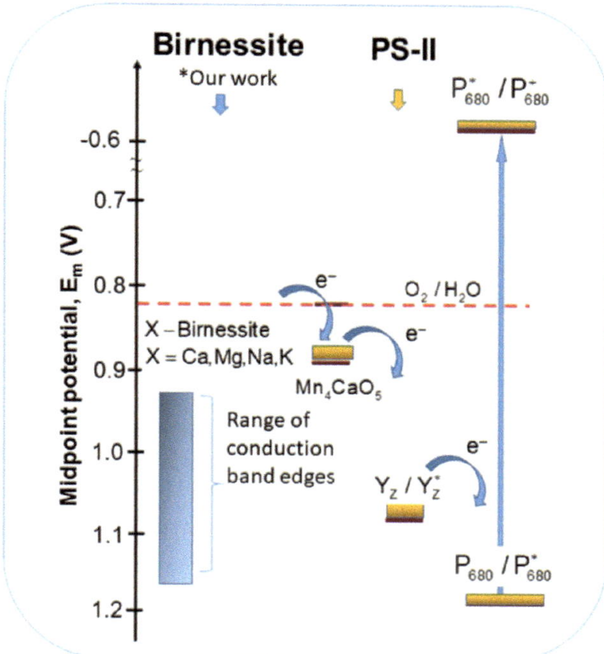

FIGURE 5 Schematic diagram of the reported range of redox potentials versus NHE of a few PS-II components and that of birnessite at pH=7.

occurs at a record low overpotential of 160–300 mV with an incident light-to-fuel conversion efficiency of 16% and a turn-over frequency (TOF) of 1000 s^{-1} [110,111]. Although no other catalysts, synthetic or otherwise, have shown oxygen evolution reaction (OER) efficacy even remotely close to that of OEC, the birnessite mineral is the closest natural analog to OEC and has the highest mass activities for water oxidation among synthetic MnO_x polymorphs with rates exceeding 1650 $mmolO_2\ mol_{Mn}^{-1} h^{-1}$ [112]. Several researchers have noted the remarkable structural and functional similarity between the two Mn structures [113,114]. Like the Kok cycle in the Mn_4CaO_5 OEC, the Mn^{II} and Mn^{III} atoms of birnessite undergo a cyclic change in the oxidative valence state, with Mn^{III} as the key intermediate [115] of water oxidation. This is further corroborated by studies [116] showing that the MnO_x cluster that reforms during photoactivation in PS-II devoid of the Mn_4CaO_5 center has a strong structural similarity to the birnessite lattice. Although no modern-day microorganisms are known to use birnessite for water oxidation, similarities have led some to propose that birnessite may have been the nature's evolutionary precursor for Mn_4CaO_5 in the initial stages of PS-II assembly [116]. The formation of a birnessite-type cluster in apo-OEC is likely due to the highly positive redox potential or U_{FB} of birnessite, which is the most positive among the various MnO_x forms. Figure 5 shows the redox potential of various components of PS-II, including the OEC (S states), relative to the conduction band position of various cation-exchanged birnessites. The equilibrium potential of the H_2O/O_2 redox reaction at a biological pH of 7 is 0.82 V *versus* the normal hydrogen electrode (NHE). To allow this reaction to be carried out at a high rate, the midpoint reduction potentials (E_m) of the Mn_4CaO_5 cluster in states S_1 to S_3 are poised at more positive values of 0.8–0.95 V (except during the transition from S_0 to S_1) and P_{680} at 1.1–1.2 V [117–119]. In fact, these positive potentials of PS-II are among the most positive values found for any biochemical process that occurs in nature and are finely modulated through electrostatic interactions between the various cofactors. Only birnessite minerals with their high electron affinities have the requisite high redox potential values (E_m of 0.9–1.4 V *versus* NHE), necessary to allow water oxidation. The relatively similar value of the redox potential of birnessite and that of the Mn_4CaO_5 cluster is the reason for the precipitation of the birnessite-type lattice in the OEC-devoid PS-II cluster. Therefore, birnessite is an important model system in the development of bioinspired OER catalysts, and being a relatively simpler analog of Ma_4CaO_5, its study of water oxidation may provide important clues to deciphering the more complex water oxidation process in PS-II.

Another consequence of the high positive redox potential of birnessite is the mechanism of water oxidation. In PS-II, the photoactivation process involves the dynamic formation of the OEC cluster to the S_0 state from membrane-bound Mn^{II} through a sequential light-induced activation step, and then subsequently oxidized to Mn^{III} and Mn^{IV}. Similarly to this photoactivation of Mn^{II}, the potential-induced water oxidation mechanism in birnessite studied in different alkali cation electrolytes shows that the water oxidation rate is governed by the rapid and reversible redox shuttle of Mn^{II} between the birnessite structure and the electrolyte, and it is this labile Mn^{II} that upon subsequent oxidation forms the key Mn^{III} intermediate, not the existing lattice Mn^{III} that is responsible for the OER process [88]. The dissolution-deposition process of Mn^{III} formation is possible due to the large electron affinity of ~6.0 eV in aqueous solution, which puts its CB well below the redox potential of the O_2/H_2O reaction. When birnessite is first immersed in the electrolyte, electron injection from water into the birnessite CB leads to the formation of an electron accumulation layer in birnessite. The localization of these electrons at the Mn^{IV} lattice sites produces Mn^{II} cations, which, under suitable pH conditions, dissolve into the electrolyte and establish equilibrium. During subsequent polarization of birnessite, dissolved Mn^{II} oxidizes at the birnessite lattice to form the Mn^{III} intermediate that subsequently catalyzes the formation of O_2. Thus, the OER activity is controlled by dynamically formed Mn^{III} through the oxidation of labile uncomplexed Mn^{II} and not the preexisting Mn^{III} in the lattice. The role of electrolyte cations (Cs^+, K^+, and Li^+) of different ionic radii and hydration strengths is to control the interlayer spacing of the basal sheets, while electrolyte anions (NO_3^-, HCO_3^-, SO_4^{2-}, PO_4^{3-}, Cl^-, and OH^-) control the extent of deprotonation of complexed Mn^{II} in the lattice. Both effects, in turn, govern the shuttling efficiency of uncomplexed Mn^{II} and its subsequent electrooxidation to the Mn^{III} intermediate.

5.2 $Fe^{II}(OH)_2$ and $Fe^{II}Fe^{III}(OH)_2$ Functional Similarity to the Multi-tasking Nitrogenase Enzyme

Nitrogen is required for all living organisms since it is an essential constituent of amino acids and nucleotides that are the basic building blocks of life. Constituting 78% of the atmosphere, the N_2 gas is the most abundant source of nitrogen on Earth, yet this N_2 is not directly accessible to most living organisms and must undergo transformation into a reduced, bioaccessible form, ammonium (NH_4^+). Enzyme nitrogenase, which is found in many prokaryotes, including cyanobacteria, rhizobacteria, aerobic heterotrophs, and phototrophs, can catalyze the energetically expensive reduction of triple bonded N_2 to NH_3 under ambient conditions. To date, three homologous nitrogenases, Mo, V, and Fe only nitrogenases, have been known, all of which are made up of two metalloproteins: a homodimeric Fe_4S_4 protein cluster reductase with a high reducing power that couples the hydrolysis of adenosine triphosphate (ATP) with electron transfer from ferredoxin to the other protein, and a second metalloprotein consisting of a heterometal center of Mo, Fe only, or V and its cofactors that house the catalytic site that converts N_2 to NH_3 using the electrons supplied by the reductase with the liberation of H_2 as a side product [120–123]. In addition to N_2, nitrogenases can also reduce several other substrates [124], such as acetylene (C_2H_2) [125], carbon monoxide (CO) [126], carbon dioxide (CO_2) [127], diazene (HN=NH) [128], and hydrazine ($H_2N–NH_2$) [129]. V-nitrogenase, but not Mo-containing nitrogenase, can reduce CO to hydrocarbons, primarily ethylene (C_2H_4), with minor amounts of ethane, propane, and methane [126,130], which is analogous to the Fischer-Tropsch industrial process. Interestingly, both $Fe^{II}(OH)_2$ and $Fe^{II}Fe^{III}(OH)_2$ show many functional similarities to nitrogenase enzymes. Schrauzer and Guth [80] reported that $Fe^{II}(OH)_2$, the Fe^{II} only counterpart of GR, can reduce molecular N_2 to hydrazine and ammonia, as well as CO to CH_4. The results of our recent investigation [131] on Fe^{II}-rich GR show that it can also reduce CO to liquid products such as acetone, in addition to its ability to convert nitrate to ammonia. Figure 6a shows the representative ^{1}H-nuclear magnetic resonance (NMR) spectrum of electrolytes of 0.1 M $KHCO_3$/K_2CO_3 containing GR powder and sparged with either N_2 or CO. The products of CO reduction (COR) are acetone, acetate, and peracetate, as determined by comparison with the reference spectrum. The *TMSP corresponds to the peak of the internal standard used for calibration and quantification. Figure 6b shows the time-dependent changes in the concentration distribution determined from the quantitative analysis of the ^{1}H-NMR spectrum taken at various time intervals and the corresponding production rate of CO reduction products detected in the electrolyte. While the acetone concentration increases with time, the production rate is seen to decrease with time (Figure 6b), which is expected given that GR functions as a reductant and provides the electron necessary for CO reduction. Consequently, slow oxidation of the material is expected. These products obtained with GR have nothing in common with the N_2 or CO reduction products of nitrogenase. However, it is striking to note that the midpoint redox potential of the [4Fe-4S] cluster (–460 to –790 mV [132,133]) of the Fe protein, the electron donors of the nitrogenase enzyme, and that of GR (–1000 mV) are among the lowest seen for biological and mineralogical species. The

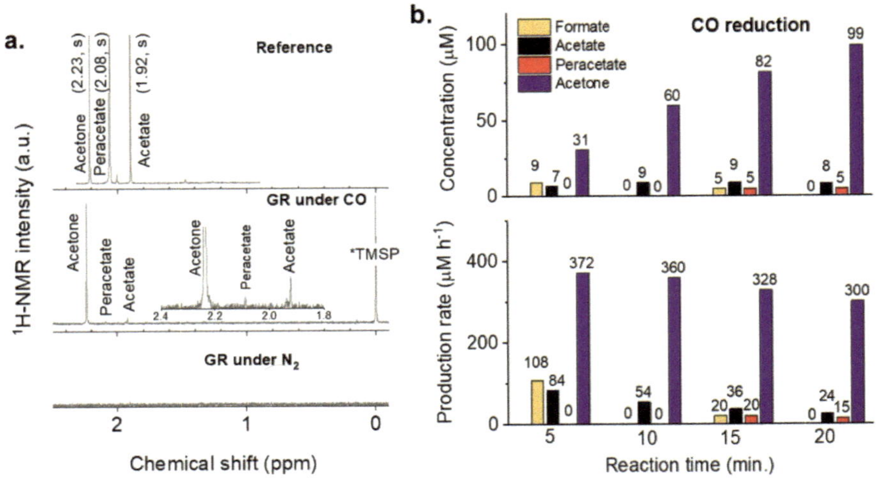

FIGURE 6 Results of CO reduction experiments with GR powder at pH = 12. (a) Representative ^{1}H-NMR spectrum of electrolytes of 0.1 M $KHCO_3/K_2CO_3$ containing GR powder and sparged with either N_2 or CO. The products of CO reduction are acetone, acetate, and peracetate, as determined by comparison with the reference spectrum. The *TMSP corresponds to the peak of the internal standard used for calibration and quantification; and (b) time-dependent changes in the production rate and concentration distribution of COR products detected in the electrolyte. Reprinted from [131]. Copyright 2024 American Chemical Society.

similarity of the extremely low potential may be the reason for the observation of some functional similarity in the reduction of carbon and nitrogen.

6 GENERAL CONCLUSIONS

The Fe and Mn minerals assume many important geochemical functions that directly and indirectly contribute to the survival of life on Earth. Among their wide range of structural diversities and functional properties, layered Fe and Mn minerals are of great importance because of their superlative properties. This chapter highlighted the unique semiconducting band structure of birnessite and GR with the highest and lowest electron affinities, respectively, of all known oxides and sulfides in the aqueous and soil environment. The chemical reactivity of these minerals was shown to be directly correlated with the extreme positions of their band edges, which also has a direct bearing on many biogeochemical functions that aid the coexistence of the mineral with a variety of microorganisms.

ACKNOWLEDGMENTS

The author thanks the graduate students who worked in her lab and whose work contributed immensely to this work.

ABBREVIATIONS AND DEFINITIONS

ε_s	permittivity of the semiconductor
$\bar{\mu}_{e-}$	electrochemical potential of electrons
Φ_S	work function
χ	electron affinity
ATP	adenosine triphosphate
C	measured capacitance
DLH	double-layered hydroxides
E_C	conduction band energy
E_G	band gap
E_F	Fermi energy
E_m	midpoint reduction potential
E_{SHE}	electron energy of the standard hydrogen electrode
E_{VAC}	vacuum energy, taken to be zero
E_V	valence band energy
EDTA	ethylenediaminetetraacetic acid
GR	green rust
HA	humic acid
k	Boltzmann constant
M-S	Mott-Schottky measurements
N_D	concentration of uncompensated charge carriers
NHE	normal hydrogen electrode
NMR	nuclear magnetic resonance
OEC	oxygen evolving complex
OER	oxygen evolution reaction
PS-II	photosystem II
PP	pyrophosphate
q	elementary charge
ROS	reactive oxygen species
SCE	standard calomel electrode
TOF	turn-over frequency
U	applied potential
U_{FB}	flat band potential

REFERENCES

1. S. Dasgupta, X. Peng, K. Ta, *Minerals* **2021**, *11*, 1324.
2. H. Dong, L. Huang, L. Zhao, Q. Zeng, X. Liu, Y. Sheng, L. Shi, G. Wu, H. Jiang, F. Li, L. Zhang, D. Guo, G. Li, W. Hou, H. Chen, *Natl. Sci. Rev.* **2022**, *9*.

3. S. Kato, K. Hashimoto, K. Watanabe, *Proc. Nat. Acad. Sci.* USA **2012**, *109*, 10042–10046.
4. S. Kato, K. Hashimoto, K. Watanabe, *Environ. Microbiol.* **2012**, *14*, 1646–1654.
5. J. E. Post, *Proc. Natl. Acad. Sci.* USA **1999**, *96*, 3447–3454.
6. D. C. Fernández-Remolar, *Iron Oxides, Hydroxides and Oxy-hydroxides.* In *Encyclopedia of Astrobiology*, M. Gargaud, W. M. Irvine, R. Amils, H. J. Cleaves, D. L. Pinti, J. C. Quintanilla, D. Rouan, T. Spohn, S. Tirard, M. Viso, Eds. Springer Berlin, Heidelberg, 2015, pp 1268–1270.
7. W. M. Landing, K. W. Bruland, *Geochim. Cosmochim. Acta* **1987**, *51*, 29–43.
8. F. T. Ling, J. E. Post, P. J. Heaney, C. M. Santelli, E. S. Ilton, W. D. Burgos, A. W. Rose, *Am. Mineral.* **2020**, *105*, 833–847.
9. J. E. Post, D. R. Veblen, *Am. Mineral.* **1990**, *75*, 477–489.
10. V. A. Drits, E. Silvester, A. I. Gorshkov, A. Manceau, *Am. Mineral.* **1997**, *82*, 946–961.
11. E. Silvester, A. Manceau, V. A. Drits, *Am. Mineral.* **1997**, *82*, 962–978.
12. B. Lanson, V. A. Drits, E. Silvester, A. Manceau, *Am. Mineral.* **2000**, *85*, 826–838.
13. H. Tan, G. Zhang, P. J. Heaney, S. M. Webb, W. D. Burgos, *Appl. Geochem.* **2010**, *25*, 389–399.
14. M. Villalobos, B. Lanson, A. Manceau, B. Toner, G. Sposito, *Am. Mineral.* **2006**, *91*, 489–502.
15. M. Villalobos, B. Toner, J. Bargar, G. Sposito, *Geochim. Cosmochim. Acta* **2003**, *67*, 2649–2662.
16. O. J. Hao, A. P. Davis, P. H. Chang, *J. Environ. Eng.* **1991**, *117*, 359–374.
17. H. Jung, T. S. Chadha, Y. Min, P. Biswas, Y.-S. Jun, *ACS Sustainable Chem. Eng.* **2017**, *5*, 10624–10632.
18. S. H. Davies, J. J. Morgan, *J. Colloid Interface Sci.* **1989**, *129*, 63–77.
19. T. Zhang, L. Liu, W. Tan, S. L. Suib, G. Qiu, F. Liu, *Environ. Sci. Technol.* **2018**, *52*, 6864–6871.
20. F. T. Ling, P. J. Heaney, J. E. Post, X. Gao, *Chem. Geol.* **2015**, *416*, 1–10.
21. D. Banerjee, H. W. Nesbitt, *Geochim. Cosmochim. Acta* **1999**, *63*, 1671–1687.
22. C. Ying, B. Lanson, C. Wang, X. Wang, H. Yin, Y. Yan, W. Tan, F. Liu, X. Feng, *Water Res.* **2020**, *187*, 116420.
23. K. P. Kong, T. B. Fischer, P. J. Heaney, J. E. Post, J. E. Stubbs, P. J. Eng, *Appl. Geochem.* **2019**, *108*, 104365.
24. H. C. B. Hansen, S. Guldberg, M. Erbs, C. B. Koch, *Appl. Clay Sci.* **2001**, *18*, 81–91.
25. H. C. B. Hansen, O. K. Borggaard, J. Sørensen, *Geochim. Cosmochim. Acta* **1994**, *58*, 2599–2608.
26. M. Erbs, H. C. Bruun Hansen, C. E. Olsen, *Environ. Sci. Technol.* **1999**, *33*, 307–311.
27. A. N. Thomas, E. Eiche, J. Göttlicher, R. Steininger, L. G. Benning, H. M. Freeman, D. J. Tobler, M. Mangayayam, K. Dideriksen, T. Neumann, *Geochem. Trans.* **2020**, *21*, 2.
28. S. C. B. Myneni, T. K. Tokunaga, G. E. Brown Jr, *Science* **1997**, *278*, 1106–1109.
29. E. J. O'Loughlin, S. D. Kelly, R. E. Cook, R. Csencsits, K. M. Kemner, *Environ. Sci. Technol.* **2003**, *37*, 721–727.
30. E. J. O'Loughlin, S. D. Kelly, K. M. Kemner, R. Csencsits, R. E. Cook, *Chemosphere* **2003**, *53*, 437–446.
31. S. H. Drissi, P. Refait, M. Abdelmoula, J. M. R. Génin, *Corros. Sci.* **1995**, *37*, 2025–2041.
32. F. Trolard, J. M. R. Génin, M. Abdelmoula, G. Bourrié, B. Humbert, A. Herbillon, *Geochim. Cosmochim. Acta* **1997**, *61*, 1107–1111.
33. M. Abdelmoula, F. Trolard, G. Bourrie, J.-M. Génin, *Hyperfine Interact.* **1998**, *112*, 235–238.
34. B. Christiansen, T. Balic-Zunic, K. Dideriksen, S. Stipp, *Environ. Sci. Technol.* **2009**, *43*, 3436–3441.

35. A. Zegeye, S. Bonneville, L. G. Benning, A. Sturm, D. A. Fowle, C. Jones, D. E. Canfield, C. Ruby, L. C. MacLean, S. Nomosatryo, *Geology* **2012**, *40*, 599–602.
36. A. Vuillemin, R. Wirth, H. Kemnitz, A. M. Schleicher, A. Friese, K. W. Bauer, R. Simister, S. Nomosatryo, L. Ordoñez, D. Ariztegui, *Geology* **2019**, *47*, 540–544.
37. T. Rennert, K. Eusterhues, V. De Andrade, K. U. Totsche, *Chem. Geol.* **2012**, *332*, 116–123.
38. J. Bearcock, W. Perkins, E. Dinelli, S. Wade, *Mineral. Mag.* **2006**, *70*, 731–741.
39. C. A. Johnson, G. Freyer, M. Fabisch, M. A. Caraballo, K. Küsel, M. F. Hochella, *Environ. Chem.* **2014**, *11*, 377–391.
40. W. C. Ghiorse, *Annu. Rev. Microbiol.* **1984**, *38*, 515–550.
41. X. Xu, Y. Li, Y. Li, A. Lu, R. Qiao, K. Liu, H. Ding, C. Wang, *Chem. Geol.* **2019**, *522*, 55–70.
42. D. R. Learman, B. M. Voelker, A. I. Vazquez-Rodriguez, C. M. Hansel, *Nat. Geosci.* **2011**, *4*, 95–98.
43. P. Refait, A.-M. Grolleau, M. Jeannin, C. Rémazeilles, R. Sabot, *Corros. Mater. Degrad.* **2020**, *1*, 198–218.
44. P. P. Stampfl, *Corros. Sci.* **1969**, *9*, 185–187.
45. H. C. B. Hansen, *Clay Miner.* **1989**, *24*, 663–669.
46. H. A. Lowenstam, *Science* **1981**, *211*, 1126–1131.
47. Y. A. Parmar, T. J. Gorby, F. G. Beveridge, N. Ferris, *Geomicrobiol. J.* **2001**, *18*, 375–385.
48. J. K. Fredrickson, J. M. Zachara, D. W. Kennedy, H. Dong, T. C. Onstott, N. W. Hinman, S.-m. Li, *Geochim. Cosmochim. Acta* **1998**, *62*, 3239–3257.
49. S. K. Chaudhuri, J. G. Lack, J. D. Coates, *Appl. Environ. Microbiol.* **2001**, *67*, 2844–2848.
50. M. Etique, F. P. A. Jorand, A. Zegeye, B. Grégoire, C. Despas, C. Ruby, *Environ. Sci. Technol.* **2014**, *48*, 3742–3751.
51. I. Halevy, M. Alesker, E. Schuster, R. Popovitz-Biro, Y. Feldman, *Nat. Geosci.* **2017**, *10*, 135–139.
52. Y.-L. Li, K. O. Konhauser, M. Zhai, *Earth Planet. Sci. Lett.* **2017**, *466*, 103–114.
53. E. Koeksoy, A. Sundman, J. Byrne, R. Lohmayer, B. Planer-Friedrich, I. Halevy, K. O. Konhauser, A. Kappler, *Geology* **2019**, *47*, 211–214.
54. J. D. Culpepper, Michelle M. Scherer, T. C. Robinson, A. Neumann, D. Cwiertny, D. E. Latta, *Environ. Sci. Process. Impacts* **2018**, *20*, 1340–1349.
55. J. Klausen, S. P. Troeber, S. B. Haderlein, R. P. Schwarzenbach, *Environ. Sci. Technol.* **1995**, *29*, 2396–2404.
56. A. Manceau, E. Silvester, C. Bartoli, B. Lanson, V. Drits, *Am. Mineral.* **1997**, *82*, 1150–1175.
57. F. T. Ling, J. E. Post, P. J. Heaney, E. S. Ilton, *Chem. Geol.* **2018**, *479*, 216–227.
58. E. A. Johnson, J. E. Post, *Am. Mineral.* **2006**, *91*, 609–618.
59. D. Banerjee, H. W. Nesbitt, *Geochim. Cosmochim. Acta* **2001**, *65*, 1703–1714.
60. T. Takashima, K. Hashimoto, R. Nakamura, *J. Am. Chem. Soc.* **2012**, *134*, 1519–1527.
61. P. Yang, S. Lee, J. E. Post, H. Xu, Q. Wang, W. Xu, M. Zhu, *Geochim. Cosmochim. Acta* **2018**, *240*, 173–190.
62. H. Zhao, M. Zhu, W. Li, E. J. Elzinga, M. Villalobos, F. Liu, J. Zhang, X. Feng, D. L. Sparks, *Environ. Sci. Technol.* **2016**, *50*, 1750–1758.
63. A. Géhin, C. Ruby, M. Abdelmoula, O. Benali, J. Ghanbaja, P. Refait, J.-M. R. Génin, *Solid State Sci.* **2002**, *4*, 61–66.
64. F. Bocher, A. Géhin, C. Ruby, J. Ghanbaja, M. Abdelmoula, J.-M. R. Génin, *Solid State Sci.* **2004**, *6*, 117–124.
65. J.-M. R. Génin, C. Ruby, A. Géhin, P. Refait, *C R Geosci.* **2006**, *338*, 433–446.

66. D. M. Sherman, *Geochim. Cosmochim. Acta* **2005**, *69*, 3249–3255.
67. K. D. Kwon, K. Refson, G. Sposito, *Geochim. Cosmochim. Acta* **2009**, *73*, 4142–4150.
68. B. A. Pinaud, Z. Chen, D. N. Abram, T. F. Jaramillo, *J. Phys. Chem. C* **2011**, *115*, 11830–11838.
69. K. P. Lucht, J. L. Mendoza-Cortes, *J. Phys. Chem. C* **2015**, *119*, 22838–22846.
70. K. D. Kwon, K. Refson, G. Sposito, *Phys. Rev. Lett.* **2008**, *100*, 146601.
71. A. Lu, Y. Li, F. Liu, Y. Liu, H. Ye, Z. Zhuang, Y. Li, H. Ding, C. Wang, *Mineral. Mag.* **2021**, *85*, 22–38.
72. A. Lu, Y. Li, H. Ding, X. Xu, Y. Li, G. Ren, J. Liang, Y. Liu, H. Hong, N. Chen, *Proc. Natl. Acad. Sci.* USA **2019**, *116*, 9741–9746.
73. A. Hankin, F. E. Bedoya-Lora, J. C. Alexander, A. Regoutz, G. H. Kelsall, *J. Mater. Chem. A* **2019**, *7*, 26162–26176.
74. H. Matsui, J. Ju, T. Odaira, N. Toyota, *J. Phys. Soc. Jpn.* **2009**, *78*, 074801.
75. A. J. Bard, R. Memming, B. Miller, *Pure Appl. Chem.* **1991**, *63*, 569–596.
76. S. Trasatti, *Pure Appl. Chem.* **1986**, *58*, 955–966.
77. C. Wang, V. Chakrapani, *ACS Earth Space Chem.* **2023**, *7*, 774–787.
78. C. Wang, N. Smieszek, V. Chakrapani, *Chem. Mater.* **2021**, *33*, 7805–7817.
79. C. Wang, V. Chakrapani, *J. Phys. Chem. C* **2023**, *127*, 23189–23198.
80. G. N. Schrauzer, T. D. Guth, *J. Am. Chem. Soc.* **1976**, *98*, 3508–3513.
81. Y. Xu, M. A. Schoonen, *Am. Mineral.* **2000**, *85*, 543–556.
82. V. Chakrapani, *Appl. Phys. Lett.* **2021**, *119* 021601.
83. W. G. Sunda, D. J. Kieber, *Nature* **1994**, *367*, 62–64.
84. C. R. Myers, K. H. Nealson, *Science* **1988**, *240*, 1319–1321.
85. C. Zerfaß, J. A. Christie-Oleza, O. S. Soyer, *Appl. Environ. Microbiol.* **2019**, *85*.
86. G. Ona-Nguema, D. Guerbois, C. Pallud, J. Brest, M. Abdelmoula, G. Morin, *Minerals* **2020**, *10*, 818.
87. M. Fonin, R. Pentcheva, Y. S. Dedkov, M. Sperlich, D. V. Vyalikh, M. Scheffler, U. Rüdiger, G. Güntherodt, *Phys. Rev. B* **2005**, *72*, 104436.
88. I. Roy, C. Wang, N. Smieszek, X. Li, L. Tsapatsaris, V. Chakrapani, *ChemSusChem* **2022**, *15*, e202200062.
89. J. J. Morgan, M. A. Schlautman, H. Bilinski, *Environ. Sci. Technol.* **2021**.
90. A. T. Stone, J. J. Morgan, *Environ. Sci. Technol.* **1984**, *18*, 450–456.
91. J. M. Harrington, D. L. Parker, J. R. Bargar, A. A. Jarzecki, B. M. Tebo, G. Sposito, O. W. Duckworth, *Geochim. Cosmochim. Acta* **2012**, *88*, 106–119.
92. J. E. Kostka, G. W. Luther, K. H. Nealson, *Geochim. Cosmochim. Acta* **1995**, *59*, 885–894.
93. A. Qian, W. Zhang, C. Shi, C. Pan, D. E. Giammar, S. Yuan, H. Zhang, Z. Wang, *Environ. Sci. Technol.* **2019**, *53*, 5768–5777.
94. A. T. Stone, *Geochim. Cosmochim. Acta* **1987**, *51*, 919–925.
95. R. E. Trouwborst, B. G. Clement, B. M. Tebo, B. T. Glazer, G. W. Luther, *Science* **2006**, *313*, 1955–1957.
96. J. K. Klewicki, J. J. Morgan, *Environ. Sci. Technol.* **1998**, *32*, 2916–2922.
97. D. Banerjee, H. W. Nesbitt, *Geochim. Cosmochim. Acta* **1999**, *63*, 3025–3038.
98. P. S. Nico, R. J. Zasoski, *Environ. Sci. Technol.* **2000**, *34*, 3363–3367.
99. M. Liang, H. Guo, W. Xiu, *Environ. Sci. Pollut. Res.* **2020**, *27*, 43769–43785.
100. S. J. Parikh, B. J. Lafferty, T. G. Meade, D. L. Sparks, *Environ. Sci. Technol.* **2010**, *44*, 3772–3778.
101. V. Q. Chiu, J. G. Hering, *Environ. Sci. Technol.* **2000**, *34*, 2029–2034.
102. G. Piantanida, A. Breskin, R. Chechik, O. Katz, A. Laikhtman, A. Hoffman, C. Coluzza, *J. Appl. Phys.* **2001**, *89*, 8259–8264.

103. S. Kim, S. Lee, K. W. Nam, J. Shin, S. Y. Lim, W. Cho, K. Suzuki, Y. Oshima, M. Hirayama, R. Kanno, J. W. Choi, *Chem. Mater.* **2016**, *28*, 5488–5494.
104. E. Yang, H. Kim, S. Kim, I. Kim, J. Kim, H. Ji, J. W. Choi, Y. Jung, *Chem. Sci.* **2018**, *9*, 433–438.
105. S. Hasegawa, K. Yasuda, T. Mase, T. Kawaguchi, *J. Catal.* **1977**, *46*, 125–131.
106. S. Zhang, J. Lv, R. Han, Z. Wang, P. Christie, S. Zhang, *Water Res.* **2021**, *196*, 117034.
107. Q. Wang, Z. Han, H. Liu, T. Chen, X. Zou, Z. Chu, J. Hu, F. Sun, H. Wang, *Chemosphere* **2023**, *341*, 139932.
108. S. B. Kanungo, K. M. Parida, B. R. Sant, *Electrochim. Acta* **1981**, *26*, 1157–1167.
109. S. B. Kanungo, *J. Catal.* **1979**, *58*, 419–435.
110. H. Dau, I. Zaharieva, *Acc. Chem. Res.* **2009**, *42*, 1861–1870.
111. H. Dau, C. Limberg, T. Reier, M. Risch, S. Roggan, P. Strasser, *ChemCatChem* **2010**, *2*, 724–761.
112. C. E. Frey, P. Kurz, *Chem. Eur. J.* **2015**, *21*, 14958–14968.
113. M. Wiechen, M. Najafpour, S. I. Allakhverdiev, L. Spiccia, *Energy Environ. Sci.* **2014**, *7*, 2203–2212.
114. M. Wiechen, I. Zaharieva, H. Dau, P. Kurz, *Chem. Sci.* **2012**, *3*, 2330–2339.
115. I. Zaharieva, M. M. Najafpour, M. Wiechen, M. Haumann, P. Kurz, H. Dau, *Energy Environ. Sci.* **2011**, *4*, 2400–2408.
116. P. Chernev, S. Fischer, J. Hoffmann, N. Oliver, R. Assunção, B. Yu, R. L. Burnap, I. Zaharieva, D. J. Nürnberg, M. Haumann, H. Dau, *Nat. Commun.* **2020**, *11*, 6110.
117. P. Geijer, F. Morvaridi, S. Styring, *Biochemistry* **2001**, *40*, 10881–10891.
118. I. Vass, S. Styring, *Biochemistry* **1991**, *30*, 830–839.
119. M. Mandal, K. Kawashima, K. Saito, H. Ishikita, *J. Phys. Chem. C Lett.* **2020**, *11*, 249–255.
120. S. Duval, K. Danyal, S. Shaw, A. K. Lytle, D. R. Dean, B. M. Hoffman, E. Antony, L. C. Seefeldt, *Proc. Nat. Acad. Sci.* USA **2013**, *110*, 16414–16419.
121. J. Kim, D. Rees, *Science* **1992**, *257*, 1677–1682.
122. M. K. Chan, J. Kim, D. Rees, *Science* **1993**, *260*, 792–794.
123. B. K. Burgess, D. J. Lowe, *Chem. Rev.* **1996**, *96*, 2983–3012.
124. L. C. Seefeldt, Z. Y. Yang, S. Duval, D. R. Dean, *Biochim. Biophys. Acta* **2013**, *1827*, 1102–11.
125. M. J. Dilworth, *Biochim. Biophys. Acta Gen. Subj.* **1966**, *127*, 285–294.
126. C. C. Lee, K. Tanifuji, M. Newcomb, J. Liedtke, Y. Hu, M. W. Ribbe, *ChemBioChem* **2018**, *19*, 649–653.
127. N. Khadka, D. R. Dean, D. Smith, B. M. Hoffman, S. Raugei, L. C. Seefeldt, *Inorg. Chem.* **2016**, *55*, 8321–8330.
128. B. M. Hoffman, D. Lukoyanov, D. R. Dean, L. C. Seefeldt, *Acc. Chem. Res.* **2013**, *46*, 587–595.
129. M. J. Dilworth, R. R. Eady, *Biochem. J.* **1991**, *277*, 465–468.
130. C. C. Lee, Y. Hu, M. W. Ribbe, *Science* **2010**, *329*, 642–642.
131. S. Lee, C. Wang, V. Chakrapani, *J. Phys. Chem. C* **2024**, *128*, 15778–15784.
132. A. Braaksma, H. Haaker, H. J. Grande, C. Veeger, *Eur. J Biochem.* **1982**, *121*, 483–491.
133. H. L. Rutledge, F. A. Tezcan, *Chem. Rev.* **2020**, *120*, 5158–5193.

8 Fe-Oxyhydroxide "Fougerite/Green Rust" Minerals Transformed (or Proto-metabolized) CO_2 and CH_4 Prior to Life's Emergence

*Orion Farr**
CNRS, CINaM (UMR 7325),
Aix-Marseille University, Marseille, France
and
CNRS, BIP (UMR 7281), Aix Marseille
University, Marseille, France

Nil Gaudu, Adriana Clouet
CNRS, BIP (UMR 7281), Aix Marseille
University, Marseille, France

Michael J. Russell
Dipartimento di Chimica, Università
degli Studi di Torino, Italy

Chloé Truong
CNRS, BIP (UMR 7281), Aix Marseille University,
Marseille, France

* Corresponding author

DOI: 10.1201/9781003459910-8

CONTENTS

ABSTRACT

The transformation of simple carbon compounds to those more complex is fundamental to life. In this analysis, we consider that the origins of these transformative abilities can be traced to entropically driven geochemical processes on the early Earth. In particular, minerals such as iron oxyhydroxide "green rusts" (GR) – naturally occurring as fougerite – are implicated as primary mineral phases within the ferruginous Archean ocean and have been found to conduct gas redox transformations of CO_2 and CH_4 into various organic products. Their precipitation in the form of mineral membranes across redox gradients at alkaline hydrothermal vents yields structures offering conditions conducive to the emergence of metabolic pathways. Firstly, these environments host dihydrogen (H_2) and methane (CH_4) as reducing agents, and Fe(III), nitrogen oxides, and carbon dioxide (CO_2) as oxidizing agents to drive reversible redox reactions (electron transfer) to be expected of early autotrophic metabolism. Secondly, their steep pH gradients favor proton transfer, which could have enabled these mineral membranes to mediate primitive chemiosmosis with the conversion of CO_2 with CH_4 to more complex carbon compounds. Such systems, inherent to the geochemical evolution of the planet, may be seen as performing the very first stages of a purely abiotic proto-metabolism.

KEYWORDS

green rust, mineral, redox, carbon dioxide, methane, metabolism, emergence of life

> *"Everything is everywhere, but, the environment selects"*
>
> — *Baas Becking, 1934*

1 A SYNTHESIS BETWEEN GEOCHEMICAL DISEQUILIBRIA AND NONEQUILIBRIUM THERMODYNAMIC PROCESSES

A new territory is opened for emergence-of-life theory exploration with the recognition that inorganic materials can mediate the synthesis and proto-metabolism of complex molecules – notably through an understanding of the various properties of the green rust/fougerite (sensu lato) system [1–13]. But how did we get here? In the half-century since Miller and Urey's electrochemical discharge experiments, a dramatic geochemical history of the ancient Archean and Hadean environments has been revealed. The revelation is one of ferruginous nutrient sequestering oceans [14], anoxic atmospheres rich in CH_4 and CO_2 gasses [15], unabated cosmic and solar irradiation impacting all surfaces [16], and intense volcanic hydrothermal vent activity and continuous meteoric impacts [17] – all challenging simple spontaneous generation hypotheses and provoking new solutions to the great "origin of life" question. Yet, contemporary research continues to follow in Miller's footsteps. In fact, reinforced by the discovery of DNA [18,19], investigations into the origin of life mainly focus on the formation of "elementary building blocks", the already complex organic molecules that then "complexified" via spontaneous interactions or wet/dry cycling into genetic material that then brought forth living organisms. Thus were born the RNA world [20–22] and numerous other prebiotic hypotheses that present lipidic [23], aromatic [24], or purely peptidic [25] origins to biochemistry.

However, these theories generally do not consider the matter of the free energy required to power living systems, as do some theories now eclipsed by the frenzied exploration of prebiotic worlds [26–28]. In opposition to the reductionist vision of the quest for the components of life, the mechanistic, energetic view considers the constraints imposed by the laws of physics: according to the second law of thermodynamics, a closed system tends toward an increase in entropy, i.e., toward greater disorder. The essence of life lies not only in its components but above all in its ability to maintain itself outside thermodynamic equilibrium, i.e., in the metabolic reactions that convert the free energy of the environment into a reduction in entropy [28,29].

Emergence of life research is therefore guided by two very distinct epistemological views, the quest for the building blocks of life – normally considered as extant organic molecules from some source or another [30,31] – and, in strong contrast, the quest for structures that could precede function and in such a way as to enable life's emergence. Our contribution here is on the latter which can provide the first step in "evolvability" [32]. It is often mentioned that minerals can be involved in prebiotic processes and redox reactions. For example, the "Iron-Sulfur World" of Wächtershäuser proposed that pyrite (FeS_2) could provide the energy needed to reduce CO_2 and participate in the formation of reduced organic compounds [33]. Once produced, it is proposed that these organic compounds could accumulate on the pyrite surfaces, catalyzing organic reactions to form more complex molecules [34]. The adsorption of organic compounds onto mineral surfaces has been particularly explored around clay mineral phases endowed with large reactive surfaces [35–37]. Interesting candidates can also be found in the oxyhydroxide group, which present minerals with interlayered structures similar to those of clays but with surfaces composed of metal cations bound to oxygen and hydroxyl groups, i.e., green rust [3,4,11,29,38,39]. Two recent papers in particular demonstrate that green rust can mediate a rich organic chemistry by oxidizing methane (CH_4) or reducing CO_2 [40,41]. Moreover, green rust minerals also share structural similarities with the reaction centers of specific metalloenzymes involved in ancient bioenergetic pathways [3]. In this chapter we propose to demonstrate how oxyhydroxide minerals, especially those of the green rust/fougèrite system, are catalytically unique in having the wherewithal to enable the first step to be made from matter to life's emergence [26].

2 DISEQUILIBRIUM GENERATING REDOX SYSTEMS ON EARLY EARTH: THE CASE OF FOUGERITE-RICH HYDROTHERMAL VENTS AS THE PRIMORDIAL CRECHE

In full context, the Universe's continual preoccupation with alleviating various levels of disequilibria all the way from the beginning of space-time [38,42] naturally would lead to the chemical evolution of a structure to act as a general free energy converter – this to relieve electro-geochemical stress that approached a single volt at the surface of the early Earth [43]! A candidate for such a structure has been proposed [44,45] to be the previously mentioned two-dimensional bi-layered mineral group known as $Fe^{2+/3+}$ oxyhydroxides, better known as green rusts [45] which today naturally occurs as Fougerite [46,47]. Fougerite, and its many related forms, is considered ubiquitous as a precipitate in the Hadean and Archean oceans – a major contributor to the first Banded Iron Formations and especially as incorporated within submarine alkaline vents [1,48–50] where life is theorized to have first emerged [43,51]. Fougerite nanocrystals, with their ready and reversible responses to redox variation between the metastable states $\sim[Fe^{2+}6Fe_2^{3+}(OH)_{16}]^{2+}\cdot[CO_3^{2-}\cdot 5H_2O]^{2-}$ and $[Fe_4^{2+}Fe_2^{3+}(OH)_{12}]^{2+}\cdot[CO_3^{2-}\cdot 3H_2O]^{2-}$, had the propensity to act as nano-electro-geochemical converters or nanoengines at these alkaline vents for redox "purposes"[3,4,52–54].

2.1 Characteristics of Green Rust

Fougerite/green rust is a bi- or double-layered (oxy)hydroxide (LDH) – Die Doppelschichtstrukturen [55–58] – comprising positively charged molecular sheets ($[Fe^{2+}/Fe^{3+}\sim X(OH)^{2-}]$) alternating with disordered but corresponding negatively charged layers, or "galleries" of anions [57]. This mineral conducts unique and versatile reactions, readily carrying out the reduction of metals (e.g., Cu, As, U), inorganic compounds (e.g., NO_3^-, ClO_4^-), and organic compounds (e.g., CT, TCE, TCM), [59–64]. In nature such reactions are observed only in diverse microbial metabolisms, yet green rust (GR) readily conducts all these electron transfer reactions, which can further be catalyzed or inhibited given modifications to its unique structure of a charged octahedrally coordinated multivalent cations (e.g., Fe, Ni, Mg, Co, Al) sandwiching interchangeable reactive anions (e.g., CO_3^{2-}, $HCOO^-$, Cl^-, SO_4^{2-}) [41,65]. We might even consider the hydrous interlayers as the harbingers of life's cytoplasm!

These high surface area (3.6–30.1 g/cm²) [66] hexagonal minerals range in size from 10 nm to 100 um and have two distinct structures based on anionic components, GR1 and GR2. Green rust 1 (GR1) [lattice parameters a = 0.3172 nm, c = 2.274 nm] is associated with planar and spherical anions, e.g., chloride $[Fe_5^{2+}Fe_2^{3+}(OH)_{14}]^{2+}[2Cl^{-}\cdot 3H_2O]^{2-}$, carbonate $[Fe^{2+}6Fe_2^{3+}(OH)_{16}]^{2+}\cdot[CO_3^{2-}\cdot 5H_2O]^{2-}$, formate $[Fe_5^{2+}Fe_2^{3+}(OH)_{14}]^{2+}[2HCOO^{-}\cdot 3H_2O]^{2-}$, as opposed to GR2's [lattice parameters a = 0.5524 nm, c = 1.1011 nm] three-dimensional hydrated anions such as sulfate $[Fe_4^{2+}Fe_2^{3+}(OH)_{12}]^{2+}\cdot[SO_4^{2-}\cdot 8H_2O]^{2-}$ with tetrahedral structures requiring two layers of water [67,68]. More complex arrangements can exist with the entry of large organic molecules of up to 24 carbon atoms [69].

A crucial quality of green rust/fougerite is that the anionic interlayers not only permit ionic migration along the hydrate layer but that these nanometric galleries can be used to promote synthesis reactions, i.e., they are effective interception and reaction galleries [1,70–72]. This is due to the LDH oxyhydroxide structure, coordination, and atomic members of GR's crystal lattice, allowing for excellent electron transfer within GR and out to other species [66], leading to impressive redox capabilities in carrying out the reduction of various compounds [73]. The multivalent octahedrally coordinated metal cation sheets that compose GR surfaces are highly catalytic toward the reduction of metals and organic and inorganic compounds [74]. GR is a natural n-type semiconductor, displaying small band gap (1.6 eV) and relatively low electron affinity (~3.2 eV) compared to other transition metal minerals, and drives its donation of electrons in many redox reactions [41]. Green rust has a pH-dependent (zpc=~8–8.6) affinity toward the adsorption and surface-layer incorporation of various metals (e.g., Ni, Zn, Cu, Si, As, W) and compounds (e.g., PO_4, NO_3^-, amino acids) [75–77]. Dense Fe-OH surface sites located on basal planes and edges preferentially adsorb molecules whereas the cations themselves can only replace Fe during synthesis or during recrystallization via Ostwald ripening [78,79]. Various metal compounds in contact with green rust minerals are found to be effectively reduced, even to their native state (e.g., Cu, As, U) under laboratory and environmental conditions, dramatically effecting their solubility and transport throughout aqueous systems [80–83]. This behavior extends to other inorganic compounds such as NO_3^- and ClO_4^- which are eventually fully converted into NH_3^- and Cl^- if left in solution [64,84]. Many carbon-bearing compounds are found to meet the same fate as even notoriously recalcitrant organic pollutants such as CCl_4, $CHCl_3$, and C_2HCl_3 are eventually reduced to ethylene, chloromethane, and CH_4 given properly doped GR

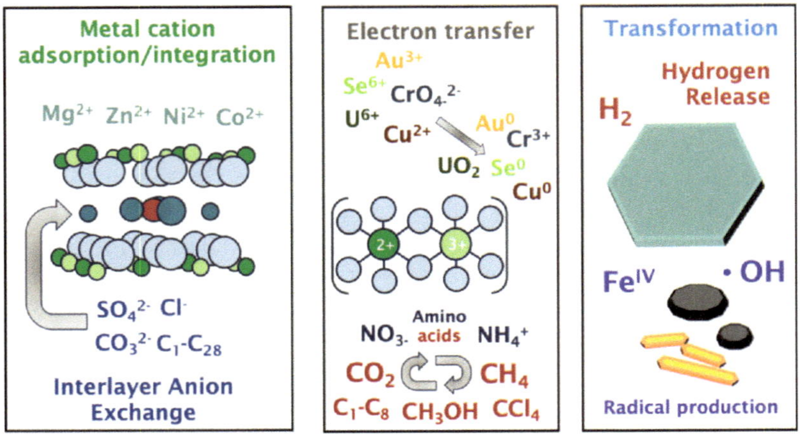

FIGURE 1 Illustrative depictions of the unique chemical and structural properties of green rust minerals. The leftmost section demonstrates the versatile compositions and arrangements of metal cations and interlayer anion species within green rust minerals. The middle section shows the variety of redox reactions carried out by green rusts upon metals, inorganic, and organic compounds. The rightmost section displays the release of radicals and H_2 following mineral transformation.

(85–88). These electron transfer properties uniquely position green rust as an abiotic mediator between geochemical disequilibria in the environment, past and present.

2.2 Resolving Geochemical Disequilibrium

The tensions available to be resolved in the Hadean were those between reductants such as hydrogen and methane emanating from the Earth, versus oxidants such as the nitrogen oxides, nitrate, and nitrite occupying a primitive atmosphere laden with the exhaust fumes of planetary formation [51,52,64,89–91] (Figure 2). Serpentinization, namely the interaction of water with olivine minerals ($MgSiO_4$ to $FeSiO_4$) occurring in AHV environments, is one of the most efficient reactions for the formation of simple organic compounds [92–94]. At high temperatures (around 300° C) it is assumed that olivine hydration generates H_2 which then reacts with CO_2 to form CH_4 and hydrocarbons in a Fischer-Tropsch Type (FTT) reaction [95,96]. At lower temperatures comparable to those found at alkaline vents (between 30 and 80° C) the production of H_2 is associated with the dissolution of olivine and to the oxidation of Fe(II) into Fe(III) [97] in a Schikorr reaction where it is usually ferrous hydroxide ($Fe(OH)_2$) that reacts with water, yielding magnetite (Fe_2O_4) and hydrogen [98] or ferric ions and hematite (a-Fe_2O_3) [99].

These serpentinization processes yield alkaline waters rich in silica and iron which can induce the growth of self-assembled mineral structures [100], among them silica/oxy-hydroxide metal bilayer membranes [101–103]. The superstructure of the idealized Hadean precipitate-mounds argued to have been the hatchery of life was comprised of the natural disequilibria converter fougerite, accompanied by

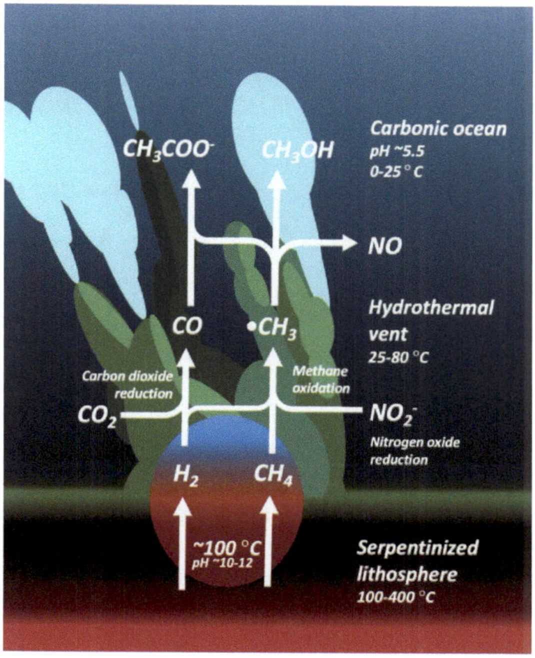

FIGURE 2 Hypothesized abiotic versions of acetogenesis and methanotrophy mediated by a Fougerite-rich alkaline hydrothermal system as described within the chapter.

greenalite $((Fe^{2+},Fe^{3+})_2\text{-}3Si_2O_5(OH)_4)$, silica ($SiO_2$), and small amounts of the layered iron monosulfide, mackinawite (FeS) [44,104] (Figure 2). Powered by the products of serpentinization, these were the entities that could theoretically drive the reduction of CO_2 *and* the oxidation of CH_4 – through the fougerite "free-energy" converters – via entirely abiotic carbon fixation pathways [52]. In such a framework, the products were the primary sources of the organic molecules to chelate trace metals as active centers to protoenzymes in the first metabolic processes [38,105–109]. Figures 2 and 3 outline such a hypothesized fougerite-catalyzed system yielding abiotic pathways for methanotrophy, given inputs of H_2, CH_4, NO, and acetogenesis given H_2 and CO_2. But for metabolism to actually take off also required an in-house specialized free energy converter, one that could continually condense and re-condense ambient phosphate to a pyrophosphate [38]. We hypothesize that fougerite could engineer such condensations as well, although few specific experiments to this effect have been undertaken [38,107,108,110,111].

This hypothesis builds on the extraordinary green rust experiments with nitrate and nitrite carried out by Hansen and colleagues [72,90,112] who demonstrated the reductions of these anions to ammonia, ammonium, and nitric oxide. Other influences were a suite of papers by Gustaf Arrhenius and coworkers who demonstrated the absorption of complex phosphates and other anions in the hydrous interlayers of green rust [1,9,48,49]. Combined, these results underpin the revised submarine

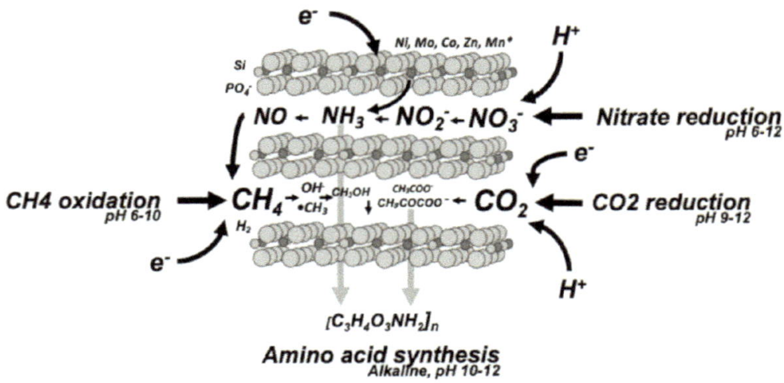

FIGURE 3 Diagram depicting the GR-mediated pathway for amino acid synthesis via the combined pathways for the reduction of CO_2 and NO_3^- and the oxidation of CH_4.

alkaline vent theory (AVT) theory for life's emergence [3,4,29,38,113] – a theory to explain the ever-sustained resolutions of the redox and disequilibria between molecular entities such as to drive nanoelectromechanical conversions for the birth of metabolism [28,29,43,114–123].

3 FIRST STEPS TOWARD METABOLISM: GREEN RUSTS MEDIATE THE REDOX OF SPECIES

The cycling of carbon on the surface of the Earth today is primarily mediated by life via cellular respiration, processing 10^3–10^6 gT of carbon in a planetary scale redox conversion of carbon species into higher and lower oxidation states utilizing highly evolved yet primitively conserved systems of electron transfer [124]. The metabolisms that conduct these reactions are mediated by enzymes with transition metal cores readily reacting with rising (>400 ppm) levels of CO_2 to form the complex sugars (e.g., DNA) that form the backbone of all of biology [125].

Paralleling this biologically mediated cycling of carbon, completely abiotic naturally forming mineral systems are found to readily conduct the same type of reactions that life exhibits [126]. Several studies of protometabolic systems where pure metals or metal-bearing minerals conduct CO_2 reduction under spontaneous, photochemical, and pressurized conditions have been recently described [127–132]. Efforts to describe fully the reactivity of minerals toward CO_2 reduction are of interest to the wider emergence of life field and its applications in society as similar mineral-mediated reactions have long been exploited by industry for their catalytic behaviors[133]. Catalytic material species (e.g., zeolites) are essential components in the processing of natural gas or the production of commodity chemicals [66]. However, these minerals, if naturally occurring, generally lack characteristics conducive to emergence of life hypotheses of which GR exemplifies.

GR exhibits a combination of properties inherent to the aforementioned materials, bearing a clay-like envelope of catalytic variable valence metal sheets containing hydrated channels of anionic reactants. For example, fougerite (naturally occurring GR) clusters and nanocrystals exchange metallic ions one for another known to be vital to life (e.g., Ni, Co, Zn with Fe) and can absorb anions into their nanometric aqueous interlayers (e.g., Cl^-, CO_3^{2-} HCO_3^-, $HCOO^-$, MoO_4^{2-}, and many organic ions). The anions in some $Fe^{2+/3+}$ hydroxysalt green rusts (carbonate, oxalate, methanoate) can act as both electron acceptor and donor in reactions that induce conformational change [134–137]. As such, GR mediates catalytic electron transfers toward organic compounds in solution, from the degradation of organics such as trichloroethylene or phenols [138] to the reduction of nitrogen oxides and the amination of carbonic or keto acids to alanine and glycine [139,140]. GR represents a natural mineral that could enable chemical disequilibria ("free" energy) conversion – a natural entity with mechanochemical potential [1,3,44,52].

Although the transformation of GR in the presence of oxidizing gasses such as O_2, CO_2, and NO has been well characterized [67,141–143], only recently have studies [40,41,144–146] explored the redox behavior of GR with CH_4, CO, and CO_2 gasses. They reveal over 20 organic compounds formed from simple interactions between the mineral and the gasses under ambient to high pressures, electrochemical impulse, and UV irradiation (Table 1). Many of the products are precursors to, or direct products of, metabolic pathways for carbon fixation (Figure 3).

One conceit in these experimental tests of protometabolic reactions is that we expect green rust minerals to act similarly to active enzymes enabling life's emergence, yet in these works, we have only been able to follow the redox reactions through the formation of stable products by analyzing solution chemistry around GR minerals or by the direct dissolution of the mineral. Thus, the full reaction pathway mediated by interlayer chemistry along with intermediates has not yet been revealed.

3.1 Experimental Evidence of $Fe^{2+/3+}$-Oxyhydroxide-Mediated Reactions with CO_2, CO, and CH_4

Reactions between GR and carbon gasses have been previously studied outside of a carbon redox context as the transformation of GR via CO_2 addition has been used to synthesize $GRCO_3$ from other GR precursor species using anion exchange [65], and CO has been reported as an oxidant for NO reduction over specially synthesized GR transformation products [143]. In solution CO_2 exists as H_2CO_3, HCO_3^-, and CO_3^{2-} dependent upon increasing solution pH, but within GR interlayers anionic exchange readily occurs with high preference given to HCO_3^- over other anions (CO_3^{2-}>SO_4^{2-}>Cl^-) [65]. GR is only stable above a pH of ~6.5 [141,147]; thus given excess CO_2 in an unbuffered solution, GR will eventually undergo an additional transformation to Fe-phases such as lepidocrocite, magnetite, and chukanovite, or to siderite. These prior experiments missed significant changes to solution chemistry during gas exposure, as Table 1 describes a variety of compounds detected following interactions between GR and gaseous CO_2, CO, and CH_4.

Given HCO_3^--rich solutions at pH >7, GR bubbled with CO_2 or CO resulted in the production of various C_1-C_5 organic species with and without electrochemical

TABLE 1
Products of GR CO_2, CO, and CH_4 Redox as Described by Recent Work

Products	CO_2 reduction	CO reduction	CH_4 oxidation	Relevant LUCA metabolism
C1				
Methanol	[41, 144]		[40, 144]	Acetogenesis, Methanogenesis, Methanotrophy, C_1 metabolism, Sulfur metabolism
Acetaldehyde			[40, 144]	Fermentation, Glycolysis
Formaldehyde			[40, 144]	Methanogenesis, Methanotrophy, C_1 metabolism
Formate	[41, 144]	[145]		Acetogenesis, Methanogenesis, Methanotrophy, C_1 metabolism, Sulfur metabolism
C2				
Ethylene			[144]	
Ethanol	[41, 144]		[144]	Fermentation
Peracetate	[41]	[145]		
Acetate	[41, 144]	[145]		Acetogenesis, Methanogenesis, Fermentation
C3				
Acetone	[41, 144]	[145]	[144]	
Pyrvuvate	[41]			Glycolysis
2-Propanol	[41]		[40, 144]	
Propanone			[40, 144]	
Glycerol	[41, 144]			Glycolysis
Malonate	[41, 144]			
Lactate	[41, 144]			Fermentation, Glycolysis
C4				
Glycolate	[41]			
Butanone			[40, 144]	
Erythritol	[41, 144]			
Dioxane	[41, 144]			
C>4				
Pentanone			[40, 144]	
Benzene			[144]	
Hydrocarbons			[40, 144]	Sulfur metabolism?

(Continued)

TABLE 1 (CONTINUED)
Products of GR CO_2, CO, and CH_4 Redox as Described by Recent Work

Products	CO_2 reduction	CO reduction	CH_4 oxidation	Relevant LUCA metabolism
C-Cl,N,S				
Chloroform			[40, 144]	
Dichloromethane			[40, 144]	
Tetrachloromethane			[40, 144]	
Nitromethane			[40, 144]	
Methylnitrate			[40, 144]	
DMSO	[41, 144, 146]			Sulfur metabolism
DMS	[144, 146]			Sulfur metabolism

stimulus [41,144,145]. While CO reduction yielded primarily acetone, the reactions with primary CO_2 led to (>100mM) production of hydroxyl-rich sugar alcohol compounds, namely erythritol, the presence of which lends further stability to GR against attack by oxidative species [148]. Surface studies indicate that CO_2 is readily adsorbed to available Fe-OH surface sites, with various C-C, C-O, and C-H bonds present following CO_2 adsorption indicating rapid transformation upon contact [41,144]. In fact, those Fe-OH bonds, specifically their length, are implicated in the reactivity of Fe-hydroxides toward catalytic reduction reactions [149]. DFT models of related mineral species such as Fe_3O_4 show very strong surface interactions between CO_2 and these Fe-OH sites common to iron minerals [150], and under these conditions, CO_2 has a predisposition to OH^--mediated reduction [151]. GR displays much higher rates of conversion (x1000 vs. gregite in $CO_2 \rightarrow$ ethanol) as compared to similar mineral species in the production of organics [41] implicates GR's unique band alignment and low electron affinity (~3.2 eV which is lower than all other similar oxide and sulfide minerals) in why GR is so prepositioned to transfer electrons. While the catalytic behavior of the surfaces is apparent, the interlayer also is implicated as a reaction site as demonstrated by the reduction of interlayer SO_4 species to DMSO in the presence of CO_2 [41,144]. Even products of CO_2 reduction by GR are further reducible by the mineral as DMSO is evidenced to be able to be reduced to DMS [144,146] and pyruvate can undergo reductive amination in the presence of NH_3 to form amino acids [140]. Thus, given a supply of CO_2 and GR fairly complex reaction pathways are realized.

In contrast to the more oxidized carbon species, the bubbling of pure CH_4 does not produce any significant organic products [40], yet XPS experiments [41,144] revealed similar patterns of C=C and C=O bonding on the surface of $GRCO_3$ following absorption of CH_4 as compared to CO_2. These signals may reflect the accumulation of carbon (e.g., ethylene, graphite, or organic phases) in a similar fashion to coke or soot formation resulting from incomplete oxidation processes. The redox active bilayers of metal cations are thought to promote reduction, not oxidation reactions under stable conditions. However, when simulating hydrothermal conditions, the pressurization of GR with CH_4 and O_2 resulted in the production of MeOH along with various other organic compounds, including trichloromethane (TCM) and dichloromethane (DCM). Farr 2023 [40] hypothesizes that GR induces the oxidation of CH_4 via the release of oxidative radicals and/or formation of Fe^{4+} sites [152,153] during its own oxidation by O_2, NO, or H_2O_2 [40,144]. This role of oxidants in the activation of methane over GR surfaces was explored where O_2, NO, and H_2O_2 were found to promote the formation of methanol, hydrocarbons, organochlorides, and organonitrates from GR mixtures under 10–50 bar CH_4 [40,144]. The reaction coincided with the sacrificial transformation of GR to various Fe-oxyhydroxide and oxide phases, providing evidence for the radical-mediated hypothesis. Further investigations reveal that CH_4 is readily photooxidized to acetone which undergoes additional reaction with UV-light or UV-induced radical species to yield compounds like benzene [144]. GR is known to be strongly photoactive and liable to transformation [154,155]; thus, these processes can be combined to yield self-catalyzing reactions [156]. Experimental evidence [144,152,157] demonstrates that the production of these hypothesized radicals and high valent (Fe^{4+}) intermediates does take place during GR oxidation. Thus, the oxidation of CH_4 over GR shares the Fe^{4+} activation mechanism utilized by the methanotrophic enzyme sMMO [158].

3.2 The Effect of GR-mediated Carbon Redox on Ancient Ferruginous Environments

Abiotic carbon redox may have greatly influenced the environmental conditions on the ancient Earth as a result of high concentrations of CO_2 and CH_4 alongside widespread ferruginous oceanic conditions during the Hadean (4.6–4 Ga), Archean (4–3.5 Ga), and early Proterozoic (3.5–2 Ga) eons [14,50,159]. These carbon species were likely reacting within the primitive oceans, carbonating the precipitating iron mineral hydroxides present [160]. High energy photons can instigate both oxidation and reduction reactions with mineral species; thus, due to a lack of significant ozone layer until the GOE, UV-C light would have catalyzed the redox processes at surface oceans bearing GR species [154,160,161], dramatically affecting the redox state of the surface ocean and the stability of any organic species present [144]. Given that the Hadean oceans leaned acidic, CO_2 reduction by GR [41] and other minerals [127, 131] (occurring pH>7) may have been inhibited at the surface and had been confined to alkaline vent sites. However, pH evolved to more alkaline conditions by the early Archean [162], so too would the conditions suitable for carbon reduction expanding from vents to open water. Methane oxidation however has been shown to occur from pH 6 to 8 and could have been catalyzed by direct UV light, radical interaction, or atmospheric NO [40,144] throughout this early period.

The chemical and biochemical evolution of the early environment could have been greatly influenced by the transformation of CO_2, CO, and CH_4 by GR minerals to volatile organic (e.g., methanol, acetone), organohalogen (e.g., $CHCl_3$, CH_2Cl_2,

CH_3Cl), and organosulfur species (e.g., DMSO, DMS) [40,144,146,160,163,164]. Today, these compounds are biologically produced and exert influence over climate forcing via cloud condensation nuclei generation [165–167]; thus, similar effects can be hypothesized to occur if produced abiotically during the Archean. This would directly influence UV-irradiation flux and thus the availability of nutrients for early metabolism in surface waters [165,168–172].

The Hadean and early Archean were subject to a solar luminosity 25–30% lower than today according to the faint young sun hypothesis [15,173]. Greenhouse gasses such as CH_4 and CO_2 are generally implicated in the warming of the early Earth above freezing temperatures; however, miscible organic molecules with lower freezing points such as methanol could have served as an antifreeze for the primitive ocean as well as those that may have or currently do exist on distant worlds such as Mars, Ceres, Enceladus, and Europa [174]. As such, the detection of exoplanet signatures corresponding to these types of compounds (e.g., methanol, dimethyl sulfide, or chloromethane) may not necessarily indicate the presence of life [175], as putative biosignatures may merely evidence a natural evolution of planetary ocean and atmospheric chemistry controlled by redox-sensitive minerals. However, the effect of such production by a GR-saturated ocean on the Archean climate is yet to be identified in the geological record, yet isotopic, morphological, or chemical evidence for such interactions may be preserved within iron, sulfur, or carbon-rich laminae present in the Eoarchean rock record [176–179].

3.3 Autochthonous Sources of Archean Oxidants for Carbon Redox

Ambient methane oxidation at the surface and at hydrothermal sites presents a route for carbon redox transformation on the early Earth. According to Farr [40] CH_4 can be oxidized to methanol given sufficiently oxidizing species. Such oxidants were previously thought to have been rare or absent during the Archean, yet evidence exists for photochemical and geochemically generated oxidizing species present prior to the GOE.

While photochemistry could have provided sufficient oxidizing power to transform Fe^{2+} to Fe^{3+} at surface waters [161,180,181], recent findings demonstrate radical production via the weathering of silicate bonds [182–184], indicating a new pathway for the oxidation of GR in a supposedly anoxic Archean. Limited radical production at the surfaces of mineral faces weathered by rocks could produce eleven times that produced by UV interactions with the atmosphere [185]. Critical interpretation of these hypotheses [186] posits that Fe^{2+} scavenging in ferruginous ocean would prevent any H_2O_2 from building up to appreciable concentrations. If so, then the interface between weathered Si-minerals and GR could be redox active and lead to runaway Fenton reactions given GR's propensity to further produce radical species. This would serve as a base for autocatalytic oxidation and reduction-based organic synthesis on itself or other mineral surfaces. Further evidence for natural radical-generating interfaces has been found at microbubble gas–liquid interfaces where strong electric fields yield hydroxyl radicals capable of oxidizing C-H bonds in CH_4, yielding products such as ethane, methanol, and formate [187,188]. Thus, the bubbling of gasses from hydrothermal sites or due to surficial gas exchange could, over geological timescales, have provided a mechanism for the abiotic transformation of the atmosphere. Another source of autochthonous formation of oxygen is that naturally occurring redox potentials present at hydrothermal environments [189] could conceivably catalyze oxygen evolution reactions (OER) yielding O_2 via water splitting over GR-related LDH minerals [190], as recently found in Fe-Mn-rich benthic polymetallic nodules [191]. Such interactions would yield a self-catalyzing oxidation process given certain concentrations and fluxes of OER-catalyzing metals (e.g., S, Ni, Mn, or Mo) [192].

Further sources of oxidants such as oxidized nitrogen species are hypothesized to have been produced through lighting and meteoritic impacts in the absence of life, and recent experimental and theoretical work places nitrogen oxides [64,91] and nitrogen oxyanions [139,193] in the position to react with GR in early marine environments. Such interactions with GR demonstrate the type of reactions initiated by NO and NO_2^- oxidizers with GR and CH_4 and imply new natural mechanisms for nitrogenated organic synthesis.

3.4 An Inferred Evolutionary Transition from Abiotic Minerals to Protoenzymes

The variety of products detected during GR-mediated carbon redox processes is described in Table 1, and certain compounds present can be linked to metabolisms associated with those of LUCA (last universal common ancestor) [194]. Many of these compounds are simple substrates for, intermediates in, or products of a variety of metabolisms associated with simple early life such as methanotrophy, methanogenesis, acetogenesis, glycolysis, fermentation, C_1 metabolism, and sulfur metabolism [52,166,195,196]. These experiments demonstrate the versatile abiotic pathways by which components of biochemical metabolism can be produced by electron transfer between green rust and carbon gasses (e.g., CO_2, CO, and CH_4). Thus, they reveal mechanistic insight into the hypothesized transition from geochemical to metabolic chemical pathways and strengthen the arguments presented by Russell, Nitschke, and Duval for a green rust–mediated transition from a purely abiotic system to ancient metabolism.

At some point under the ferruginous conditions that dominated the Eoarchean, geochemistry transitioned into biochemistry [196–198] and the links between the analogous structures and catalytic ability inherent in redox-sensitive minerals and metallo-protein reaction centers invoke hypotheses relating to a mineral origin of enzymes [52,106,199]. Extant biochemistry displays various examples of bio-mineral relationships, such as magnetite serving as microbial electron transporters [200] or within the P-amended ferrihydrite core of ferritin [201]. Further parallels can be drawn between enzymatic metal cofactors and organic ligands with green rust's capacity to be modified by metallic and organic species, notably increasing its reactivity in the presence of certain metals [60,86,202–205] or its increased stability in the presence of inorganic and organic moieties [147,157].

In the case of CO_2 reduction under alkaline conditions, carbonate green rust is able to produce C_2-C_4 hydroxyl-rich compounds such as glycerol in high quantities (Figure 3) [41]. Glycerol and similar hydroxyl-rich compounds are known to retard the oxidation of green rust indefinitely [206,207]. Shielding crystals from physical oxidation by hydroxyl coating or producing anti-oxidant heterocycles (e.g., quinones) [208] could constitute innate forms of dynamic self-preservation in autocatalytic chemical systems. Furthermore, organic interactions are known to modify coordination chemistry of metals [209] and modify surfaces [210–213] further augmenting the reactivity and capabilities of the green rust system. Various additions to green rust, such as metals (e.g., Ni, Co, Cu, Zn, Mo), can also modify the reaction mechanism and products formed during CH_4 oxidation or CO_2 reduction. Sorbed inorganic species such as silicon or phosphate are known to protect GR from oxidation [81,147], and the presence of adsorbed phosphate species specifically is known to increase the release of radicals during oxidation [157] possibly via the prevention of resorption by the mineral itself, thus granting the protected green rust with catalytic oxidative properties.

Given appropriate polymerization [214,215], GR morphologically and functionally begins to resemble (Figure 4C) a metallic center of an ancestral enzyme or protein surrounded by cofactor ligands [105]. This system evolves from an innate chemical structure that preceded its function [216].

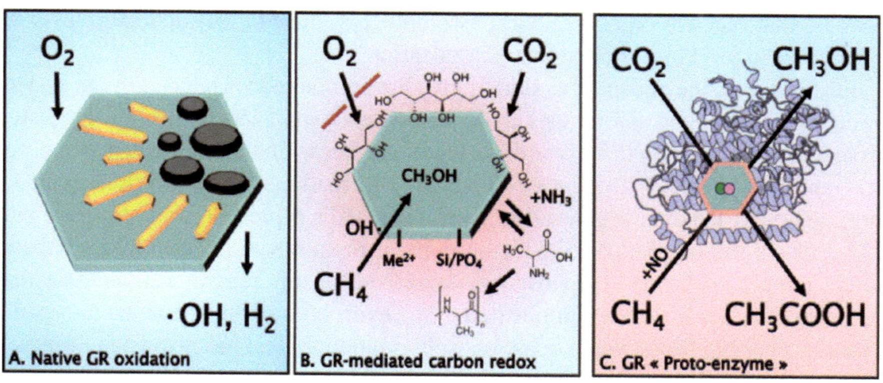

FIGURE 4 Demonstration of GR's response to external redox species at different levels of complexity. In section A, GR is liable to oxidative transformation in the presence of O_2. In B, this no longer occurs due to the production of protective glycols resulting from spontaneous CO_2 reduction. Many reactions can be nonspecifically catalyzed at this level of complexity. In slide C, due to specific selection pressures the organic molecules complexified into a hypothetical polyamino ligand that selects for only the oxidation of CH_4 to CH_3OH, thus resembling the extant enzymatic function of sMMO. Red and blue coloration indicate relative levels of reducing and oxidizing conditions, respectively.

In the diagram (Figure 4B) this activated green rust mineral is held to spontaneously convert CO_2 into protective glycolic sugar alcohol species [41,144] and pyruvate. This pyruvate can then be reductively aminated to alanine [140] (Figure 3). These amino acid products can be used as a feedstock for further organic synthesis, or be used to reactivate GR that has been deactivated by Si/P adsorption [211] by competing with blocked surface sites. The next step in uncovering this hypothesized primordial metabolism would be the polymerization and functionalization of glycols and amino acids to more complex proteins and on to an entropically driven formation of organic membranes attached to various mineral surfaces [214,215,217,218]. This enzyme-like function of GR is hinted to occur with the spontaneous hydration of key components of the rTCA cycle such as phosphoenolpyruvate [219]. From there, the autocatalytic production of polymerized amino acid ligands both protects the mineral cofactor from oxidation and increases its reactivity toward certain ions. This is implied in Figure 4C which shows GR surrounded by ligands (like those of sMMO) for the catalysis of CH_4 oxidation. Here the diagram demonstrates how GR – perhaps subject to a redox gradient – resembles a protoenzyme in form *and* function [220]. The reaction centers of biochemical enzymes share certain structural features in common with reactive mineral species like GR and given the appropriate conditions whereby GR – modified with various metals, inorganic compounds, and polyamino acid chains – could have its reactivity selectively tuned much like an enzyme, to service specific reaction mechanisms, such as CH_4 oxidation to methanol. The relationship between inorganic proto-enzymes as described and the initial biochemical enzymes to emerge from similar environments is not known, but it is possible that reactive organo-mineral assemblies produced [221] or served as [222] the templates for their production and replication.

Furthermore, the phosphorylation of glycolic compounds produced in the system by metals to form lipid membranes [223,224] presents a pathway for this GR-mediated proto-metabolic system to further drive reactions, encapsulate and concentrate products, and preserve itself from environmental perturbations. Observed liposome-like morphological features preserved in Eoarchean BIFs represent possible fossil evidence of simple organic [225–227] or mineral membranes [228]. Lipid synthesis would require pathways for phosphorylation, and in this regard, it is notable that Fe-oxyhydroxides have been implicated in the hydrolysis of phosphate from organic phosphates [229]. Moreover, a quasi-phosphorylation system has also been observed in GR and other mineral systems to form pyrophosphates [111,230,231], or in their absence via condensation [232]. If these polyphosphates could be transferred to the glycolic and amino acid products, then pathways to lipid membrane and nucleotide synthesis would arise from very simple starting conditions. In the functionalization of these membranes, GR and other minerals have been demonstrated to integrate within liposomes (Figure 5) [233], opening the door to the chemiosmotic exploitation [234] of proton gradients to do work, a necessary hallmark of life [29].

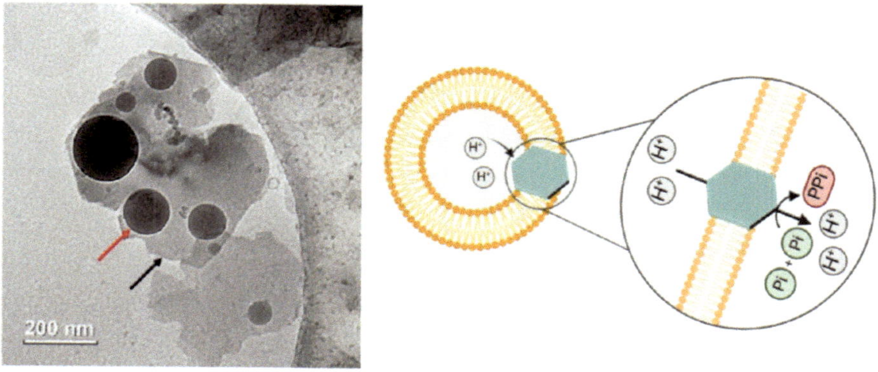

FIGURE 5 (Left) Cryo-transmission electron microscopy image of phosphatidylcholine liposomes (red arrows) of 100–200 nm in diameter adsorbed at the surface of μm-sized crystals of layered-double-hydroxide (LDH) hydroxycarbonate green rust $\{[Fe^{2+}_{(1\text{-}x)}\, Fe^{3+}_{x}\, (OH^{-})_2]^{x+}\, [CO_3^{2-}{}_{(x/2)}\, H_2O_{(y)}]^{x-}\}$ (black arrows). Republished from Gaudu et al. [233] with permission. (Right) Diagram depicting green rust integration within a lipid membrane and its putative conduction of an abiotic chemiosmosis via H^+ transport and phosphorylation of inorganic phosphate.

4 MINERAL MEMBRANES AS PRIMITIVE ANALOGS OF CHEMIOSMOTIC SYSTEMS

4.1 Metabolism and Chemiosmosis

The physico-chemical phenomenon of chemiosmosis underpinning all metabolic reactions is mediated by metallo-enzymes which couple the resolution of a redox gradient, constituted by the presence of electron donors and acceptors, to the generation of a pH gradient across a membrane through coordinated electron and proton transfer (Figure 6). This pH gradient powers phosphate-condensing enzymes which couple the translocation of protons through the membrane to the condensation of phosphate, thereby forming increased amounts of phosphorylated over non-phosphorylated forms. This in turn constitutes a thermodynamic disequilibrium as "electrochemical free energy" in the form of potential energy-dissipating exergonic dephosphorylation reactions.

In living organisms, this mechanism has the effect of massively increasing concentrations of condensed phosphates from a transmembrane redox disequilibrium, thus constituting an energy-dissipative system. Indeed, the second law of thermodynamics allows for a local decrease in entropy (i.e., generation of order, such as a thermodynamic disequilibrium) if compensated by a global increase in entropy (i.e., getting closer to thermodynamic equilibrium). Applied to the current bioenergetic systems that involve electron transport chains, this allows the thermodynamically unfavored (i.e., endergonic) concentration of condensed phosphates to reach orders of magnitude above their chemical equilibrium (i.e., an astonishing decrease in entropy) by compensation for the exergonic resolution of the redox and pH gradients

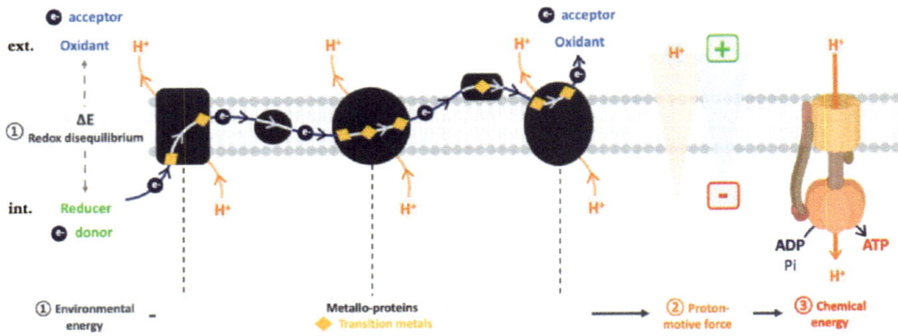

FIGURE 6 General scheme based on the functioning of chemiosmosis in current organisms. (1) An environmental redox disequilibrium is solved by metallo-proteins coupling electron and proton transfer across the membrane, thus generating (2) a proton-motive force (PMF) comprising a pH gradient and a membrane electric potential. This PMF powers a phosphate-condensing enzyme such as the ATPase, through the spontaneous translocation of protons down gradient, so forcing the phosphorylation of ADP. The increased out-of-equilibrium concentrations of ATP over ADP then constitute a free energy reservoir in the form of potential energy-dissipating exergonic dephosphorylation reactions (i.e., "chemical free energy").

across the membrane through electron and proton transfer, respectively, as a dissipative counterpart (i.e., the global increase in entropy).

To make the connection with organic chemistry, it is the tremendous displacement of the thermodynamic equilibrium toward condensed phosphates over less- or un-condensed ones (e.g., ATP over ADP, or even pyrophosphate (PPi) over orthophosphate (Pi)) that makes the dephosphorylation/hydrolysis of the condensed forms thermodynamically favorable (i.e., exergonic). Some endergonic reactions that would not occur spontaneously, for instance, those involved in biochemical pathways leading to the condensation of organic macromolecules in an aqueous system, become thermodynamically favorable when coupled to these exergonic returns to equilibrium.

Given its ubiquity in current organisms, this mechanism must have existed in one form or another since – and indeed enabling – life's emergence [113]. Primitive mineral-mediated systems coupled electron and proton transfer may have presaged this development via organic or inorganic structures (Figure 4) [3,233,235]. Such systems would inevitably form from geochemical redox disequilibria.

4.2 Electron and Proton Transfers across Mineral Membranes

The Alkaline Hydrothermal Vent (AHV) environment is a mineral precipitate-based inorganic system that, involving the green rust/fougerite, provides analogous features to the ones observed in current chemiosmosis, thus being a relevant candidate for hosting a primitive form of this mechanism. Both naturally occurring and lab-simulated alkaline vent systems have been shown to harbor strong redox gradients, even reflected in electrical currents across the chimney wall [101,140,236–238].

Such redox gradients are constituted by chimney wall-mediated spatial separation of reducing compounds (i.e., electron donors) such as CH_4, H_2, H_2S, or S_2 in the alkaline fluid on the internal side of the membrane and oxidizing compounds (i.e., electron acceptors) on the external acidic ocean such as CO_2, NO_3^-, or O_2 in modern conditions [113,239]. The display of electric currents at the surface of these vents tends to show that electrons are transferred from the reducers in the alkaline fluid toward the oxidants in the seawater through metal sulfide and oxide catalysis within the chimney wall [237].

At this point the analogy of such inorganic metal-based cross-compartment electron-transferring systems with metallo-enzymes performing cross-membrane electron transport is clear. However, the enzymes themselves exploit this electron transfer to associate with proton transfer across membranes, which is still scarcely explored in alkaline vent chimneys despite the presence of steep pH gradients across the chimney wall, i.e., the mineral membrane. These pH gradients occur naturally due to the separation of pH 10 to 11 alkaline internal fluids and the pH ~6 external acidic ocean, creating an intermembrane differential of 4 to 5 pH units [240]. The magnitude of this pH gradient has been discovered to be critical in the catalysis of cross-membrane electron transfer associated with H_2S oxidation and CO_2 reduction, leading to the synthesis of formate in lab-produced Fe(Ni)S precipitates [127]. However, in this system the pH gradient is not involved as an out-of-thermodynamic equilibrium free-energy to power chemical reactions through proton translocation across the membrane as in bioenergetic systems, but rather as a facilitator of H_2S oxidation and CO_2 reduction on the internal and external sides of the membrane, respectively. While the existence of such a phenomenon in hydrothermal systems is criticized as a theoretical impossibility – i.e., to build an energy-dissipating pH gradient across dielectric barriers of such thickness – Weingart and collaborators [241] and Moller and colleagues [242] demonstrate that such pH gradients do form in micro-compartments of smaller size within the overall chimney wall. Hye-Eun Lee and her collaborators have shown that the bc layer of the plate-like layered nanocrystals of brucite ($Mg(OH)_2$), another LDH mineral, is arranged in a concentric manner radiating outwards from (i.e., geometrically normal to) the deep-sea micro- to nano-channels or chimneys bearing alkaline fluid [238].

Such preferential crystal orientations may be associated with the movement of charged particles along an ionic concentration gradient, e.g., a pH gradient [238,243,244] thus forming transverse channels for ions to be vectorially transported from one side of a vent wall to the other. Moreover, LDHs such as brucite or similar FeS like mackinawite are known to adsorb a wide variety of ions [245,246], which can locally influence the surface charge of the AHV precipitates, thus increasing the selectivity toward certain ions. This effect was proposed to occur throughout the chimney wall as protons made their rapid way down the pH gradient, a transport that is converted into electric current across the mineral precipitate [238]. However, LDH adsorption properties change with pH as a result of protonation/deprotonation dynamics and competition with OH^- [246]. In fact, a mechanism involving proton transport across LDH interlayers rather than adsorption would be more consistent with the data. For example, LDH green rust minerals can be partially or totally deprotonated when structural hydroxides (-OH) are deprotonated to oxides ($-O^-$), a phenomenon related to the redox gradient [134]. In this scenario, electrons would be driven by the redox disequilibrium (i.e., from the most reducing regions in the crystals to the most oxidizing edges) similar to charge hopping, thanks to the progressive and successive oxidation of iron sites alongside the interlayers [136]. Since the interstitial anions charge-compensate the oxidized Fe^{3+} sites, they would move accordingly. Additionally, the hydroxo bridges will preferentially protonate when linked to newly reduced iron, thus simulating proton displacement alongside electron transfer, in a process similar to the "Grotthuss shuttling" mechanism of translocation that has been considered in the bioenergetically relevant transmembrane pyrophosphatase (TmPPase) [4,247].

This primitive system of inorganic membranes that react with and transport electrons, protons, and carbon compounds across steep pH and redox gradients begins to resemble a primitive ancestor to life.

5 LIFE AS AN EMERGENT PROPERTY OF GEOCHEMISTRY

To conclude this chapter we iterate upon the Alkaline Vent Theory (AVT) or Fougerite Alkaline Vent Theory (FAVT) for life's emergence (see chapter 9 by Russell) and connect new findings on GR carbon redox to a global geochemical perspective (Figure 7) [3,4,7,235,248].

Situated within a Hadean or Eoarchean hydrothermal vent pore, the interface between cool oxidized acidic seawater and hot alkaline reduced effluent results in the precipitation of dissolved metal ions as silicates, sulfates, sulfide, and oxyhydroxide nano- to micro-minerals. Such a network of Fe, Ni, Mg, Zn, Co, Mn, Mo, and W precipitates, loosely bound by intermolecular forces, serve as a semipermeable membrane separating the redox disequilibria inherent to the outer and inner environments [101,111,113]. Here, fougerite goes to work. As a metastable state of iron, fougerite varies in its redox state and even comes in and out of existence as suitable pH fronts advance and retreat due to oscillatory variations in pressure, redox, and pH at these active sites [249]. GRs transform to magnetite and goethite as a result of heat and redox reactions, or to greenalite and vivianite where Si or P fluxes are high [222,250], so affecting the fate of dissolved cations and anions [81,251,252]. Carbon dioxide and nitric oxides dissolved in the ocean meet and interact with inhaled methane from the effluent in micro-pores at the base of hydrothermal mounds and are thereby transformed by GR and/or FeS to formate, acetate, malonate, succinate, glycols, and amino acids [40,41,127,131,140,144]. The transport of compounds introduced or produced within the hydrothermal vents relies on the affinity or charge of the mineral species present, which in turn depends on pH. With transformation comes the production of H_2 and radicals, introducing new players in the synthesis and modification of produced or adsorbed species [152,53].

The synthesis and accumulation of organic anions and compounds within vent pores would likely lead to the polymerization of amino acids and sugars [215,217,253], thus yielding simple peptidic matter. The mineral membranes are thus rendered near impermeable and thereby highly selective as to what can enter and exit

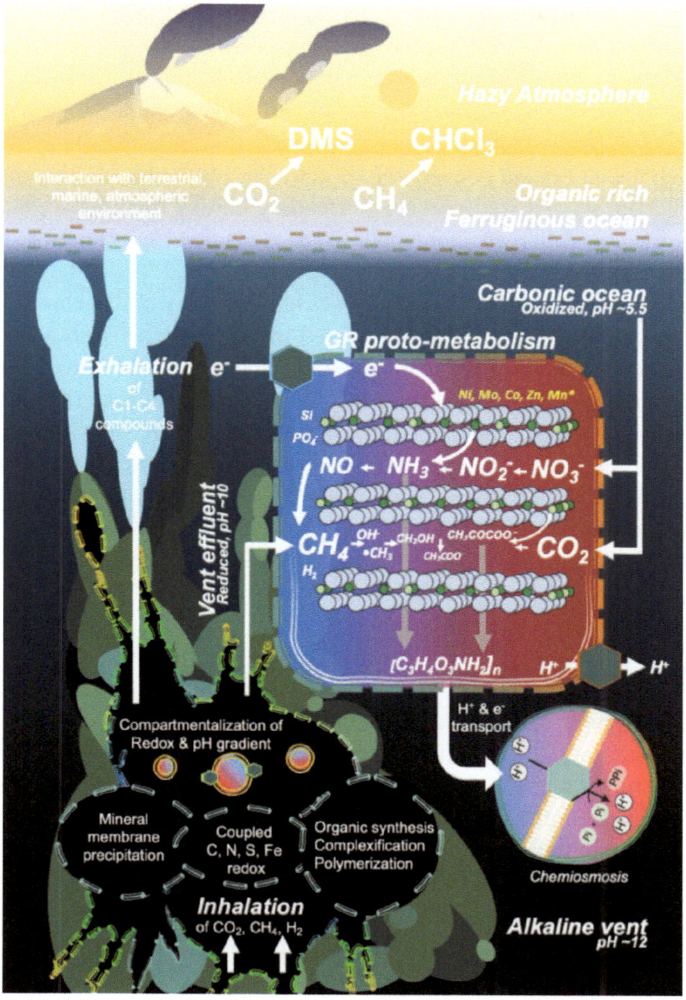

FIGURE 7 An illustrative diagram depicting the various chemical reaction pathways mediated by GR redox chemistry at early hydrothermal vent environments. The *inhalation* of dissolved and gaseous species from serpentinization processes naturally leads to redox disequilibria with cold oxidized sea water at these mineral mounds, precipitating reactive mineral species like GR. These minerals conduct C, N, S, and Fe redox which leads to the formation of simple organic compounds which complexify and polymerize in high surface area mineral pores. Here GR's particular characteristics, namely anionic transport and organic integration, are shown to cross organic membranes and conduct H^+ and e^- transport, leading to local disequilibria in a kind of protometabolism. This pH and redox gradient developing into a chemiosmotic system capable of phosphorylation is hypothesized to be necessary for life to emerge. Beyond this, the products yielded from redox reactions (e.g., DMS, $CHCl_3$) undergo *exhalation* from the vent environment out into the ocean and interact with surface ocean and atmospheric chemistry.

these primitive chemical garden-like cells, so influencing the encapsulated reaction networks within [96,254–257]. Protons are transported across membranes via green rust interlayers, yielding primitive chemiosmotic systems [113,234]. Electrons are transported through the semiconducting mineral layers, connecting chemiosmotic reactors distributed throughout the hydrothermal vents via ramified mineral plumbing [199]. This vent system converts NO_3^- and CO_2 into reduced compounds while CH_4 and H_2 convert oxidized species, portending energy metabolism and, eventually, life – life that emerges spontaneously as a result of geochemical thermodynamic disequilibria.

The point at which geochemistry becomes biochemistry is still elusive. However, the reactions between green rusts and related mineral species with a methane and carbon dioxide-rich atmosphere that exhale organic compounds in a manner mimicking biochemistry [40,41,131,144,160] suggest profound mineralogical forcing over the evolution of the Early Earth's oceans, atmosphere, and early biosphere. Extrapolating these findings, such reactions may not be limited to Earth as hydrothermal activity is thought to occur on Mars, Ceres, Europa, Enceladus, and many other icy bodies within our solar system [258–264]. The discovery of such systems would imply a universal nature to the emergence of life mediated by redox-sensitive mineral chemistry.

ACKNOWLEDGMENTS

We thank Wolfgang Nitschke and Simon Duval for their efforts to uncover a mineral-mediated emergence of life, Christian Ruby and Francois Guyot for their expertise in mineral chemistry, and Daniel Ferry and Olivier Grauby for their continued experimental support and advice. This work received support from the French Government under the France 2030 investment plan, as part of the initiative d'Excellence d'Aix Marseille Université (grant no. A*MIDEX -AMX-21-PEP-039) as well as a grant from the French Agence Nationale pour la Recherche (grant no. ANR-22-CE30-0035-01).

ABBREVIATIONS AND DEFINITIONS

AVT alkaline vent theory
BIF banded iron formation
CT carbon tetrachloride
DCM dichloromethane
DMS dimethyl sulfide
DMSO dimethyl sulfoxide
FAVT fougerite alkaline vent theory
FTT Fischer-Tropsch type
GEO great oxidation event
GR Green Rust
LDH layered double hydroxide
LUCA last universal common ancestor

OER oxygen evolution reactions
TCE trichloroethylene
TCM trichloromethane
XPS X-ray photoelectron spectroscopy

REFERENCES

1. G. O. Arrhenius, *Helv. Chim. Acta.* **2003**, *86* (5), 1569–1586.
2. G. Arrhenius, B. Sales, S. Mojzsis, T. Lee, *J. Theor. Biol.* **1997**, *187* (4), 503–522.
3. S. Duval, F. Baymann, B. Schoepp-Cothenet, F. Trolard, G. Bourrié, O. Grauby, E. Branscomb, M. J. Russell, W. Nitschke, *Interface Focus.* **2019**, *9* (6), 16–20.
4. S. Duval, E. Branscomb, F. Trolard, G. Bourrié, O. Grauby, V. Heresanu, B. Schoepp-Cothenet, K. Zuchan, M. J. Russell, W. Nitschke, *Appl. Clay Sci.* **2020**, *195*, 105737.
5. X.-L. Huang, J. R. Harmer, G. Schenk, G. Southam, *Front. Chem.* **2024**, *12*.
6. M. Preiner, J. C. Xavier, A. Do Nascimento Vieira, K. Kleinermanns, J. F. Allen, W. F. Martin, *Interface Focus.* **2019**, *9* (6).
7. W. Martin, M. J. Russell, *Phil. Trans. Roy. Soc. B.* **2007**, *362* (1486), 1887–1925.
8. G. Arrhenius, *NASA space life sciences symposium, Washington DC.* **1987**, 260–261.
9. G. Arrhenius, *Clay and clay minerals and the origin of life.* **1986**, 97–104.
10. G. Arrhenius, *Fundamental studies and the future of science.* **1984**, 301–319.
11. A. G. Cairns-Smith, A. J. Hall, M. J. Russell, *Origins Life Evol. Biosphere* **1992**, *22* (1–4), 161–180.
12. C. F. Brunk, C. R. Marshall, *Life* **2024**, *14* (2), 226.
13. X.-L. Huang, *New. J. Chem.* **2022**, *46* (32), 15273–15291.
14. S. W. Poulton, D. E. Canfeld, *Elements* **2011**, *7* (2), 107–112.
15. D. C. Catling, K. J. Zahnle, *Sci. Adv.* **2020**, *6* (9).
16. M. Tedetti, R. Sempéré, *Photochem. Photobiol.* **2006**, *82* (2), 389.
17. N. J. Tosca, B. M. Tutolo, *Geochim. Cosmochim. Acta.* **2023**, *352*, 51–68.
18. J. D. Watson, F. H. C. Crick, *Nature* **1953**, *171* (4356), 737–738.
19. R. E. Franklin, R. G. Gosling, *Nature* **1953**, *171* (4356), 740–741.
20. W. Gilbert, *Nature* **1986**, *319* (6055), 618–618. https://doi.org/10.1038/319618a0.
21. L. E. Orgel, *J. Theor. Biol.* **1986**, *123* (2), 127–149.
22. R. Krishnamurthy, N. V. Hud, *Chem. Rev.* **2020**, *120* (11), 4613–4615.
23. D. Segré, D. Ben-Eli, D. W. Deamer, D. Lancet, *Origins Life Evol. Biosphere* **2001**, *31* (1/2), 119–145.
24. J. Groen, D. W. Deamer, A. Kros, P. Ehrenfreund, *Origins Life Evol. Biosphere* **2012**, *42* (4), 295–306.
25. B. M. A. G. Piette, J. G. Heddle, *Trends. Ecol. Evol.* **2020**, *35* (5), 397–406.
26. E. Branscomb, M. J. Russell, *Biochim. Biophys. Acta Bioenerg.* **2013**, *1827* (2), 62–78.
27. I. Prigogine, R. Lefever, *Synergetics* **1973**, 124–135.
28. E. Schrödinger, *What Is life? The physical aspect of the living cell*, Cambridge University Press, London, UK, **1944**.
29. W. Nitschke, B. Schoepp - Cothenet, S. Duval, K. Zuchan, O. Farr, F. Baymann, F. Panico, A. Minguzzi, E. Branscomb, M. J. Russell, *Electrochem. Sci. Adv.* **2022**.
30. S. Arrhenius, *Die Umschau.* **1903**, *7*, 481.
31. F. Hoyle, C. Wickramasinghe, *Comets-A vehicle for panspermia*, in *Comets and the Origin of Life*, Springer Netherlands, Dordrecht **1981,** 227–239.
32. R. Dawkins, *Climbing Mount Improbable, Harmondsworth, UK, Viking,* **1998**, WW Norton & Company
33. C. Huber, G. Wächtershäuser, *Science* **1997**, *276* (5310), 245–247.
34. G. Wächtershäuser, *Prog. Biophys. Mol. Biol.* **1992**, *58* (2), 85–201.

35. J. D. Bernal, *Pro. Phys. Soc.* **1949**, *62* (10), 597.
36. A. Brack, In *Developments in Clay Science.* **2013**, *5*, 507–521.
37. J. P. Ferris, A. R. Hill, R. Liu, L. E. Orgel, *Nature* **1996**, *381* (6577), 59–61.
38. M. J. Russell, W. Nitschke, E. Branscomb, *Phil. Trans. Roy. Soc. B.* **2013**, *368* (1622).
39. M. J. Russell, W. Nitschke, *Astrobiol.* **2017**, *17* (10), 1053–1066.
40. O. Farr, N. Gaudu, G. Danger, M. J. Russell, D. Ferry, W. Nitschke, S. Duval, *J. R. Soc. Interface.* **2023**, *20* (206). https://doi.org/10.1098/rsif.2023.0386.
41. S. Lee, C. Wang, V. Chakrapani, *ChemRxiv.* **2023**.
42. W. Nitschke, M. J. Russell, *J. Cosmol. Astropart. Phys.* **2010**, 3200–3216.
43. M. J. Russell, A. J. Hall, *J. Geol. Soc. London.* **1997**, *154* (3), 377–402.
44. M. J. Russell, *Front. Microbiol.* **2023**, *14*. https://doi.org/10.3389/fmicb.2023.1145915.
45. M. Russell, *Life* **2018**, *8* (3), 35.
46. F. Trolard, J.-M. R. Génin, M. Abdelmoula, G. Bourrié, B. Humbert, A. Herbillon, *Geochim. Cosmochim. Acta.* **1997**, *61* (5), 1107–1111.
47. F. Feder, F. Trolard, G. Klingelhöfer, G. Bourrié, *Geochim. Cosmochim. Acta.* **2005**, *69* (18), 4463–4483.
48. G. Arrhenius, B. Gedulin, S. Mojzsis, In *Chemical Evolution and Origin of Life.* **1993**, 25–50.
49. S. J. Mojzsis, R. Krishnamurthy, G. Arrhenius, in *Cold Spring Harbor Monograph Series, 37*, **1999**.
50. I. Halevy, M. Alesker, E. M. Schuster, R. Popovitz-Biro, Y. Feldman, *Nat. Geosci.* **2017**, *10* (2), 135–139.
51. M. J. Russell, D. Daia, A. J. Hall, in *Thermophiles: The keys to the molecular evolution and the origin of life?* Eds: J. Wiegel, A. W. W. Michael, CRC Press **1998**.
52. W. Nitschke, M. J. Russell, *Phil. Trans. R. Soc. B Bio.* **2013**, *368* (1622), 20120258.
53. M. J. Russell, L. M. Barge, R. Bhartia, D. Bocanegra, P. J. Bracher, E. Branscomb, R. Kidd, S. McGlynn, D. H. Meier, W. Nitschke, T. Shibuya, S. Vance, L. White, I. Kanik, *Astrobiol.* **2014**, *14* (4), 308–343.
54. S. Duval, K. Zuchan, F. Baymann, B. Schoepp-Cothenet, E. Branscomb, M. J. Russell, W. Nitschke, *Minerals and emergence of life*, in *Metals, microbes, and minerals - The biogeochemical side of life*, Eds P. M. H. Kroneck, M. E. Sosa-Torres, De Gruyter, Berlin, **2021**, 135–157.
55. M. J. Russell, A. J. Hall, A. R. Mellersh, On the dissipation of thermal and chemical energies on the early earth, in *Natural and laboratory-Simulated thermal geochemical processes*, Springer, Dordrecht, The Netherlands **2003**, 325–388.
56. W. Buser, *Helv. Chim. Acta.* **1954**, *37* (7), 2334–2344.
57. J. D. Bernal, *Clay Miner.* **1959**, *4* (21), 15–30.
58. R. Allmann, *Fortschr. Miner.* **1971**, *48*, 24–30.
59. M. Etique, A. Zegeye, B. Grégoire, C. Carteret, C. Ruby, *Water Res.* **2014**, *62*, 29–39.
60. J. Choi, B. Batchelor, C. Won, J. Chung, *Chemosphere* **2012**, *86* (8), 860–865.
61. J. Jönsson, D. M. Sherman, *Chem. Geol.* **2008**, *255* (1–2), 173–181.
62. A. Onoguchi, G. Granata, D. Haraguchi, H. Hayashi, C. Tokoro, *R. Soc. Open. Sci.* **2019**, *6* (4).
63. L. N. Døssing, K. Dideriksen, S. L. S. Stipp, R. Frei, *Chem. Geol.* **2011**, *285* (1–4), 157–166.
64. S. Buessecker, H. Imanaka, T. Ely, R. Hu, S. J. Romaniello, H. Cadillo-Quiroz, *Nat. Geosci.* **2022**, *15* (12), 1056–1063.
65. M. I. Agnel, S. Grangeon, F. Fauth, E. Elkaïm, F. Claret, M. Roulet, F. Warmont, C. Tournassat, *Environ. Sci. Technol.* **2020**, *54* (2), 851–861.
66. M. Usman, J. M. Byrne, A. Chaudhary, S. Orsetti, K. Hanna, C. Ruby, A. Kappler, S. B. Haderlein, *Chem. Rev.* **2018**, *118* (7), 3251–3304.

67. L. Legrand, L. Mazerolles, A. Chaussé, *Geochim. Cosmochim. Acta.* **2004**, *68* (17), 3497–3507.
68. C. Ruby, A. Géhin, M. Abdelmoula, J. M. R. Génin, J. P. Jolivet, *Solid State Sci.* **2003**, *5* (7), 1055–1062.
69. K. B. Ayala-Luis, C. B. Koch, H. C. B. Hansen, *Appl. Clay. Sci.* **2010**, *50* (4), 512–519.
70. V. Rives, M. Angeles Ulibarri, *Coord. Chem. Rev.* **1999**, *181* (1), 61–120.
71. P. M. L. Bonin, M. S. Odziemkowski, E. J. Reardon, R. W. Gillham, *J. Solution. Chem.* **2000**, *29* (10), 1061–1074.
72. H. C. B. Hansen, O. K. Borggaard, J. Sørensen, *Geochim. Cosmochim. Acta.* **1994**, *58* (12), 2599–2608.
73. W. Yao, J. Zhang, K. Gu, J. Li, J. Qian, *Environ. Pollut.* **2022**, *304*.
74. C. Ruby, C. Upadhyay, A. Géhin, G. Ona-Nguema, J. M. R. Génin, *Environ. Sci. Technol.* **2006**, *40* (15), 4696–4702.
75. Y. L. Li, K. O. Konhauser, M. Zhai, *Earth. Planet. Sci. Lett.* **2017**, *466*, 103–114.
76. J. M. Bearcock, W. T. Perkins, N. J. G. Pearce, *J. Hazard. Mater.* **2011**, *190* (1–3), 466–473.
77. E. J. O'Loughlin, S. D. Kelly, K. M. Kemner, R. Csencsits, R. E. Cook, *Chemosphere* **2003**, *53* (5), 437–446.
78. L. H. G. Chaves, J. E. Curry, D. A. Stone, M. D. Carducci, J. Chorover, *Rev. Bras. Cienc. Solo.* **2009**, *33* (5), 1115–1123.
79. R. Guilbaud, M. L. White, S. W. Poulton, *Geochim. Cosmochim. Acta.* **2013**, *108*, 141–153.
80. E. J. O'Loughlin, S. D. Kelly, R. E. Cook, R. Csencsits, K. M. Kemner, *Environ. Sci. Technol.* **2003**, *37* (4), 721–727.
81. J. P. H. Perez, D. J. Tobler, L. G. Benning, *Environ. Sci. Process. Impacts.* **2024**.
82. K. Inoue, K. Shinoda, S. Suzuki, Y. Waseda, *Mater. Trans.* **2008**, *49* (9), 1941–1946.
83. S. Yan, M. I. Boyanov, B. Mishra, K. M. Kemner, E. J. O'Loughlin, *Environ. Sci. Technol.* **2018**, *52* (8), 4601–4609.
84. Z. Zhang, W. Yan, O. Messan, J. Fang, W. A. Jackson, *ACS Earth Space Chem.* **2021**, *5* (8).
85. M. Erbs, H. C. B. Hansen, C. E. Olsen, *Environ. Sci. Technol.* **1999**, *33* (2), 307–311.
86. E. J. O'loughlin, M. I. Boyanov, K. M. Kemner, D. R. Burris, *ChemRxiv,* **2022**.
87. W. Lee, B. Batchelor, *Environ. Sci. Technol.* **2002**, *36* (24), 5348–5354.
88. J. Ai, H. Chr. Bruun Hansen, K. Dideriksen, D. J. Tobler, *Chem. Eng. Jl.* **2022**, *446*.
89. M. J. Russell, A. J. Hall, A. J. Boyce, A. E. Fallick, *Econ. Geol.* **2005**, *100* (3), 419–438.
90. H. Hansen, *Appl. Clay Sci.* **2001**, *18* (1–2), 81–91.
91. M. L. Wong, B. D. Charnay, P. Gao, Y. L. Yung, M. J. Russell, *Astrobiol.* **2017**, *17* (10), 975–983.
92. A. Neubeck, N. T. Duc, D. Bastviken, P. Crill, N. G. Holm, *Geochem. Trans.* **2011**, *12* (1), 6.
93. M. Preiner, J. C. Xavier, F. L. Sousa, V. Zimorski, A. Neubeck, S. Q. Lang, H. Chris Greenwell, K. Kleinermanns, H. Tüysüz, T. M. McCollom, N. G. Holm, W. F. Martin, *Life* **2018**, *8* (4).
94. C. J. Lissenberg, A. M. McCaig, S. Q. Lang, P. Blum, N. Abe, W. J. Brazelton, R. Coltat, J. R. Deans, K. L. Dickerson, M. Godard, B. John, F. Klein, R. Kuehn, K. Lin, H. Liu, E. Lopes, T. Nozaka, A. Parsons, V. Pathak, M. Reagan, J. Robare, I. Savov, E. Schwarzenbach, O. Sissmann, G. Southam, F. Wang, C. Wheat, L. Anderson, S. Treadwell, *Science (1979).* **2024**, *385* (6709), 623–629.
95. T. M. McCollom, B. S. Lollar, G. Lacrampe-Couloume, J. S. Seewald, *Geochim. Cosmochim. Acta.* **2010**, *74* (9), 2717–2740.

96. C. Konn, J. L. Charlou, J. P. Donval, N. G. Holm, F. Dehairs, S. Bouillon, *Chem. Geol.* **2009**, *258* (3–4), 299–314.
97. A. Neubeck, N. T. Duc, H. Hellevang, C. Oze, D. Bastviken, Z. Bacsik, N. G. Holm, *Planet. Space Sci.* **2014**, *96*, 51–61.
98. G. Schikorr, *Zeitschrift für Elektrochemie und angewandte physikalische Chemie.* **1929**, *35* (2), 65–70.
99. K. K. Aligizaki, M. R. de Rooij, D. D. Macdonald, *Cem. Concr. Res.* **2000**, *30* (12), 1941–1945.
100. J. M. García-Ruiz, E. Nakouzi, E. Kotopoulou, L. Tamborrino, O. Steinbock, *Sci. Adv.* **2017**, *3* (3).
101. F. Glaab, M. Kellermeier, W. Kunz, E. Morallon, J. M. García-Ruiz, *Angew. Chem. Int. Ed.* **2012**, *51* (18), 4317–4321.
102. L. M. Barge, Y. Abedian, M. J. Russell, I. J. Doloboff, J. H. E. Cartwright, R. D. Kidd, I. Kanik, *Angew. Chem. Int. Ed.* **2015**, *54* (28), 8184–8187.
103. R. Saladino, G. Botta, B. M. Bizzarri, E. Di Mauro, J. M. Garcia Ruiz, *Biochemistry* **2016**, *55* (19), 2806–2811.
104. L. M. White, R. Bhartia, G. D. Stucky, I. Kanik, M. J. Russell, *Earth Planet. Sci. Lett.* **2015**, *430*, 105–114.
105. L. I. Gutierrez-Rus, G. Gamiz-Arco, J. A. Gavira, E. A. Gaucher, V. A. Risso, J. M. Sanchez-Ruiz, *Mol. Biol. Evol.* **2023**, *40* (6).
106. X.-L. Huang, *iScience* **2024**, 109555.
107. E. J. Milner-White, M. J. Russell, *Biol. Direct.* **2008**, *3* (1), 3.
108. E. J. Milner-White, M. J. Russell, *Origins Life Evol. Biospheres.* **2005**, *35* (1), 19–27.
109. W. Nitschke, S. E. McGlynn, E. J. Milner-White, M. J. Russell, *Biochim Biophys Acta Bioenerg.* **2013**, *1827* (8–9), 871–881.
110. E. J. Milner-White, M. J. Russell, *Genes (Basel).* **2011**, *2* (4), 671–688.
111. Q. Wang, L. M. Barge, O. Steinbock, *Chem. Eur. J.* **2019**, *25* (18), 4732–4739.
112. H. Chr. B. Hansen, C. B. Koch, H. Nancke-Krogh, O. K. Borggaard, J. Sørensen, *Environ. Sci. Technol.* **1996**, *30* (6), 2053–2056.
113. W. Nitschke, M. J. Russell, *J. Mol. Evol.* **2009**, *69* (5), 481–496.
114. J. J. R. Fraústo da Silva, R. J. P. Williams, *The biological chemistry of the elements*, Oxford University Press, Oxford **2001**.
115. F. Baymann, E. Lebrun, M. Brugna, B. Schoepp–Cothenet, M. Giudici–Orticoni, W. Nitschke, *Philos. Trans. R. Soc. Lond. B Biol. Sci.* **2003**, *358* (1429), 267–274.
116. B. Pignataro, *Ideas in chemistry and molecular sciences: Where chemistry meets life*, John Wiley & Sons **2010**.
117. L. Cademartiri, *Ange. Chem. Int. Ed.* **2011**, *50* (5), 989–990.
118. G. Arrabito, Y. Aleeva, V. Ferrara, G. Prestopino, C. Chiappara, B. Pignataro, *J. Funct. Biomater.* **2020**, *11* (2).
119. A. Cottrell, *Contemp. Phys.* **1979**, *20* (1), 1–10.
120. R. D. Astumian, *Acc. Chem. Res.* **2018**, *51* (11), 2653–2661.
121. E. Branscomb, T. Biancalani, N. Goldenfeld, M. Russell, *Phys Rep.* **2017**, *677*, 1–60.
122. A. I. Brown, D. A. Sivak, *Chem Rev.* **2020**, *120* (1), 434–459.
123. M. A. Caraballo, M. P. Asta, J. P. H. Perez, M. F. Hochella, *Gondwana Res.* **2022**, *110*, 283–304.
124. P. Falkowski, R. J. Scholes, E. Boyle, J. Canadell, D. Canfield, J. Elser, N. Gruber, K. Hibbard, P. Högberg, S. Linder, F. T. Mackenzie, B. Moore, T. Pedersen, Y. Rosenthal, S. Seitzinger, V. Smetacek, W. Steffen, *The global carbon cycle: A test of our knowledge of earth as a system*, Vol. 290 **2000**.
125. P. G. Falkowski, T. Fenchel, E. F. Delong, *Science (1979)* **2008**, *320* (5879), 1034–1039.
126. R. de Graaf, Y. De Decker, V. Sojo, R. Hudson, *Chem. Eur. J.* **2023**, *29* (53).

127. R. Hudson, R. de Graaf, M. S. Rodin, A. Ohno, N. Lane, S. E. McGlynn, Y. Yamada, R. Nakamura, L. Barge, D. Braun, V. Sojo, *Proc. Natl. Acad. Sci.* **2020**, *117* (37), 22873–22879.
128. T. Beyazay, K. S. Belthle, C. Farès, M. Preiner, J. Moran, W. F. Martin, H. Tüysüz, *Nat. Commun.* **2023**, *14* (1), 570.
129. S. A. Rauscher, J. Moran, *Ange. Che. Int. Ed.* **2022**.
130. S. J. Varma, K. B. Muchowska, P. Chatelain, J. Moran, *Nat. Ecol. Evol.* **2018**, *2* (6), 1019–1024.
131. M. Grosch, M. T. Stiebritz, R. Bolney, M. Winkler, E. Jückstock, H. Busch, S. Peters, A. F. Siegle, J. van Slageren, M. Ribbe, Y. Hu, O. Trapp, C. Robl, W. Weigand, *ChemSystemsChem.* **2022**, *4* (5).
132. V. Vinogradoff, V. Leyva, E. Mates-Torres, R. Pepino, G. Danger, A. Rimola, L. Cazals, C. Serra, R. Pascal, C. Meinert, *Earth Planet. Sci. Lett.* **2024**, *626*, 118558.
133. M. Prakash, J. M. Weber, L. E. Rodriguez, R. Y. Sheppard, L. M. Barge, *Int. J. Astrobiol.* **2022**, *21* (6), 423–440.
134. J. M. R. Génin, R. Aïssa, A. Géhin, M. Abdelmoula, O. Benali, V. Ernstsen, G. Ona-Nguema, C. Upadhyay, C. Ruby, *Solid State Sci.* **2005**, *7* (5), 545–572.
135. C. Ruby, A. Géhin, R. Aissa, J. M. R. Génin, *Corros. Sci.* **2006**, *48* (11), 3824–3837.
136. M. C. F. Wander, K. M. Rosso, M. A. A. Schoonen, *J. Physl. Chem. C.* **2007**, *111* (30), 11414–11423.
137. H. Antony, L. Legrand, A. Chaussé, *Electrochim. Acta.* **2008**, *53* (24), 7146–7156.
138. R. Matta, K. Hanna, S. Chiron, *Sep. Purif. Technol.* **2008**, *61* (3), 442–446.
139. L. M. Barge, E. Flores, J. M. Weber, A. A. Fraeman, Y. L. Yung, D. VanderVelde, E. Martinez, A. Castonguay, K. Billings, M. M. Baum, *Geochim. Cosmochim. Acta.* **2022**, *336*, 469–479.
140. L. M. Barge, E. Flores, M. M. Baum, D. G. V. Velde, M. J. Russell, *Proc. Natl. Acad. Sci.* USA **2019**, *116* (11), 4828–4833.
141. C. Ruby, M. Abdelmoula, S. Naille, A. Renard, V. Khare, G. Ona-Nguema, G. Morin, J. M. R. Génin, *Geochim. Cosmochim. Acta.* **2010**, *74* (3), 953–966.
142. A. Yoshino, S. Fujieda, K. Shinoda, S. Suzuki, *ISIJ Int.* **2013**, *53* (5), 894–899.
143. P. Wang, X. Mou, Y. Li, W. Shen, *Green Energy and Environ.* **2021**.
144. O. Farr, *Mineral spirits and the redox transformations of CO2 & CH4 by Fe2+/3+oxyhydroxide "green rust" minerals*, Aix-Marseille Université, Marseille, France **2024**.
145. S. Lee, C. Wang, V. Chakrapani, *J. Phys. Chem. C.* **2024**, *128* (38), 15778–15784.
146. O. Farr, *An abiotic gaia?: $Fe^{2+/3+}$ oxyhydroxide "Green rust" mineral mediated CO_2 reduction and CH_4 oxidation yields climate forcing agents dimethyl sulfide (DMS) and chloroform (CCl_3), in prep*, **2025**.
147. F. Bocher, A. Géhin, C. Ruby, J. Ghanbaja, M. Abdelmoula, J. M. R. Génin, *Solid State Sci.* **2004**, *6* (1), 117–124.
148. H. C. B. Hansen, I. F. Poulsen, *Clays Clay Miner.* **1999**, *47* (3), 312–318.
149. H. Li, J. Deng, Q. Jia, L. Zhu, L.-Z. Huang, *Water Res.* **2024**, 121589.
150. D. Santos-Carballal, A. Roldan, N. Y. Dzade, N. H. De Leeuw, *Phil. Trans. R. Soc. A.* **2018**, *376* (2110).
151. M. T. C. Martins-Costa, M. Ruiz-Lopez, *Chem. Eur. J.* **2024**.
152. L. Fang, L. Xu, J. Deng, S. Gao, L. Z. Huang, *Chem. Eng. J.* **2021**, *414*, 128780.
153. Z. Li, M. Li, B. Tan, N. Du, Q. Zhang, C. Li, Y. Zhang, J. Li, J. Li, *Environ. Res.* **2022**, *208*.
154. C. Wang, V. Chakrapani, *J. Phys. Chem. C.* **2023**, *127* (48), 23189–23198.
155. W. Liu, *Photogeochemistry of minerals under anoxic conditions*, Rutgers University, USA **2021**.

156. X. He, L. Zhang, J. Chen, H. Liu, Y. Su, H. Li, Y. Cao, P. O. Dral, C. Wang, *Inorg. Chem.* **2023**, *62* (26), 10343–10350.
157. L. Z. Huang, X. Zhang, R. Liu, L. Fang, *Chem. Eng. J.* **2022**, *429.*
158. D. A. Kopp, S. J. Lippard, *Curr. Opin. Chem. Biol.* **2002**, *6* (5), 568–576.
159. F. Thibon, J. Blichert-Toft, H. Tsikos, J. Foden, E. Albalat, F. Albarede, *Earth. Planet. Sci. Lett.* **2019**, *506*, 360–370.
160. L. Ernst, U. Barayeu, J. Hädeler, T. P. Dick, J. M. Klatt, F. Keppler, J. G. Rebelein, *Nat. Commun.* **2023**, *14* (1), 4364.
161. M. S. Dodd, H. Wang, C. Li, M. Towner, A. R. Thomson, J. F. Slack, Y. sheng Wan, F. Pirajno, C. Manikyamba, Q. Wang, D. Papineau, *Earth Planet. Sci. Lett.* **2022**, *584.*
162. I. Halevy, A. Bachan, *Science* **2017**, *355*, 1069–1071.
163. J. Hädeler, G. Velmurugan, R. Lauer, R. Radhamani, F. Keppler, P. Comba, *J. Am. Chem. Soc.* **2023**, *145*, 24590–24602.
164. F. Althoff, K. Benzing, P. Comba, C. McRoberts, D. R. Boyd, S. Greiner, F. Keppler, *Nat. Commun.* **2014**, *5*, 4205.
165. J. E. Lovelock, L. Margulis, *Tellus A: Dyn MeteorolOceanogr.* **1974**, *26* (1–2), 2.
166. S. M. Ledford, L. K. Meredith, *J. Mol. Evol.* **2024**.
167. S. Kloster, J. Feichter, E. Maier-Reimer, K. D. Six, P. Stier, P. Wetzel, *Biogeosci.* **2006**, *3* (1), 29–51.
168. Q. Li, R. P. Fernandez, R. Hossaini, F. Iglesias-Suarez, C. A. Cuevas, E. C. Apel, D. E. Kinnison, J. F. Lamarque, A. Saiz-Lopez, *Nat. Commun.* **2022**, *13* (1).
169. O. Farr, J. Hao, W. Liu, N. Fehon, J. R. Reinfelder, N. Yee, P. G. Falkowski, *Proc. Natl. Acad. Sci.* USA **2023**, *120* (30).
170. J. Hao, D. A. Sverjensky, R. M. Hazen, *Earth Planet. Sci. Lett.* **2017**, *471*, 148–159.
171. J. Hao, A. Knoll, F. Huang, J. Schieber, R. Hazen, I. Daniel, *Geochim. Cosmochim. Acta*, **2020**, 360–377.
172. J. Hao, W. Liu, J. L. Goff, J. A. Steadman, R. R. Large, P. G. Falkowski, N. Yee, *Sci. Adv.* **2022**, *8.*
173. C. Sagan, G. Mullen, *Science* **1972**, *177* (4043), 52–56.
174. I. Egea-González, C. P. Mckay, J. E. Hallsworth, A. Jiménez-Díaz, J. Ruiz, *rXiv.* **2024**.
175. C. Malaterre, I. L. ten Kate, M. Baqué, V. Debaille, J. L. Grenfell, E. J. Javaux, N. Khawaja, F. Klenner, Y. J. Lara, S. McMahon, K. Moore, L. Noack, C. H. L. Patty, F. Postberg, *Astrobiol.* **2023**, *23* (11), 1213–1227.
176. Z. Guo, D. Papineau, J. O'Neil, H. Rizo, Z.-Q. Chen, X. Qiu, Z. She, *Nat. Commun.* **2024**, *15* (1), 5679.
177. M. S. Dodd, D. Papineau, Z. B. She, C. Manikyamba, Y. S. Wan, J. O'Neil, J. A. Karhu, H. Rizo, F. Pirajno, *Earth Planet. Sci. Lett.* **2019**, *512*, 163–174.
178. L. Sun, M. Lechte, X. Shi, X. Zhou, L. Zhou, H. Fang, B. Xie, M. Wu, D. Tang, *Am. Mineralo.* **2022**, *107* (5), 970–984.
179. H. Oduro, S. Ono, F. Alrasheed, D. L. Eldridge, *Geochem, Geophys, Geosys.* **2023**, *24* (10).
180. D. C. Mauzerall, *Origins and evolution of biosphere*, **1990**, *20*, 293-302
181. N. X. Nie, N. Dauphas, R. C. Greenwood, *Earth Planet. Sci. Lett.* **2017**, *458*, 179–191.
182. G. P. Anipsitakis, D. D. Dionysiou, *Environ. Sci. Technol.* **2004**, *38* (13), 3705–3712.
183. H. He, X. Wu, H. Xian, J. Zhu, Y. Yang, Y. Lv, Y. Li, K. O. Konhauser, *Nat. Commun.* **2021**, *12* (1).
184. G. Zhao, M. Tan, B. Wu, X. Zheng, R. Xiong, B. Chen, A. Kappler, C. Chu, *Environ. Sci. Technol.* **2023**.
185. H. He, X. Wu, J. Zhu, M. Lin, Y. Lv, H. Xian, Y. Yang, X. Lin, S. Li, Y. Li, H. H. Teng, M. H. Thiemens, *Proc. Natl. Acad. Sci.* USA **2023**, *120* (13).
186. W. H. Koppenol, H. Sies, *Redox. Biol.* **2023**, 103012.

187. S. Jin, H. Chen, X. Yuan, D. Xing, R. Wang, L. Zhao, D. Zhang, C. Gong, C. Zhu, X. Gao, Y. Chen, X. Zhang, *JACS Au.* **2023**, *3* (6), 1563–1571.
188. J. Li, J. Xu, Q. Song, X. Zhang, Y. Xia, R. N. Zare, *Chem. Sci.* **2024**.
189. N. Kitadai, R. Nakamura, M. Yamamoto, S. Okada, W. Takahagi, Y. Nakano, Y. Takahashi, K. Takai, Y. Oono, *Commun. Chem.* **2021**, *4* (1).
190. Y. Ou, L. P. Twight, B. Samanta, L. Liu, S. Biswas, J. L. Fehrs, N. A. Sagui, J. Villalobos, J. Morales-Santelices, D. Antipin, M. Risch, M. C. Toroker, S. W. Boettcher, *Nat. Commun.* **2023**, *14* (1).
191. A. K. Sweetman, A. J. Smith, D. S. W. de Jonge, T. Hahn, P. Schroedl, M. Silverstein, C. Andrade, R. L. Edwards, A. J. M. Lough, C. Woulds, W. B. Homoky, A. Koschinsky, S. Fuchs, T. Kuhn, F. Geiger, J. J. Marlow, *Nat. Geosci.* **2024**.
192. N. Kitadai, S. Maruyama, *Geosci. Front.* **2018**, *9* (4), 1117–1153.
193. A. L. Ducluzeau, R. van Lis, S. Duval, B. Schoepp-Cothenet, M. J. Russell, W. Nitschke, *Trends Biochem. Sci.* **2009**, *34* (1), 9–15.
194. N. Glansdorff, Y. Xu, B. Labedan, *Biol. Direct.* **2008**, *3* (1), 29.
195. L. Schwander, M. Brabender, N. Mrnjavac, J. L. E. Wimmer, M. Preiner, W. F. Martin, *Biochim. Biophys. Acta Bioenerg.* **2024**, *1865*, 149224.
196. J. L. E. Wimmer, A. D. N. Vieira, J. C. Xavier, K. Kleinermanns, W. F. Martin, M. Preiner, *Microorganisms* **2021**, *9* (2), 1–16.
197. J. E. Johnson, T. M. Present, J. S. Valentine, *Proc. Natl. Acad. Sci.* USA **2024**, *121* (38).
198. T. W. Lyons, C. J. Tino, G. P. Fournier, R. E. Anderson, W. D. Leavitt, K. O. Konhauser, E. E. Stüeken, *Nat Rev Microbiol.* **2024**.
199. A. D. Goldman, J. M. Weber, D. E. LaRowe, L. M. Barge, *Proc. Natl. Acad. Sci.* USA **2023**, *120* (34).
200. F. Liu, A. E. Rotaru, P. M. Shrestha, N. S. Malvankar, K. P. Nevin, D. R. Lovley, *Environ. Microbiol.* **2015**, *17* (3), 648–655.
201. A. García-Prieto, J. Alonso, D. Muñoz, L. Marcano, A. Abad Díaz De Cerio, R. Fernández De Luis, I. Orue, O. Mathon, A. Muela, M. L. Fdez-Gubieda, *Nanoscale* **2016**, *8* (2), 1088–1099.
202. E. J. O'Loughlin, D. R. Burris, *Minerals* **2022**, *12* (2), 138.
203. Y. Zhang, S. Wei, Y. Lin, G. Fan, F. Li, *ACS Omega* **2018**, 12778–12787.
204. A. N. Thomas, E. Eiche, J. Göttlicher, R. Steininger, L. G. Benning, H. M. Freeman, D. J. Tobler, M. Mangayayam, K. Dideriksen, T. Neumann, *Geochem. Trans.* **2020**, *21* (1).
205. O. Farr, E. J. Elzinga, N. Yee, *Geochem. Trans.* **2022**, *23* (1), 3.
206. J.-M. R. Génin, P. Refait, G. Bourrié, M. Abdelmoula, F. Trolard, *Appl. Geochem.* **2001**, *16* (5), 559–570.
207. Y. Sun, G. Jiang, R. Jiang, S. Lan, Y. Li, Y. Yan, Q. Zhang, *J. Agro-Environ. Sci.* **2023**, *42* (2), 393–402.
208. A. J. Anschutz, R. L. Penn, *Geochem. Trans.* **2005**, *6* (3), 60–66.
209. Q.-Q. Jia, X.-J. Zhang, L. Zhu, L.-Z. Huang, *Water Res.* **2024**, 121342.
210. W. Yin, L. Huang, E. B. Pedersen, C. Frandsen, H. C. B. Hansen, *J. Colloid. Interface. Sci.* **2017**, *497*, 429–438.
211. W. Yin, J. Ai, L. Z. Huang, D. J. Tobler, H. C. B. Hansen, *Environ. Sci. Technol.* **2018**, *52* (14), 7876–7883.
212. M. R. Hooks, P. Webster, J. M. Weber, S. Perl, L. M. Barge, *Langmuir.* **2020**, *36* (21), 5793–5801.
213. A. Borrego-Sánchez, C. Gutiérrez-Ariza, C. I. Sainz-Díaz, J. H. E. Cartwright, *Langmuir.* **2022**, *38* (34), 10538–10547.
214. J. F. Lambert, *Orig. Life Evol. Biosph.* **2008**, *38* (3), 211–242.
215. U. Shanker, B. Bhushan, G. Bhattacharjee, Kamaluddin, *Orig. Life Evol. Biosph.* **2012**, *42* (1), 31–45.

216. M. L. Wong, C. E. Cleland, D. Arend, S. Bartlett, H. J. Cleaves, H. Demarest, A. Prabhu, J. I. Lunine, R. M. Hazen, *Pro. Natl. Acad. Sci.* USA **2023**, *120* (43).
217. H. Ogasawara, A. Yoshida, E. I. Imai, H. Honda, K. Hatori, K. Matsuno, *Orig. Life Evol. Biosph.* **2000**, *30* (6), 519–526.
218. B. T. Burcar, L. M. Barge, D. Trail, E. B. Watson, M. J. Russell, L. B. McGown, *Astrobiol.* **2015**, *15* (7), 509–522.
219. J. Zimmerman, B. Atalay, J. Moran, *ChemRxiv,* **2024**.
220. D. P. Henriques Pereira, J. Leethaus, T. Beyazay, A. do Nascimento Vieira, K. Kleinermanns, H. Tüysüz, W. F. Martin, M. Preiner, *FEBS J.* **2022**, *289* (11), 3148–3162.
221. I. Mamajanov, G. D. Cody, *Philos. Trans. R. Soc. A.* **2017**, *375* (2109).
222. B. Rasmussen, J. R. Muhling, W. W. Fischer, *Astrobiol.* **2020**, *21* (2), 1–14.
223. K. T. Madanan, Y. Li, V. J. Boide-Trujillo, D. A. Russell, C. Bonfio, *ChemRxiv.* **2024**.
224. M. Aleksandrova, F. Rahmatova, D. A. Russell, C. Bonfio, *J. Am. Chem. Soc.* **2023**, *145* (47), 25614–25620.
225. D. Kanaparthi, M. Lampe, B. Zhu, T. Boesen, A. Klingl, P. Schwille, T. Lueders, *Elife* **2024**.
226. M. S. Dodd, D. Papineau, T. Grenne, J. F. Slack, M. Rittner, F. Pirajno, J. O'Neil, C. T. S. Little, *Nature* **2017**, *543* (7643), 60–64.
227. S. Jordan, M. Van Zuilen, Z. Martins, N. Lane, *bioRxiv.* **2023**.
228. S. McMahon, *Proc. R. Soc. B.* **2019**, *286* (1916).
229. A. R. Klein, J. J. Basinski, A. Niyitanga Manzi, C. X. Cheng, M. Sanchez, W. Thongsomboon, L. Aristilde, *Environ. Chem. Lett* **2024**.
230. M. Akouche, M. Jaber, M. C. Maurel, J. F. Lambert, T. Georgelin, *Angew. Chem. Int. Ed.* **2017**, *56* (27), 7920–7923.
231. L. M. Barge, I. J. Doloboff, M. J. Russell, D. Vandervelde, L. M. White, G. D. Stucky, M. M. Baum, J. Zeytounian, R. Kidd, I. Kanik, *Geochim. Cosmochim. Acta.* **2014**, *128*, 1–12.
232. Y. Hirakawa, T. Kakegawa, Y. Furukawa, *Sci. Rep.* **2024**, *14* (1).
233. N. Gaudu, O. Farr, G. Ona-Nguema, S. Duval, *Biochimie* **2023**, *215*, 100–112.
234. D. Milshteyn, G. Cooper, D. Deamer, *Sci. Rep.* **2019**, *9* (1).
235. W. Nitschke, O. Farr, N. Gaudu, C. Truong, F. Guyot, M. J. Russell, S. Duval, *Life* **2024**, *14* (5), 607.
236. R. Nakamura, T. Takashima, S. Kato, K. Takai, M. Yamamoto, K. Hashimoto, *Angew. Chem. Int. Ed.* **2010**, *49* (42), 7692–7694.
237. M. Yamamoto, R. Nakamura, T. Kasaya, H. Kumagai, K. Suzuki, K. Takai, *Angew. Chem. Int. Ed.* **2017**, *56* (21), 5725–5728.
238. H.-E. Lee, T. Okumura, H. Ooka, K. Adachi, T. Hikima, K. Hirata, Y. Kawano, H. Matsuura, M. Yamamoto, M. Yamamoto, A. Yamaguchi, J.-E. Lee, H. Takahashi, K. T. Nam, Y. Ohara, D. Hashizume, S. E. McGlynn, R. Nakamura, *Nat. Commun.* **2024**, *15* (1), 8193.
239. G. Pillot, Ó. Santiago, S. Kerzenmacher, P.-P. Liebgott, *Life* **2023**, *13* (2), 356.
240. H.-E. Lee, T. Okumura, H. Ooka, K. Adachi, T. Hikima, K. Hirata, Y. Kawano, H. Matsuura, M. Yamamoto, M. Yamamoto, A. Yamaguchi, J-E. Lee, K. T. Nam, Y. Ohara, D. Hashizume, S. McGlynn, R. Nakamura, *Chemrxiv.* **2023**.
241. M. Weingart, S. Chen, C. Donat, V. Helmbrecht, W. D. Orsi, D. Braun, K. Alim, *ChemRxiv.* **2023**.
242. F. M. Möller, F. Kriegel, M. Kieß, V. Sojo, D. Braun, *Angew. Chem. Int. Ed.* **2017**, *56* (9), 2340–2344.
243. D. Velegol, A. Garg, R. Guha, A. Kar, M. Kumar, *Soft Matt.* **2016**, *12* (21), 4686–4703.
244. S. Shim, *Chem. Rev.* **2022**, *122* (7), 6986–7009.
245. I. M. Ahmed, M. S. Gasser, *Appl. Surf. Sci.* **2012**, *259*, 650–656.

246. Z. zhu Yang, J. jing Wei, G. ming Zeng, H. qing Zhang, X. fei Tan, C. Ma, X. cheng Li, Z. hao Li, C. Zhang, *Coord. Chem. Rev.* **2019**, *386*, 154–182.
247. A. A. Baykov, *Front. Plant. Sci.* **2020**, *11.*
248. F. Trolard, S. Duval, W. Nitschke, B. Ménez, C. Pisapia, J. Ben Nacib, M. Andréani, G. Bourrié, *Earth. Sci. Rev.* **2022**, *225.*
249. H. Elderfield, A. Schultz, *Annu. Rev. Earth. Planet. Sci.* **1996**, *24*, 191–224.
250. S.-F. Xiong, S.-Y. Jiang, J.-X. Zhao, P.-P. Niu, Y. Ma, X.-Y. Bai, *Precambrian. Res.* **2023**, *393*, 107088.
251. A. Zegeye, S. Bonneville, L. G. Benning, A. Sturm, D. A. Fowle, C. A. Jones, D. E. Canfield, C. Ruby, L. C. MacLean, S. Nomosatryo, S. A. Crowe, S. W. Poulton, *Geology* **2012**, *40* (7), 599–602.
252. Y. Xiong, R. Guilbaud, C. L. Peacock, M. D. Krom, S. W. Poulton, *Geochim. Cosmochim. Acta.* **2023**, *351*, 139–151.
253. B. Ménez, C. Pisapia, M. Andreani, F. Jamme, Q. P. Vanbellingen, A. Brunelle, L. Richard, P. Dumas, M. Réfrégiers, *Nature* **2018**, *564* (7734), 59–63.
254. B. Hazra, M. Prasad, S. Das, R. Mandal, A. Sardar, N. Dewangan, P. K. Tarafdar, *Langmuir* **2023**, *39* (48), 17031–17042.
255. T. M. Mccollom, G. Ritter, B. R. T. Simoneit, *Orig. Life Evol. Biosph.* **1999**, *29* (153–166).
256. G. Purvis, L. Šiller, A. Crosskey, J. Vincent, C. Wills, J. Sheriff, C. Xavier, J. Telling, *Commun. Earth Environ.* **2024**, *5* (1).
257. D. He, X. Wang, Y. Yang, R. He, H. Zhong, Y. Wang, B. Han, *Pro. Natl. Acad. Sci.* **2021**, 118, 1–7.
258. J. M. Weber, E. C. Czaplinski, B. L. Henderson, L. M. Barge, J. C. Castillo-Rogez, R. Hodyss, *ACS Earth Space Chem.* **2024**.
259. L. M. Barge, G. P. Fournier, *Astrobiology* **2024**, *24* (3), 328–338.
260. C. Liu, W. Xu, Z. Zhang, K. Robinson, M. Lau, F. 放 Huang 黄, F. 方 Huang 黄, C. R. Glein, J. Hao, *Astrophys. J.* **2024**, *971* (1), 51.
261. H. E. Maynard-Casely, M. L. Cable, M. J. Malaska, T. H. Vu, M. Choukroun, R. Hodyss, *Am. Mineralo.* **2018**, *103* (3), 343–349.
262. M. J. Russell, A. E. Murray, K. P. Hand, *Astrobiology* **2017**, *17* (12), 1265–1273.
263. S. M. MacKenzie, S. P. D. Birch, S. Hörst, C. Sotin, E. Barth, J. M. Lora, M. G. Trainer, P. Corlies, M. J. Malaska, E. Sciamma-O'Brien, A. E. Thelen, E. Turtle, J. Radebaugh, J. Hanley, A. Solomonidou, C. Newman, L. Regoli, S. Rodriguez, B. Seignovert, A. G. Hayes, B. Journaux, J. Steckloff, D. Nna-Mvondo, T. Cornet, M. Y. Palmer, R. M. C. Lopes, S. Vinatier, R. Lorenz, C. Nixon, E. Czaplinski, J. W. Barnes, E. Sittler, A. Coates, *Planet. Sci. J.* **2021**, *2* (3), 112.
264. J. R. Michalski, E. Z. N. Dobrea, P. B. Niles, J. Cuadros, *Nat. Commun.* **2017**, *8* (1), 15978.

9 Fougerite: Free Energy Converter for Life's Conception

Michael J. Russell
Dipartimento di Chimica, Università degli Studi di Torino, Torino, Italy
michaeljrussell80@gmail.com

CONTENTS

DOI: 10.1201/9781003459910-9

ABSTRACT

The unique yet diverse electro-mechanochemical functionalities of the double-layer oxy-hydroxide "fougerite" – the mineral name of green rust – are outlined with respect to what we call the "fougerite submarine alkaline vent theory" (FAVT) for the emergence of energy-metabolism that eventuated in life. Fougerite, we maintain, is the "found object" that allowed a moderate temperature submarine alkaline hydrothermal system to generate an ordered autogenic energy-metabolism through the conversion of environmental free energy into a host of topologically focused chemical disequilibria. When rapidly precipitated as chemical gardens in the company of syngenetically generated organic molecules at such a hydrothermal alkaline/acidic redox boundary, fougerite nanocrystals act as variable-valence inorganic macromolecules. In this milieu their most notable functions arise from the formation of an osmotic barrier with strikingly "biological" properties such as those supporting steep vectorially directed redox and ionic gradients. Indeed, fougerite's bilaterally active pH, redox-sensitive, and flexible cationic layers are demonstrably capable of a variety of protometabolic-like free energy (strictly "disequilibria") conversions – providing both the structures required to support the relevant geochemically mediated driving gradients and the nanoengines needed to convert those disequilibria to life's essential "starter kit" of internal disequilibria: notably the individual reduction of carbonate to formate and on through to acetate, pyruvate, glyoxalate and, converted to magnetite, to ⋜C18 fatty acids; the reduction of nitrogen oxides to ammonium and of pyruvate and glyoxalate to alanine and glycine, and, conversely, the oxidation of methane to a methyl group and ⋜C4 organics. Most importantly, fougerite may be instrumental in converting the environmental free energy into high polyphosphate to phosphate ratios thereby furnishing a topologically confined chemical disequilibrium able to drive a host of otherwise endergonic processes, just as is the case in extant life. Thus the bilayers of fougerite offer the fairly well-ordered confined channels performing the order-from-order conversions at the hydrothermal vent which render the emergence of life compatible with the second law of thermodynamics.

KEYWORDS

abiogenesis, disequilibria, fougerite-alkaline-vent-theory (favt), free-energy-converter, green rust, second law of thermodynamics

> *Life ... operates by utilising the free energy of its environment. It is a free energy converter ...*
>
> Frederick George Donnan 1929, page 1397

1 MINERAL-MEDIATED AUTOGENIC THEORIES FOR LIFE'S EMERGENCE: AN OVERVIEW

The prevalent mindset on the origin of life was, and still is, the notion that it first requires the presence of a mixture of supposedly energy-rich organic molecules (nucleotides, amino acids, lipids, etc.) from which life will (more or less) spontaneously onset [1–4]. The bulk of research into life's origin therefore attempts to redraw the prebiotic synthesis pathways for these organic molecules. Traditionally (going back to Stanley Miller's groundbreaking experiments), these approaches react the supposed principal abiotic precursors for carbon-, nitrogen-, oxygen-, etc., based molecules, that is, carbon dioxide, ammonia, methane, hydrogen, cyanide and so on, to form simple organic ones, generally at a primeval pond or pool on otherwise dry land. The activation energies impeding these mixtures of (mostly metastable) precursors to attain thermodynamic equilibrium with the products (the organic molecules) are overcome by excitations such as electrical discharge, UV-light, ionizing radiation, heat or bolide impact. Yet by 1929, Donnan had warned that "Organic molecules containing only carbon, hydrogen, oxygen, nitrogen, sulphur, and phosphorus do not reign supreme in the chemical and physical mechanism of the living cell" [5]. And even before that, in 1911 Stéphane Leduc – a committed Darwinist – argued that "The chain of life is of necessity a continuous one, from the mineral at one end to the most complicated organism at the other" [6, page xv].

However, even most of the mineral-mediated theories for life's origin still subscribe to the general organic scheme, envisioning the described "organics-forming-reactions" to benefit from the catalytic power of transition metals and clusters thereof, adsorbed on certain minerals. The tendency of transition metals to decrease activation barriers (the very definition of catalysis) does away with (or at least strongly reduces) the above-mentioned need of energetic excitations to activate the precursors. Additionally, involving minerals in prebiotic synthesis of organic molecules acknowledges the fact that a vast number of metabolic reactions in extant life are indeed catalyzed by metalloenzymes and thereby renders the envisioned prebiotic reaction schemes more "life-like". We will, in the following, describe the dominant hypotheses following the "prebiotic chemistry"-roadmap [7] before we come back to questioning the actual foundations of the described mindset.

Abundant CO_2, carbonate and bicarbonate – to this day the carbon substrates of the biosphere – made up much of the early inclement volatisphere [8–11]. Armed with this geological knowledge of the early Earth – far removed from the reduced and clement atmosphere supposed by Harold Urey [12] and so many others – Leduc [6], Goldschmidt [13] and Bernal [14] suggested that life originated autogenically from CO_2 (or carbonic acid) at the surfaces or interfaces of various minerals acting as mediators or catalysts. Such autogenic theories received a boost at the beginning of the 1980s with the discovery of acidic high temperature submarine sulfidic hydrothermal chimneys – "Black Smokers" – and their associated microbiomes on the East Pacific Rise in 1979 [15–25].

For example, Günter Wächtershäuser, calling especially upon Karl Stetter's discoveries of autotrophic hyperthermophilic prokaryotes at these black smokers

[19,26–29], favored Fe^{2+} + H_2S at an ancient sulfidic volcanic hot spring as the CO_2 reductant in a pyrite-(FeS_2)-catalyzed reaction for the "origin" of life [30–33]. In contrast Michael Russell and collaborators, inspired by their geological discovery of moderate temperature and much smaller centimeter-wide fossil pyrite hydrothermal chimneys that fed a 350 Myr-old submarine base metal sulfide and associated barite deposit in Ireland [34–38], coupled with their then ongoing research into syngenetic links between serpentinization and magnesite mineralization in Alpine-type ultramafic complexes [39,40], opted for moderate temperature submarine alkaline springs bearing H_2 into a Hadean carbonic ocean as their preferred site for the "emergence" of life (Figure 1). Such submarine springs were presumed to have precipitated the iron sulfide chimneys and chemical gardens, bolstered by syngenetically generated organic molecules at the vent site [36,41–43]. These same authors considered such composite fine tubes and chimneys to be capable of redox- and pH-gradient-driven hydrogenations of carbonic acid in what became known as the submarine alkaline vent theory for life's emergence (AVT) [41,44].

This "minerobiolization" model [51] came full circle when it was found that the supergiant hydrothermal base-metal sulfide orebody discovered at Navan, Ireland, in 1970, was shown to have required bacteriogenic sulfide for at least 95% of its ore in a process we have called "bacteriometallogenesis" [38,64,65].

In the early days of hot spring theories for life's emergence there was a sense of determinism about the expected first pathways to an autotrophic metabolism [66] – that the journey to life would begin at the acetyl coenzyme-A pathway (ACp) prior to its segue into the reverse but incomplete tricarboxylic acid cycle/pathway (rTCAp) – a pathway terminating at succinate and glyoxylate [30,46,67,68]. That FeS and Fe_3S_4 were the assumed precursors to pyrite in both the AVT [41,69], and Wächtershäuser's [33] "iron-sulfur world" paper found further support from Eck and Dayhoff's [70] much earlier suggestion that FeS, then evolving to ancestral ferredoxins through a single highly reduced Fe_4S_4 center sequestered first by 4-mer peptide, were the first "biological" catalysts. Wächtershäuser had previously referred to Eck and Dayhoff's iron sulfide precursor to ferredoxin in his iron-sulfur world paper [33] noting sulfur's capacity to form iron-sulfur clusters with both redox and non-redox functions, often under anaerobic conditions [71–74]. On a different tack Russell and Hall [46] and Milner-White and Russell [75] drew attention to the fact that the active centers of ferredoxins share conformational affinities with atomic structures of greigite and mackinawite, as nested in short peptides.

2 CAVEATS RAISED AGAINST MINERAL-MEDIATED SCENARIOS IN GENERAL AND AGAINST THE AVT IN PARTICULAR

2.1 THERMODYNAMICS AND THE ROLE OF FREE-ENERGY CONVERTERS

Research on the emergence of life is replete with scenarios trying to distinguish themselves from others by using antagonistic terms. "Building-block-first" hypotheses compete with "metabolism-first" ones, or organics-based theories are seen to be opposed to their mineral/metal-based "alternatives". Organics-based scenarios are

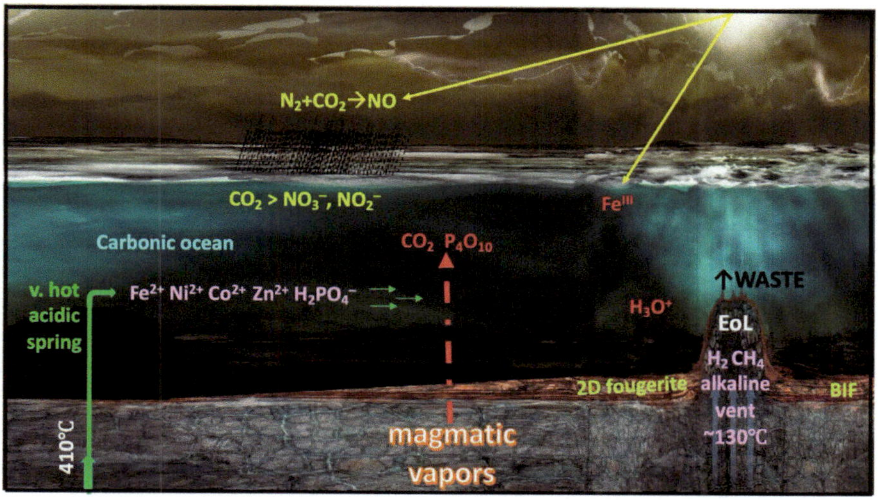

FIGURE 1 The putative Hadean disequilibria at ~4.2Ga [45–50] resulting in life's emergence according to the submarine Fougerite Alkaline Vent Theory (FAVT) [44, 51–54]. Metals dissolved in acidic ~400°C hydrothermal solutions associated with mantle plumes and delivered to the carbonic Hadean Ocean remain supersaturated as a result of general and continuous degassing of CO_2 with subsidiary CH_4 at the ocean floor [17, 55–58]. Precipitation of metals occurs chiefly where H_2- and CH_4-bearing alkaline (pH~10.5) fluids, emitted from a ~130°C vent at a distance from magmatic intrusions, meet this somewhat acidic ocean water [38, 55, 59] (but see Shibuya and collaborators, for a variation on this scenario [60]). This bottom-up view of a thermophilic emergence of life [61] meets a favorable top-down finding that a reconstruction of ancestral proteins with their extant counterparts suggests primitive life had an alkaline habitat [62]. Further support for alkaline conditions stems from a study using "microbial fossils" and isotope records under a new cross-bracing implementation [62, 63]. From [38] with permission.

often equated with building-blocks-first models, and metabolism-first theories are considered to necessarily be based on metals/minerals. However, these antagonisms frequently obscure the tangled connections between all these scenarios.

As reviewed recently [76], the view that life originated spontaneously out of an agglomerate of building blocks is severely at odds with thermodynamic principles and in particular with the second law of thermodynamics, a fact masterly laid out in Erwin Schrödinger's book *What Is Life* [77]. In short, the ordering (entropy decrease, or as Gibbs would have it, "decrease in *mixed-up-ness*" [78]) inherent to life cannot come about by itself (as envisioned by the above-mentioned "primordial soup" scenarios) but must be extracted from an environmental bath of low entropy (i.e., some sort of environmental "order"). Such environmental "order" corresponds to thermodynamic disequilibria which often occur in the form of "Gibbs- (or alternatively Helmholtz-) free energy". Schrödinger's message – grounded in the physical statistics of Ludwig Boltzmann and Frederick G. Donnan's statistical mechanics with an emphasis on the contrast in arrangements of atoms between living organisms and non-living matter [79,80] – was that, to be in line with the second law of thermodynamics, life's ordering must be continually driven by environmental free energy – his "negentropy" (see page 74 in [77]), cf. electrochemical disequilibria [81].

Unfortunately, this message is often misunderstood as saying that the fundamental processes of life only need to feature negative free energies, in other words: to be exergonic. This vision is completely inappropriate and demonstrably wrong. Exergonic reactions take a system toward thermodynamic equilibrium which is precisely the antithesis of what life does! For example, in the Miller/Urey experiments, highly out-of-equilibrium concentrations of inorganic precursor molecules are exergonically reacted to achieve thermodynamic equilibrium yielding very small quantities of more complex organic molecules (nota bene: the free energy of this reaction resides in the non-equilibrium concentrations of precursors and products, the energy – not free energy – of the electric discharge only serves to overcome the activation barriers and thus speeds up the reaction). Analogously, an abiotic version of the Wood-Ljungdahl pathway producing biomass from an exergonic reaction of H_2 and CO_2 was proposed [67] to have engendered life as discussed in more detail below. Again, this reaction can only take the precursors (H_2 and CO_2) and the products (methane/acetate and organics) to their thermodynamic equilibrium. Life, by contrast, takes the precursors and products to extremely out-of-equilibrium concentrations and constantly keeps them there. For a living entity, a return to thermodynamic equilibrium simply spells death!

Schrödinger's crucial step forward consisted in recognizing life as a subsystem embedded in its out-of-equilibrium environment with this subsystem's ordering driven by processes converting part of the environmental low entropy into ordering of the subsystem. Life thus is "continually sucking orderliness from its environment" – all while simultaneously taking the entire system closer to global equilibrium through order-from-order conversions [77,82,83].

What makes life possible therefore is neither the availability of organic building blocks nor the exergonicity of imaginary "metabolic" reaction schemes [84], but the existence of free energy converting processes that "steal" part of the initial low entropy of the environment to build up life's ordering [85]. In extant life, this disequilibrium-converting mechanism is called "energy-metabolism" and to deduce the conditions producing the emergence of life we need to understand where energy-metabolism comes from [76,86,87]. Indeed, the whole thrust of "origin-of-life" research may have taken a more productive route had scientists from the earliest days not overlooked Frederick G. Donnan's realization that life was a "free energy converter" [5].

In this context, it is important to remember that "metabolism" isn't a synonym of "energy-metabolism". In biology, metabolism is subdivided into three sub-processes: (a) energy metabolism, (b) build-up metabolism (anabolism) and (c) degrading metabolism (catabolism). The vast majority of "metabolism-first" scenarios are actually (exergonic) anabolism-first ones! This is true for both Wächtershäuser's pyrite hypothesis [31] and for the older versions of the AVT (as will be discussed in the next section). They therefore fundamentally resemble the "primordial soup" scenarios in that they imagine exergonic anabolic processes to spontaneously engender life's entropy-decreasing order formation with the only difference of involving metals and minerals in the catalytic reactions taking the system to equilibrium.

The perceived competition between "organics- (i.e., building blocks) first" and "metabolism-first" hypotheses thus appears misguided in the light of the second law which attributes the penultimate primacy to energy metabolism, that is, disequilibrium conversion. Anabolism and catabolism – *per se* extremely endergonic processes – become exergonic only under the conditions of an energized cell (i.e., featuring extremely strongly out-of-equilibrium ATP/ADP-ratios, an order channeled off from the environmental order via energy-metabolism).

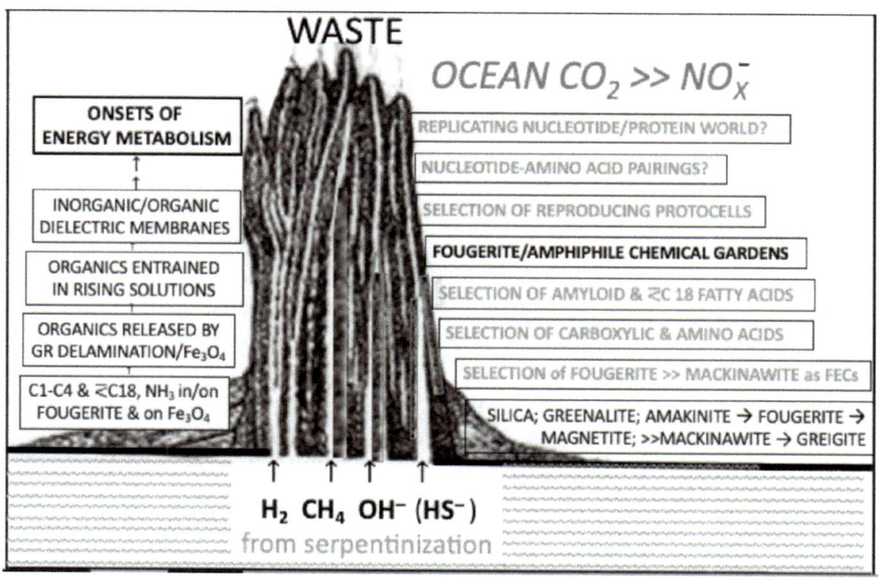

FIGURE 2 The submarine alkaline hydrothermal mound acts as a large-scale free-energy-converter and selection column which produces all manner of increasing – "evolved" – complexity with time [92–97]. Initially the mound comprises a myriad of narrow inter-braided chemical garden compartments composed of amorphous silica, amakinite, fougerite, greenalite, magnetite and minor mackinawite [99–104]. Of these amakinite, fougerite, magnetite and mackinawite are "selected" [96, 98] as free energy converters and/or conductors to produce amino acids, disordered peptides, amyloid and long-chain fatty acids above veinlets of microcrysts of magnetite converted through hydrothermal zone refinement over a matter of days from earlier precipitates of disordered magnetite and fougerite [38, 61, 88, 103–112]. The various polymers so produced are, in turn, subjected to further selection. The fougerite itself is a natural battery rechargeable at the vent from electrons stemming from hydrothermal H_2 [113–119]. To be compared with Branscomb and Russell [92] and Figure 2 in Cartwright and Russell [93]. Recast from [38] and [61].

2.2 How Does All This Affect the AVT?

From its very inception, the AVT seemed to eschew the above laid-out conflict with the second law since it was seen as closely emulating life's ubiquitous form of free energy conversion, chemiosmosis. One of the founding observations in favor of AVT was indeed the fact that the redox- and pH-gradients generated at the mineral-walled voids of the hydrothermal chimneys by the encounter of reduced, alkaline vent fluid with oxidized, mildly acidic ocean water, intriguingly resembled the situation of cellular life as it converts free energy via chemiosmosis [88] (Figure 2). Just as the cytoplasm of the majority of living cells is more reducing and slightly more alkaline in the majority of cases than its environment (however see alkalophilic bacteria [89]), it was argued that the interior of the mineral bubbles would be mainly fed by the alkaline fluid highly charged in the reductants H_2 and CH_4, while the exterior would be kept at lower pH and fed with oxidants (nitrogen oxyanions and CO_2) as a result of contact with ocean waters [46,68,81,88,90,91]. These perceived similarities suggested the possibility of a spontaneous transition from the natural, abiotic conditions of the alkaline hydrothermal vent to living entities, a hypothesis further detailed via the proposal that an abiotic version of the Wood-Ljungdahl pathway of energy conversion and CO_2-fixation may have smoothly morphed into the first type of biological metabolism [67,68]. The reduction of CO_2 by H_2 as part of the Wood-Ljungdahl pathway was perceived as an exergonic reaction scheme bound to occur abiotically and the biological take-over mainly implied speeding-up of individual reaction steps.

However, as pointed out above, the essence of a living system is not that it exergonically tends toward thermodynamic equilibrium but that it employs a process converting the free energy driving the overall exergonic reaction into a different sort of free energy and more specifically into a chemical disequilibrium (in extant life in the form of far-from-equilibrium ATP/ADP ratios) driving the extremely endergonic processes of anabolism and catabolism and enabling life's ordering processes to occur. In the older versions of the AVT, such free energy converting processes were viewed as a later invention by fledgling life whereas Schrödinger's analysis tells us that they ineluctably must be at the very basis of life's emergence. These versions of the AVT therefore also (just like Wächtershäuser's) are anabolism-first rather than energy-metabolism-first constructs and therefore fail the litmus-test of the second law.

2.3 ... And Then the Devil Also Lies in Some Details ...

The earliest formulations of the AVT suffered from a number of other problems and inconsistencies, some of which were obvious from the start but shrugged off, others pointed out more recently.

(a) While the environmental pH- and redox-gradients provided by the setting of alkaline hydrothermal vents (i.e., reducing, alkaline fluids emanating from the serpentinization of peridotitic rock at the base of the chimneys and flowing toward their tips versus mildly acidic, oxidizing ocean waters percolating into the chimneys from their flanks pulled by the Venturi-effect) will build up macroscopically over the mound, it is not obvious how this should translate into a microscopic gradient between inside and outside of individual bubbles. Instead, it is more likely that mineral compartments located further inwards will be exposed to increasingly reducing and alkaline fluid while those toward the periphery of the mound will encounter acidulous and oxidizing conditions (Figure 3). The transition between those extremes is expected to be smooth, and it is not clear how cell-like conditions might arise in individual compartments. Nevertheless, at certain locales in the mound, pH gradients may exist between the two sides of mineral barriers that are said to assist, for example, in CO_2-reducing reactions.

(b) In early versions of the AVT, the mineral walls of bubble-like voids in the mound were straightforwardly likened to the chemiosmotic barrier of a living cell. It has since been pointed out [124] that the dielectric properties of these two barriers differ fundamentally. The orders-of-magnitude greater thickness of mineral walls observed in alkaline hydrothermal chimneys as compared to a bioenergetic membrane implies a much shallower pH- and electric field-gradient in the mineral settings. Worse, these mineral barriers in general are semi-conducting, suggesting an extremely rapid dissipation of any potential electrostatic field induced by asymmetric redox reactions on different sides of the barrier. In response to this criticism, an early coating of the mineral bubbles' walls by either lipidic membranes [125,126] or polypeptides [127] was proposed.

(c) This inference of an early emergence of organics-based dielectric barriers entails a new conundrum. Whereas (uncharged) dissolved gases such as H_2, CO_2, and CH_4 readily diffuse through such barriers, the passage of cations and anions will become substantially impeded. The inner volume of the voids therefore cannot be replenished anymore with the high pH-fluid and so any imagined [68] pH-driven anabolic reactions will quickly stall due to equilibration of proton activities on both sides of the barrier. Nick Lane's group has therefore proposed that an enzyme system functioning analogously to the Rnf-complex may already very early on have actively maintained the proton gradient [130]. The author of the present article has, by contrast, resorted to considering the proton gradient as a secondary disequilibrium resulting from anisotropic oxidation and reduction reactions inside vs. outside of the dielectric barrier (see below).

So how might we rescue the original AVT from the onslaught of criticism from the very diverse angles detailed above? We hold that modifications injected into the initial AVT over the last decade do indeed respond to these criticisms and allow us to put forward a fully second-law-compliant scenario for life's emergence in submarine alkaline hydrothermal settings. These modifications were motivated by input from palaeo-geochemistry, mineralogy, and biology as will be detailed below, and

FIGURE 3 Sketch of a "community" of discrete polymeric protocells decorated with variable valence fougerite nanocrysts (the green, brown and red "laths" illustrate the redox trend) and mackinawite (yellow). These nanocrysts decorating the polymeric cell membrane margins act as disequilibria or free energy converters and conductors [112,122] (cf. Figures 2 and 6). From [64] and [123] with permission.

basically all gravitate around the consideration of a crucial role for the mineral fougerite. They have been individually laid out in a number of recent articles [87,123,127] while here I will try and propose the synthesis of all these modifications into an updated version of the AVT.

3 THE INVOLVED MINERALS: REVISITING THE INITIAL GEOCHEMICAL SETTING

As mentioned in Section 9.1, the inception of the AVT was inspired by our discovery of ~350 million-year-old fossil hydrothermal pyrite chimneys in a Pb-Zn-$BaSO_4$ open pit ore deposit, at and through which metals in acid solution must have been precipitated on meeting alkaline seawater bearing bacteriogenic sulfide [35]. We assumed the initial precipitates to have comprised mackinawite and greigite that only later transformed into pyrite [36,41–43]. These pH conditions were a reverse of those to be expected in the Hadean-Eoarchaean times where ocean water was carbonic while moderate temperature (~130 °C) hydrothermal solutions emanating from serpentinization of peridotitic ocean floor would have been alkaline. Yet I had carelessly assumed that sulfide would still dominate hydroxide in such solutions. However, when Randall Mielke and collaborators [52] attempted to mimic a serpentinization reaction expected to deliver alkaline fluid to a putative carbonic ocean, they failed to produce sufficient hydrothermal hydrogen bisulfide (HS^-) to generate FeS/Fe_3S_4 chimneys alone; the results speaking "against the possibility that the first compartments of life consisted purely of sulfides" [52]. Instead chimneys and compartments "would be composed of carbonate, green rust (GR), silicates, or silica itself" [52,131], though pure carbonate minerals were an unlikely primary precipitate in the Hadean [132].

Green rust is a double-layered iron(II/III)-oxyhydroxide studied in its synthetic form and as a corrosion product by mineralogists for more than half a century [103,133–137]. In 1997, green rust was found in natural environments and named as "fougerite" by its discoverer Fabienne Trolard and her collaborators [138]. Fougerite sensu lato [137] is now considered as the likely precursor to the magnetite comprising so much of the Hadean banded iron formation in Quebec and in the Eoarchean of Greenland [139–143]. This telling link with Banded Iron Formations (BIFs) – the oldest known chemical sediments on Earth – renders scenarios for the emergence of life limited only to iron sulfides also untenable (Figure 1).

Further evidence for an ineluctable role of green rust/fougerite comes from the analysis of modern-day analogs of the alkaline hydrothermal vents put forward as hatcheries of life in the Hadean. The Lost City Alkaline vents (pH 9–11 and ≤96 °C) discovered in the year 2000 at a depth of 750 meters in the northern North Atlantic [144,145] were the first empirical support for the AVT which previously was only deduced from our studies of sedimentary exhalative magnesite ($MgCO_3$) deposits in the Alpine serpentinite belts eventually published by Fallick and colleagues [39] and Zedef and collaborators [40]. The very first, though ephemeral precipitate at Lost City spires, is valence-invariant brucite – a Mg analogue (as ~$Mg[OH]_2$) of variable-valence fougerite ($Fe^{2+}{}_4Fe^{3+}{}_2(OH)_{12}CO_3{\cdot}3H_2O$). Brucite, only insoluble at high pH, is rapidly replaced by carbonate on meeting ocean waters [145,146]. Brucite is the type-structure for all the hydroxyl clay derivatives where cations fill trioctahedra [134,147]. Fougerite is just such a mineral, but one with the particular advantage, in this context, of wide valence tunability through the diadochic incorporation of various transition metal cations in the "brucite" layers facing the internal galleries [148]. Indeed, judging from the experiments of Helmbrecht and her collaborators [149] the ferrous iron homolog of brucite ($Fe[OH]_2$), i.e., white rust or amakinite, was likely the first precipitate, rapidly oxidized by ocean water to fougerite with the evolution of hydrogen. Under the conditions of an iron-rich Hadean ocean, fougerite/GR is brucite's equivalent, and it is almost inescapable to assume the chimneys in Hadean alkaline hydrothermal vents to have contained significant amounts of green rust, in line with results on artificial systems [52,99,131,149].

However, another modern analogue for the putative fougerite submarine alkaline vent theory is the discovery, in 2010, of the Shinkai Seep Field in the Southern Mariana Trench [150]. Although the alkaline fluid exhaling from serpentinizing peridotite there is cool (~2°C), the depths at 5,861 to 5,550 m, i.e., well below the carbon compensation depth of ~4,500 m, are more in line with expectations of fougerite submarine AVT-hosted, deep sea alkaline vents in the Hadean [99,132,151]. The chimneys so generated there comprise >80% of the layered magnesium hydroxide, brucite ($Mg[OH]_2$) [147,150,152]. The subsidiary carbonates, calcite and aragonite, are likely ephemeral as the vent system lies below the carbonate compensation depth [132,153] just as siderite is generally missing in comparable banded-iron formation sites in the earliest ocean.

Ryuhei Nakamura and colleagues [147,152,154–160] have made detailed explorations and examinations of such brucite chimneys. These studies culminated in the realization that the orientation of the layered brucites lay normal to alkaline flow

[147,152] and therefore could act as inorganic membranes. This research at extant springs, including those at black smokers, was onboarded to the RIKEN laboratories for examination and experimentation in a concerted effort to eventually mimic these natural geological, geochemical, and geophysical conditions to the point where deep understanding could inform emergence-of-life theory. Indeed Lee, H.-E. and her collaborators [147,152] demonstrate that brucite membranes preside over a natural osmotic energy conversion available at modern vents, thus adding to the portfolio of sustainable geo-electrochemical free energies hypothesized for GR/fougerite chimneys in the early ocean.

We therefore embraced GR /fougerite – albeit dosed with sulfide and trace metals – as the mineral without which life could not have begun [44,51,162]. We consider that a wide apron of GR /fougerite was precipitated around the alkaline vents in the Hadean across the ocean floor surviving as banded iron formations (BIFs) to the present day [62,163]. The inclusion of fougerite/GR in the specifics of the alkaline hydrothermal vent theory allows resolution of most of the aforementioned objections to this scenario for life's emergence. We call this "evolved" form of the AVT the "Fougerite-Alkaline Vent Theory" or FAVT.

4 GREEN RUST/FOUGERITE, EXPOSED

Green rust is a member of the class of layered minerals frequently termed the hydrotalcite supergroup [137]. In hydrotalcite-type minerals, a 2D lattice of mixed valence ($2^+/3^+$) metal cations, bridged by hydroxide groups, form the positively charged mineral layer. Stacks of these layers are charge-balanced by an aqueous layer of interstitial anions. These minerals feature a sandwiched alternation of positively charged metal-hydroxide lattices and negatively charged interlayer anions. They are therefore referred to as layered double hydroxides (LDHs) or occasionally also as double-layered hydroxides (DLHs), the latter term emphasizing the alternation between two distinct 2D leaves. Hydrotalcite proper contains Al^{3+} and Mg^{2+} as metal cations in the hydroxide layer. Green rust distinguishes itself from most members of the hydrotalcite group by the fact that its sole metal cation is the redox-active iron ion and that a range of ratios of Fe^{2+} to Fe^{3+} allows the formation of thermodynamically stable forms of the mineral. In the range of $1/3 < Fe^{3+}/Fe_{total} < 2/3$, GR is considered to keep its structural integrity for some time before the onset of degradation. Indeed, we speculate that rapid rhythmic coupled oscillations of redox states between 1/3 and 2/3 Fe^{3+}/Fe_{total} states will homeodynamically self-organize along the fougerite layers, perhaps in a way broadly comparable to that of the plant flavan-3-ol (±)-catechin complexes when coordinated with Fe^{2+} or Fe^{3+} [164–166].

Figure 4b and b' illustrates the case of equal amounts of Fe^{2+} and Fe^{3+}. The interlayer anion in the case of Figure 4b is nitrate, i.e., a "flat" molecule. Under these circumstances, the height of the interstitial galleries is roughly 8 Å. Other "2D"-type anions such as carbonate and chloride [169,170] yield comparable layer arrangements and the resulting mineral is called green rust 1 (GR1). In the presence of bulky (usually tetrahedral) anions such as sulfate or orthophosphate, two distinct layers of interstitial anions are incorporated and the gallery height increases to about 12

Å [Christian Ruby, personal communication 2024] (Figure 4b'). These forms are referred to as GR2 [103]. Incorporation of even bulkier anions has been observed and gallery heights of up to 40 Å have been described [171]. At higher oxidation states (i.e., $Fe^{3+} > 2/3\ Fe_{total}$), GR is observed to convert into magnetite or lepidocrocite on a timescale of hours. However, on shorter timescales (minutes), green rust (see Figure 4c) keeps its layered structure even at high oxidation states as evidenced by its XRD diffractogram [172]. The fully oxidized form is called "ferric" GR (Figure 4d). An analogous behavior is observed at high reduction states (i.e., $Fe^{3+}/Fe_{total} << 1/3$): over longer timescales, highly reduced GR dissolves while the layered GR structure persists long enough to even be characterized as a distinct mineral called amakinite or white rust [170] (Figure 4a). In strongly oxidized GR, the increasing charge imbalance is compensated by increasing deprotonation of the hydroxy-bridges (see Figure 4c and d), thereby rendering the layer less positively charged. This redox-softness afforded by protonation/deprotonation of the oxy-bridges is the reason why

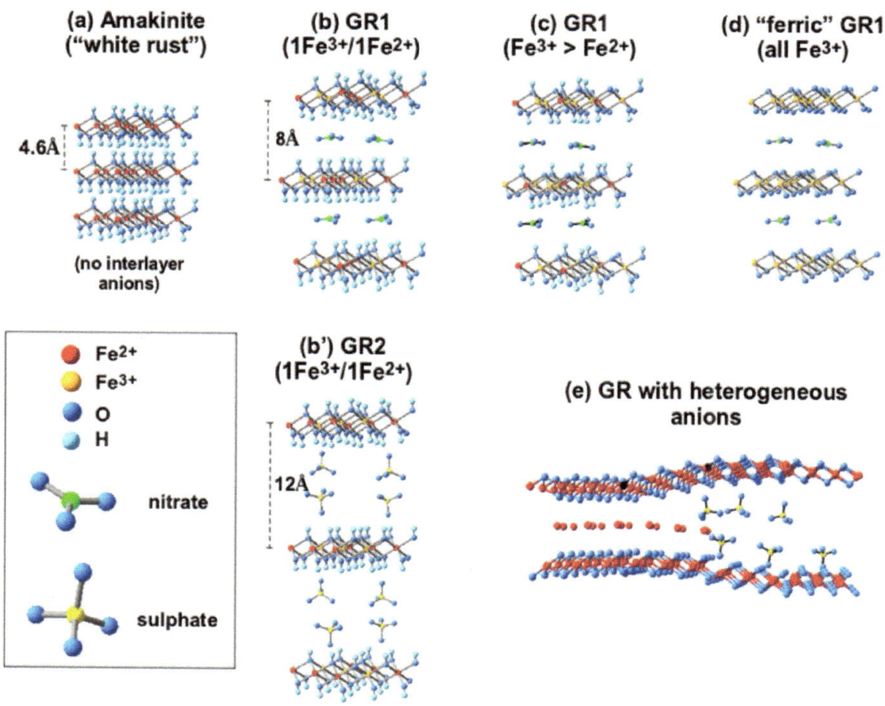

FIGURE 4 Iron hydroxide or white rust (cf. brucite) is the ephemeral mineral amakinite (a) with only a vestige of an interlayer. 4(b to e) are the GR1 oxyhydroxides that comprise equal amounts of Fe^{2+} and Fe^{3+} and host the planar molecule nitrate (and/or carbonate and chloride) in their 8 Å interlayer. 4(b') is host to the "doubled up" tetrahedral anions sulfate or phosphate in a 12 Å interlayer. 4(c) and (d) are more oxidized versions though with similar gallery heights. 4(e) demonstrates a green rust comprising both GR1- and GR2-type domains that show its potential to act as a peristaltic pump and to produce condensation pressure [167,168].

GR is commonly referred to as an Fe^{2+}/Fe^{3+} oxyhydroxide-type LDH. At very high reduction states, the interlayer is expelled since the global charge of the iron-hydroxide layer approaches zero. In this state, between-layer stabilization is achieved by interlayer hydrogen bridges. The stacking distance between layers shrinks to 4.6 Å (Figure 4a). The overall structure of highly reduced GR strongly resembles that of the abovementioned mineral brucite (Section 2). Brucite contains only magnesium (2+) as cation, and the metal-hydroxide layer is therefore charge-neutral and no interlayer anions are required for stabilization.

While the mineralogists tend to focus on "pure" (i.e., homogeneous) cases with well-defined crystallographic parameters to which they can attribute specific mineral names (e.g., amakinite for the fully reduced, brucite-like form), the juxtaposition in Figure 4 intends to emphasize that these pure forms are only extreme cases of a continuum of degrees of reduction (Fe^{2+}/Fe^{3+} ratios) and interlayer dimensions. Rather than obeying the intuitive notion of a mineral as a solid crystalline structure, green rust really is a soft, "breathing", adaptive structural entity highly sensitive and reactive to environmental conditions such as redox poise and the nature of solutes. Results by Guerbois and her colleagues [173] furthermore suggest that individual nanocrystals of GR can contain both GR1- and GR2-type domains (as illustrated in Figure 4e), a property which will be crucial to the present version of the FAVT scenario to be detailed in Section 7.

All the above-described forms of GR are synthetic minerals with well-defined types and ratios of all involved atoms. Fougerite is the GR-type mineral first found in natural settings by Fabienne Trolard [174] and is a much more messy entity. While it has been shown to feature all characteristic parameters of GR in XRD and Mössbauer spectroscopy, it contains variable amounts of other di- and trivalent metal ions such as nickel and aluminum as well as a range of distinct interstitial anions.

5 GREEN RUST AND ITS ANALOGIES TO BIOLOGICAL (AND IN PARTICULAR BIOENERGETIC) ENTITIES

5.1 GR Is a Structural and Functional Analog of Di-iron Hydroxylase Enzymes

Structural affinities between the catalytic metal clusters in certain enzymes and specific minerals (mainly greigite and mackinawite) have been pointed out in the past [46,162,175–178]. The ineluctable conclusion from geochemical evidences that GR must have been a major component of Hadean alkaline hydrothermal vents raised the question of whether comparable similarities also existed for the structural arrangement of atoms in the GR lattice. Indeed, the coordination pattern of GR's iron sites by μ-hydroxo bridges is strongly reminiscent of the basic structure of the di-iron centers in certain hydroxylases. Hydroxylases are enzymes transforming C-H functions in hydrocarbons into alcohols (C-OH) via activated oxygen, the latter being most frequently produced through reduction by the strong biological electron donor NADH (reduced nicotinamide adenine dinucleotide).

The recognition that methane represented a substantial ingredient in vent fluids in Lost City-type alkaline hydrothermal systems [179], together with the results of phylogenetic analyses suggesting methanotrophy to be an ancient bioenergetic energy-metabolism [86,162], drew attention to the initiating step of the methanotrophic pathway, i.e., the O_2-radical-mediated oxidation of CH_4 to CH_3OH:

$$CH_4 + O_2 + 2e^- + 2H^+ \rightleftharpoons CH_3OH + H_2O \quad (1)$$

Of course, this is a standard hydroxylation reaction and the enzyme performing this step in many methanotrophic prokaryotes dutifully is a member of the superfamily of di-iron hydroxylases [123,180, 181]. Guided by the structural similarities (Figure 5a), it was therefore proposed that GR may (among its many other catalytic capacities to be discussed below) also be able to perform the methane monooxygenase reaction [87,123]. While this proposal remained wild speculation for several years, Orion Farr and colleagues have recently shown that GR does indeed convert methane to methanol, methyl chloride, and di- and trichloromethane (as well as longer hydrocarbons) with significant yields at room temperature and at a pressure of 10 bars [111,182]. Future experiments need to test whether NO can replace molecular oxygen as hydroxylating agent to satisfy the predictions of Nitschke and Russell [91].

5.2 GR as a Chemiosmotic Device

Considering the features of GR detailed in Section 4, a tantalizing inference imposes itself: GR features a host of properties generally observed in chemiosmotic systems:

(a) Redox flexibility and electron transfer. GR can cycle through a continuum of redox states from all-Fe^{2+} (fully reduced) to all-Fe^{3+} (fully oxidized) (Figure 5b). Whereas the long-term (i.e., > hours) stability of the mineral is only maintained in the range of 1/3 to 2/3 Fe^{3+}, GR sustains full and repeated redox cycling for several minutes without loss of the layered structure [Ievgen Mazurenko and Simon Duval, personal communication]. In the range of mixed-valence states, GR is a semiconductor based on charge-hopping between ferrous and ferric sites [113]. This conductivity is dimensionally restrained with electron transfer rates within layers substantially exceeding those between layers [113].

(b) One of the most fundamental principles rendering chemiosmosis possible is the coupling of redox states and protonation/deprotonation equilibria [184]. Just as in bioenergetic redox centers, the increasing oxidation of GR is associated with decreasing p*K*-values of the proton on the μ-hydroxo-bridges (Figure 5b) eventually resulting in full deprotonation (μ-oxo-bridges) in ferric (i.e., fully oxidized) GR [103,135]. Inevitably, electron transfer through an iron-oxyhydroxy layer will entail concomitant modification of pK-values and therefore pull a proton along the electron transfer path.

(c) As seen in Figure 4, the thickness of the interstitial space is flexible and depends both on the average oxidation state of the Fe-oxyhydroxy layers

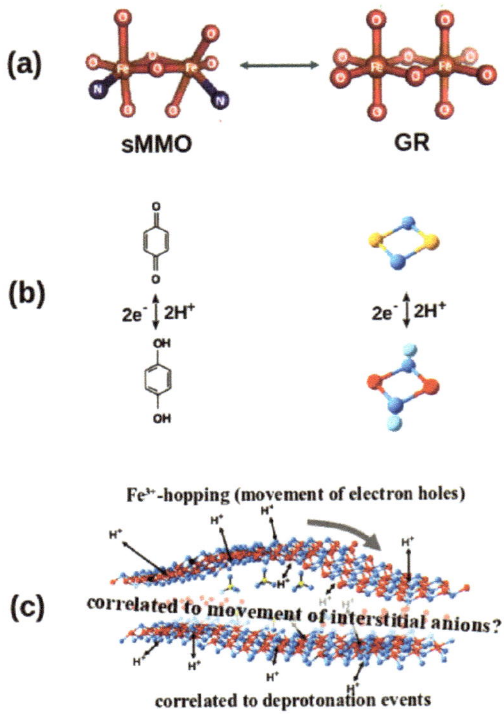

FIGURE 5 Comparisons of metal ion coordination between (a) green rust GR and soluble methane monooxygenase sMMO [111]; (b) quinone + $2e^-$ + $2H^+ \rightleftharpoons$ quinol, cf., $[Fe_2^{3+}O_2^{2-}]^{2-}$ + $2e^-$ + $2H^+ \rightleftharpoons [Fe_2^{2+}OH^-_2]^{2+}$; (c) Fe^{3+}-hole hopping argued here to be correlated with movement of interstitial anions accompanied by deprotonation events [113,183]. Noteworthy here is an evolutionary fit between GR and sMMO – an initial "first order" configuration of matter that persisted (and still does persist in the absence of a chelate) unless and until coopted by a kinetically favorable and more stable "second-order" protein configuration [96].

and on the steric nature of the (charge-compensating) interstitial anions. In fully reduced GR (Figure 4a), the effective charge of the Fe-hydroxy layers is basically nil and no interstitial anions are required for charge compensation. Consequently, the interstitial anions are expulsed and the height of the interstitial space shrinks to 4 Å. At higher oxidation states, however, these interstitial galleries will accommodate a wide variety of anions, and the height is mainly controlled by the 3D structures of these anions. Heights of galleries ranging from 8 Å ("point-like" or "flat" anions such as chloride, hydroxide, carbonate, nitrate as in GR1) through 12 Å (sulfate and likely phosphate GR2 [Christian Ruby, personal communication]) up to half a nanometer (when long chain hydrocarbons become sorbed [171]) have been observed. While so far only homogeneous cases have been analyzed, it is very likely that different types of anions can simultaneously be present in nanocrystalline GR, resulting in wavy features of the Fe-oxyhydroxy layers and allowing the formation of polaron-like structures as proposed previously [44,118,127,185]. Electron/proton migration within the layers may entail these polarons and result in redox-controlled transport of the encaged anions as suggested by the scheme in Figure 5c [44,114].

6 GREEN RUST: A CATALYST FOR ALL SEASONS

The idea that GR/fougerite was a requirement of AVT as a generator of organic molecules had been introduced in Russell et al. [186] as a contributor to "a natural fluidized- or flat-bed reactor with dialytic capability" underlying active submarine alkaline springs and chimneys. So, fully alerted to the requirement of GR demonstrated by Mielke et al, [52] we scoured the literature for further information. And there in plain sight were the extensive contributions of Hans Christian Bruun Hansen's group in Copenhagen, Denmark, as well as Jean-Marie Génin's in Nancy, France. In these mineralogy laboratories, a wealth of information on GR's catalytic versatilities has been accumulated over several decades. Even before this, Gustav Arrhenius' group had investigated [187,188] his "dysoxic environments" model for synthetic pathways to life involving GR as life's primordial catalyst [187,189–195]. However, the Arrhenius model is more given to explaining a path to a nucleotide world involving GR as a condensation catalyst of "prebiotic" organic molecules of uncertain origin [189] rather than as an enabler of the earliest pathways of both energy-metabolism and anabolism [44,51]. Incidentally, it should be mentioned that Huber and Wächtershäuser [196] had shown $Fe(OH)_2$ (white rust) to be just as effective as FeS in reductively aminating carboxylates to amino acids. In the following I will list some of the catalytic functionalities of GR most pertinent to the FAVT.

6.1 Nitrate Reduction to Ammonia

The anabolic synthesis of both amino acids and nucleotides requires the availability of reduced nitrogen compounds. GR has been shown to deliver the most reduced inorganic nitrogen compound, ammonia, directly from the most oxidized nitrogen oxyanion, nitrate. The remarkable 8-electron reduction of nitrate to ammonia with reduced GR (fougerite) (Fe^{2+}:Fe^{3+} = 3:1) between 15 and 50 °C and pH 7 to 8.5 is a "chemical reaction that competes with biochemistry" [103,137,197–203]. For example, with GR_{Cl} and GR_{carb} the following "sacrificial" reactions are recorded [103,198]:

$$Fe_{4.5}{}^{2+}Fe_{1.5}{}^{3+}(OH)_{12}Cl_{1.5} + {}^{5}/_{16}NO_3{}^{-} \rightleftharpoons 2Fe^{2+}Fe_2{}^{3+}O_4 + {}^{3}/_{2}Cl^{-} + {}^{5}/_{16}NH_4{}^{+} + {}^{14}/_{16}H^{+} + {}^{79}/_{16}H_2O \quad (2)$$

$$Fe_4{}^{2+}Fe_2{}^{3+}(OH)_{12}CO_3 + {}^{3}/_{8}NO_3{}^{-} + {}^{3}/_{4}H^{+} \rightleftharpoons 5FeOOH + Fe^{2+} + CO_3{}^{2-} + {}^{3}/_{8}NH_4{}^{+} + {}^{25}/_{8}H_2O \quad (3)$$

In these experiments the pH is held between 7 and 8.5 as the OH^{-} ions in the GR/fougerite are deprotonated during the reaction [103,135,136,198]. GR acts as a "sacrificial catalyst" in these reactions since it is oxidized to magnetite. Whether the reaction could proceed without the loss of GR/fougerite integrity is an invitation to further experimentation at alkaline-vent conditions, notionally:

$$3Fe_4^{2+}Fe_2^{3+}(OH)_{12}Cl_2 + NO_3^- \rightleftharpoons 3Fe_2^{2+}Fe_4^{3+}O_2(OH)_{10}Cl_2 + NH_4^+ + 2H^+ \quad (4)$$

implying that the electrons lost to production of ammonia are replenished from the vent fluid [204].

6.2 Nitrite Reduction to Produce Nitrogen Oxides and Reactive Intermediates

Guerbois and her collaborators [173] use fougerite/$GR_{carbonate}$ but replace the interlayer carbonate with nitrite to produce $GR_{nitrite}$. This is a new type of GR (fougerite) – different to, although structurally affine with, GR $1_{nitrate}$ [133]. Products at neutral pH are different too, featuring volatile NO, N_2O, and N_2, with NO as the intermediate as the reaction progresses to nitrous oxide [173,205,206], e.g.:

$$[Fe_3^{2+}Fe_3^{3+}(OH)_{10}(OOH)CO_3] + 3NO_2^- \rightleftharpoons 6FeOOH + 3NO + 2H^+ + HCO_3^- + 3H_2O \quad (5)$$

$$[Fe_3^{2+}Fe_3^{3+}(OH)_{10}(OOH)CO_3] + 3NO \rightleftharpoons 6FeOOH + 1.5N_2O + HCO_3^- + H^+ + 1.5H_2O \quad (6)$$

$$[Fe_3^{2+}Fe_3^{3+}(OH)_{10}(OOH)CO_3] + N_2O \rightleftharpoons 6FeOOH + 1.5N_2 + H^+ + HCO_3^- + 1.5H_2O \quad (7)$$

In this reaction cascade, the generation of NO is of particular interest to the FAVT, since NO may replace O_2 in the above-described hydroxylation reaction of methane to methanol [111] and furthermore assist in producing nitrogen-bearing aromatic molecules [179] (Figure 5).

6.3 CO_2-Reduction to Sugars, Acetone and Alcohols

Lee and collaborators [207] demonstrate the "sacrificial " reduction at STP of CO_2 in a 0.5 M $KHCO_3$ solution mixed with $GR_{carbonate}$ powder. At pH=7.7 products include acetate (CH_3COO^-) with minor 2-propanol ($CH_3CH_2CH_2OH$) and DMSO (dimethyl sulfoxide, $[CH_3]_2SO$) while at pH 9.5 acetate with minor acetone (CH_3COCH_3), ethanol (CH_3CH_2OH) and pyruvate ($CH_3COCOOH$) are generated. The $GR_{carbonate}$ dissolves during these reactions leading to some products significantly more reduced than those to be expected of the acetyl CoA and reverse TCA cycles as well as those produced by electrocatalytic over-potentials [207]. It is notable that no long-chain fatty acids are apparently produced in their system. However, Purvis and collaborators [112] have synthesized long-chain fatty acids up to length C18 in alkaline vent conditions (90 °C and 16 bar) over magnetite, a common product of fougerite alteration [110].

6.4 White Rust (Amakinite)/$GR_{CHLORIDE}$ as an Amine Transferase or Dehydrogenase to Aminate Pyruvate to Alanine and Glyoxylate to Glycine: The Path to Amino Acids

Huber and Wächtershäuser [196] have demonstrated the reductive amination of alpha-keto acids such as pyruvic acid ($CH_3COCOOH$) into alanine ($CH_3CH(NH_2)COOH$) in the presence of white rust ($Fe(OH)_2$) – a reaction also promoted by alkaline conditions and moderate temperatures (pH ~ 9.3, T = 50 °C):

$$CH_3COCOOH + NH_3 + 2Fe(OH)_2 \rightleftharpoons CH_3CH(NH_2)COOH + 2FeO(OH) + H_2O \quad (8)$$

Barge and her collaborators [208] also demonstrate the reductive amination of pyruvate to alanine at 70 °C within fougerite/GR_{Cl} sediment in alkaline conditions with Fe^{2+}:Fe^{3+} poised at 50%,

$$2CH_3COCOO^- + 2NH_3 + 2Fe_3^{2+}Fe_3^{3+}(OH)_{10}(OOH)Cl_2 \rightleftharpoons$$
$$2CH_3CH(NH_2)COO^- + 2Fe_2^{2+}Fe_4^{3+}O_2(OH)_{10}Cl_2 + H_2O \quad (9)$$

But in their chimney experiments only lactate was recovered, perhaps because of the short duration of the experiment and the resulting small size of the chimney. Alternatively alanine may have been immobilized in the fougerite interlayers [208]. It is incumbent upon us to repeat such chimney experiments as dispositive tests of the FAVT!

Barge and colleagues [209] also synthesized glycine (perhaps the first amino acid formed at the vent and the first to be coded [210]) through the reductive amination of glyoxylate, an entity possibly supplied via disproportionation of oxaloacetate, itself produced by the carboxylation of pyruvate [211,212]:

$$2(OHCCOO^-) + 2NH_3 + H_2 + 2Fe_3^{2+}Fe_3^{3+}(OH)_{10}(OOH)Cl_2 \rightleftharpoons$$
$$2NH_2CH_2COO^- + 2Fe_2^{2+}Fe_4^{3+}O_2(OH)_{10}Cl_2 + 2H_2O \quad (10)$$

Also, notwithstanding the difficulty of its synthesis, Grégoire and collaborators [213] show the introduction of aspartate into a fougerite analog, a chloride-bearing MgAl double-layer oxyhydroxide, in alkaline conditions where it readily condenses to di-aspartate. And aspartate is the amino acid from which other "early" amino acids – asparagine, methionine, threonine, lysine and isoleucine – are derived [214].

A further aspect to consider regarding the synthesis of amino acids within the fougerite interlayers is their effect in the weakening of the electrostatic attraction between the layers that can lead to delamination [215–217], a mechanism enhanced were they to be condensed there to peptides [218–220] as discussed next.

6.5 From Amino Acids to Peptides?

A critical support for the FAVT would be a demonstration of amino acid condensation to functional biomimetic peptides within hydrous fougerite [75,221–224]. Of course, a substantial lowering of water activity might be expected to drive the condensation of amino acid monomers to peptides. Indeed, as in the general context of sprayed or electro-sprayed microdroplets in air and using either phosphoric acid or oxazolidinone ions as condensing agents, both Wang and collaborators [225] and Holden et al. [226] demonstrate dipeptide formation. However, Muñoz-Santiburcio and Marx [218] caution that such soft interfaces are not comparable to conditions to be found between the relatively hard interfaces within the 001 cleavage planes of layered minerals where water activity, its acidity and the degrees of freedom are all low. In this respect Muñoz-Santiburcio and Marx [219] make the theoretical case, using ab initio molecular and metadynamic techniques, for the abiotic dimerization of glycine within the 001 interlayers of 2D mackinawite (~FeS). Confined water conditions are compared between those using (i) unprimed glycines, (ii) glycine zwitterions and (iii) glycine with N-carboxyanhydride, and all are found to result in dimerization. These authors assume extrapolation to polypeptide formation would be favored in these same circumstances. Muñoz-Santiburcio and Marx [219] also remark on the physical similarities to be expected within fougerite interlayers. Hansma [220], though favoring a mica interlayer as substrate, reaches similar conclusions and predicts that alanine would be condensed to di- and tri-alanine and through to polyalanine – an idea awaiting experimental test.

Using iron, nickel sulfide surfaces, CO as condensation agent and methane thiol or H_2S as catalyst at 100°C and a pH of ~9.1, Huber and Wächtershäuser [227] have generated around 10% of glycylglycine and phenylalanylphenylalanine from their respective monomers. For best results solutions were buffered with brucite $Mg(OH)_2$ [196], a common precipitate at modern submarine alkaline vents [144,147,152,153,228,229]. For evidence of more effective condensation of amino acids, Ferris et al. [230] have demonstrated the polymerization to peptides of up to 50 amino acid monomers on the Mg-rich smectite clay montmorillonite and consider the potential length to be "enormous". Of course, the greater the length, the more firmly the polymer would be surface-bound. To be functional the peptide would need to be prized off such a surface or off the internal surfaces of a smectic mineral revealed through delamination [231]. Zhai and collaborators [232] demonstrate how, in alkaline conditions, HPO_4^{2-} biomimetically detaches the hexapeptide SGAGKT from a mica surface to form SGAGKT-HPO_4^{2-} in a classical P(phosphate)-loop, a loop that fortuitously discriminates against toxic arsenate.

Such generations of uncoded polymers would provide peptides to supply a portion of the first organic coats to charged native inorganic nanocrystals, anions and cations. They might even have ended up generating the first dielectric barriers initiating chemiosmosis-like processes as discussed below (Section 7). Even the single amino acid has some use [233] while di- and tri-peptides have some functionality [234–236]. And 5-mer to 10-mer peptides would give the flexibility required for earliest metabolic functions of protoenzymes. In support of the first stage of such a sequence of events achiral and heterochiral structures have been shown to sequester phosphate in natural conditions, an analog to the P-loop in the tri- and pyro-phosphatases [223,237]. Beyond pyrophosphate, 5-mer peptides are also able to sequester a variety of other anions as well as certain cations in such "peptide nests" [221,237,238]. In this context, the more relevant cations $Ni^{1+/2+}$, $Co^{1+/2+}$ and Zn^{2+} also form such complexes spontaneously in alkaline solution through the displacement of protons from the amide nitrogens comprising the main chain of peptide. The "ATCUN motif" is one such, named [234] for the fact that the amino terminal of the 4-mer peptide was seen to sequester either copper or nickel in the core of its flat structure much in the way of corrinoids [239]. Thus these complexes also constitute likely precursors to the active centers of certain metalloenzymes. For example, we note that the acetyl coenzyme-A pathway [23, 66] involves a tetrapyrrole in the corrinoid ring with a similar electronic structure [221,222,235,236]. The iron-sulfur centers $[Fe_2S_2](RS)_4$, $[Fe_3S_4](RS)_3$ and $[Fe_4S_4](RS)_4$ could also be bound in such loops [75].

However, the 50-mer plus "Ferris peptides" would likely have further functionalities. Judging from Alva and colleagues' [240] comparative studies and Romero Romero and collaborators' [241] computations, some 55-mer peptides could bind a variety of phosphate-containing ligands as well as Fe-S, and even RNA and single-strand DNA, perhaps in multiple P-loops. Moreover, given certain sequences, some peptides may have been self-replicating [242].

7 OUTLINE OF THE PRESENT STATUS OF THE FAVT

Taking into account all the points of criticism listed in Section 2 and taking full advantage of the physicochemical/electrochemical/synthetic properties of fougerite/green rust as detailed in Sections 4, 5 and 6, the following scenario is proposed: Already at the hot base of the mound, simple organic molecules are produced by mineral-mediated reactions [112,243,244]. The involved minerals are likely mainly fougerite [111,152,245] and magnetite [110–112,245] but with associated mackinawite and greigite [175,186,246,247]. These organic molecules may be condensed [248,249] to more complex structures such as peptides while relatively intricate organics (e.g., heteroatomic aromatics) may be synthesized early on in fougerite interlayers [123]. The continuous upward flow of the alkaline fluid will distribute these organics throughout the mound (Figure 2) where they will accumulate due to adsorption on mineral surfaces and eventually produce an organic coating to the internal voids of the mound [112,123] (Figures 3, 6). Many of these "organic bubbles" will end up as topologically insulated voids in which "inside" and "outside" are separated by barriers featuring the dielectric properties of proteinaceous material, that is, resembling the situation in extant living cells [76]. However, it needs to be emphasized that so far there is nothing remotely life-like to be found in such structures (cf. [250]).

Following Schrödinger's insight and the layout of the chemiosmotic mechanism in extant life [76], one is forced to search for disequilibria-converting processes exploiting the presence of these dielectric barriers and how they might convert redox tensions into phosphate/polyphosphate disequilibria, just as in extant life

[44,153,251,252]. The nature of the redox tensions is straightforward given the composition of the reducing fluids on the inside resulting from serpentinization and of the surrounding ocean waters on the outside (Figures 2 and 6). The alkaline vent fluids are highly charged in the reductants H_2 and CH_4. As dissolved (electrically uncharged) gases, both dihydrogen and methane can pass through the described dielectric barrier essentially unimpeded (Figure 6). The ocean waters with their high concentrations of dissolved CO_2 and carbonate together with likely significant amounts of nitrogen oxyanions [253–255] (mainly nitrate [256] as nitrite will be readily further reduced by the ubiquitous Fe^{2+} in kinetically uninhibited 1-electron reactions). These charged nitrogen oxyanions as well as carbonate cannot cross the dielectric barrier and will therefore remain excluded from the internal spaces (Figure 6). However, the very weak oxidant CO_2 can and may thus be available as a carbon source for organosynthetic processes within the pre-cellular structures. The diffusional barrier to the strong oxidants together with permeability toward the reduced gases straightforwardly produces a redox disequilibrium over the barrier similar to that encountered in extant chemiosmotic cellular life.

Certain nano-crystalline minerals such as fougerite and mackinawite will be entrapped within the pre-cellular voids and will oxidize the reducing gases to produce electrons. The salient feature of the FAVT is the prediction that a number of fougerite along with subordinate mackinawite nano-crystals (possibly beforehand coated in organic material) will integrate into the dielectric barrier (Figures 3 and 6) and serve as conductors permitting the electric gradient to be dissipated.

Judging from experimental analogues [152] the organic monomers and polymers would have been produced in the fougerite interlayers in a process perhaps leading to some delamination and also from amphiphiles synthesized on magnetite [112,131, 208,209,215–217]. Desorption of these organic polymers would be partly triggered by the vibrations accompanying hydrothermal flow and partly by zone refining. Thus freed, these amphiphiles were entrained in rising hydrothermal solutions to be redistributed and aggregated in fougerite garden tubes developing higher in the mound where, given the higher ratio of oxidized ocean water to highly reduced hydrothermal solution emanating from below, reductive stress would have been more limited [38,75,88,104,166,186,221–223,257–261]. Were nucleotides to be also generated in fougerite interlayers, they would tend to be trapped by, and attach to, the amyloid agglomerations [112,130,149,259,260,262–269].

However, this dissipation (i.e., electron transfer from inside to outside) will be accompanied by a transfer of protons (Figure 6) as a result of the electron/proton-coupling properties of green rust discussed in Sections 4 and 5.2. This process perfectly mimics the first step of the chemiosmotic mechanism, that is, conversion of a redox gradient into a cation-concentration and electrostatic gradient across the membrane.

As previously speculated [44,127] and as shown in Figure 4b, the reported properties of GR-fougerite strongly suggest the possibility that the phosphate/pyrophosphate equilibrium of bulk water will be displaced to higher amounts of pyrophosphate in the interstitial galleries of fougerite as a result of the much lower water activity in these confined spaces. And as furthermore proposed by Branscomb

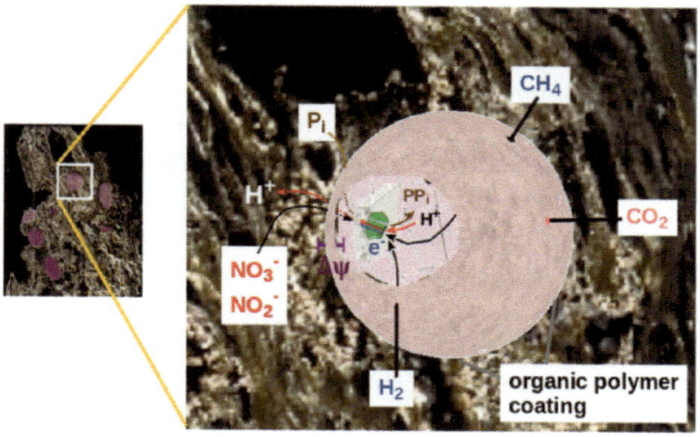

FIGURE 6 Illustration of a hypothetical primeval cell generated within ramifying chemical garden tubules and bubbles with organic coating (Farr et al. [182]). Comparable structures have been realized in several papers over the last decades [128,129,233–235,250–256]. In this case it is superimposed over a brucite garden produced at the Shinkai seep field [147,153]. The putative cell is a dielectrically insulated organic compartment that hosts fougerite nanocrysts protruding on both sides of the barrier (Figure 3) [76,111,117,128–131]. Hydrogen and CH_4 are introduced from the alkaline hydrothermal fluids while inorganic monophosphate P_i, CO_2, NO^-_3 and NO^-_2 derive from the acidulous ocean as in Figure 3 [91]. Oxides are reduced as reductants are oxidized while phosphate is condensed in the interlayers of fougerites acting as free energy converters. Note that the electrons and protons are transported through the fougerite layers [87,127]. The underlying micrograph of a brucite garden was kindly provided by Ryuhei Nakamura and Hye-Eun Lee.

and Russell [270], these pyrophosphate anions with their higher overall negative charge may be selectively and vectorially entrained by electron-hole transfers within the Fe-oxyhydroxy layers and thus transported within GR2-like polaronic structures (Figure 5c) toward the interior of the pre-cellular structures (Figure 6). Thereby the two-stage disequilibrium conversion of chemiosmosis resulting from the redox tensions through the membrane potential will be attained to finally produce phosphate/pyrophosphate-based chemical disequilibria (Figure 6).

While the impetus for designing the above scenario lay in the observation that a number of the firmly established properties of GR nanocrystals are reminiscent of reaction schemes in the chemiosmotic process, other properties of GR crucial to the validity of this proposal require future investigation. Some are likely but as yet undemonstrated whereas others are plausible but unexplored. In the following section we will briefly list the most pressing experimental questions to be addressed in the framework of the FAVT.

8 EXPERIMENTATION AND TESTABLE PREDICTIONS

Conductivity: The complete transformation of GR into magnetite, lepidocrocite and others via oxidative attack from the nanocrystals' edges shows that charge transfer within a single crystal must necessarily exist. Hole-hopping within the iron-oxyhydroxy layers has been argued to rationalize these kinds of charge transfer and the very, to extremely high, rates for this transfer have been predicted based on ab initio molecular orbital calculations on idealized systems containing single Fe^{3+} sites in a bath of Fe^{2+} layers [113]. However, in actualistic GR nanocrystals, conductivity certainly will strongly depend on a number of intrinsic parameters such as degree of reduction (i.e., Fe^{2+}/Fe^{3+} ratio), type of interstitial anions and pH (influencing the tendency of hydroxide bridges to deprotonate as addressed below). Empirical data with respect to these parameters are sorely missing.

Effective (bulk) pK values of the hydroxy-bridges: Mineralogical investigations of "ferric GR" (Figure 4d) have found that the hydroxy-bridges deprotonate in the presence of very high Fe^{3+}/Fe^{2+} ratios. This demonstrates that increasing the degree of oxidation of the iron sites must decrease the p*K* values of the protons on these bridges, as is to be expected for any redox-linked protonatable site [183]. The absolute (at a certain degree of GR oxidation) and relative (i.e., strength of dependence on changes in degree of oxidation) p*K* values are presently unknown but are of paramount importance to assess the potential degree of coupling of electron- to proton-transfer through an individual GR nanocrystal.

Condensation of phosphates to polyphosphates: The mass-action principle demands that the P_i to PP_i equilibrium be displaced (toward higher amounts of PP_i) in the low-water-activity interstitial galleries with respect to bulk water. While this effect appears ineluctable, it has not yet been demonstrated and is even less quantified in GR (or other LDHs). Thus, respective analyses would be extremely helpful for assessing the scenario of Section 7. However, to effectively generate far-from-equilibrium ratios of PP_i to P_i, the condensed polyphosphates need to be (more or less) selectively transported to the interior of the voids bounded by dielectric barriers as shown in Figure 6. Such selective transport would occur if hopping of the positive hole located on the Fe^{3+} sites pulled the interstitial anions (*nota bene*: the pyrophosphate anion is more highly negatively charged than P_i at a given pH) along and therefore ultimately deliver them to the bubbles' interior. Once again, this prediction is plausible but requires experimental validation.

The above-listed experiments are certainly easily defined on paper, but in the lab, GR is obviously blighted by redox instability severely hampering experimental approaches. A remedy at least concerning the physical aspects (e.g., quantifying the displacement of the P_i/PP_i ratio) may consist in turning to the mixed-valence hydrotalcite ($[Mg_6{}^{2+}AI_2{}^{3+}(OH)_{16}][CO_3.4H_2O]$) as a stable structural analog – stable because this 2D mineral is immune to both oxidation and reduction [193]. Moreover, the carbonate form of hydrotalcite is exceptionally stable because of the high affinity of carbonate to the magnesium ion [271]. Another solution to this issue of GR's susceptibility to rapidly oxidize may have been resolved for certain experimental requirements by Tahawy and colleagues [272] who synthesized a super-stable GR using a one-pot solvothermal reaction using ferric chloride and glycerol [273]. Finally, the electrochemical generation of GR/fougerite also makes for stability [274]. While electrochemical and/or microfluidic approaches have shown their worth in sulfide and silicate membrane experimentation [120,121,157,204,275–282], to date only Lee and collaborators [207] and Helmbrecht and colleagues [149] have undertaken similar experiments with GR.

9 CONCLUSIONS

With the industrial revolution/evolution as a guide we already know that to function life needs – always needed – appropriately convertible free energies, particular forms of materials to hand, toolboxes, construction kits, switches, engines and waste disposal facilities [5,46,76–80,277–290] – always with Erwin Schrödinger's proviso ([77] page 84) that all purely mechanical events seem to follow distinctly and directly the "order-from-order" principle. To make such analogies appropriate to scale Stephen Mann directs us to see "the living cell as an integrated self-regulating machine run by nanoscale miniaturization", and "that this specific level of miniaturization is critical for the emergence and sustainability of the cell" ([291] p. 5308). But how to get started?

The AVT from its very inception solved the problem of the environmental free energies that needed to be converted into the low entropy condition characterizing life. The redox gradients resulting from oxidizing, acidulous ocean waters encountering the strongly reduced, alkaline hydrothermal fluids produced by serpentinization provided a type of Gibbs free energy intriguingly similar to that driving extant life on planet Earth. However, the second, equally important part of Schrödinger's message, that is, the necessity for a system converting the environmental disequilibrium into life's ordering, was not dealt with in the previous versions of the AVT and was relegated to later, more evolutionarily evolved, steps toward life as we find it today. Instead, the catalytic properties of minerals present in the hydrothermal mound were considered to fundamentally solve the problem of the inanimate-to-life transition by speeding up exergonic reactions resembling anabolic processes in certain extant prokaryotes. As detailed in Section 2.1, this view is thermodynamically untenable. Including the mineral fougerite (green rust) and analogues thereof [292,293] into the scenario provides the toolbox [184], more precisely the "engine" [270], theoretically able to convert environmental disequilibria into chemical disequilibria, right from the beginning [76]. In this contribution I have provided a detailed version of a scenario (the "Fougerite-AVT") that tries to remedy previous criticism of the AVT and aims at a full observance of Schrödinger's strictures. All aspects of this scenario are experimentally testable, and I have tried to lay out directions for future research.

Of course, the really hard question remains open; how does the system, initially well-governed by functional fougerite segue, through the processes of selection, to an algorithmically dictated digital system capable of taking each and every entropically steeper step involved in persistent evolution [87,88,91,286–289]. In this regard at least we find solace in paraphrasing Darwin [299, p. 514] "It is no valid objection to the theory of fougerite functionality at a submarine alkaline vent that science as yet throws no light on the far higher problem of the … origin of the genetic code".

ACKNOWLEDGMENTS

Wolfgang Nitschke and Elbert Branscomb were my constant advisors, guides and companions in this venture. I am also most grateful to Simon Duval, Orion Farr, Nil Gaudu, Ryuhei Nakamura and Julyan Cartwright for their experimental insights and discussions, and thank John Allen, Stuart Bartlett, Jason Feldman, Elizabeth Jagger, Adrian Ponce and Michael Wong for keeping my mind well stirred.

ABBREVIATIONS

Å	Ångström unit
ADP	adenosine di-phosphate
ATP	adenosine tri-phosphate
AVT	alkaline hydrothermal vent theory
CoA	coenzyme A
DLH	double-layered hydroxide
FAVT	fougerite alkaline hydrothermal vent theory
GR	Green Rust
$GR_{Cl/carb}$	Green Rust with chloride/carbonate anions in the interlayer
LDH	layered double hydroxide
NADH	reduced nicotinamide adenine dinucleotide
P_i	inorganic phosphate
PP_i	pyrophosphate
RIKEN	Institute of Physical and Chemical Research/Japan
STP	standard temperature and pressure
TCA	tricarboxylic acid cycle
XRD	X-ray diffraction

REFERENCES

1. D. D. Sasselov, J. P. Grotzinger, J. D. Sutherland, *Sci. Adv.* **2022**, *6*, eaax3419.
2. M. Ferus, P. Rimmer, G. Cassone, A. Knížek, S. Civiš, S. J. E. Šponer, H. Saeidfirozeh, P. Kubelík, R. Dudžák, L. Petera, L. Juha, A. Pastorek, A. Křivková, M. Krůs, *Astrobiology* **2020**, *20*, 1476–1488.
3. K. Kobayashi, J.-i. Ise, R. Aoki, M. Kinoshita, K. Naito, T. Udo, B. Kunwar, J.-i. Takahashi, H. Shibata, H. Mita, H. Fukuda, Y. Oguri, K. Kawamura, Y. Kebukawa, V. S. Airapetian, *Life* **2023**, *13*, 1103.
4. Y. Oba, T. Koga, Y. Takano, N. O. Ogawa, N. Ohkouchi, K. Sasaki, H. Sato, D. P. Glavin, J. P. Dworkin, H. Naraoka, S. Tachibana, H. Yurimoto, T. Nakamura, T. Noguchi, R. Okazaki, H. Yabuta, K, Sakamoto, T. Yada, M. Nishimura, A. Nakato, A. Miyazaki, K. Yogata, M. Abe, T. Okada, T. Usui, M. Yoshikawa, T. Saiki, S. Tanaka, F. Terui, S. Nakazawa, S.-i. Watanabe, Y. Tsuda, *Nat. Commun.* **2023**, *14*, 1292.
5. F. G. Donnan, *J. Soc. Dyers Colour* **1929**, *45*, 1387–1398.
6. S. Leduc, *Mechanism of Life,* Rebman Ltd, 1911.

7. G. Horneck, N. Walter, F. Westall, J. L. Grenfell, W. F. Martin, F. Gomez, S. Leuko, N. Lee, S. Onofri, K. Tsiganis, R. Saladino, E. Pilat-Lohinger, E. Palomba, J. Harrison, F. Rull, C. Muller, G. Strazzulla, J. R. Brucato, P. Rettberg, M. T. Capria, *Astrobiology* **2016**, *16*, 201–243.
8. H. T. De La Beche, *Researches in theoretical geology*, FJ Huntington & Company, 1837.
9. F. Darwin, *The life and letters of Charles Darwin*, Vol. 3, John Murray, London/UK, 1888.
10. W. W. Rubey, *Geol. Soc. Am. Bull.* **1951**, *62*, 1111–1148.
11. D. C. Catling, K. J. Zahnle, *Sci. Adv.* **2020**, *6*, eaax1420.
12. H. C. Urey, *Proc. Natl. Acad. Sci.* USA **1952**. *38*, 351–363.
13. V. M. Goldschmidt, *New Biology* **1952**, *12*, 97–105.
14. J. D. Bernal, in *Aspects of the Origin of Life,* Ed M. Florkin, Pergamon Press, New York, 1960, pp. 30–45.
15. F. N. Spiess, K. C. Macdonald, T. Atwater, R. Ballard, A. Carranza, D. Cordoba, C. Cox, V. M. Diaz Garcia, J. Francheteau, J. Guerrero, J. Hawkins, R. Haymon, R. Hessler, T. Juteau, M. Kastner, R. Larson, B. Luyendyk, J. D. Macdougall, S. Miller, W. Normark, J. Orcutt, C. Rangin, *Science* **1980**, *207,* 1421–1433.
16. R. Hekinian, M. Fevrier, J. L. Bischoff, P. Picot, W. C. Shanks, *Science* **1980**, *207,* 1433–1444.
17. J. B. Corliss, J. A. Baross, S. E. Hoffman, *Corvallis, Oregon State University, School of Oceanography, Spec. Pub.* **1980**, 80–87.
18. J. B. Corliss, J. A. Baross, S. E. Hoffman, *Oceanologica Acta* (Special issue) **1981**, 59–69.
19. O. Kandler, K. O. Stetter, *Zentralbl. Bakteriol. Parasitenkd. Infektionskr. Hyg. Abt. 1 Orig. Teil* C **1981**, *2,* 111–121.
20. F. Fischer, W. Zillig, K. O. Stetter, G. Schreiber, *Nature* **1983**, *301*, 511–513.
21. G. Fuchs, in *Autotrophic bacteria,* Eds H. G. Schlegel, B. Bowien, Science Tech Publishers, Madison/WI 1989, pp. 365–382.
22. H. W. Jannasch, M. J. Mottl, *Science* **1985**, *229*, 717–725.
23. H. G. Wood, *FASEB J.* **1991**, *5*, 156–163.
24. S. W. Ragsdale, *Crit. Rev. Biochem. Mol. Biol.* **1991**, *26*, 261–300.
25. L. G. Ljungdahl, in *Acetogenesis*, Ed. H. L. Drake, Chapman and Hall, New York, 1994, pp. 63–87.
26. K. O. Stetter, *Nature* **1982**, 300, 258–260.
27. K. O. Stetter, *Syst. Appl. Microbiol.* **1988**, *10*, 172–173.
28. K. O. Stetter, H. König, E. Stackebrandt, *Syst. Appl. Microbiol.* **1983**, *4*, 535–551.
29. K. O., Stetter, G. Lauerer, M. Thomm, A. Neuner, *Science* **1987**, *236*, 822–824.
30. G. Wächtershäuser, *Microbiol. Rev.* **1988**, *52*, 452–484.
31. G. Wächtershäuser, *System. Appl. Microbiol.* **1988**, *10*, 207–210.
32. G. Wächtershäuser, *Orig. Life Evol. Biosph.* **1990**, *20*, 173–176.
33. G. Wächtershäuser, *Prog. Biophys. Mol. Biol.* **1992**, *58*, 85–201.
34. R. C. L. Larter, A. J. Boyce, M. J. Russell, *Miner Depos.* **1981**, *16*, 309–318.
35. A. J. Boyce, J. L. Coleman, M. J. Russell, *Nature* **1983**, *306*, 545–550.
36. M. J. Russell, in *Proc. Seventh IAGOD Symposium,* Ed E. Zachrisson, Schweizerbartsche Verlagsbuchhandlung, 1988, pp.183–190.
37. M. J. Russell, A. J. Hall, A. G. Cairns-Smith, P. S. Braterman, *Nature* **1988**, *336*, 117.
38. M. J. Russell, in *Irish-type deposits around the world*, Eds C. J. Andrew, M. W. Hitzman, G. Stanley, *Irish Assoc. Econ. Geol. Dublin* 2023, 45–94.
39. A. E. Fallick, M. Ilich, M. J. Russell, *Econ. Geol.* **1991**, *86*, 847–861.
40. V. Zedef. M. J. Russell, A. J. Hall, A. E. Fallick, *Econ. Geol.* **2000**, *95*, 429–446.

41. M. J. Russell, A. J. Hall, D. Turner, *Terra Nova* **1989**, *1*, 238–241.
42. M. J. Russell, A. J. Hall, A. P. Gize, *Nature* **1990**, *344*, 387.
43. M. J. Russell, R. M. Daniel, A. J. Hall, *Terra Nova* **1993**, *5*, 343–347.
44. M. J. Russell, W. Nitschke, E. Branscomb, *Philos. Trans. R. Soc. Lond. B Biol. Sci.* **2013**, *368*, 20120254.
45. Y. W. Yamagata, H. Watanabe, M. Saitoh, T. Namba, *Nature* **1991**, *352*, 516–519.
46. M. J. Russell, A. J. Hall, *J. Geol. Soc. Lond.* **1997**, *154*, 377–402.
47. M. J. Russell, A. J. Hall, *Geol. Soc. America Memoir* **2006**, *198*, 1–32.
48. J. O'Neil, R. Carlson, J. Paquette, D. Francis, *Precamb. Res.* **2012**, *20–21,* 23–44.
49. M. J. Russell, A. Ponce, *Life* **2020**, *10*, 291.
50. I. N. Bindeman, J. O'Neil, *Earth Planet. Sci. Lett.* **2022**, *586*, 117539
51. M. J. Russell, L. M. Barge, R. Bhartia, D. Bocanegra, P. J. Bracher, E. Branscomb, R. Kidd, S. E. McGlynn, D. H. Meier, W. Nitschke, T. Shibuya, S. Vance, L. M. White, I. Kanik, *Astrobiology* **2014**, *14*, 308–343.
52. R. E. Mielke, M. J. Russell, P. R. Wilson, S. McGlynn, M. Coleman, R. Kidd, I. Kanik, *Astrobiology* **2010**, *10*, 799–810.
53. L. M. White, T. Shibuya, S. D. Vance, L. E. Christensen, R. Bhartia, R. Kidd, A. Hoffmann, G. D. Stucky, I. Kanik, M. J. Russell, *Astrobiology* **2020**, *20*, 307–326.
54. D. Papineau, Z. She, M. S. Dodd, F. Iacoviello, J. F. Slack, E. Hauri, P. Shearing, C. T. S. Little, *Sci. Adv.* **2022**, *8,* eabm2296.
55. G. Macleod, C. Mckeown, A. J. Hall, M. J. Russell, *Orig. Life Evol. Biosph.* **1994**, *2,* 19–41.
56. E. L. Shock, *Orig. Life Evol. Biosph.* **1992**, *22*, 67–107.
57. S. A. Marakushev, O. V. Belonogova, *Biol. Bull.* **2021**, *48,* 519–529.
58. H. Ueda, T. Shibuya, *Fortschr. Mineral.* **2021**, *11*, 389.
59. S. S. S. Cardoso, J. H. E. Cartwright, *Proc. R. Soc. A* **2017**, *473*, 20170387.
60. T. Shibuya, , T. Komiya, , K. Nakamura, K. Takai, S. Maruyama, *Precambrian Res.* **2010**, *182*, 230–238.
61. M. J. Russell, N. T. Arndt, *Biogeosciences* **2005**, *2*, 97–111.
62. T. Fujikawa, T. Sasamoto, F. Zhao, A. Yamagishi, S. Akanuma, *Sci. Rep.* **2024**, *14*, 398.
63. E. R. Moody, S. Alvarez-Carretero, T. Mahendrarajah, J. W. Clark, H. Betts, N. Dombrowski, L. L. Szánthó, R. A. Boyle, S. Daines, X. Chen, N. Lane, Z. Yang, G. A. Shields, G. J. Szöllősi, A. Spang, D. Pisani, T. A. Williams, T. M. Lenton, P. C. J. Donoghue, *Nat. Ecol. Evol.* **2024**. https://doi.org/10.1038/s41559-024-02461-1.
64. M. J. Russell, *Front. Microbiol.* **2023**, *14*, 1083.
65. A. E. Fallick, J. H. Ashton, A. J. Boyce, R. M. Ellam, M. J. Russell, *Econ. Geol.* **2001**, *96*, 885–890.
66. D. Schwartzman, C. H. Lineweaver, *Temperature, Biogenesis, and Biospheric Self Organization,* Springer Berlin Heidelberg, 2005, pp. 207–221.
67. M. J. Russell, W. Martin, *Trends Biochem. Sci.* **2004**, *24*, 358–363.
68. W. Martin, M. J. Russell, *Phil. Trans. R. Soc. B* **2007**, *362,* 1887–1926.
69. A. G. Cairns-Smith, A. J. Hall, M. J. Russell, "Mineral theories of the origin of life and an iron sulfide example", *Marine Hydrothermal Systems and the Origin of Life: Report of SCOR Working Group 91*, Springer, Dordrecht/Netherlands, 1992, pp. 161–180.
70. R. V. Eck, M. O. Dayhoff, *Science* **1966**, *152*, 363–366.
71. L. Que, J. R. Anglin, M. A. Bobrick, A. Davison, R. H. Holm, *J. Am. Chem. Soc.* **1974**, *96,* 6042–6048.
72. B. A. Averill, W. H. Orme-Johnson, in *Metal Ions in Biological Systems,* Vol 7., Ed H. Sigel, Marcel Dekker, New York. 1978, pp. 127–183.
73. D. C. Yoch, R. P. Carithers, *Microbiol. Rev.* **1979**, *43*, 384–421.
74. H. Beinert, *JBIC* **2000**, *5*, 2–15.

75. E. J. Milner-White, M. J. Russell, *Orig. Life Evol. Biosph.* **2005**, *35*, 19–27.
76. W. Nitschke, O. Farr, N. Gaudu, C. Truong, F. Guyot, M. J. Russell, S. Duval, *Life* **2024**, *14*, 607.
77. E. Schrödinger, *What is Life? The Physical Aspect of the Living Cell*, Cambridge University Press, London, 1944.
78. J. W. Gibbs, The Scientific Papers, Thermodynamics, Longmans, Green, and Co., London, 1906, pp. 1, xxviii + 434.
79. F. G. Donnan, *Scientia* **1918**, *24*, 282–288.
80. F. G. Donnan, *Annual Report of the Smithsonian Institution* **1929**, 309–321.
81. W. Nitschke, M. J. Russell, *J. Mol. Evol.* **2009**, *69*, 481–496.
82. E. D. Schneider, J. J. Kay, *Math. Comp. Model.* **1994**, *19*, 25–48.
83. E. Branscomb, 2023, preprint. https://arxiv.org/submit/5240040/pdf.
84. W. Martin, K. Kleinermanns, *The Geochemical Origin of Microbes*, CRC Press, 2024.
85. E. Branscomb, T. Biancalani, N. Goldenfeld, M. J. Russell, *Phys. Rep.* **2017**, *677*, 1–60.
86. H. Morowitz, E. Smith, *Complexity* **2007**, *13*, 51–59
87. S. Duval, K. Zuchan, F. Baymann, B. Schoepp-Cothenet, E. Branscomb, M. J. Russell, W. Nitschke, in Vol 21 of *Metals in Life Sciences*, Eds P. Kroneck, M. E. Sosa Torres, Walter de Gruyter, Berlin, 2021, pp. 136–157.
88. M. J. Russell, R. M. Daniel, A. J. Hall, J. Sherringham, *J. Mol. Evol.* **1994**, *39*, 231–243.
89. G. Mamo, *Alkaliphiles in Biotechnology*, 2020, Springer Heidelberg, pp. 85–133.
90. W. Martin, M. J. Russell, *Phil. Trans. R. Soc. B* **2003**, *358*, 59–85.
91. W. Nitschke, M. J. Russell, *Philos. Trans. R. Soc. Lond. B* **2013**, *368*, 20120258.
92. E. Branscomb, M. J. Russell, "*Why Frankenstein theories for the emergence of life ('warm little ponds,''prebiotic soups,''energized assemblages of building blocks,'...) cannot in principle be correct*", *NASA Astrobiological Institute Principal Investigators Meeting*, University of Illinois, Urbana-Champaign/IL, 2017.
93. J. H. Cartwright, M. J. Russell, *Interface Focus* **2019**, *9*, 20190104.
94. P. Knoll, B. C. Batista, A. McMahon, O. Steinbock, *ACS Earth Space Chem.* **2022**, *6*, 2644–2650.
95. M. Weingart, S. Chen, C. Donat, V. Helmbrecht, W. D. Orsi, D. Braun, K. Alim, *Sci. Adv.* **2023**, *9*, eadi1884.
96. M. L. Wong, C. E. Cleland, D. Arend, S. Bartlett, H. J. Cleaves, H. Demarest, A. Prabhu, J. I. Lunine, R. M. Hazen, *Proc. Natl Acad. Sci.* USA **2023**, *120*, p.e2310223120.
97. S. J. Bartlett, P. Beckett, *Interface Focus* **2019**, 9, 20190058.
98. R. K. Szilagyi, R. Hanscam, E. M. Shepard, S. E. McGlynn, *Interface Focus* **2019**, *9*, 20190071.
99. M. J. Russell, *Life* **2018**, *8*, 35.
100. M. R. Hooks, P. Webster, J. M. Weber, S. Perl, L. M. Barge, *Langmuir* **2020**, *36*, 5793–5801.
101. B. Rasmussen, J. R. Muhling, W. W. Fischer, *Astrobiology* **2021**, *21*, 246–259.
102. S. Testón-Martínez, T. Huertas-Roldán, P. Knoll, L. M. Barge, C. I. Sainz-Díaz, J. H. Cartwright, *Phys. Chem. Chem. Phys.* **2023**, *25*, 30469–30476.
103. J.-M. R. Génin, R. Aïssa, A. Géhin, M. Abdelmoula, O. Benali, V. Ernstsen, G. Ona-Nguema, C. Upadhyay, C. Ruby, *Solid State Sci.* **2005**, *7*, 545–572.
104. R. E. Mielke, K. J. Robinson, L. M. White, S. E. McGlynn, K. McEachern, R. Bhartia, I. Kanik, M. J. Russell, *Astrobiology* **2011**, *11*, 933–950.
105. C. M. Rufo, Y. S. Moroz, O. V. Moroz, J. Stöhr, T. A. Smith, X. Hu, W. F. DeGrado, I. V. Korendovych, *Nat. Chem.* **2014**, *6*, 303–309.
106. J. Greenwald, M. P. Friedmann, R. Riek, *Angew. Chem. Int. Ed.* **2016**, *55*, 11609–11613.
107. Y. Sonzogni, A. H. Treiman, S. P. Schwenzer, *JMPS* **2017**, *112*, 59–76.
108. A. Pohorille, M. A. Wilson, G. Shannon, *Life* **2017**, *7*, 23.

109. C. P. J. Maury, *Cell. Mol. Life Sci.* **2018**, *75*, 1499–1507.
110. T. Asimakidou, A. Makridis, S. Veintemillas-Verdaguer, M. P. Morales, I. Kellartzis, M. Mitrakas, G. Vourlias, M. Angelakeris, K. Simeonidis, *Chem. Eng. J.* **2020**, 124593.
111. O. Farr, N. Gaudu, G. Danger, M. J. Russell, D. Ferry, W. Nitschke, S. Duval, *J. R. Soc. Interface Focus* **2023**, *20*, 20230386.
112. G. Purvis, L. Šiller, A. Crosskey, J. Vincent, C. Wills, J. Sheriff, C. Xavier, J. Telling, *Commun. Earth Environ.* **2024**, *5*, 30.
113. M. C. F. Wander, K. M. Rosso, M. A. A. Schoonen, *J. Phys. Chem. C* **2007**, *111*, 11414–11423.
114. J. E. Katz, X. Zhang, K. Attenkofer, K. W. Chapman, C. Frandsen, P. Zarzycki, K. M. Rosso, R. W. Falcone, G. A. Waychunas, B. Gilbert, *Science* **2012**, *337*, 1200–1203.
115. M. Yamamoto, R. Nakamura, K. Oguri, S. Kawagucci, K. Suzuki, K. Hashimoto, K. Takai, *Angew. Chem. Int. Ed.* **2013**, *52*, 10758–10761.
116. M. Yamamoto, R. Nakamura, T. Kasaya, H. Kumagai, K. Suzuki, K. Takai, *Angew. Chem.* **2017**, *129*, 5819–5822.
117. L. M. Barge, Y. Abedian, M. J. Russell, I. J. Doloboff, J. H. Cartwright, R. D. Kidd, I. Kanik, *Angew. Chem. Int. Ed. Engl.* **2015**, *54*, 8184–8187.
118. H. Ooka, S. E. McGlynn, R. Nakamura, *ChemElectroChem.* **2019**, *6*, 1316–1323.
119. M. J. Russell, *Life* **2021**, *11*, 429.
120. R. Vasiliadou, N. Dimov, N. Szita, S. F. Jordan, N. Lane, *Interface Focus* **2019**, *9*, 20190073.
121. R. Hudson, R. de Graaf, M. Strandoo Rodin, A. Ohno, N. Lane, S. E. McGlynn, Y. M. A. Yamada, R. Nakamura, L. M. Barge, D. Braun, V. Sojo, *Proc. Natl. Acad. Sci.* USA **2020**, *117*, 22873–22879.
122. S. Zhang, T. Holmes, C. Lockshin, A. Rich, *Proc. Natl. Acad. Sci.* USA **1993**, *90*, 3334–3338.
123. S. Duval, F. Baymann, B. Schoepp-Cothenet, F. Trolard, G. Bourrie, O, Grauby, E. Branscomb, M. J. Russell, W. Nitschke, *Interface Focus* **2019**, *9*, 20190063.
124. J. B. Jackson *J. Mol. Evol.* **2016**, 83: 1–11.
125. N. Lane, J. F. Allen, W. Martin, *BioEssays* **2010**, *32,* 271–80.
126. N. Lane, *BioEssays* **2017**, *39*, 1600217.
127. S. Duval, E. Branscomb, F. Trolard, G. Bourrié, O. Grauby, V. Heresanu, B. Schoepp-Cothenet, K. Zuchan, M. J. Russell, W. Nitschke, *Appl. Clay Sci.* **2020**, 105737.
128. Y. Ding, J. H. E. Cartwright, S. S. S. Cardoso, *Interface Focus* **2019**, *9*, 20190064.
129. Y. Ding, S. S. S. Cardoso, J. H. E. Cartwright, *Geobiology* **2024**, *22*, e12611.
130. V. Sojo, B. Herschy, A. Whicher, E. Camprubi, N. Lane, *Astrobiology* **2016**, *16*, 181–197.
131. L. M. Barge, J. P. Jones, J. J. Pagano, E. Martinez, J. Bescup, *ACS Earth Space Chem.* **2020**, *4*, 1663–1669.
132. F. Gäb, C. Ballhaus, J. Siemens, A. Heuser, M. Lissner, T. Geisler, D. Garbe-Schönberg, *Geochim. Cosmochim. Acta* **2017**, *214*, 209–225.
133. J. D. Bernal, D. R. Dasgupta, A. L. Mackay, *Clay Miner. Bull.* **1959**, *4*, 15–30.
134. P. S. Braterman, Z. P. Xu, F. Yarberry, *Handbook of Layered Materials* **2004**, 373–474.
135. J.-M. R. Génin, C. Ruby, C. Upadhyay, *Solid State Sci.* **2006**, *8,* 1330–1343.
136. J.-M. R. Génin, C. Ruby, *Solid State Sci.* **2008**, *10*, 244–259.
137. S. J. Mills, A. G. Christy, J. M. Génin, T. Kameda, F. Colombo, *Mineral. Mag.* **2012**, *76*, 1289–1336.
138. F. Trolard, J.-M. R. Génin, M. Abdelmoula, G. Bourrie, B. Humbert, A. Herbillon, *Geochim. Cosmochim. Acta* **1997**, *61*, 1107–1111.
139. P. W. U. Appel, *Precamb. Res.* **1980**, *11*, 73–87.

140. I. Halevy, M. Alesker, E. M. Schuster, R. Popovitz-Biro, Y. Feldman, *Nat. Geosci.* **2017**, *10*, 135–139.
141. M. S. Dodd, D. Papineau, T. Grenne, J. F. Slack, M. Rittner. F. Pirajno, J. O'Neil, C. T. S. Little, *Nature* **2017**, *543*, 60–64.
142. M. S. Dodd, H. Wang, C. Li, M. Towner, A. R. Thomson, J. F. Slack, Y.-s. Wan, F. Pirajno, C. Manikyamba, Q. Wang, D. Papineau, *Earth Planet. Sci. Lett.* **2022**, *584*, 117469.
143. N. J. Tosca, C. Z. Jiang, B. Rasmussen, J. Muhling, *Free Radic. Biol. Med.* **2019**, *140*, 138–153.
144. D. S. Kelley, J. A. Karson, D. K. Blackman, G. L. Früh-Green, D. A. Butterfield, M. D. Lilley, E. J. Olson, M. O. Schrenk, K. K. Roe, G. T. Lebon, *Nature* **2001**, *412*, 145–149.
145. K. A. Aquino, G. L. Früh-Green, J. Rickli, S. M. Bernasconi, S. Q. Lang, M. D. Lilley, D. A. Butterfield, *Geochim. Cosmochim. Acta* **2022**, *332*, 239–262.
146. K. A. Aquino, G. L. Früh-Green, S. M. Bernasconi, T. R. Bontognali, A. Foubert, S. Q. Lang, *Geochem. Geophys. Geosyst.* **2024**, *25*, e2023GC011010.
147. H.-E. Lee, T. Okumura, H. Ooka, K. Adachi, T. Hikima, K. Hirata, Y. Kawano, H. Matsuura, M. Yamamoto, M. Yamamoto, A. Yamaguchi, J.-E. Lee, K. T. Nam, Y. Ohara, D. Hashizume, S. E. McGlynn, R. Nakamura, *Nat. Commun.* **2024**, *15*, 8193
148. P. Refait, P. Bauer, A. A. Olowe, J.-M. R. Genin, *Hyperfine Interact.* **1990**, *57*, 2061–2066.
149. V. Helmbrecht, M. Weingart, F. Klein, D. Braun, W. D. Orsi, *Geobiology* **2023**, *21*, 758–769.
150. Y. Ohara, M. K. Reagan, K. Fujikura, H. Watanabe, K. Michibayashi, T. Ishii, R. J. Stern, I. Pujana, F. Martinez, G. Girard, J. Ribeiro, M. Brounce, N. Komori, M. Kino, *Proc. Natl. Acad. Sci.* USA **2012**, *109*, 2831–2835.
151. J. W. Morse, F. T. Mackenzie, *Geochemistry of Sedimentary Carbonates*, Amsterdam, Elsevier, 1990, p. 707.
152. H.-E. Lee, M. J. Russell, R. Nakamura, *Chem. Eur.* **2024**, 2, e202400038.
153. T. Okumura, Y. Ohara, R. J. Stern, T. Yamanaka, Y. Onishi, H. Watanabe, C. Chen, S. H. Bloomer, I. Pujana, S. Sakai, T. Ishii, K. Takai, *Geochem. Geophys. Geosystems* **2016**, *17*, 3775–3796.
154. A. Yamaguchi, M. Yamamoto, K. Takai, T. Ishii, K. Hashimoto, R. Nakamura, *Electrochim. Acta* **2014**, *141*, 311–318.
155. A. Yamaguchi, Y. Li, T. Takashima, K. Hashimoto, R. Nakamura, in *Solar to Chemical Energy Conversion: Theory and Application*, Eds K. Fujii, S. Nakamura, Springer, Cham/Switzerland, 2016, 213–228.
156. A. Yamaguchi, K. Arai, A. N. E. Aisnada, J. E. Lee, N. Kitadai, R. Nakamura, M. Miyauchi, *J. Phys. Chem. C* **2022**, *126*, 2772–2779.
157. N. Kitadai, R. Nakamura, M. Yamamoto, K. Takai, Y. Li, A. Yamaguchi, A. Gilbert, Y. Ueno, N. Yoshida, Y. Oono, *Sci. Adv.* **2018**, *4*, eeao7265
158. N. Kitadai, R. Nakamura, M. Yamamoto, K. Takai, N. Yoshida, Y. Oono, *Sci. Adv.* **2019**, *5*, eaav7848.
159. N. Kitadai, R. Nakamura, M. Yamamoto, S. Okada, W. Takahagi, Y. Nakano, Y. Takahashi, K. Takai, Y. Oono, *Commun. Chem.* **2021**, *4*, 37.
160. S. A. Sanden, R. K. Szilagyi, Y. Li, N. Kitadai, S. M. Webb, T. Yano, R. Nakamura, M. Hara, S. E. McGlynn, *Dalton Trans.* **2021**, *50*, 11763–11774.
161. J. E. Lee, A. Yamaguchi, H. Ooka, T. Kazami, M. Miyauchi, N. Kitadai, R. Nakamura, *Chem. Commun.* **2021**, *57*, 3267–3270.
162. W. Nitschke, S. E. McGlynn, E. J. Milner-White, M. J. Russell, *Biochim. Biophys. Acta* **2013**, *1827*, 871–881.
163. M. J. Russell, *Trans. Instn. Min. Metall., Appl. Earth Sci.Sect B* **1975**, *84*, B128–133.

164. D. Lloyd, M. A. Aon, S. Cortassa, *Sci. World* **2001**, *1*, 133–145.
165. L. Kubicova, G. Bachmann, W. Weckwerth, V. Chobot, *Cells* **2022**, *11*, 958.
166. L. Zhang, M. Zhang, X. Guo, D. Gan, Y. Ye, Y. Zhao, J. Ying, *Chem. Commun.* **2024**, *60*, 2748–2751.
167. R. D. Astumian, *Proc.Natl.Acad.Sci.* USA **2018**, *115*, 9405–9413.
168. M. I. Agnel, S. Grangeon, F. Fauth, E. Elkaïm, F. Claret, M. Roulet, F. Warmont, C. Tournassat, *Environ. Sci. Technol.* **2020**, *54*, 851–861.
169. F. Trolard, *Clays and Clay Miner.* **2007**, *55*, 451–451.
170. F. Trolard, S. Duval, W. Nitschke, B. Ménez, C. Pisapia, J. B. Nacib, M. Andréani, G. Bourrié , *Earth-Sci. Rev.* **2022**, *225*, 103910.
171. K. B. Ayala-Luis, C. B. Koch, H. C. B. Hansen, *Appl. Clay Sci.* **2010,** *50*, 512–519.
172. C. Ruby, M. Abdelmoula, S. Naille, A. Renard, V. Khare, G. Ona-Nguema, G. Morin, J.-M. R. Génin, *Geochim. Cosmochim. Acta* **2010**, *74*, 953–966.
173. D. Guerbois, G. Ona-Nguema, G. Morin, M. Abdelmoula, A. M. Laverman, J. M. Mouchel, K. Barthelemy, F. Maillot, J. Brest, *Environ. Sci. Technol.* **2014**, *48*, 4505–4514.
174. F. Trolard, *C R Geosci.* **2006**, *338*, 1158–1166.
175. L. M. Barge, I. J. Doloboff, M. J. Russell, D. VanderVelde, L. M. White, G. D. Stucky, M. M. Baum, J. Zeytounian, R. Kidd, I. Kanik. *Geochim. Cosmochim. Acta* **2014**, *128*, 1–12.
176. X. L. Huang, *New J. Chem.* **2022**, *46*, 15273–15291.
177. X. L. Huang, J. R. Harmer, G. Schenk, G. Southam, *Front. Chem.* **2024**, *12,* 1349020.
178. R. Szilagyi, *ChemRxiv* **2024**. https://doi.org/10.26434/chemrxiv-2024-q3v7b.
179. M. J. Russell, W. Nitschke, *Astrobiology* **2017**, *17*, 1053–1066.
180. D. A. Kopp, S. J. Lippard *Curr. Opin. Chem. Biol.* **2002**, *6*, 568–576.
181. R. Banerjee, Y. Proshlyakov, J. D. Lipscomb, D. A. Proshlyakov, *Nature* **2015**, *518*, 431–434.
182. O. Farr, N. Gaudu, A. Clouet, M. J. Russell, C. Truong, chapter 8 of this book
183. V. Alexandrov, K. M. Rosso, *J.Chem. Phys.* **2014**, *140*, 234701.
184. W. Nitschke, B. Schoepp-Cothenet, S. Duval, K. Zuchan, O. Farr, F. Baymann, F. Panico, A. Minguzzi, E. Branscomb, M. J. Russell, *Electrochem. Sci. Adv.* **2022**, e2100192.
185. E. Branscomb, M. J. Russell, *Interface focus* **2019**, *9*, 20190061.
186. M. J. Russell, A. J. Hall, A. R. Mellersh, in *Natural and laboratory-simulated thermal geochemical processes*, Ed R. Ikan, Kluwer, Dordrecht, 2003, pp. 325–388.
187. G. Arrhenius, in *Clay and Clay Minerals and the Origin of Life,* Eds A. G. Cairns-Smith, H. Hartman, Cambridge University Press, 1986, pp. 97–104.
188. G. Arrhenius, B. Sales, S. Mojzsis, T. Lee, *J. Theoret. Biol.* **1997**, *187*, 503–522.
189. G. Arrhenius, in *Fundamental Studies and the Future of Scienc*e, Ed C. Wickramasinghe, University College Cardiff Press, Cardiff UK, 1984, pp. 301–319.
190. G. Arrhenius, "*Interaction of hydrous minerals with bioorganic precursor molecules*", *NASA Space Life Sciences Symposium,* Washington DC June 21–26, 1987, 260–261.
191. G. O. Arrhenius, *Helv. Chim. Acta* **2003**, *86*, 1569–1586.
192. K. Kuma, W. Paplawsky, B. Gedulin, G. Arrhenius, *Orig. Life Evol. Biosph.* **1989**, *19*, 573–601.
193. S. Pitsch, A. Eschenmoser, B. Gedulin, S. Hui, G. Arrhenius, *Orig. Life Evol. Biosph.* **1995**, *25*, 297–334.
194. S. J. Mojzsis, G. Arrhenius, K. D. McKeegan, T. M. Harrison, A. P. Nutman, C. R. L. Friend, *Nature* **1996**, *384*, 55–59.
195. R. Krishnamurthy, G. Arrhenius, A. Eschenmoser, *Orig. Life Evol. Biosph.*, **1999**, *29*, 333–354.

196. C. Huber, G. Wächtershäuser, *Tetrahedron Lett.* **2003**, *44*, 1695–1697.
197. H. C. B. Hansen, C. B. Koch, H. Nancke-Krogh, O. K. Borggaard, J. Sørensen, *Environ. Sci. Technol.* **1996**, *30*, 2053–2056.
198. H. C. B. Hansen, S. Guldberg, M. Erbs, C. B. Koch, *Appl. Clay Sci.* **2001**, *18*, 81–91.
199. H. C. B. Hansen, C. B. Koch, *Clay Min.* **1998**, *33*, 87–101.
200. J.-M. Génin, S. J. Mills, A. G. Christy, O. Guérin, A. J. Herbillon, E. Kuzmann, G. Ona-Nguema, C. Ruby, C. Upadhyay, *Mineral. Mag.* **2014**, *78*, 447–465.
201. J.-M. Génin, A. Christy, E. Kuzmann, S. Mills, C. Ruby, *Hyperfine Interact.* **2014**, *226*, 459–482.
202. F. Trolard, G. Bourrié, in *Clay minerals in nature: their characterization, modification and application*, Eds M. Valaskova, G. S. Martynkova, InTech, Rijeka, Croatia, 2012, pp. 171–188.
203. F. Picardal, *Front. Microbio.* **2012**, *3*, 112.
204. R. Nakamura, T. Takashima, S. Kato, K. Takai, M. Yamamoto, K. Hashimoto, *Angew. Chem.* **2010**, *122*, 7858–7860.
205. M. J. Kampschreur, R. Kleerebezem, W. W. de Vet, M. C. van Loosdrecht, *Water Res.* **2011**, *45*, 5945–5952.
206. S. Buessecker, H. Imanaka, T. Ely, R. Hu, S. J. Romaniello, H. Cadillo-Quiroz, *Nat. Geosci.* **2022**, *15*, 1056–1063,
207. S. Lee, C. Wang, V. Chakrapani, **2023**. https://doi.org/10.26434/chemrxiv-2023-bch3f-v2.
208. L. M. Barge, E. Flores, M. M. Baum, D. G. VanderVelde, M. J. Russell, *Proc. Natl. Acad. Sci.* USA **2019**, *116*, 4828–4833.
209. L. M. Barge, E. Flores, D. G. VanderVelde, J. M. Weber, M. M. Baum, A. Castonguay, *J. Geophys. Res. Planets* **2020**, *125*, e2020JE006423.
210. H. S. Bernhardt, W. M. Patrick, *J. Mol. Evol.* **2014**, *78*, 307–309.
211. A. Bar-Even, A. Flamholz, E. Noor, R. Milo, *Biochim. Biophys. Acta Bioenerg.* **2012**, *1817*, 1646–1659.
212. A. Bar-Even, *Biochem.* **2016**, *55*, 3851–3863.
213. B. Grégoire, H. C. Greenwell, D. G. Fraser, *ACS Earth Space Chem.* **2018**, *2*, 852–862.
214. S. A. Harrison, W. L. Webb, H. Rammu, N. Lane, *Life* **2023**, *13*, 1177.
215. S. V. Prasanna, P. V. Kamath, *Ind. Eng. Chem. Res.* **2009**, *48*, 6315–6320.
216. Z. Ferencz, *Mechanochemical Preparation and Structural Characterization of Layered Double Hydroxides and their Amino Acid-Intercalated Derivatives*, PhD Thesis, University of Szeged, Hungary, 2016.
217. Z. Yin, K. Dideriksen, M. Abdelmoula, C. Ruby, F. M. Michel, M. J. Bjerrum, H. C. B. Hansen, *Appl. Clay Sci.* **2019**, *170*, 86–96.
218. D. Muñoz-Santiburcio, D. Marx, *Phys. Rev. Lett.* **2017**, *119*, 056002.
219. D. Muñoz-Santiburcio, D. Marx, *Chem. Sci.* **2017**, *8*, 3444–3452.
220. H. G. Hansma, *Science* **2020**, *2*, 88.
221. E. J. Milner-White, M. J. Russell, *Biol. Direct* **2008**, 3, 745–761.
222. E. J. Milner-White, M. J. Russell, *Genes* **2011**, *2*, 671–688.
223. A. Bianchi, C. Giorgi, P. Ruzza, C. Toniolo, E. J. Milner-White, *Proteins* **2012**, *80*, 1418–1424.
224. A. Levin, T. A. Hakala, L. Schnaider, G. J. Bernardes, E. Gazit, T. P. Knowles, *Nat. Rev. Chem.* **2020**, *4*, 615–634.
225. W. Wang, L. Qiao, J. He, Y. Ju, K. Yu, G. Kan, C. Guo, H. Zhang, J. Jiang, *J. Phys. Chem. Lett.* **2021**, *12*, 5774–5780.
226. D. T. Holden, N. M. Morato, R. G. Cooks, *Proc. Natl. Acad. Sci.* USA **2022**, *119*, e2212642119.
227. C. Huber, G. Wächtershäuser, *Science* **1998**, *281*, 670–672.

228. V. T. Marteinsson, J. K. Kristjánsson, H. Kristmannsdóttir, M. Dahlkvist, K. Sæmundsson, M. Hannington, S. K. Pétursdóttir, A. Geptner, P. Stoffers, *Appl. Environ. Microbiol.* **2001**, *67*, 827–833.
229. R. Price, E. S. Boyd, T. M. Hoehler, L. M. Wehrmann, E. Bogason, H. Þ. Valtýsson, J. Örlygsson, B. Gautason, J. P. Amend, *Geology* **2017**, *45*, 1135–1138.
230. J. P. Ferris, A. R. Hill, R. Liu, L. E. Orgel, *Nature* **1996**, *381*, 59–61.
231. T. Hibino, *Chem. Mat.* **2004**, *16*, 5482–5488.
232. H. Zhai, L. Qin, W. Zhang, C. V. Putnis, L. Wang, *Environ. Sci. Tech.* **2018**, *52*, 10472–10479.
233. S. F. Jordan, I. Ioannou, H. Rammu, A. Halpern, L. K. Bogart, M. Ahn, R. Vasiliadou, J. Christodoulou, A. Maréchal, N. Lane, *Nature Commun.* **2021**, *12*, 5925.
234. C. Harford, B. Sarkar, *Acc. Chem. Res.* **1997**, *30*, 123–130.
235. B. Kandemir, L. Kubie, Y. Guo, B. Sheldon, K. L. Bren, *Inorg. Chem.* **2016**, *55*, 1355–1357.
236. J. L. Alvarez-Hernandez, J. W. Han, A. E. Sopchak, Y. Guo, K. L. Bren, *Am. Chem. Soc. Energy Lett.* **2021**, *6*, 2256–2261.
237. J. D. Watson, E. J. Milner-White, *J. Mol. Biol.* **2002**, *315*, 183–191.
238. J. D. Watson, E. J. Milner-White, *J. Mol. Biol.* **2002**, *315*, 199–207.
239. A. Eschenmoser, *Angew. Chem. Int. Ed. Engl.* **1988**, *27*, 5–39.
240. V. Alva, J. Söding, A. N. Lupas, *eLife* **2015**, *4*, e09410.
241. M. L. Romero Romero, F. Yang, Y. R. Lin, A. Toth-Petroczy, I. N. Berezovsky, A. Goncearenco, W. Yang, A. Wellner, F. Kumar-Deshmukh, M. Sharon, D. Baker, G. Varani, D. S. Tawfik, *Proc. Natl. Acad. Sci. U. S. A.* **2018**, *115*, E11943-E11950.
242. D. H. Lee, J. R. Granja, J. A. Martinez, K. Severin, M. R. Ghadri, *Nature* **1996**, *382*, 525–552
243. T. Shibuya, M. J. Russell, K. Takai, *Geochim. Cosmochim. Acta* **2016**, *175*, 1–19.
244. M. Preiner, K. Igarashi, K. B. Muchowska, M. Yu, S. J. Varma, K. Kleinermanns, M. K. Nobu, Y. Kamagata, H. Tüysüz, J. Moran, W. F. Martin, *Nat. Ecol. Evol.* **2020**, *4*, 534–542.
245. M. Usman, J. M. Byrne, A. Chaudhary, S. Orsetti, K. Hanna, C. Ruby, A. Kappler, S. B. Haderlein, *Chem. Rev.* **2018**, *118*, 3251–3304.
246. B. Herschy, A. Whicher, E. Camprubi, C. Watson, L. Dartnell, J. Ward, J. R. G. Evans, N. Lane, *J. Mol. Evol.* **2014**, *79*, 213–227.
247. L. M. White, R. Bhartia, G. D. Stucky, I. Kanik, M. J. Russell, *Earth Planet. Sci. Lett.* **2015**, *430*, 105–114.
248. G. Matrajt, D. Blanot, *Amino Acids* **2004**, *26*, 153–158.
249. V. Erastova, M. T. Degiacomi, D. G. Fraser, H. C. Greenwell, *Nature Comm.* **2017**, *8*, 2033.
250. D. Kanaparthi, F. Westall, M. Lampe, B. Zhu, T. Boesen, B. Scheu, A. Klingl, P. Schwille, T. Lueders, **2021** *bioRxiv.*
251. L. Fumagalli, A. Esfandiar, R. Fabregas, S. Hu, P. Ares, A. Janardanan, Q. Yang, B. Radha, T. Taniguchi, K. Watanabe, G. Gomila, K. S. Novoselov, A. K. Geim, *Science* **2018**, *360*, 1339–1342.
252. A. O. M. Holmes, A. Kalli, A. Goldman, *Front. Mol. Biosci.* **2019**, *6*, 132.
253. A.-L. Ducluzeau, R. van Lis, S. Duval, B. Schoepp-Cothenet, M. J. Russell, W. Nitschke, *Trends Biochem. Sci.* **2009**, *34*, 9–15.
254. M. L. Wong, B. D. Charnay, P. Gao, Y. L. Yung, M. J. Russell, *Astrobiology* **2017**, *17*, 975–983.
255. L. M. Fifer, M. L. Wong, *Astrobiology* **2024**, *24*, 590–603.
256. A. Johansen, E. Camprubi, E. van Kooten, H. J. Hoeijmakers, *Astrobiology* **2024**.
257. E. Shock, P. Canovas, *Geofluids* **2010**, *10*, 161–192.

258. S. E. McGlynn, I. Kanik, M. J. Russell, *Philos. Trans. R. Soc. A* **2012**, *370*, 3007–3022.
259. B. T. Burcar, L. M. Barge, D. Trail, E. B. Watson, M. J. Russell, L. B. McGown, *Astrobiology* **2015**, *15*, 509–522.
260. S. D. Fried, K. Fujishima, M. Makarov, I. Cherepashuk, K. Hlouchova, *J. R. Soc. Interface* **2022**, *19*, 20210641.
261. N. Gaudu, O. Farr, G. Ona-Nguema, S. Duval, *Biochemie* **2023**, *215*, 100–112.
262. T. M. McCollom, J. S. Seewald, *Chem. Rev.* **2007**, *107*, 382–401.
263. D. Yang, S. Peng, M. R. Hartman, T. Gupton-Campolongo, E. J. Rice, A. K. Chang, Z. Gu, G. Q. Lu, D. Luo, *Sci. Rep.* **2013**, *3*, 3165.
264. B. Herschy, A. Whicher, E. Camprubi, N. Lane, *Astrobiology* **2016**, *16*, 181–197.
265. S. F. Jordan, H. Rammu, I. N. Zheludev, A. M. Hartley, A. Maréchal, N. Lane, *Nat. Ecol. Evol.* **2019**, *3*, 1705–1714.
266. D. He, X. Wang, Y. Yang, R. He, H. Zhong, Y. Wang, B. Han, F. Jin, *Proc. Natl. Acad. Sci.* USA **2021**, *118*, e2115059118.
267. S. K. Rout, R. Cadalbert, N. Schröder, J. Wang, J. Zehnder, O. Gampp, T. Wiegand, P. Güntert, D. Klingler, C. Kreutz, A. Knörlein, J. Hall, J. Greenwald, R. Riek, *J. Am. Chem. Soc.* **2023**, *145*, 21915–21924.
268. S. Tagami, P. Li, *Development, Growth & Differentiation* **2023**, *65*, 167–174.
269. S. Holler, S. Bartlett, R. J. Löffler, F. Casiraghi, C. I. S. Diaz, J. H. Cartwright, M. M. Hanczyc, *Proc. Natl. Acad. Sci.* USA **2023**, *120*, e2300491120.
270. E. Branscomb, M. J. Russell, *Biochim. Biophys. Acta-Bioenergetics* **2013**, *1827*, 62–78.
271. J. Arulraj, *Intercalation of organic anions and intracrystalline reactions in anionic clays*, Ph.D. Thesis, Manipal University, Manipal, India, 2013.
272. R. Tahawy, E. Doustkhah, E. S. A. Abdel-Aal, M. Esmat, F. E. Farghaly, H. El-Hosainy, N. Tsunoji, F. I. El-Hosiny, Y. Yamauchi, M. H. N. Assadi, Y. Ide, *Appl. Catal. B: Environmental* **2021**, *286*, 119854.
273. V. P. Aguirre, S. Jocic, P. Webster, C. Buser, J. A. Moss, L. M. Barge, Y. Tang, Y. Guo, M. M. Baum, *ACS Earth Space Chem.* **2020**, *5*, 40–54.
274. H. A. Moreno, D. L. Cocke, J. A. Gomes, P. Morkovsky, J. Parga, E. Peterson, C. Garcia, *Electrochem. Soc. Trans.* **2007**, *3*, 67.
275. V. Sojo, A. Ohno, S. E. McGlynn, Y. M. Yamada, R. Nakamura, *Life* **2019**, *9*, 16.
276. Q. Wang, L. M. Barge, O. Steinbock, *Chemistry–Euro J.* **2019**, *25*, 4732–4739.
277. T. Altair, L. G. Borges, D. Galante, H. Varela, *Life* **2021**, *11*, 777.
278. A. Bagger, W. Ju, A. S. Varela, P. Strasser, J. Rossmeisl, *ChemPhysChem.* **2017**, *18*, 3266–3273.
279. A. Borrego-Sánchez, C. M. Gutiérrez-Ariza, C. I. Sainz-Díaz, J. H. E. Cartwright, *Langmuir* **2022**, *38*, 10538–10547.
280. M. Fracchia, A. Visibile, E. Ahlberg, A. Vertova, A. Minguzzi, P. Ghigna, S. Rondinini, *ACS Appl. Energy Materials* **2018**, *1*, 1716–1725.
281. A. Al-Shameri, M. C. Petrich, K. Junge Puring, U. P. Apfel, B. M. Nestl, L. Lauterbach, *Angew. Chem. Int. Ed.* **2020**, *59*, 10929–10933.
282. F. Panico, D. Nava , M. Fracchia, A. Bergomi, P. Fermo, S. Pellegrino, A. Minguzzi, M. J. Russell, A. Vertova, *ACS Sustain. Chem. Eng.* **2024**, *In review.*
283. J. Frausto da Silva, R. J. P. Williams, *The biological chemistry of the elements: The inorganic chemistry of life.* Oxford University Press, 2001.
284. F. Baymann, E. Lebrun, M. Brugna, B. Schoepp-Cothenet, M.-T. Giudici-Orticoni, W. Nitschke, *Phil. Trans. R. Soc. London Series B* **2003**, *358*, 267–274.
285. B. Pignataro, *Ideas in Chemistry and Molecular Sciences: Where Chemistry Meets Life.* John Wiley & Sons, 2010.
286. G. Arrabito, A. Bonasera, G. Prestopino, A. Orsini, A. Mattoccia, E. Martinelli, B. Pignataro, P. G. Medaglia, *Crystals* **2019**, *9*, 361.

287. A. Cottrell, *Contemp. Phys.* **1979**, *20*, 1–10.
288. R. D. Astumian, *Proc. Natl. Acad. Sci.* **2005**, *102*, 1843–1847.
289. R. D. Astumian, *Phys. Chem.* **2007**, *9*, 5067–5083.
290. A. I. Brown, D. A. Sivak, *Chem. Rev.* **2019**, *120*, 434–459.
291. S. Mann, *Angew. Chem. Int. Ed.* **2008**, *47*, 5306–5320.
292. S. Boyd, K. Ganeshan, W. Y. Tsai, T. Wu, S. Saeed, D. E. Jiang, N. Balke, A. C. T. van Duin, V. Augustyn, *Nat. Mater.* **2021**, *20*, 1689–1694.
293. A. G. de Herrera, T. Markert, F. Trixler, *Commun. Chem.* **2023**, *6*, 69.
294. M. Barbieri, *Biosemiotics* 2**012**, *5*, 297–299.
295. M. Barbieri, *Biosystems* **2018**, *164*, 1–10.
296. R. G. Endres, *Scientific. Rep.* **2017**, *7,* 1–13.
297. D. L. Abel, *Why is Abiogenesis Such a Tough Nut to Crack?* Preprint, 2023. https://doi.org/10.20944/preprints202312.0231.v1.
298. D. L. Gonzalez, S. Giannerini, R. Rosa. *Interface Focus* **2019**, *9*, 20190038
299. C. Darwin, *The Origin of Species*; A Mentor Book, New York, NY, USA.

Index

A

B

C

D

E

N

O

P

Q

R

S